T0192517

MICELLAR CATALYSIS

SURFACTANT SCIENCE SERIES

FOUNDING EDITOR

MARTIN J. SCHICK
1918–1998

SERIES EDITOR

ARTHUR T. HUBBARD
Santa Barbara Science Project
Santa Barbara, California

ADVISORY BOARD

DANIEL BLANKSCHTEIN
Department of Chemical
Engineering
Massachusetts Institute
of Technology
Cambridge, Massachusetts

S. KARABORNI
Shell International Petroleum
Company Limited
London, England

LISA B. QUENCER
The Dow Chemical Company
Midland, Michigan

JOHN F. SCAMEHORN
Institute for Applied Surfactant
Research
University of Oklahoma
Norman, Oklahoma

P. SOMASUNDARAN
Henry Krumb School of Mines
Columbia University
New York, New York

ERIC W. KALER
Department of Chemical
Engineering
University of Delaware
Newark, Delaware

CLARENCE MILLER
Department of Chemical
Engineering
Rice University
Houston, Texas

DON RUBINGH
The Procter & Gamble Company
Cincinnati, Ohio

BEREND SMIT
Shell International
Oil Products B.V.
Amsterdam, The Netherlands

JOHN TEXTER
Strider Research Corporation
Rochester, New York

MICELLAR CATALYSIS

Mohammad Niyaz Khan
University of Malaysia
Kuala Lumpur, Malaysia

CRC Press
Taylor & Francis Group
Boca Raton London New York

CRC Press is an imprint of the
Taylor & Francis Group, an **informa** business

A TAYLOR & FRANCIS BOOK

CRC Press
Taylor & Francis Group
6000 Broken Sound Parkway NW, Suite 300
Boca Raton, FL 33487-2742

First issued in paperback 2020

© 2007 by Taylor & Francis Group, LLC
CRC Press is an imprint of Taylor & Francis Group, an Informa business

No claim to original U.S. Government works

ISBN-13: 978-0-367-57781-0 (pbk)
ISBN-13: 978-1-57444-490-2 (hbk)

This book contains information obtained from authentic and highly regarded sources. Reprinted material is quoted with permission, and sources are indicated. A wide variety of references are listed. Reasonable efforts have been made to publish reliable data and information, but the author and the publisher cannot assume responsibility for the validity of all materials or for the consequences of their use.

No part of this book may be reprinted, reproduced, transmitted, or utilized in any form by any electronic, mechanical, or other means, now known or hereafter invented, including photocopying, microfilming, and recording, or in any information storage or retrieval system, without written permission from the publishers.

For permission to photocopy or use material electronically from this work, please access www.copyright.com (http://www.copyright.com/) or contact the Copyright Clearance Center, Inc. (CCC) 222 Rosewood Drive, Danvers, MA 01923, 978-750-8400. CCC is a not-for-profit organization that provides licenses and registration for a variety of users. For organizations that have been granted a photocopy license by the CCC, a separate system of payment has been arranged.

Trademark Notice: Product or corporate names may be trademarks or registered trademarks, and are used only for identification and explanation without intent to infringe.

Library of Congress Cataloging-in-Publication Data

Khan, Mohammad Niyaz.
  Micellar catalysis / Mohammad Niyaz Khan.
    p. cm. -- (Surfactant science series ; 133)
  Includes bibliographical references and index.
  ISBN 1-57444-490-5 (alk. paper)
  1. Catalysis. 2. Micelles. I. Title. II. Series: Surfactant science series ; v. 133.

QD505.K53 2006
541'.395--dc22   2006040127

Visit the Taylor & Francis Web site at
http://www.taylorandfrancis.com

and the CRC Press Web site at
http://www.crcpress.com

Preface

Catalysis has been known since the 14th century, but the mechanistic aspects of catalytic processes at the molecular level are far from fully understood. It is almost certain that Nature controls the fascinating selectivity and specificity of all biochemical and nonbiochemical reactions through catalytic processes. Much of the work on catalysis lies in the domain of homogeneous and heterogeneous catalysis. Micellar-mediated reactions are characterized as catalytic processes involving microheterogeneous catalysis. Kinetic studies, and hence mechanistic studies, of micellar-mediated reactions at the molecular level started only in the late 1950s. But the lack of understanding of the highly dynamic micellar structure and polarity of the micromicellar reaction environment has resulted in the development of many kinetic models for micellar-mediated reactions. Although the applicability of these models is not sufficiently general, each has increased our understanding of the highly dynamic structural features of micelles, as well as the complex mechanistic aspects of their effects on reaction rates.

This book strongly reflects my own philosophy of science. Much of the problem with learning science arises from a lack of appreciation of the fundamental conceptual aspects of science. For example, it is much easier to understand, and consequently to appreciate, how — rather than why — a certain event occurs in the scientific domain. Furthermore, every such event poses three basic challenges to scientists: (1) to discover a new event to understand, and to comprehend (2) how and (3) why such an event occurs. The complexity, and consequently the challenge to appreciation increases as we move in turn from the first to the second to the third challenge. In the scientific domain, concepts and theories are advanced to rationalize experimental observations. But sometimes the problem is that the same experimental observations can be almost equally well rationalized in terms of different concepts and theories. These alternative concepts and theories cannot be discredited entirely, because they help develop more refined and coherent concepts and theories. To the best of my ability, I have tried to keep these ideas in mind while writing this book.

The physicochemical properties of a micelle, as well as the forces responsible for its formation and highly dynamic structural features, are described in Chapter 1. Chemical forces or molecular interactions responsible for catalytic effects and general mechanisms — also called general theories of catalytic processes — are described in some detail in Chapter 2. General mechanisms or various kinetic models for micelle-catalyzed reactions are discussed in Chapter 3, and Chapter 4 discusses the effects of micelles on the rate constants of unimolecular, solvolytic, bimolecular, and intramolecular organic reactions; the effects of mixed aqueous-organic solvents on the micellar binding constants of solubilizates; and micelles as modifiers of reaction rates. The effects of mixed micelles on the rates of unimolecular, solvolytic, and bimolecular reactions have been described in Chapter 5. A rather brief description of the effects of metallomicelles and induced metallomicelles on reaction rates is given in Chapter 6, and Chapter 7 discusses the fundamental principles of kinetics and practical kinetics, which are often needed in kinetic data analysis.

My debts to my postdoctoral supervisor, the late Professor Jack Hine of Ohio State University, Columbus, are enormous. My research thinking has been highly influenced by the work of a few individuals such as Professors J. Hine, M.L. Bender, W.P. Jencks, T.C. Bruice, C.A. Bunton, A.J. Kirby, F.M. Menger, R. Breslow, M.I. Page, and A. Williams. I would like to extend my sincere gratitude to Professor Arthur Hubbard of the Santa Barbara Science Project, California, who suggested that I write this book. I wish to express my sincere thanks to all my coworkers, who directly and indirectly contributed to this work. I am also grateful to Bayero University, Kano, Nigeria; National Science Council for R&D, IRPA, Malaysia; Academy Sciences of Malaysia for Scientific Advancement Grant Allocation (SAGA), and the University of Malaya for generous financial support of research in the area covered by this book. Last, but certainly not least, I want to acknowledge my wife's help at every step of the way.

Mohammad Niyaz Khan
Kuala Lumpur, Malaysia

About the Author

Mohammad Niyaz Khan is a professor at the Department of Chemistry, University of Malaya, Malaysia. After obtaining his Ph.D. in chemistry from Aligarh Muslim University (1975), India, Dr. Niyaz carried out postdoctoral work with Jack Hine at the Department of Chemistry (November 1975–December 1977) and Louis Malspeis at the College of Pharmacy (January 1978–August 1980), the Ohio State University, Columbus. He was on the faculty of science of Bayero University, Kano, Nigeria, from late 1980 through late 1990 before joining the chemistry department of University Putra Malaysia as a visiting research fellow. Since late 1993, he has been a faculty member at the Department of Chemistry, University of Malaya. He was a member of the Editorial Advisory Board of the *International Journal of Chemical Kinetics* (1996–1998), the *Indian Journal of Chemistry* (1998), and *Malaysian Journal of Science* (2001–2004). His research has centered around two areas: (1) kinetics and mechanisms of homogeneous catalysis and intramolecular organic reactions, and (2) kinetics and mechanisms of micellar catalysis of organic reactions.

Contents

1 Micelles

1.1 NORMAL MICELLES

Almost all significant or breakthrough scientific inventions were met with skepticism from the world scientific community in the initial stages. This is natural, for the human brain appears to be conditioned to deny any concept it first encounters that is not supported by convincing and rational reasoning or proof. However, this does not appear to be an absolute rule, because the human brain does accept, knowingly or unknowingly, concepts based on mere intuition. A suggestion by McBain regarding the formation of molecular aggregates in an aqueous solution of surfactant at the turn of the 20th century was treated with incredulity by the leading scientists of the time. McBain pointed out in a lecture to the Royal Society of London that unusual solution properties of an aqueous solution of surfactant above a critical surfactant concentration could be explained by surfactant aggregation under such conditions.[1] A leading physical chemist chairing the meeting responded to this idea with two words, "Nonsense, McBain."[1] But, everyone today who is familiar with colloidal chemistry accepts the notion that surfactant molecules, above a critical surfactant concentration, do aggregate in aqueous solvents. In 1936, Hartley[2] introduced the term *amphipathy* to describe the unusual properties of aqueous solutions of soap and detergent molecules:

> This unsymmetrical duality of affinity is so fundamental a property of paraffin-chain ions, being directly responsible for all the major peculiarities of paraffin-chain salts in aqueous solutions, that it is worthwhile, if only for emphasis, to give it a special name. The property is essentially the simultaneous presence (in the same molecule) of separately satisfiable sympathy and antipathy for water. I propose, therefore, to call this property *amphipathy* — the possession of both feelings.

A solute molecule is said to dissolve in water solvent when the absolute value of *solvation energy* (i.e., energy released owing to formation of solvation shell around the solute molecule) becomes equal to or larger than the energy required to form a cavity in the water solvent for embedment of the solute molecule. A surfactant (surface-active) or amphiphile molecule consists of both hydrophobic (lipophilic or lipid-loving) and hydrophilic (lipophobic or lipid-hating) molecular segments. When the hydrophobic segment becomes considerably larger than voids in the three-dimensional structural network of water solvent, and the solvation energy for a single molecule is not sufficient to counterbalance the energy needed to form a cavity for embedding the molecule, then surfactant molecules begin to aggregate, because it is energetically less expensive or easier to form a larger cavity than a relatively smaller one. However, because the surfactant molecules contain both hydrophilic segments (called *head groups*) and lipophilic segments (generally long methylene chains [C_nH_{2n+1} with n > 8] called tails), packing of head groups and tails during the formation of a micelle involves different energetics in terms of molecular interactions. Thus, the shape and size of a micelle depend upon the energetics of the interaction between adjacent hydrophilic head groups as well as adjacent lipophilic tails. Surfactants are often classified on the basis of an empirical scale called *hydrophilic–lipophilic balance* (HLB) *number*, which gives a simple index for the molecular balance of surfactant at an oil–water interface.[3]

Based on characteristic physical properties of head groups, surfactants have been categorized as cationic (if the head groups are cationic), anionic (if the head groups are anionic), zwitterionic (if the head groups possess both cationic and anionic sites), and nonionic (if the head groups are nonionic). The increase in the concentration of a particular surfactant in an aqueous solvent reveals a sudden change in various aqueous surfactant solution properties such as surface tension, equivalent conductivity, solubilization, osmotic pressure, turbidity, self-diffusion, magnetic resonance, UV-visible/fluorescence spectra of solutes, and reaction rates above a sharp surfactant concentration (Figure 1.1). Such changes in various physical properties of an aqueous solution of the surfactant are attributed to the formation of aggregates of surfactant molecules above a critical surfactant concentration, which is termed as *critical micelle concentration* (CMC), because these surfactant molecular aggregates are called *micelles*. The word *micelle*, which is now also called *normal micelle*, was introduced by McBain in 1913 to describe the formation of molecular aggregates in aqueous solution of soaps.[4] The surfactant molecules below CMC are believed to remain as monomers. However, the concept of surfactant molecules as monomers below CMC is considered to be an oversimplified situation.[5] A recent report provides fluorescence spectral evidence in favor of premicellar aggregates.[6]

Although these micelles cannot be seen with naked eyes or detected directly by UV-visible spectral studies (because the sizes of these micelles are significantly smaller than the wavelength of the electromagnetic radiation covering its UV-visible region), there is convincing indirect experimental evidence, including common logic and intuition, in favor of aggregates or micelles formation in aqueous solutions of surfactants.[7] Aqueous solutions of micelle-forming surfactants scatter light, and this property provided compelling evidence for the formation of surfactant aggregates or micelles.[8]

1.1.1 CRITICAL MICELLE CONCENTRATION (CMC) AND ITS DETERMINATION

Critical micelle concentration (CMC) of a surfactant is defined as the optimum aqueous concentration of the surfactant at which micelles begin to form under a specific reaction condition. Extensive research on the physicochemical behavior

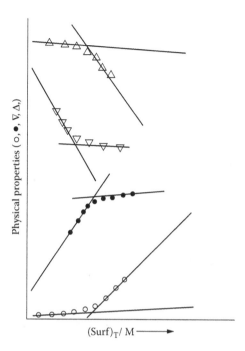

Physical properties $(\circ, \bullet, \triangledown, \triangle_i)$

$(Surf)_T / M \longrightarrow$

FIGURE 1.1 Change in physical properties of micellar aqueous solution as a function of total surfactant concentration, $[Surf]_T$. The point of intersection of two linear lines in each plot represents the critical micelle concentration (CMC).

of aqueous organized molecular assemblies during the last few decades reveal that CMC of a surfactant is an extremely important parameter in view of its importance in determination and optimization of various characteristic properties of micelles, such as micellar stability and binding affinity/binding constant of a solubilizate as well as surfactant use in facial cleansers, shampoo, and baby-care products.[9,10] Various physicochemical properties of micellar solutions are distinctly different from those of nonmicellar solutions (aqueous solution containing monomers) of the same surfactant, and this behavior of an aqueous solution of a surfactant forms the basis for various physicochemical methods to determine the value of CMC. These methods include the following:

1. Surface tension
2. Ionic conductance
3. Vapor pressure osmometry
4. Static light scattering
5. Dynamic light scattering
6. Refractive index
7. Dye solubilization
8. Dye micellization
9. Molecular adsorption

10. Diffusion coefficient
11. Viscosity
12. Partial molal volume
13. Sound velocity

The details of most of these methods are found in books[11] and review articles.[12]

The structure of a micelle is highly dynamic as well as sensitive to additives; because of these characteristic properties of micelles, the physicochemical methods produce slightly different CMCs of a surfactant.[8a,13,14] Although solubilization of substances containing chromophores or fluorescence for determination of CMC has been criticized because incorporation of solubilizate into a micelle can change the CMC values,[11c,15] the use of such a method is now common.[16] For instance, because of the report[17] that the luminescent probe (pyrene) could be used to determine the CMC of surfactant solutions, the so-called pyrene 1:3 ratio method has become one of the most popular methods for the determination of CMC[18] in pure[19] and mixed surfactant systems.[20–26] However, attempts have been made to compare CMC values obtained from different physicochemical methods.[19,27–36]

Apart from conventional techniques, a few more new techniques have recently been introduced to determine CMC of a surfactant solution. For instance, Gosh et al.[37] have determined CMC of surfactant solutions by the use of hyper-Rayleigh scattering (HRS), which is sensitive to micellar size and shape changes. This technique has also provided indirect evidence for the presence of premicellar aggregates of much lower aggregation numbers compared to those of micelles. The CMC value obtained by linear Rayleigh scattering is slightly higher than that obtained by HRS, which is attributed to the known fact that CMC values do vary a little depending upon the method employed for its determination. The CMC values of anionic, cationic, and nonionic surfactants have been determined by a new application of nuclear track microfilters,[38] and these CMC values have been found to be comparable with the literature values. A spectrophotometric estimation of CMC of ethoxylated alkyphenol surfactants has been shown to be feasible through the so-called absorbance deviation method.[39] The CMC estimates are satisfactory and, thus, the method may be an alternative for alkylphenol nonionic surfactants. A new method of CMC determination has been reported, which is based upon the decrease in the water structure in the critical range of micelle formation.[40] The decrease in water structure is measured by the significant increase in the rate of disintegration of gelatin microcapsules.

Kinetic studies on micellar-catalyzed reactions form the backbone of this book, and a quantitative analysis of kinetic data requires knowledge of the exact value of CMC of the micellar solution under strict reaction conditions of kinetic runs. However, most kinetic studies on micellar-catalyzed reactions have used CMC values determined by the usual physical methods under experimental conditions not strictly similar to those of kinetic runs. Although it is not impossible, it is extremely difficult from a practical point of view to obtain CMC values under reaction kinetic conditions by using any of the various physicochemical methods. Broxton et al.[41] have advanced a so-called kinetic graphical method to determine

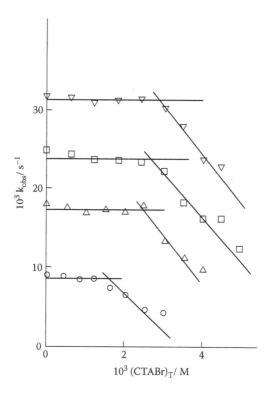

FIGURE 1.2 Plots showing the dependence of pseudo-first-order rate constants, k_{obs}, for methanolysis of ionized phenyl salicylate upon total concentration of cetyltrimethylammonium bromide, $[CTABr]_T$ at 25°C (O), 35°C (Δ), 40°C (□), and 45°C (∇). The point of intersection of two linear lines in each plot represents the CMC. (Data from Khan, M.N., Arifin, Z. Kinetics and mechanism of intramolecular general base-catalyzed methanolysis of ionized phenyl salicylate in the presence of cationic micelles. *Langmuir* **1996**, *12*(2), 261–268.)

the CMC value under reaction kinetic conditions. This method involves the plot of rate constant (k_{obs}) vs. total concentration of surfactant ($[Surf]_T$) within the $[Surf]_T$ range (covering values both below and above CMC). The point of intersection of two linear plots of k_{obs} vs. $[Surf]_T$ drawn just below and above CMC gives the value of CMC. Such plots obtained under different reaction conditions for intramolecular general base-catalyzed methanolysis of ionized phenyl salicylate in the presence of cetyltrimethylammonium bromide (CTABr) surfactant are shown in Figure 1.2.[42] This graphical method has been used to determine CMC in a number of kinetic studies.[43–46]

The graphical methods, including the kinetic graphical method, cannot produce very reliable values of CMCs if the inflection point in the plot, which gives the value of CMC, is not very sharp. Under such circumstances, a kinetic iterative method is expected to give a more reliable value of CMC. The kinetic iterative method involves the fitting of experimentally determined first-order or pseudo-first-order rate constants (k_{obs}) to the kinetic equation, Equation 1.1, derived based upon a pseudophase model of micelle.[47] In Equation 1.1, $[D_n] = [Surf]_T - CMC$, $k_W = k_{obs}$ at $[D_n] = 0$, k_M is the

$$k_{obs} = \frac{k_W + k_M K_S [D_n]}{1 + K_S [D_n]} \qquad (1.1)$$

rate constant for the reaction occurring in the micellar pseudophase, and K_S is the micellar binding constant of reactant S. If k_{obs} represents pseudo-first-order rate constants, then the assumption in the derivation of Equation 1.1 is that the micellar binding constants of all reactants except S are nearly zero. The unknown parameters, k_M and K_S, as well as the least squares, Σd_i^2 (where $d_i = k_{obs\,i} - k_{calcd\,i}$ with $k_{obs\,i}$ and $k_{calcd\,i}$ representing experimentally determined and calculated first-order rate constants at the i-th total concentration of surfactant, respectively), are calculated from Equation 1.1 at a given (presumed) value of CMC using the nonlinear least-squares technique. In a given set of experimental conditions, the value of k_W is determined experimentally by carrying out the kinetic run at $[Surf]_T = 0$, provided surfactant concentration below CMC does not affect k_{obs}. The magnitudes of the least squares, Σd_i^2, are determined at different given (presumed) values of CMC, and the specific CMC value at which the Σd_i^2 value turns out to be minimum is considered the best kinetic CMC value. The values of CMC of CTABr, obtained by kinetic graphical method and kinetic iterative method under different reaction conditions for CTABr micellar-catalyzed methanolysis of ionized phenyl salicylate, are not significantly different from each other (Table 1.1).[42] The kinetic iterative method has been used to determine CMC of surfactants under a variety of kinetic reaction conditions.[48,49]

Recently, attempts have been made to predict CMC values of nonionic,[50–52] cationic,[53] anionic,[54] and different classes of surfactants[55] based upon a quantitative structure property relationship (QSPR) study.

1.1.2 Effects of Additives on CMC

Effect of additives on CMC of an aqueous solution of a surfactant depends on the nature of interaction of additive with the following: (1) hydrophobic segment of surfactant, (2) hydrophilic segment of surfactant, (3) monomers of surfactant, which are in fast equilibrium with micelles, and (4) water molecules in aqueous pseudophase. Molecular interaction between interacting molecules may involve some or all of the following interactions: dipole–dipole, ion–dipole, ion–ion, van der Waals/dispersion forces, and hydrogen bonding. Energetically favorable interactions between additive and micellized surfactant molecules will increase the stability of micelle, which will, in turn, cause the decrease in CMC.

Depending on the nature of the additive and surfactant (nonionic, anionic, cationic, and zwitterionic), all possible interactions between the additive and micellized surfactant molecules may not be equally effective, and some of them may

TABLE 1.1
Effect of Methanol Content on Critical Micelle Concentration (CMC) of CTABr Obtained by Kinetic Iterative Method and Kinetic Graphical Method

Methanol/% v/v	10^4 CMC/M[a]	10^4 CMC/M[b]
10	1.8	1.7
15	2.9	2.8
20	4.6	4.3
25	7.8	7.9
30	18.0	17.7
35	35.0	34.0
40	43.0	43.0
45	48.0	48.5
50	60.0	65.0

Note: CMC value obtained from Khan, M.N., Arifin, Z. Kinetics and mechanism of intramolecular general base-catalyzed methanolysis of ionized phenyl salicylate in the presence of cationic micelles. *Langmuir* **1996**, *12*(2), 261–268.

[a] Determined by the kinetic iterative method.
[b] Determined by the kinetic graphical method.

oppose each other for micellar stability or instability. For example, a more hydrophilic cation and anion can cause an energetically favorable interaction with the respective anionic and cationic head groups of ionic micelles. Such an interaction will reduce the electrostatic repulsion between head groups of ionic micelles, which, in turn, will increase the stability of ionic micelles and consequently decrease the value of CMC. However, moderately or highly hydrophilic ions, because of their high charge density, promote water structure (i.e., in the presence of these ions, water molecules in water solvent become more structured). It is obvious that energy needed to form a cavity to accommodate a micellar aggregate (micelle) should be larger in a more structured than that in a relatively less structured water solvent. Thus, the water-structure-forming ability of hydrophilic ions is expected to increase the energetically favorable hydrophobic interactions between hydrophobic segments of micellized surfactant molecules and consequently increase the micellar stability. It is now apparent that the effect of hydrophilic ions on CMC of an ionic surfactant is the combined effects of these ions on electrostatic repulsive interaction between ionic head groups and hydrophobic attractive interaction between tails (hydrophobic segments) of the ionic surfactant molecules.

On the other hand, although water-structure-breaking cations and anions decrease the electrostatic repulsion between respective anionic and cationic head groups of ionic micelles and thus increase the stability of micelles, these ions decrease the stability of ionic micelles by decreasing the energy of hydrophobic interactions between hydrophobic segments of micellized surfactants because of the lower energy required to form a cavity to embed a micellar aggregate in a less-structured compared to that in a more or moderately structured water solvent. The water-structure-breaking tetraalkylammonium ions are known to cause salting-in effect in aqueous solution of hydrophobic molecules.[56] Thus, the effects of water-structure-breaking cations and anions on CMC of respective anionic and cationic surfactants are governed by two seemingly opposing effects of these ions on the stability of ionic micelles.

Both nonelectrolyte and electrolyte additives cause dehydration of nonionic micellar aggregates in aqueous solvents.[57-65] The dehydration effect of these additives facilitates the expansion of the hydrophobic microenvironment of nonionic micelles and consequently increases the stability of micelles. This interaction is similar to what is known as *salting-out effect* of hydrophilic ions. Ions with a large charge density increase the water-structure and hence, the increase in the concentration of water-structure-forming hydrophilic ions is expected to cause nonlinear decrease in CMC of nonionic surfactants.

The effects of water-structure-breaking ions on the CMC of an aqueous solution of a nonionic surfactant depend not only on the intrinsic characteristics of these ions to break the water structure but also on the extent of hydrophobicity of these ions (such as tetraalkylammonium ions). Although the water-structure-breaking ability of moderately hydrophobic ions decreases the stability of nonionic micelles, the hydrophobic interactions between moderately hydrophobic ions and micellized surfactant molecules are most likely to increase the stability of nonionic micelles. The interaction between polar head groups of nonionic micelles and additives (moderately hydrophobic ions) may, depending upon the structural features of the head groups and the type of charge on additive ions, increase the stability of nonionic micelles. Thus, it is very difficult to predict *a priori* the effects of ionic additives on CMC of nonionic micelles.

Nonpolar hydrocarbon molecules (additives) are expected to interact weakly with hydrophobic segments of ionic and nonionic micellized surfactants involving weak dispersion forces. However, based upon the assumption of the ideal mixing of the solubilizate with micellized surfactant, Mukerjee[66] concluded that the CMC of the micelle-forming surfactant is reduced by a factor of $1 - X$ where X stands for the mole fraction of the solubilizate (i.e., micellized additive) in a micelle. It is also stressed that additives such as ethanol or dioxane at high concentrations reduce hydrophobic interactions and cause the CMC to increase through a medium effect. Thus, the plot of CMC vs. [additive] show a minimum.[66] The uncharged solubilized additive molecules increase the micellar volume and thus cause the reduction in the density of the head groups, which, in turn, increases the stability of micelle.

Study on the effects of nonelectrolyte additives, 2-butoxyethanol, poly(ethylene glycol), and glucose, on various physicochemical aspects of aqueous solutions of two nonionic silicone surfactants based on poly(dimethylsiloxane)-*graft-*

polyethers, reveals that the increase in the concentrations of additives decreases the CMC values and increases dehydration of micellized surfactants.[67] The decrease in CMC due to the presence of nonelectrolyte additives is attributed to the dehydration of micellized surfactant molecules. The dehydration effects of these additives increase the attractive hydrophobic interaction between hydrophobic segments of micellized surfactant molecules. Fine details of the dehydration effects of both nonelectrolyte and electrolyte additives on micellized surfactant molecules are not well understood. However, it is widely believed that the dehydration of micellized surfactants due to the presence of such additives causes the transfer of water molecules from the solvation shell of micellized molecules to the bulk-water solvent. In view of this held opinion, the systematic increase in hydration of the micelles with increase in temperature in the presence of 2-butoxyethanol, poly(ethylene glycol), and glucose additives is considered to be unusual. This unusual observation is explained in terms of a proposal that the dehydration of the micelles may follow a preferential pattern in which water may be lost successively from the inner core to the fringe and to the outer shell parts. It is further postulated that the dehydration under these conditions expels more water from within the micelles to the outer corona.[67]

Effects of four different amine additives (ethylenediamine, diethylenetriamine, triethylenetetramine, and tetraethylenepentamine) on CMC of CTABr and sodium dodecyl sulfate (SDS) in buffered aqueous solutions of pH 7.00 show a decrease in CMC with increase of amine concentrations and the decreasing effects follow this sequence: tetraethylenepentamine > triethylenetetramine > diethylenetriamine > ethylenediamine. The decreasing effect on CMC of an amine is almost the same for both CTABr and SDS surfactants. These results are explained in terms of polar and hydrogen bond interactions between additive and micellized surfactant molecules.[68] The nature of the counterion (Br^-, F^-, HO^-, and SO_4^{2-}) reveals little effects on the CMC of CTABr and tetradecyltrimethylammonium bromide (TTABr), whereas the presence of ionic and nonionic aromatic substrates exhibit a significant lowering effect and insignificant effect on CMC, respectively.[41b]

Effects of different volume contents of methanol,[42] ethanediol,[69] and acetonitrile[46] on CMC of CTABr in mixed aqueous solvent containing 2×10^{-4} M anionic phenyl salicylate are described in terms of the following empirical relationship:

$$CMC = CMC_0 \exp(\theta X) \qquad (1.2)$$

where X represents volume content of organic cosolvent, $CMC_0 = CMC$ at X = 0, and θ is an empirical constant whose value is the measure of the ability of organic cosolvent to increase CMC of micelle-forming surfactant. The values of CMC of CTABr obtained under different reaction conditions for organic cosolvent, 1,2-ethanediol, methanol, and acetonitrile are summarized in Table 1.2. The values of CMC_0 and θ have been calculated from Equation 1.2 by the use of the nonlinear least-squares technique and the results obtained are summarized

TABLE 1.2
Effects of Organic Cosolvent (X) and Temperature on Critical Micelle Concentration (CMC) of CTABr, Obtained by Kinetic Graphical Method

[X]/% v/v	Temp/°C	10^4 CMC/M	10^4 CMC_{calcd}/M^a	10^4 CMC_{calcd}/M^b	10^4 CMC_{calcd}/M^c
\multicolumn{6}{c}{X = CH₃CNᵈ}					

[X]/% v/v	Temp/°C	10^4 CMC/M	10^4 CMC_{calcd}/M^a	10^4 CMC_{calcd}/M^b	10^4 CMC_{calcd}/M^c
			$X = CH_3CN^d$		
5	35	3.0	3.5		
8	35	5.2	5.7		
10	35	7.3	7.9		
15	35	19.0	17.8		
20	35	40.0	40.3		
			$X = HOCH_2CH_2OH^e$		
15	30	1.6	1.1		
20	30	2.4	1.8		
25	30	2.7	3.0		
30	30	4.3	4.9		
40	30	13.1	12.9		
50	30	34.0	34.0		
20	25	2.1	1.5	2.0	2.0
20	30	2.4	1.8	2.4	2.4
20	35	2.8	2.3	3.0	3.0
20	40	3.7	3.2	3.7	3.6
20	45	4.5	4.5	4.5	4.4
30	25	4.0	4.4	3.5	3.6
30	30	4.3	4.8	4.7	4.8
30	35	6.0	6.4	6.3	6.3
30	40	8.4	8.8	8.3	8.2
30	45	11.0	11.0	10.9	10.7
40	25	12.6	12.5	11.8	11.8
40	30	13.1	13.0	14.7	14.6
40	35	18.0	17.9	18.2	18.0
40	40	24.0	23.9	22.3	22.1
40	45	26.4	26.4	27.2	27.3
			$X = CH_3OH^f$		
10	30	1.7	1.1		
15	30	2.8	2.2		
20	30	4.3	4.4		
25	30	7.9	8.7		
30	30	17.7	17.2		
35	30	34.0	34.1		
40	30	43.0	g		
45	30	48.5	g		
50	30	65.0	g		

TABLE 1.2 *(Continued)*
Effects of Organic Cosolvent (X) and Temperature on Critical Micelle Concentration (CMC) of CTABr, Obtained by Kinetic Graphical Method

[X]/% v/v	Temp/°C	10^4 CMC/M	10^4 CMC$_{calcd}$/M[a]	10^4 CMC$_{calcd}$/M[b]	10^4 CMC$_{calcd}$/M[c]
10	25	1.6	1.2	1.4	1.5
10	30	1.7	1.2	1.7	1.8
10	35	2.1	1.3	2.1	2.1
10	40	2.2	1.8	2.5	2.5
10	45	3.2	2.8	3.0	2.9
20	25	4.1	4.3	3.6	3.8
20	30	4.3	4.6	4.5	4.6
20	35	5.2	5.6	5.6	5.6
20	40	6.8	7.0	6.9	6.8
20	45	8.8	9.0	8.6	8.3
30	25	16.0	16.0	16.5	16.1
30	30	17.7	17.7	19.4	19.0
30	35	25.0	25.0	22.7	22.4
30	40	27.6	27.6	26.4	26.5
30	45	29.2	29.2	30.6	31.2

[a] Calculated from Equation 1.2 with [X], CMC$_0$, and θ summarized in Table 1.2 and Table 1.3.

[b] Calculated from Equation 1.10 with T, and ΔH_m^0, ΔS_m^0 summarized in Table 1.2 and Table 1.7, respectively.

[c] Calculated from Equation 1.15 with T, and A, B summarized in Table 1.2 and Table 1.8, respectively.

[d] CMC values are obtained from Reference 46 where reaction mixture for each kinetic run contained 10% v/v CH$_3$OH and 0.01 M NaOH.

[e] CMC values are obtained from Reference 69 where reaction mixture for each kinetic run contained 2% v/v CH$_3$CN and 0.01 M NaOH.

[f] CMC values are obtained from Reference 42 where reaction mixture for each kinetic run contained 2% v/v CH$_3$CN and 0.01 M NaOH.

[g] The values of CMC at ≥ 40% v/v CH$_3$OH did not fit to Equation 1.2.

in Table 1.3. The extent of reliability of the fit of observed data to Equation 1.2 is evident from CMC$_{calcd}$ values summarized in Table 1.2 and from the standard deviations associated with the calculated parameters, CMC$_0$ and θ (Table 1.3).

The effect of mixed water-tetrahydrofuran (THF) and water-acetonitrile on CMC of CTABr and SDS have been studied at different temperatures ranging from 25 to 40°C.[70] These CMC values as listed in Table 1.4 were found to fit to Equation 1.2. The least-squares calculated values of unknown parameters, CMC$_0$ and θ, are summarized in Table 1.3. Although the values of CMC$_0$ are not very different from CMC obtained at [X] = 0 for X = CH$_3$CN (Table 1.4), for X =

TABLE 1.3
Values of Fitting Parameters, CMC_0 and θ, Calculated from Equation 1.2 for CTABr in the Presence of Various Organic Cosolvent (X) at Different Temperatures

X	Surfactant	Temp/°C	10^4 CMC_0/M	10^2 θ/(% v/v)$^{-1}$
CH_3CN[a]	CTABr	35[b]	1.54 ± 0.19[c]	16.3 ± 0.7[c]
		35[d]	0.30 ± 0.06	16.3 ± 0.7
$HOCH_2CH_2OH$[a]		25	0.19 ± 0.10	10.5 ± 1.4
		30[e]	0.26 ± 0.04	9.7 ± 0.3
		30[f]	0.24 ± 0.13	10.0 ± 1.4
		35	0.29 ± 0.10	10.3 ± 0.9
		40	0.43 ± 0.10	10.0 ± 0.6
		45	0.78 ± 0.01	8.8 ± 0.1
CH_3OH[a]		25	0.32 ± 0.10	13.0 ± 1.0
		30[g]	0.29 ± 0.04	13.6 ± 0.5
		30[h]	0.31 ± 0.11	13.5 ± 1.2
		35	0.28 ± 0.13	15.0 ± 1.6
		40	0.45 ± 0.08	13.7 ± 0.6
		45	0.87 ± 0.12	11.7 ± 0.5
CH_3CN[i]		25	9.70 ± 0.87	6.09 ± 0.23
		30	11.0 ± 0.9	5.87 ± 0.21
		35	12.5 ± 1.0	5.65 ± 0.21
		40	13.8 ± 1.1	5.51 ± 0.20
	SDS[i]	25	85.7 ± 3.7	6.75 ± 0.20
		30	104 ± 4	5.98 ± 0.24
		35	104 ± 3	5.98 ± 0.24
		40	115 ± 7	5.69 ± 0.35
THF[i]	CTABr	25	23.5 ± 1.1	10.2 ± 0.3
		30	26.2 ± 1.8	9.77 ± 0.40
		35	28.4 ± 2.5	9.42 ± 0.50
		40	31.1 ± 2.6	9.01 ± 0.48
	SDS[i]	25	16.7 ± 0.4	3.70 ± 0.17
		30	18.7 ± 0.3	3.00 ± 0.12
		35	20.0 ± 1.1	2.56 ± 0.45
		40	22.8 ± 0.6	1.89 ± 0.21
$HOCH_2CH_2OH$[i]	CTABr	25.2	5.2 ± 0.9	4.18 ± 0.35
	TTABr	25.2	17.9 ± 6.2	3.92 ± 0.76
	DTABr	25.2	95.6 ± 14.2	2.68 ± 0.34
	CTABr	25.2	5.9 ± 0.2	4.43 ± 0.10
	TTABr	25.2	28.4 ± 3.2	2.53 ± 0.39
	DTABr	25.2	125 ± 8	1.89 ± 0.23

[a] CMC values are listed in Table 1.2.
[b] The values of CMC_0 and θ were calculated from Equation 1.2 with X = % v/v content of CH_3CN only.
[c] Error limits are standard deviations.
[d] The values of CMC_0 and θ were calculated from Equation 1.2 with X = % v/v content of CH_3CN + 10% v/v CH_3OH.
[e] The values of CMC_0 and θ were calculated from Equation 1.2 with X range 15–50% v/v.
[f] The values of CMC_0 and θ were calculated from Equation 1.2 with X range 20–40% v/v.
[g] The values of CMC_0 and θ were calculated from Equation 1.2 with X range 10–35% v/v.
[h] The values of CMC_0 and θ were calculated from Equation 1.2 with X range 10–30% v/v.
[i] CMC values are listed in Table 1.4.

TABLE 1.4
Effects of Organic Cosolvent (X) and Temperature on Critical Micelle Concentration (CMC) of Surfactants Obtained by Different Methods

Temp/°C =	25		30		35		40	
	10^4	10^4	10^4	10^4	10^4	10^4	10^4	10^4
[X]/% v/v	CM	$CM_{calcd}{}^a$	CM	$CM_{calcd}{}^a$	CM	$CM_{calcd}{}^a$	CM	$CM_{calcd}{}^a$
X = CH_3CN^b				CTABr				
0	9.4		9.6		9.8		10.0	
4	11.0	12.4	13.0	13.9	15.5	15.6	18.5	17.2
8	16.0	15.8	18.8	17.6	22.2	19.6	26.0	21.4
12	21.0	20.1	23.0	22.2	26.0	24.6	28.0	26.7
16	25.0	25.7	28.0	28.1	31.0	30.8	34.0	33.3
20	31.0	32.8	34.0	35.6	36.0	38.6	39.0	41.5
24	39.0	41.9	43.0	45.0	46.0	48.4	49.0	51.8
28	50.0	53.4	52.0	56.9	55.0	60.7	58.0	64.6
32	77.0	68.2	80.0	72.0	84.0	76.1	87.0	80.6
44	140.0	141.6	144.4	145.6	149.0	149.9	156.0	156.1
56	226.0	c	226.0	c	232.0	c	240.0	c
64	484.0	c						
				SDS				
0	82		85		88		90	
4	120	112	136	133	140	133	155	144
8	148	147	156	168	162	168	176	181
12	192	193	208	214	215	214	225	227
16	250	252	262	272	268	272	275	285
20	320	330	335	346	348	345	365	358
24	440	433	440	440				
THF^b				CTABr				
0	9.4		9.6		9.8		10.0	
4	33.0	35.4	35.0	38.6	36.0	41.4	18.5	17.2
8	53.0	53.3	58.0	57.1	63.0	60.3	26.0	21.4
12	80.0	80.2	83.0	84.4	86.0	87.9	28.0	26.7
16	124.0	120.6	130.0	124.8	134.0	128.2	34.0	33.3
20	180.0	181.6	182.0	184.4	184.0	186.8	39.0	41.5
25	240.0	c	241.0	c	241.0	c	49.0	51.8
				SDS				
0	82		85		88		90	
4	192	193	212	211	224	222	248	246
8	225	225	238	238	250	245	265	265
12	265	261	265	268	260	272	280	286
16	300	302	304	302	308	302	312	308
$HOCH_2CH_2OH$	CTABr		TTABr		DTABr			
0^d	9.25		36.2		145			
20	13.1	12.1	45.9	39.2	173	163		
35	21.5	22.6	63.0	70.5	231	244		

Continued.

TABLE 1.4 *(Continued)*
Effects of Organic Cosolvent (X) and Temperature on Critical Micelle
Concentration (CMC) of Surfactants Obtained by Different Methods

Temp/°C =	25		30		35		40	
	10^4	10^4	10^4	10^4	10^4	10^4	10^4	10^4
[X]/% v/v	CM	CM_{calcd}^a	CM	CM_{calcd}^a	CM	CM_{calcd}^a	CM	CM_{calcd}^a
	CTABr		TTABr		DTABr			
50	42.6	42.3	129	127	369	365		
12e	11.5	10.1	40.0	38.5	160	157		
25	16.0	17.9	51.0	53.5	195	201		
35	28.5	27.8	70.0	68.9	245	242		

Note CM = CMC with units expressed in M.

ᵃ Calculated from Equation 1.2 with [X], CMC_0, and θ summarized in Table 1.3 and Table 1.4.
ᵇ CMC values are obtained from Reference 70.
ᶜ The value of CMC did not fit to Equation 1.2.
ᵈ CMC values at 25.2°C are obtained from Reference 71.
ᵉ CMC values at 25.2°C are obtained from Reference 72.

THF, the CMC_0 values are nearly 3-fold larger than the corresponding values of CMC at [X] = 0 (Table 1.4). The sharp increase in CMC compared to that predicted by Equation 1.2 (at initial very low contents of THF) may be the consequence of the difference in orientation effect of the organic cosolvent CH_3CN and THF in the mixed aqueous solvent. The values of CMC of CTABr, TTABr, and dodecyl-trimethylammonium bromide (DTABr), obtained at 20, 35, and 50% w/w ethylene glycol (EG)[71] as well as 12, 25, and 35% w/w EG[72] (Table 1.4) have been treated with Equation 1.2, and the least-squares-calculated values of CMC_0 and θ are summarized in Table 1.3. Although CMC_{calcd} values are comparable with the corresponding experimentally determined CMC values, the values of CMC at [EG] = 0 (Table 1.4) are 50 to 100% larger than the corresponding CMC_0 values for the data treatment using CMC values obtained within 20 to 50% w/w EG, whereas such a large discrepancy is not so evident for the data treatment using CMC values obtained within 12 to 35% w/w EG (Table 1.3). The most obvious reason for this discrepancy is the absence of data points below 20% w/w EG and perhaps inclusion of a data point at 50% w/w EG in the data treatment with Equation 1.2. It appears from Table 1.2 and Table 1.4 that the CMC values obtained within a large range of organic cosolvent cannot fit satisfactorily to Equation 1.2.

The increase in the concentration of tetra-*n*-butylammonium bromide (*n*-Bu$_4$NBr) from 0.0 to 6.0×10^{-3} *M* decreased CMC of aqueous solution of sodium dodecyl benzenesulfonate (SDBS) from 2.5×10^{-3} to 4.7×10^{-4} *M* (Table 1.5).[73] These observed data fit reasonably well to Equation 1.2 where X = [*n*-Bu$_4$NBr]. The calculated values of CMC_0 and θ are summarized in Table 1.6. The value of CMC_0 is not significantly different from the experimentally determined CMC

TABLE 1.5

Effects of [n-Bu$_4$NBr] and [Urea] on Critical Micelle Concentration (CMC) of Surfactants

[X]/M	Surfactant	Temp/°C	10^3 CMC/M	10^3 CMC$_{calcd}$/M[b]
n-Bu$_4$NBr[a]	SDBS			
0		30	2.5	
0.002		30	1.28	1.27
0.004		30	0.71	0.75
0.006		30	0.47	0.44
Urea[c]	DDPI			
0		45	6.70	
3.4		45	11.8	12.1
5.9		45	17.1	16.6
8.0		45	21.3	21.5
0[d]		25	4.75	
0.96		25	5.75	5.92
3.4		25	9.10	8.86
5.9		25	13.3	13.4
0[d]		45	5.63	
0.96		45	7.10	7.32
3.4		45	11.0	10.7
5.9		45	15.7	15.8

[a] CMC values obtained from Reference 70.
[b] Calculated from Equation 1.2 with [X] = [n-Bu$_4$NBr] and CMC$_0$ = 2.2×10^{-3} M and $\theta = -265$ M^{-1} as mentioned in the text.
[c] CMC values obtained from Reference 73.
[d] Surfactant solution contained 0.001-M Na$_2$S$_2$O$_3$.

value (2.5×10^{-3} M) at [n-Bu$_4$NBr] = 0.[73] The increase in [urea] increases CMC of dodecylpyridinium iodide (DDPI) in water at 25 and 45°C.[74] These data fit to Equation 1.2 where X = [urea] and the least-squares calculated values of CMC$_0$ and θ are shown in Table 1.6. The CMC of SDS micelles decreases upon addition of salt and this decrease follows the expression[75]

$$\log(CMC) = -A_1 - A_2 \log(CMC_0 + [Na^+]_{added}) \qquad (1.3)$$

where [Na$^+$]$_{added}$ is the molar concentration of added salt, $A_1 = 3.6$, and $A_2 = 0.678$.

1.1.3 THERMODYNAMICS OF MICELLE FORMATION

The thermodynamic formulation of micellization is based on the assumption that micelles exist in equilibrium with micelle-forming surfactant monomers as expressed by Equation 1.4 for nonionic surfactants.[76]

TABLE 1.6
Values of Fitting Parameters, CMC$_0$, and θ Calculated from Equation 1.2 for Different Surfactants in the Presence of n-Bu$_4$NBr and Urea

X	Surfactant	Temp/°C	10^3 CMC$_0$/M	θ/M^{-1}
n-Bu$_4$NBr[a]	SDBS	30	2.16 ± 0.17[b]	−265 ± 25[b]
Urea[a]	DDP	45	7.94 ± 0.73	0.125 ± 0.013
		25[c]	5.06 ± 0.26	0.165 ± 0.010
		45[c]	6.30 ± 0.32	0.156 ± 0.011

[a] CMC values are listed in Table 1.5.
[b] Error limits are standard deviations.
[c] Values of CMC$_0$ and θ are obtained in the presence of 0.001-M Na$_2$S$_2$O$_3$.

$$\{(n-N)/N_A\} \text{ monomer} \underset{}{\overset{K_M}{\rightleftharpoons}} (N/rN_A) \text{ micelle} \qquad (1.4)$$

where n represents total number of surfactant molecules, N is the total number of surfactant molecules used up in the formation of (N/r) number of micelles where r is the mean aggregation number of micelles, and N$_A$ is Avogadro's number. Concentration equilibrium constant K$_M$ may be expressed as

$$K_M = \frac{[\text{micelle}]^x}{[\text{monomer}]^y} \qquad (1.5)$$

where x = N/rN$_A$ and y = (n − N)/N$_A$. Thermodynamically, free energy of micelle formation, ΔG_m^0, is defined as $\Delta G_m^0 = -RT \ln K_M$, which, when combined with Equation 1.5, leads to Equation 1.6

$$\Delta G_m^0 = -RT\{x \ln[\text{micelle}] - y \ln[\text{monomer}]\} \qquad (1.6)$$

where R and T have their usual meanings. Under limiting conditions such as lim [micelle] → 0, N → 0 and y ln [monomer] → ln [CMC], Equation 1.6 is reduced to Equation 1.7

$$\Delta G_m^0 = RT \ln[\text{CMC}] \qquad (1.7)$$

Similar arguments can lead to Equation 1.8 for ionic univalent surfactants where β is the fraction of charges

$$\Delta G_m^0 = RT (1 + \beta) \ln[\text{CMC}] \qquad (1.8)$$

of micellized univalent surfactant ions neutralized by micelle-bound univalent counterions. Emerson and Holtzer[77] have shown that ΔG_m^0 represents the free energy for the addition of a single monomer to a micelle with most probable size, if the micelles are polydisperse.

Based on the assumption used in the derivation of Equation 1.7 and Equation 1.8, Zana[78] derived an equation, Equation 1.9, for the relationship between CMC of a surfactant in solution and its free energy of micellization, ΔG^0_M, for a general type of ionic surfactant. In Equation 1.9, i and j represent respective number of charged groups of valency z_s and number of alkyl chains connected by some spacer

$$\Delta G_m^0 = RT[(1/j) + \beta\ (i/j)\,|\,(z_s/z_c)\,|\,]\ \ln[CMC] + RT[(i/j)\,|\,(z_s/z_c)\,|\,\beta\ \ln(i/j)\,|\,(z_s/z_c)\,|\, - (\ln j)/j] \tag{1.9}$$

groups of an ionic surfactant A_j^{izs}, z_c is the valency of the counterion X^{zc}, and $\beta = p\,|\,z_c\,|\,/Ni\,|\,z_s\,|$ with N and p representing respective number of surfactant ions and counter ions in a micelle. In case $i = j = |\,z_c\,| = |\,z_s\,| = 1$, Equation 1.9 reduces to Equation 1.8.

Although Equation 1.7 and Equation 1.8 have very often been used for the calculation of free energy of micellization of nonionic and ionic univalent surfactants, respectively, there is a so-called mathematical paradox in the derivation of Equation 1.7 and Equation 1.8. For instance, $\lim N \to 0$, $K_M \to 0$ which means K_M is no longer constant under such limiting conditions and therefore Equation 1.7 and Equation 1.8 are no longer real equations for the calculation of free energy of micellization.

Equation 1.8 can be expressed in terms of enthalpy (ΔH_m^0) and entropy (ΔS_m^0) of micellization of ionic univalent surfactants as expressed by Equation 1.10.

$$[CMC] = \exp(\Delta H_m^0/((1 + \beta)RT))\ \exp(-\Delta S_m^0/((1 + \beta)R)) \tag{1.10}$$

The values of [CMC] at different temperatures as summarized in Table 1.2 were used to calculate ΔH_m^0 and ΔS_m^0 from Equation 1.10 by the use of nonlinear least-squares method. The calculated values of ΔH_m^0 and ΔS_m^0 at different contents of organic cosolvents ($HOCH_2CH_2OH$ and CH_3OH) are shown in Table 1.7. A reasonably good fit of experimentally determined CMC values to Equation 1.10 is evident from the CMC_{calcd} values summarized in Table 1.2.

Jolicoeur and Philip[79] used an alternative approach to calculate ΔH_m^0 from observed data (CMC vs. temperature). They found that CMC values of several nonionic and ionic surfactants obtained at different temperatures fit to the following empirical third-degree polynomial

$$\ln[CMC] = A + BT + CT^2 + DT^3 \tag{1.11}$$

where A, B, C, and D are empirical constant parameters. Gibbs–Helmholtz equation $d\ln K/dT = \Delta H^0/RT^2$ and Equation 1.7 lead to Equation 1.12

TABLE 1.7
Values of ΔH_m^0 and ΔS_m^0, Calculated from Equation 1.10 for CTABr in the Presence of Different Organic Cosolvent (X)

(1 + β)=	1.8		1.5		1.2		10^{10}
X/% v/v	$-\Delta H_m^0$	$-\Delta S_m^0$	$-\Delta H_m^0$	$-\Delta S_m^0$	$-\Delta H_m^0$	$-\Delta S_m^0$	$\Sigma d_i^{2\ a}$
20[b]	13.96 ± 1.07[c]	16.3 ± 3.4[c]	11.64 ± 0.89[c]	13.6 ± 2.9[c]	9.31 ± 0.71[c]	10.9 ± 2.3[c]	6.165
30	19.53 ± 1.52	37.0 ± 4.9	16.3 ± 1.26	30.9 ± 4.0	13.02 ± 1.01	24.7 ± 3.2	51.81
40	14.14 ± 1.85	23.4 ± 5.9	11.79 ± 1.54	19.5 ± 4.9	9.43 ± 1.23	15.6 ± 3.9	673.7
10[d]	12.46 ± 2.52	10.2 ± 8.1	10.38 ± 2.10	8.5 ± 6.7	8.31 ± 1.68	6.8 ± 5.4	16.94
20	14.84 ± 1.74	21.4 ± 5.6	12.36 ± 1.45	17.8 ± 4.6	9.89 ± 1.16	14.3 ± 3.7	56.32
30	10.44 ± 1.93	12.1 ± 6.2	8.70 ± 1.61	10.1 ± 5.2	6.96 ± 1.28	8.1 ± 4.1	1182

Note: The units for ΔH_m^0 and ΔS_m^0 are kcal mol^{-1} and cal K^{-1} mol^{-1}, respectively.

[a] Least-squares value (Σd_i^2 where $d_i = CMC_i - CMC_{calcd\ i}$ with CMC_i and $CMC_{calcd\ i}$ representing respective experimentally determined and calculated critical micelle concentration at the i-th temperature) remained unchanged with change in β.

[b] X = $HOCH_2CH_2OH$.

[c] Error limits are standard deviations.

[d] X = CH_3OH.

$$\Delta H_m^0 = -RT^2(d\ ln[CMC]/dT) \tag{1.12}$$

Equation 1.11 and Equation 1.12 give Equation 1.13, which was used to calculate ΔH_m^0 by using the values of B, C, and D calculated from Equation 1.11.[79]

$$\Delta H_m^0 = -RT^2(B + 2CT + 3DT^2) \tag{1.13}$$

The values of ΔG_m^0 and ΔS_m^0 were calculated from Equation 1.7 and Equation 1.14, respectively.

$$\Delta S_m^0 = (\Delta H_m^0 - \Delta G_m^0)/T \tag{1.14}$$

The values of CMC summarized in Table 1.2 could not fit to Equation 1.11 with [CMC] expressed on molar concentration scale. Instead, these CMC values fit to Equation 1.15.

$$ln[CMC] = A + BT \tag{1.15}$$

where [CMC] expressed on molar concentration scale. The critical micelle concentrations of chlorhexidine digluconate in aqueous solution were found to fit to Equation 1.11 with D = 0 and CMC in mole fraction units.[80] Recently, Kim and Lim[81] used Equation 1.8 and the empirical linear relationship between ΔH_m^0 and

TABLE 1.8
Values of A and B, Calculated from Equation 1.15
for CTABr in the Presence of Various Organic
Cosolvents (X) at Different Temperatures

X/% v/v	−A/M	10^2 B/M K^{-1}	Temp range/K
20[a]	20.2 ± 0.9[b]	3.91 ± 0.30[b]	298–318
30	24.0 ± 1.7	5.39 ± 0.55	298–318
40	19.2 ± 1.7	4.17 ± 0.55	298–318
10[c]	18.6 ± 1.9	3.29 ± 0.61	298–318
20	19.7 ± 1.6	3.97 ± 0.52	298–318
30	16.3 ± 1.7	3.29 ± 0.56	298–318

[a] $X = HOCH_2CH_2OH$.
[b] Error limits are standard deviations.
[c] $X = CH_3OH$.

ΔS_m^0 (i.e., enthalpy–entropy compensation for micellization) to derive an equation similar to Equation 1.15 with an additional term C/T on the right-hand side.

The values of unknown empirical constants A and B were calculated from Equation 1.15 by using the least-squares method and these calculated values are summarized in Table 1.8. The reasonably good fit of observed data to Equation 1.15 is evident from CMC_{calcd} values (Table 1.2) and from the standard deviations associated with the calculated values of A and B (Table 1.8). The values of B at different contents of organic cosolvents (X) were used to calculate ΔH_m^0 from Equation 1.16,

$$\Delta H_m^0 = -RT^2(1 + \beta)B \qquad (1.16)$$

which is similar to Equation 1.13 except with the respective absence and presence of additional multiple C and D terms and $(1 + \beta)$, which appears because of the univalent cationic nature of micelle-forming surfactant CTABr. The reported value of β for CTABr micelle in aqueous solution with the absence of any additive is in the range 0.7 to 0.9.[82] But the presence of the significant amounts of organic cosolvent additives such as 1,2-ethanediol and methanol are expected to decrease β value. Degree of CTABr micellar ionization, α (=1 − β), at 25.2°C increases from 0.21 to 0.29 with increase in the content of EG from 0 to 50% w/w in mixed aqueous solvents.[71] In the absence of experimentally determined β values under the experimental conditions of study reported in Reference 69 and Reference 42, arbitrary values of β as 0.8, 0.5, and 0.2 have been considered for the calculation of ΔG_m^0 and ΔH_m^0 from Equation 1.8 and Equation 1.16, respectively. The values ΔS_m^0 at different T were calculated from Equation 1.14. The calculated values of ΔG_m^0, ΔH_m^0, and ΔS_m^0 at different values of X and T are summarized in Table 1.9.

TABLE 1.9
Values of ΔG_m^0, ΔH_m^0 and ΔS_m^0, Calculated from Equation 1.8, Equation 1.16, and Equation 1.14, Respectively, for CTABr in the Presence of Various Organic Cosolvents (X) at Different Temperatures

$(1 + \beta)=$		1.8			1.5			1.2		
X	Tem	$-\Delta G_m^0$	$-\Delta H_m^0$	$-\Delta S_m^0$	$-\Delta G_m^0$	$-\Delta H_m^0$	$-\Delta S_m^0$	$-\Delta G_m^0$	$-\Delta H_m^0$	$-\Delta S_m^0$
20[a]	25	9.03	12.42	11.4	7.52	10.35	9.5	6.02	8.28	7.6
	30	9.03	12.84	12.6	7.53	10.70	10.5	6.02	8.56	8.4
	35	9.01	13.27	13.8	7.51	11.06	11.5	6.01	8.84	9.2
	40	8.85	13.70	15.5	7.37	11.42	12.9	5.90	9.13	10.3
	45	8.76	14.14	16.9	7.30	11.78	14.1	5.84	9.43	11.3
30	25	8.34	17.12	29.5	6.95	14.27	24.6	5.56	11.41	19.6
	30	8.40	17.70	30.7	7.00	14.75	25.6	5.60	11.80	20.5
	35	8.17	18.29	32.8	6.81	15.24	27.4	5.45	12.19	21.9
	40	7.93	18.89	35.0	6.61	15.74	29.2	5.29	12.59	23.3
	45	7.75	19.49	36.9	6.46	16.24	30.8	5.17	13.00	24.6
40	25	7.12	13.24	28.6	5.93	11.04	17.1	4.74	8.83	13.7
	30	7.19	13.69	21.4	5.99	11.41	17.9	4.80	9.13	14.3
	35	6.96	14.15	23.3	5.80	11.79	19.4	4.64	9.43	15.6
	40	6.75	14.61	25.1	5.63	12.18	20.9	4.50	9.74	16.7
	45	6.75	15.08	26.2	5.63	12.57	21.8	4.50	10.05	17.5
10[b]	25	9.32	10.45	3.8	7.76	8.71	3.2	6.21	6.97	2.5
	30	9.41	10.80	4.6	7.84	9.00	3.8	6.27	7.20	3.1
	35	9.33	11.16	6.0	7.77	9.30	5.0	6.22	7.44	4.0
	40	9.43	11.53	6.7	7.86	9.61	5.6	6.29	7.69	4.4
	45	9.15	11.90	8.6	7.63	9.92	7.2	6.10	7.93	5.8
20	25	8.31	12.61	14.4	6.93	10.51	12.0	5.54	8.41	9.6
	30	8.40	13.04	15.3	7.00	10.86	12.7	5.60	8.69	10.2
	35	8.33	13.47	16.7	6.94	11.22	13.9	5.55	8.98	11.1
	40	8.16	13.91	18.4	6.80	11.59	15.3	5.44	9.27	12.2
	45	8.00	14.36	20.0	6.67	12.00	16.7	5.33	9.57	13.3
30	25	6.86	10.45	12.0	5.72	8.71	10.0	4.57	6.97	8.0
	30	6.87	10.80	13.0	5.72	9.00	10.8	4.58	7.20	8.7
	35	6.60	11.16	14.8	5.50	9.30	12.3	4.40	7.44	9.9
	40	6.60	11.53	15.8	5.50	9.61	13.1	4.40	7.69	10.5
	45	6.64	11.90	16.5	5.53	9.92	13.8	4.43	7.93	11.0

X	$(1 + \beta)$	$-\Delta G_m^{0\,c}$	$-\Delta H_m^{0\,c}$	$-\Delta S_m^{0\,c}$
20[a]	1.8	8.94 ± 0.12[d]	13.27 ± 0.68[d]	14.0 ± 2.2[d]
30	1.8	8.12 ± 0.27	18.3 ± 0.94	33.0 ± 3.0
40	1.8	6.95 ± 0.20	14.15 ± 0.73	24.9 ± 2.7
20	1.5	7.45 ± 0.10	11.06 ± 0.57	11.7 ± 1.8
30	1.5	6.77 ± 0.23	15.25 ± 0.78	27.5 ± 2.5
40	1.5	5.80 ± 0.17	11.80 ± 0.61	19.4 ± 2.0
20	1.2	5.96 ± 0.08	8.85 ± 0.45	9.4 ± 1.5
30	1.2	5.41 ± 0.18	12.2 ± 0.63	22.0 ± 2.0

TABLE 1.9 *(Continued)*
Values of ΔG_m^0, ΔH_m^0 and ΔS_m^0, Calculated from Equation 1.8, Equation 1.16, and Equation 1.14, Respectively, for CTABr in the Presence of Various Organic Cosolvents (X) at Different Temperatures

	$(1 + \beta)$	$-\Delta G_m^{0\ c}$	$-\Delta H_m^{0\ c}$	$-\Delta S_m^{0\ c}$
40	1.2	4.64 ± 0.14	9.44 ± 0.48	15.6 ± 1.6
10[b]	1.8	9.33 ± 0.11	11.17 ± 0.57	5.9 ± 1.9
20	1.8	8.24 ± 0.16	13.48 ± 0.69	17.0 ± 2.3
30	1.8	6.71 ± 0.14	11.17 ± 0.57	14.4 ± 1.9
10	1.5	7.77 ± 0.09	9.31 ± 0.48	5.0 ± 1.6
20	1.5	6.87 ± 0.13	11.24 ± 0.59	14.1 ± 1.9
30	1.5	5.59 ± 0.12	9.31 ± 0.48	12.0 ± 1.6
10	1.2	6.22 ± 0.07	7.45 ± 0.38	4.0 ± 1.3
20	1.2	5.49 ± 0.11	8.98 ± 0.46	11.3 ± 1.5
30	1.2	4.48 ± 0.09	7.45 ± 0.38	9.6 ± 1.2

Note: The units for X, Tem, ΔH_m^0, and ΔS_m^0 are % v/v, °C, kcal mol^{-1}, and cal K^{-1} mol^{-1}, respectively.

[a] X = HOCH$_2$CH$_2$OH.
[b] X = CH$_3$OH.
[c] Mean value of those obtained within the temperature range 25 to 45°C.
[d] Error limits are standard deviations.

It is evident from Table 1.7 and Table 1.9 that although the approaches of calculating of ΔH_m^0 using Equation 1.10 and Equation 1.16 are technically different, both approaches produced nearly the same ΔH_m^0 values under the same experimental conditions. The values of thermodynamic parameters listed in Table 1.9 show a continuous increase in ΔG_m^0 with increase in the contents of HOCH$_2$CH$_2$OH from 20 to 40% v/v and CH$_3$OH from 10 to 30% v/v at constant or slightly decreasing values of β under such conditions. But the variation of ΔH_m^0 and ΔS_m^0 with % v/v contents of HOCH$_2$CH$_2$OH and CH$_3$OH reveals U-shaped plots with minimum at 30% v/v HOCH$_2$CH$_2$OH and 20% v/v CH$_3$OH (Table 1.7 and Table 1.9). Such plots may be ascribed to the enthalpy–entropy compensation for micellization or aqueous solubilization of amphipathic molecules. Enthalpy–entropy compensation behavior is believed to appear owing to the fact that the molecular interactions between solubilizate and solvent molecules that decrease or increase ΔH_m^0 must also decrease or increase ΔS_m^0, thus causing a linear relationship between ΔH_m^0 and ΔS_m^0 with change in the solvent.

The micellization of various cationic surfactants including CTABr in pure aqueous solvent is predominantly entropy controlled (i.e., ΔS_m^0 values are positive) and ΔH_m^0 values are very close to zero or slightly negative or positive.[79] But contrary to these results in pure aqueous solvents, the micellization of CTABr in mixed H$_2$O-HOCH$_2$CH$_2$OH and H$_2$O-CH$_3$OH seems to be enthalpy controlled, and negative ΔS_m^0 values (Table 1.7 and Table 1.9) show an unfavorable entropy effect on micellization. Enthalpy-controlled rather than entropy-controlled micellization has been observed for SDS and cetylpyridinium chloride (CPC) in mixed H_2O-CH_3CN.[83] However, depending upon the mole fraction of organic cosolvent, both enthalpy-controlled and entropy-controlled micellization of SDS and CTABr is noted in mixed H_2O-CH_3CN and H_2O-THF.[70] The switch from predominantly entropy-controlled to predominantly enthalpy-controlled micellization of ionic surfactant with change in solvent from pure water to mixed water–organic cosolvent could be attributed to the effects of organic cosolvent additive on a three-dimensional structural network of water as well as solvation shells of surfactant monomers and micelles. The three-dimensional structural network of water is crumbled in the presence of significant amounts of alkanols and acetonitrile.[84] Thus, it becomes energetically easier to form a cavity to embed surfactant molecules in mixed water–alkanol or water–acetonitrile solvents than in a pure water solvent. Furthermore, compared to water molecules, methanol, 1,2-ethanediol, and acetonitrile molecules are expected to interact more strongly with cationic surfactant molecules. These two types of effects of molecular interactions between surfactant and solvent molecules result in negative ΔH_m^0 and ΔS_m^0 as obtained (Table 1.7 and Table 1.9).

Enthalpy–entropy compensation for micellization in aqueous solutions for a large number of nonionic and ionic surfactants has been observed although with some significant scattering of observed data from the correlation straight line.[79] The study on the effects of temperature and pure methanol (MeOH), ethanol (EtOH), n-propanol (n-PrOH), isopropanol (i-PrOH), n-butanol (n-BuOH), isobutanol (i-BuOH), tert-butanol (t-BuOH), 1,2-ethanediol (EtdOH), 1,2-propanediol (PrdOH), and 1,2,3-propanetriol (PrtOH) on CMC and thermodynamic parameters of micellization of 1-hexadecylpyridinium bromide reveals the following: (1) for alcohols, the plots of the CMC vs. temperature display U-shaped curves with minimum at 30°C, whereas in pure water, the CMC increases nonlinearly over the temperature range (10–50°C) studied; (2) CMC value varies in the order H_2O < n-BuOH < i-BuOH < t-BuOH < PrtOH < n-PrOH < i-PrOH < PrdOH < EtOH < EtdOH < MeOH; (3) the values of β decrease as the temperature increases; (4) whereas the drop in the ΔG_m^0 values over the given temperature range is only 1 to 2 kJmol^{-1}, the lowering of ΔH_m^0 and ΔS_m^0 is much more pronounced; and (5) enthalpy–entropy compensation plots give compensation temperatures, which lie in the range 289 to 313 K.[85] An enthalpy–entropy compensation effect has been observed with an isostructural temperature (T_c) of 300 K for both the micellization and interfacial adsorption of polyoxyethylene (10) lauryl ether ($C_{12}E_{10}$) in water.[86]

Isothermal titration microcalorimetry has been used to determine directly the enthalpy of micellization (ΔH_m^0) of 1-alkyl-4-n-dodecylpyridinium surfactants with several counterions in aqueous solutions.[87] The values of ΔH_m^0 for all these micelle-forming cationic surfactants vary within -1.7 to -22.4 kJ/mol at a constant temperature of 30°C. The increase in temperature from 30 to 50°C decreases ΔH_m^0 by nearly 8 to 13 kJ/mol for all surfactants of the study.[87] Recently, the enthalpies of micellization of two cationic surfactant series have been determined by calorimetry.[88]

1.1.4 STRUCTURE OF MICELLES

The structural aspects of micelles have been studied by an unusually wide variety of techniques such as x-rays, NMR, ESR, fluorescence, light scattering, calorimetry, kinetics, and techniques involving viscosity, conductivity, and surface tension measurements. Despite all this attention, micelles have managed to elude detailed understanding of their structural behavior and, as such, the structure of a micelle still remains a controversial topic. Attempts have been made from time to time to present a clearer picture of micellar structure, but they give no better than a crude and very approximate picture of the highly dynamic and extremely complex micellar structure. Among the best-known attempts are due to Hartley (Hartley model, 1935),[89] Menger (Menger model, 1979),[90] Dill and Flory (Dill–Flory model, 1980,1981,1984),[91–93] Fromherz (Fromherz model, 1981),[94] and Butcher and Lamb (Butcher–Lamb model, 1984).[95] The basis of all these models is the same, the hydrophobic region being wholly or partially encased by a hydrophilic region. These models are perhaps the best representation of micellar structure within the domain of experimental observations, obtained until the year of the proposed models, on the micellar-modulated reaction rates and physicochemical properties of solubilizates of diverse features in terms of structure, polarity, and ionic character as well as micelle-forming surfactants. Rusanov et al.[96] have proposed nanostructural models of micelles and premicellar aggregates. Two nanostructural models of molecular aggregates have been discussed: (1) the classical drop model, which assumes flexibility of hydrocarbon chains of molecules and their full immersion into the hydrocarbon core of a molecular aggregate, and (2) a quasi-drop model, which assumes partial outcropping of the chains in the strain-free state from the core.

It seems that atoms and molecules and their building blocks — fundamental particles — tend to attain the lowest energy state through a least energy path under a specific condition or environment. These atoms and molecules can smartly, perfectly, and effectively discriminate energy paths and energy states, which differ in respective energy barrier and energy state by one or less than one kcal/mol in a chemical or physical transformation from one state to another. However, we do not have any perfect theoretical model that could be used to predict very accurately (as atoms and molecules could determine) the lowest energy path and lowest energy state in a chemical or physical transformation from one state to the other involving a larger number of atoms or even larger number of molecules such as in micelle formation. I still remember the Evans memorial lecture delivered by the late Professor M. J. S. Dewar at the chemistry department of the Ohio State University, Columbus, Ohio, in 1978–1979. At the end of the talk, a person in the audience posed a question to Dewar: "Do you share the view of a well-known theoretical chemist that in the future, synthetic organic chemists could know the practical feasibility or nonfeasibility of various reaction steps of a multistep organic synthesis prior to the start of the synthesis?" In response, Dewar displayed again the first transparency of his talk, which contained a series of assumptions and empirical approaches involved in several theoretical models that are commonly used in the so-called molecular mechanics, and replied gently in a few words: "I don't think so."

The formation of a micelle and similar molecular aggregates involve various types of weak molecular interactions, and the theoretical evaluation of such weak molecular interaction energies for highly dynamic micellar structure is bound to contain a significant amount of uncertainty. For instance, the dynamics of hydrogen bonds among water molecules themselves and with the polar head groups (PHG) at a micellar surface have been investigated by long molecular dynamics simulations.[97] The lifetime of the hydrogen bond between a PHG and a water molecule is found to be much longer than that between any two water molecules, and it is shown that water molecules can remain bound to the micellar surface for > 100 psec, that is, for much longer than their average lifetime. The activation energy for such a transition from the bound to a free state for the water molecules is estimated to be ≈ 3.5 kcal/mol.[97] These results predict that the rate of a reaction between a free water molecule and a substrate at the micellar surface should be slower than the rate of the same reaction in the bulk water solvent, provided the fraction of free water molecules is much lower than that of PHG-bound water molecules in microreaction environment at the micellar surface. However, the rate of reaction of H_2O with ionized methyl salicylate (MS^-) remains essentially unchanged with change in [SDS] from 0 to 0.4 M at 0.03 M NaOH,[98] whereas the SDS micellar binding constant of MS^- is 6 M^{-1}.[99]

1.1.4.1 Effects of Concentration of Micelle-Forming Surfactant and Temperature

It is known that size of a micelle increases with increasing concentrations of surfactants.[100] Micelles are believed to be approximately spherical at surfactant concentration not very much greater above the CMC. As the surfactant concentration increases, the number of aggregates increases with an average constant size until, above a second critical concentration, referred to as the *second critical micellar concentration*, the micelles start to increase their average size and form cylindrical, rodlike, threadlike, or disklike aggregates.[101] Porte et al.[102] have argued that the sphere-to-rodlike transition of micellar structure must involve an energy barrier, and this idea is necessary to explain the presence of a second CMC. The direct experimental evidence for the existence of such a barrier and techniques to measure its energy as well as transition temperature (T_u) are provided by differential scanning calorimetry.[103,104]

The progress in the cryogenic transmission electron microscopic (cryo-TEM) technique allowed Bernheim–Grosswasser et al.[105] to show using cryo-TEM images the distinct coexistence of small spheroidal aggregates along with much larger rodlike micelles for aqueous cationic Gemini surfactant solution at concentrations between ~ 5 and 25 CMC. The proposal of the presence of a second

CMC led to the emergence of a molecular theory of the sphere-to-rodlike transition in dilute micellar solutions taking into account the molecular arrangement as a function of local curvature of the micelles.[106] This theoretical model predicts the presence of a second CMC as well as the coexistence of small aggregates along with much larger micelles. The use of small-angle-x-ray scattering (SAXS), static light scattering (SLS), and dynamic light scattering (DLS) measurements on salt-free aqueous solutions of a cationic Gemini surfactant {ethanediyl-1,2-bis(dodecyl dimethylammonium bromide)} reveal strong evidence for the coexistence of small, intermediate, and very large micellar sizes in the whole range of concentration of the study.[107] These observations are in good agreement with cryo-TEM observations on the same micellar system.[105] However, small-angle-neutron scattering (SANS) and DLS data indicate spherical structures of micelles of octadecyltrimethylammonium chloride and ammonium dodecyl sulfate at concentrations as high as 300 mM.[108] Compact packing of surfactant molecules in micelles is observed with increase in the total concentration of surfactant.

The early investigations on the so-called sphere-to-rodlike transitions were all concerned with the ionic surfactants SDS,[109] cetylpyridinium bromide,[110] tetradecyltrimethylammonium chloride,[111] and CTABr.[112–114] The sphere-to-rodlike transition occurs in a broad range of concentrations. The concentration at which micellar transition occurs seems to be technique dependent. A micellar transition of CTABr has been reported in the concentration range from 0.05 to 0.34 mol/kg by using viscosity, Rayleigh light scattering, ^{81}Br NMR, SAXS, and other techniques.[115–121] For nonionic n-alkyl (C_m)-poly-(ethylene glycol) (E_n)ethers, the increase in the molecular weight of molecular aggregates has been deduced from osmotic measurements ($C_{12}E_8$ and $C_{16}E_9$),[122] and from light-scattering studies ($C_{12}E_6$,[123] $C_{14}E_8$,[124] and $C_{12}E_5$[125]). A series of SANS experiments on the structure of water–nonionic surfactant (C_mE_n) systems ($C_mE_n = C_{12}E_6$, $C_{12}E_5$, $C_{10}E_4$, C_8E_5, and C_8E_3) accompanied by complementary ultralow shear experiments and depolarized light scattering reveal: (1) the trend of a sphere-to-rodlike transition with increasing temperature, (2) this trend depends on the nature of the surfactant and somewhat on concentration, and (3) surfactants with a relatively high ratio of m and n form rodlike micelles even at low temperatures. For instance, $C_{12}E_5$ and C_8E_4 form rodlike micelles, whereas C_8E_5 and $C_{12}E_6$ form spherical micelles at low temperatures.[126]

Theories advanced by Drye and Cates[127] and Bohbot et al.[128] predict the formation of networks of branched rodlike micelles. It is difficult to find experimental evidence to support the formation of multiconnected branched micellar networks instead of simple rodlike micelles. The only available technique, at the moment, to unambiguously distinguish between branched and unbranched micelles is cryo-TEM.[129–132] The use of cryo-TEM in the study of the micellar growth and network formation in aqueous solutions of $C_{12}E_5$ (0.5 wt%) shows the gradual formation of the multiconnected branched micellar networks from disconnected polymer-like micelles as temperature is raised from 8 to 29°C.[133] At very low temperatures and surfactant concentrations, the coexistence of short disconnected rodlike and spherical micelles is observed. The effect of temperature on micellar growth is explained through temperature effect on the end-cap energy of the micelles E_c (= the difference in the free energy of adding surfactant molecules to the cylindrical core as compared to adding surfactant molecules to the two spherical end-caps of the micelle), where E_c increases linearly with temperature.[133] The gradual formation of micellar networks is not only temperature dependent but also concentration dependent, which is consistent with theory.[127,128,134] At a very low temperature of 2°C, the coexistence of only spheroidal and rodlike micelles is observed even at a concentration as high as 1.5 wt%. The transition from unbranched to branched micelles depends on the difference between the junction and end-cap energies, E_j and E_c, respectively.[134] SLS and DLS experiments show that the mean micelle contour length, L, increases as $c^{0.5}$ where c represents surfactant concentration.[133]

1.1.4.2 The Random Micelle Aggregation Model for Sphere-to-Rodlike Transition

The random micelle aggregation model is based on the assumption that the growth of the micelles is brought about by the association or fusion of complete micelles, and this model is among the first models that have been applied to describe the sphere-to-rodlike transition.[135] The model is characterized by the chemical processes expressed as

$$M_i + M_j \rightleftarrows M_{i+j} \quad i, j = 1, 2, 3, \ldots \tag{1.17}$$

where M_i and M_j denote rodlike micelles consisting of a fused assembly of i and j small spheroidal micelles, respectively, acting as basic units. Each chemical process in the scheme expressed by Equation 1.17 is in principle characterized by an equilibrium constant, K_{ij}, and the corresponding rate constants k_{ij} for forward and k_{-ij} for backward processes. The model, in its simplest form, assumes equal equilibrium constants for each equilibrium step, that is, $K_{11} = K_{12} = K_{21} = K_{22} = K_{23} = \ldots = K$.

Thermodynamically, the random micelle aggregation model (Equation 1.17) is equivalent to the model of stepwise incorporation of only small spheroidal micelles, which is expressed as

$$M_{m-1} + M_1 \rightleftarrows M_m \quad \text{for } m = 3, 4, \ldots \tag{1.18}$$

where M_1 represents the intact small spheroidal micelle. This model also assumes equal equilibrium constants for each step, that is, $K_{21} = K_{31} = K_{41} = \ldots = K$.

Recently, Ilgenfritz et al.[136] have studied the energetics and dynamics of the growth of $C_{14}E_8$ and $C_{16}E_8$ surfactant micelles in the aqueous isotropic phase using the differential scanning calorimetric method, temperature-jump (TJ) relaxation kinetics, stop-flow (SF) kinetics, transient electric birefringence (TEB) dynamics, and viscosity measurements. The calorimetric data show that the transition from spherical to rodlike structures is an endothermic process by 2.1 kJ/mol of surfactant in the case of $C_{16}E_8$ and a similar value for $C_{14}E_8$. The cause of transition from spherical to rodlike structure is believed to lie in a loss of hydration of the polar ethoxy chain, leading to a change of its conformation and responsible for the observed barrier of 2.1 kJ/mol of surfactant molecule for transition. The loss of hydration would also lead to a reduced effective head-group area, leading to a larger packing parameter, P, which favors rodlike structure. In a simple picture of packing constraints put forward by Israelachvili,[137] the values of P (= v/a_0l, v = volume of hydrophobic part of the surfactant molecule, l = length of hydrophobic chain, and a_0 = head group area of a surfactant molecule) indicate the structural features of the surfactant aggregates. For example, for spherical micelles, $0 < P \leq 1/3$; for rodlike micelles, $1/3 < P \leq 1/2$; for vesicles and planar bilayer, $1/2 < P \leq 1$, and for reverse micelles, $P > 1$. The transition occurs in a narrow temperature range for both $C_{14}E_8$ and $C_{16}E_8$ surfactants; thus, the process must be highly cooperative.

The relaxation kinetic data on micellar growth obtained by TJ method (with observation of scattered and transmitted light), SF method (monitoring the dissociation of rodlike micelles), and the stress relaxation dynamic (monitoring TEB) are explained by using random micelle aggregation model and not by the stepwise incorporation of single surfactant molecule into preexisting micelles. At the transition temperature, the dynamics of the structural transition for $C_{14}E_8$ is characterized by association and dissociation rate constants of 4.4×10^6 M^{-1} sec^{-1} and 1.8×10^3 sec^{-1}, respectively, compared to the dynamics of $C_{16}E_8$ surfactant with the corresponding respective rate constants 8.5×10^6 M^{-1} sec^{-1} and 29 sec^{-1}. The rate constants of micellar fusion increase, whereas those of micellar scission decrease with increasing temperature. The reaction step of dissociation or fission with apparent negative activation energy cannot be considered an elementary step with single activation barrier. It could be interpreted in terms of a fast equilibrium chemical process preceding the rate-determining step. The authors have suggested a few alternative possibilities for the apparent negative activation energy for dissociation of rodlike micelles of $C_{14}E_8$ and $C_{16}E_8$ surfactants.[136] The interesting finding of nearly 100-fold larger values of both association and dissociation rate constants for $C_{14}E_8$ compared to the corresponding rate constants for $C_{16}E_8$ remains to be explained.

1.1.4.3 Effects of Additives

Cationic micelles, especially CTABr, grow with the increase in the concentration of counterions. But, such micellar growth is less important with cetyltrimethylamooinium chloride (CTACl).[138] It is widely believed that micellar growth due to increasing concentration of counterions is accompanied by micellar structural changes from spherical to rodlike or disklike.[139–144] The effects of various organic anions (altogether 24 anions) on micellar properties of CTABr and N-cetylpyridinium bromide (CPBr), examined by solvatochromic probe betaine dye (4-

(2,4,6-triphenylpyridinium-1-yl)-2,6-diphenylphenolate), led to the qualitative conclusion that the presence of organic anion additives caused sphere-to-rodlike micellar transition.[145]

Generally, the sphere-to-rodlike ionic micellar transition is achieved by the appropriate conditions of surfactant concentration, temperature, and presence of counterions. However, Kumar et al.[146-148] have shown that not only inorganic salts but nonionic organic compounds also such as n-alcohols, amines, and aromatic hydrocarbons are also potential candidates for such micellar structural changes. It has been proposed that interfacial partitioning of organic additives causes micellar growth whereas interior solubilization of nonionic organic additives produces swollen micelles.[146-148] Certain nonionic organic additives cause the breakdown of rodlike micelles into spherical micelles.[149] The sphere-to-rodlike micellar transition appears to get enhanced when nonionic organic additives and inorganic salts[146,147,150,151] or ammonium[152]/quaternary ammonium salts[153-156] are added simultaneously. In an attempt to study the effects of K^+ and Na^+ ions on CTABr micellar transition from spherical to rodlike structures in the presence of n-octylamine, it has been found that the effect of KCl on such micellar structural transition is much stronger than NaCl.[157] Potassium chloride additives are shown to screen out only coulombic forces between CTABr micelles whereas van der Waals interactions between micelles remain independent of KCl.[158]

SLS and DLS experiments on the aqueous micellar solutions of mixed surfactants SDS and cocoamidopropyl betaine (CAPB) show that the mixed micelles undergo a sphere-to-rodlike transition at total concentrations of surfactants as low as 10 mM and at molar fraction 0.8 of CAPB, whereas mixed micelles of CAPB and sodium laureth sulfate (sodium dodecyl-trioxyethylene sulfate, SDP3S) undergo from sphere-to-rodlike transition at higher total surfactant concentration of about 40 mM and at the same molar fraction of CAPB.[159] This difference in the

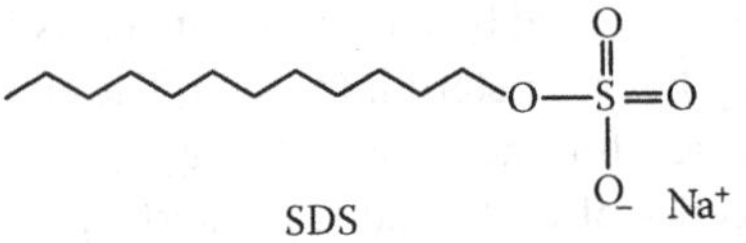

SDS

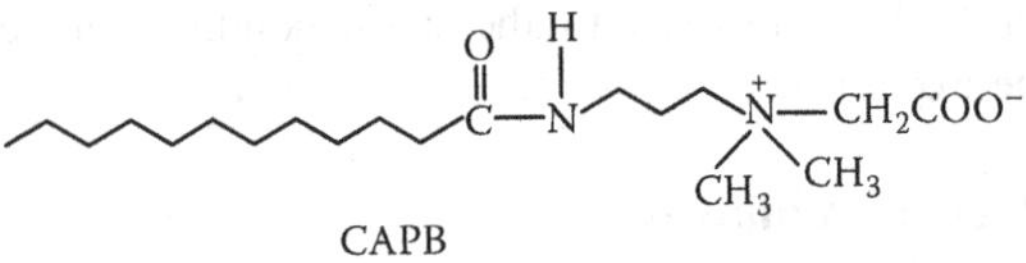

CAPB

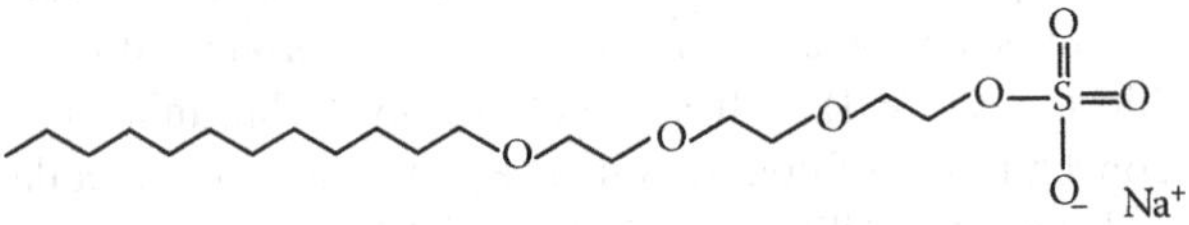

SDP$_3$S

transition concentrations for SDS and SDP3S is explained in terms of relative size of head groups of SDS and SDP3S. The bulkier SDP3S head group leads to a larger mean area per molecule in the micelles containing SDP3S and, hence, to smaller spontaneous radius of curvature of the micelles, which, in turn, leads to a less favored transition from spherical to rodlike micelles.

The mechanistic details of the sphere-to-rodlike micellar structural transition in aqueous surfactant solutions caused by variations in the concentrations of surfactant, additives, and temperature are not fully understood at the moment. Mere speculative explanations for such fascinating structural transitions of molecular aggregates may be extracted from the literature, which are described as follows:

1. The decrease of the content of water of hydration of head groups of nonionic and ionic micelles appears to cause predominance of van der Waals attractive interaction among surfactant molecules or energetically favorable micelle–micelle interaction.[160] Thus, the factors that could cause dehydration of micellar head groups may facilitate sphere-to-rodlike transition of micelles.

2. The substantial SDS micellar growth due to increase in [n-Bu$_4$NBr] is attributed to the presence of both cationic nitrogen and hydrophobic butyl chains of n-Bu$_4$N$^+$. Some of these butyl chains of n-Bu$_4$N$^+$ may penetrate into the micellar interior owing to hydrophobic interactions.[161] Under such situations, icebergs on the butyl chains penetrating toward the micellar interior will break down, thereby increasing the entropy of the system. This entropy increase may be a driving force for the micellar growth.[144] This explanation predicts that alkane hydrocarbon solubilization by nonionic micelles should lead to micellar growth. Furthermore, the long-held concept of iceberg hydration shells of aliphatic hydrocarbons in aqueous solvent is currently under debate.[162]

3 The positive charge of n-Bu$_4$N$^+$ is expected to decrease the effective charge of the anionic micelle, whereas the butyl chains may penetrate the micellar surface with a concomitant removal of water molecules from the micellar surface clefts, which could be responsible for micellar growth.[163]

4. n-Alkylamines such as n-octylamine molecules are solubilized in CTABr micelles through electrostatic and hydrophobic effects with amine groups left on the surface of the micelle. The electrostatic and hydrophobic interactions between micellar solubilized n-octylamine and micelles decrease the intramicellar coulombic repulsive forces and increase the hydrophobic forces among the monomers of the micelle and thus favor micellar growth.[157]

It seems certain that the existence of micellar aggregates (whether spherical, rodlike, or any other shape) is governed by the delicate balance between energetically unfavorable interactions between polar or ionic head groups of monomers and van der Waals attractive and repulsive interactions between hydrophobic segments of monomers in a micellar aggregate. At present, there is no easy, precise, and completely flawless experimental or theoretical technique that could be used to determine very accurately this delicate balance of attractive and repulsive molecular interactions in the formation of a micellar aggregate in aqueous solvent. Thus, any factor that facilitates micelle formation or micellar growth must either reduce the repulsive interactions between head groups of monomers or increase the attractive van der Waals interactions between hydrophobic segments of monomers in a micellar aggregate. The factor that causes both the reduction in repulsive interactions and increase in van der Waals attractive interactions simultaneously is considered to be the most efficient promoter for micelle formation and micellar growth. The structural features of additives are very important in regulating these opposing forces for the formation of micellar aggregates.

1.1.4.4 Effects of Specific Additives

Cationic surfactants in the presence of some specific counterions with relatively strong micellar binding affinity form micellar solutions, which can exhibit unusual physicochemical properties such as viscoelasticity.[164–169] An extensively studied viscoelastic system is the mixture of a cationic surfactant such as CTABr and sodium salicylate.[164,168,170–174] Apart from salicylate ion, a few more organic anions have also been found to cause rodlike/wormlike micelle formation in the presence of cationic surfactants.[175–183] These rodlike micelles show some characteristics of the so-called living polymer system.[184,185] The term "living" (or "equilibrium") polymer is used for linear macromolecules that can break and recombine. The length of these highly flexible micelles range from 100 to 200 nm. These micelles are composed of rigid regions of length ranging from 10 to 20 nm, which can break and recombine.

Spinnable, viscoelastic solutions of tetradecyltrimethylammonium salicylate ($C_{14}TAS$) consist of short rodlike micelles at < 25°C.[172] The addition of sodium salicylate (SS) additive lengthens rodlike micelles to 800 to 1600 nm in 0.1-M SS. Further addition of SS diminishes the micellar size as evident from the presence of only spherical micelles at 1.0-M SS. Electrophoretic mobility experiments indicate that with addition of SS, net micellar surface charge changes from positive to negative through neutral at ~ 0.1-M SS.[172] Rehage and Hoffmann[186] reported that the viscosity of aqueous solution of cetylpyridinium chloride (CPC) containing SS rises sharply until slightly above a 1:1 molar ratio of CPC/SS, and then the viscosity drops off drastically. On continued addition of SS, the viscosity rises again until the CPC/SS ratio reaches about 1:6 and then drops off again with continued addition of SS. Even today, no satisfactory explanations are available for these interesting unexplained observations.

Cationic micellar growth in salt-free aqueous solutions of CTA2,6-dichlorobenzoate, CTA2-chlorobenzoate, CTA3,5-dichlorobenzoate, CTA4-chlorobenzoate, and CTA3,4-dichlorobenzoate has been investigated by the use of SANS and rheology measurements.[177a] The aqueous solutions of CTA2,6-dichlorobenzoate and CTA2-chlorobenzoate reveal the formation of short rodlike micelles and entangled polymer-like micelles only at surfactant concentrations of > 70 mM and ≥ 200 mM, respectively. But the other three surfactants CTA3,5-dichlorobenzoate, CTA4-chlorobenzoate, and CTA3,4-dichlorobenzoate form highly viscoelastic aqueous solutions containing entangled wormlike micelles at surfactant concentrations much lower than 70 mM.[177a] The aqueous solutions of mixed nonionic surfactants, polyoxyethylene cholestryl ether (ChEO$_{15}$), and alkanoyl-N-methylethanolamide (NMEA-n) at 25°C exhibit a viscoelastic micellar solution of entangled wormlike micelles as suggested by rheological measurements.[187] The SLS and SANS study of nonionic C$_{16}$E$_6$ surfactant micelles doped with small amounts of ionic surfactant C$_{16}$SO$_3$Na reveals the presence of giant polymer-like micelles.[188] It has also been demonstrated that the aqueous solutions of nonionic surfactants doped with variable amounts of ionic surfactants can be used to investigate the influence of electrostatic interactions on micelle formation and micellar growth.

The aqueous solutions of cetyltrimethylammonium chloride, CTACl (5 mM) and 2-ClC$_6$H$_4$COONa, 3-ClC$_6$H$_4$COONa, and 4-ClC$_6$H$_4$COONa (each at 12.5 mM) studied by drag reduction, shear and extensional rheometry, and cryo-TEM reveal the presence of viscoelastic rodlike micellar networks with 4-ClC$_6$H$_4$COONa whereas only nonviscoelastic spherical micelles with 2-ClC$_6$H$_4$COONa are present at 20°C.[189] However, cryo-TEM images of aqueous solutions of CTACl (5 mM) + 3-ClC$_6$H$_4$COONa (12.5 mM) at 20°C show rodlike micellar networks in some pictures and vesicles in others. These authors[189] postulate that these different images of the same sample are because of variations in the level of shear the samples were subjected to during preparation. This is an important observation, which reminds one that direct micellar structural information obtained through cryo-TEM images are sample preparation technique dependent. Furthermore, Almgren et al.[190] have shown that a variety of structures, from lacelike to multiconnected threadlike aggregates, of SDS micelles revealed by cryo-TEM examination are formed at the air–solution interface and not in the bulk solution. Thus, caution should be exercised when the micellar structure obtained through cryo-TEM images is claimed to be similar to the micellar structure of the same aqueous surfactant solution without being subjected to the experimental conditions of cryo-TEM measurements. The fact that the presence of 2- ClC$_6$H$_4$COONa and 3-ClC$_6$H$_4$COONa, as well as 4-ClC$_6$H$_4$COONa, causes respective spherical and rodlike micellar networks is explained qualitatively in terms of specific positions of Cl substituent in the benzene ring of benzoate ion. The hydrophobic chlorine in the 2-Cl counterion must reside in an energetically unfavorable environment of aqueous phase and, hence, only spherical micelles are formed. The Cl groups in the 3-Cl and 4-Cl counterions reside into energetically favorable nonpolar hydrocarbon core of the micelles and, hence, stable elongated micelles are formed.

The micellar chain model for the origin of the viscoelasticity in dilute surfactant solutions containing specific additives is advanced based on the FT NMR
study on the effects of addition of SS, sodium m-chlorobenzoate (SMCB), and
sodium m-hydroxybenzoate (SMHB) to micellar solutions of CTABr.[166] The
carboxylic group protrudes out of the micellar surface in the case of SS and
SMCB, whereas it tilts toward the micellar surface in the case of SMHB. The
viscoelasticity exhibited in the cases of SS and SMCB is attributed to their unique
orientation on the micellar surface, which allows the formation of a chain of
micellar beads, with the additive acting as a bridge. The tendency of SS to form
dimmers is argued to have a synergistic effect on the formation of micellar
chain.[166,191] The presence of 0.5 M SS in decyltrimethylammonium bromide
micellar solution causes extensive micellar growth, whereas micellar growth
remains insignificant in the presence of 0.5 M SMCB and SMHB.[171b]

The formation of rodlike micelles from spherical micelles of alkylpyridinium
surfactants in aqueous solution is strongly dependent on the following: (1)
structure and hydrophobicity of counterion, (2) hydrophobicity and the type of
substituent as well as the substituent pattern of the aromatic ring counteri-
ons/additives, (3) the microenvironment of the counterion (substituent) in the
Stern region, and (4) the structure of the surfactant monomer (i.e., the surfactant
cation) where head group effects are proposed to be the main driving force for
sphere-to-rodlike transition.[176b]

One of the peculiar features of mixed water–surfactant solutions is the
inverted phase diagram, which shows a first-order phase separation as the temperature is increased, in contrast to conventional phase separations that are usually
observed as the temperature is decreased. Zilman et al.,[192] by using a general
theory of phase separation in equilibrium networks, have shown that the unusual
inverted temperature dependence of the phase behavior of the wormlike micelles
has its origin in the temperature dependence of the spontaneous curvature of the
surfactant layers comprising the micelles. At low temperature, the preferred
spontaneous curvature is high, favoring the semispherical end caps and long, unbranched rodlike micelles. As the temperature is increased, the spontaneous
curvature decreases, which favors branching and, eventually, phase separation.
These researchers have also shown theoretically that below a certain temperature,
T_r, where the junctions become sufficiently numerous compared with the end
caps, the relaxation time related to the approach to equilibrium for chain breaking
and coalescence is determined mainly by the former and reassembly via end-cap
fusion and is almost independent of the junction kinetics. However, for temperatures greater than T_r, the relaxation mechanism arises from the kinetics of
junction formation, and the end caps are negligible.

Although some qualitative explanations for the formation of flexible rodlike
micelles are found in the recent literature, it is not fully clear at the molecular
level how such micelles are formed and why they are formed in the presence of
some specific additives, whereas in the presence of other additives of the same
class of compounds, such micelles do not form.

1.1.5 Micellar Location of the Solubilizate

In view of the amphipathic nature and structural features of micelle-forming surfactant molecules, it is not illogical to believe that there should be a continuous and not an abrupt change in hydrophobicity/hydrophilicity of the micellar medium with continuous change in the distance from the micellar surface to the interior of the micelle. This behavior of micellar medium is expected to remain unchanged with a change in the shape and size of a micelle, because the change in micellar shape and size does not bring a significant change in the micellar surfactant molecular arrangement. The notion of continuous decrease in hydrophilicity with increasing distance from exterior to interior or core of a micelle is supported indirectly by some experimental and theoretical studies.[193–196] Micellar solubilization of a solubilizate is governed by mainly ionic or polar and van der Waals interactions between micellar medium and solubilizate and, consequently, the micellar location of a solubilizate is governed by these molecular interactions. Depending upon the specific molecular structure of the surfactant and solubilizate molecules, hydrogen bonding, and specific cation–π[197,198] interaction between solubilizate and micelle may also contribute to the specific micellar location of the solubilizate. As the micellar medium characteristics remain essentially unchanged with a change in micellar shape and size, the micellar location of a solubilizate is expected to be independent of micellar shape and size.

Various types of physicochemical studies aimed at investigating the micellar environment reveal that it is inhomogeneous in terms of various solution properties such as polarity/dielectric constant, water concentration, water structure, viscosity, and ionic strength (for ionic micelles only). Some studies[199–201] suggest that the micellar core is a relatively dense aliphatic medium, but the size of this medium is difficult to ascertain. Much of the characteristic properties of micellar medium are obtained by the use of optical probes, which are sparingly soluble in water but highly soluble in hydrophobic solvents.[202] Mishra et al.[203,204] have mapped the polarity of micellar environment with varied $E_T(30)$ values for the localization of some cationic cyanine-dye probes of varied hydrophobicity. From the studies of the electronic and emission spectra of these cationic dyes, they proposed that a micelle has a hydrophobic force field, and the dyes of varying hydrophobicity are localized in various pockets of the field. Multicompartmental micelles with distinct hydrophobic microdomains having different properties have been realized via micellar polymers bearing hydrocarbon and fluorocarbon fragments.[205]

The use of a probe molecule to study the properties of micellar medium requires the knowledge of the micellar binding site of the probe molecule, with absolute certainty under a specific condition, which is almost impossible to achieve by any experimental or theoretical technique. Pyrene, a highly hydrophobic water insoluble hydrocarbon molecule, has been used to determine the polarity of micellar medium,[206] but its precise micellar binding site is not known with absolute certainty. For instance, pulsed Fourier transform NMR and chemical shift analyses used to determine the dynamic solubilization site of the pyrene in cationic micellar solutions show that pyrene is solubilized in the interior of the

CTABr micelles.[207] But other studies, which used different experimental techniques, conclude that pyrene does not reside in the micellar interior; instead, it resides in a relatively polar micellar environment, i.e., interfacial region.[208,209] Microviscosities of the six micelles (SDS, CTABr, 2-phenyldodecanesulfonate, 3-phenyldodecanesulfonate, 4-phenyldodecanesulfonate, and 6-phenyldodecanesulfonate) have been obtained from the average reorientation times of two probes (2,5-dimethyl-1,4-dioxo-3,6-diphenylpyrrolo[3,4-c]pyrrole and coumarin 6) in micelles, and it has been observed that both the probes sense almost identical microviscosity for a single micelle.[210]

The seems to be no fully convincing and ambiguity-free technique to determine directly the exact and precise location of a solubilizate in the micelle. One- and two-dimensional ^{1}H NMR studies of mixed counterion surfactant systems, CTABr/CTAY (Y = 3,5-dichlorobenzoate ion and 2,6-dichlorobenzoate ion), indicate that dichlorobenzoate counterions intercalate among the surfactants' head groups and 3,5-dichlorobenzoate counterions insert farther into the interface of their rodlike micelles than the 2,6-dichlorobenzoate counterions insert into the interface of their spherical micelles.[177b] The orientational assignments of TTABr micellized 1-naphthoate, 2-naphthoate, 1-hydroxy-2-naphthoate, and 3-hydroxy-2-naphthoate ions have been studied by the use of ^{1}H NMR technique where the aromatic induced chemical shifts of the alkyl chain methylene protons demonstrate the deep penetration into the palisade layer by these anions.[211] The probable sites of SDS micellar incorporation of solubilizate molecules (phenol, 4-methylphenol, 4-allyl-2-methoxyphenol, anisole, 4-methylanisole, 4-propenylanisole, 1,8-cineole, and limonene) studied by proton NMR are found to depend upon hydrophobicity and structural features of solubilizates.[212] Phenolic compounds, being the most hydrophilic among the present solubilizates, reside at the hydrophilic/hydrophobic boundary of micelle–water interface, whereas aromatic methoxy and aliphatic compounds, being relatively more hydrophobic in nature, reside inside the micellar core.

The presence of salts in ionic micellar solutions appears to change the expected solubilization sites of organic additives.[150,151,153,154] Viscosity measurement studies of SDS micellar solutions containing organic additives, R_4NBr (R = n-C_4H_9), show that, with an appropriate R_4NBr concentration present in combination with a surfactant, change of the solubilization site from micellar interior to the interfacial region for even nonpolar compounds (such as aliphatic hydrocarbons) is possible.[156] The effects of counterions on the solubilization of benzene in cationic micelles, studied by ^{1}H, ^{13}C, and ^{14}N NMR spectroscopy, are explained in terms of a qualitative model in which benzene displaces H_2O from the clefts, which characterize the surface of the micellar aggregates.[213]

The solubilizates of different effective hydrophobicity are expected to occupy different microregions of a micelle. The approximate assignment of the location of a solubilizate in the micelle depends very much upon the type of technique used for this purpose and the interpretation of the results obtained. For example, benzene molecules are concluded to exist close to the interface between hydrocarbon and water based upon ^{1}H NMR chemical shift measurements in aqueous solution of CTABr.[214] The increase in the benzene concentration causes benzene molecules to enter the micellar interior after the micellar interface becomes saturated. The solubilization of benzene in CTABr and SDS micellar solutions has been studied by Fendler and Patterson[215] with pulse radiolysis and these authors arrived at the same conclusion as in Reference 214 for CTABr micelles, but in SDS micelles, benzene is always located in the interior. The UV spectral study on the solubilization of benzene in CTABr and SDS micelles concludes that in both micellar systems, benzene molecules were predominantly located in the interior of the micelles.[216] On the other hand, Mukerjee and Cardinal,[217,218] using the UV spectral technique, conclude that the micellar environment of solubilized benzene molecules in SDS and CTACl micelles is more polar at low value of [benzene] than at high value of [benzene]. Experimental evidence and theoretical considerations have been used to conclude that, depending upon benzene concentration, benzene molecules may be located both at the interface and in the interior of micelles of cetylpyridinium chloride, SDS, dodecylammonium chloride, and sodium diamylsulfosuccinate.[219] Studies on solubilization of benzene in DTABr micelles carried out by solubility, tracer diffusion coefficient, electrical conductivity, viscosity, and UV absorbance measurements, show that micelles swell owing to solubilization, and benzene molecules reside in the central core.[220]

1.1.6 KINETICS OF MICELLIZATION AND MICELLAR SOLUBILIZATION

Kinetic study of the reaction rates provides, perhaps, the most refined, detailed, and complete mechanism for chemical transformation from the reactant state to stable product state. But the interpretation of kinetic data, within the framework of a reaction mechanism, requires utmost care because sometimes haste and chemically/physically illogical interpretation of kinetic data lead to a reaction mechanism that later turns out to be wrong. A micelle or a micellar aggregate constitutes an inhomogeneous microreaction environment, which is highly dynamic, in the sense that its constituents (surfactant monomers) are in rapid equilibrium with surfactant monomers in aqueous phase. Thus, in a strict sense, a micelle or a micellar aggregate is not a separate phase like aqueous phase although it does provide microreaction medium, which is called *pseudophase*, in which micellar-mediated reactions occur. Kinetic studies on the rates of formation and breakup of a micelle under a variety of experimental conditions are expected to provide valuable information on energetics and mechanism, which control not only the stability but also shape, size, and rotational or spinning behavior[221] of a micelle. Various practical applications of micelles (such as adjuvant, microreaction environment, foaming, wetting, emulsification, solubilization, and detergency) are intimately connected to stability, shape, and size of a micelle. The importance of micellar kinetics in relation to technological processes is elegantly highlighted elsewhere in the literature.[222]

In the very early stage of our understanding about the fascinating properties of aqueous solutions of various types of surfactants, the notion of the presence of equilibrium between monomers of surfactant molecules (i.e., CMC) and micelles has been unequivocally established by various experimental approaches. The presence of an equilibrium between CMC and micelles is a strong indicator of the eminent fact that micelles are dynamic; i.e., they constantly form and break up in aqueous solutions. But studies on the kinetics of the formation and breakup of micelles could not be carried out until the advent of fast reaction kinetic techniques or chemical relaxation spectrometry, a term coined by Eigen to describe the experimental techniques that enable the study of the kinetics of fast reactions in solution whose lifetimes range from 1 to 10^{-10} sec.

1.1.6.1 Kinetics of Micellization

The rates of micellization in aqueous solution have been studied extensively since the late 1960s by using fast reaction kinetic methods (also known as *chemical relaxation methods*) such as temperature-jump (TJ), stopped flow (SF), pressure jump (PJ), and ultrasonic absorption spectrometry (UAS). These methods are briefly discussed by Fendler and Fendler[223] and thoroughly reviewed by Lang and Zana[224]. Just as any physical method used to monitor the reaction rate by observing the variation of a certain physical property of the reaction mixture as a function of reaction time, chemical relaxation methods provide the variation of a certain physical property of the sudden perturbed equilibrium system as a function of time for restoration of equilibrium. These observed data simply give the relaxation time or rate constants for the restoration of equilibrium of the micellar system that has been slightly perturbed by the use of a specific chemical relaxation method. A necessary requirement in the use of the chemical relaxation methods is that the perturbation of micellar equilibrium must be extremely mild. The most attractive mechanistic model for micelles formation is the so-called law of mass action. In 1935, Goodeve[225] pointed out that the simultaneous coming together of many micelle-forming monomer molecules to produce a micelle (Equation 1.19) is an improbable process for perturbation of micellar equilibrium caused by the use of chemical relaxation methods.

The observed relaxation times, obtained experimentally from the kinetic study of rates of formation and breakup of micelles under a variety of experimental conditions, can lead to the mechanism for micelle formation as expressed by Equation 1.20, which involves the addition of a monomer surfactant molecule to the existing aggregates of different aggregation numbers.

$$nA_1 \underset{k_{-1}^{nA1}}{\overset{k_n^1}{\rightleftharpoons}} A_n \tag{1.19}$$

$$A_1 + A_{n-1} \underset{k_{-1}^{A1,An-1}}{\overset{k_2^{1,n-1}}{\rightleftharpoons}} A_n \quad \text{where } n = 2,3,4,\ldots \tag{1.20}$$

In Equation 1.19 and Equation 1.20, A_1 represents the monomer of surfactant molecule, A_n is an aggregate (micelle) made up of n monomers of surfactant molecules, and k_n^1 and $k_2^{1,n-1}$ are respective n-th and second-order rate constants for aggregation of n monomers and single monomer with aggregate A_{n-1}, whereas k_{-1}^{nA1} and $k_{-1}^{A1,An-1}$ represent respective first-order rate constants for the exit of n and one monomer from A_n. The equilibrium processes represented by Equation 1.19 and Equation 1.20 are distinguished as micellar dissolution and dissociation/disintegration, respectively.

Although the mechanism of micelle formation is still incompletely understood, a generally accepted mechanism for such a complex dynamic process is represented by Equation 1.20. The stability of aggregate A_n (n = 2, 3, 4, ...) is governed by both polar/ionic and van der Waals attractive as well as repulsive forces. The rate constants, $k_2^{1,n-1}$, stand for diffusion-controlled association processes and, hence, the magnitudes of $k_2^{1,n-1}$ lie within 10^9 to 10^{10} M^{-1} sec^{-1} for all values of n ≥ 2. But the values of $k_{-1}^{A1,An-1}$ are expected to decrease with an increase in n. The kinetic requirement for the multiple-equilibrium processes as expressed by Equation 1.20 necessitate that $k_{-1}^{A1,An-1} \gg k_2^{1,n-1}[A_1]$ (where n = 2, 3, 4, ...). This kinetic condition reveals that the growth in size of A_n should cease when the inequality $k_{-1}^{A1,An-1} \gg k_2^{1,n-1}[A_1]$ is changed to $k_{-1}^{A1,An-1} \ll k_2^{1,n-1}[A_1]$. Thus, this kinetic analysis predicts that (1) $k_2^{1,n-1}$-step is the rate-determining step when n is equal to the aggregation number of the most stable aggregate (i.e., micelle) and (2) $[A_2]/[A_1] < [A_3]/[A_2] < [A_4]/[A_3] < ... < [A_n]/[A_{n-1}]$ because $k_2^{1,1} \approx k_2^{1,2} \approx k_2^{1,3} \approx ... \approx k_2^{1,n-1}$ and $k_{-1}^{A1,A1} > k_{-1}^{A1,A2} > k_{-1}^{A1,A3} > ... > k_{-1}^{A1,An-1}$.

As the multiple-equilibrium model (Equation 1.20) contains a large number of equilibrium constants, drastic simplifying assumptions about the relations between them must be made in order to derive a relationship between experimentally determined relaxation times and rate constants of micelle formation as expressed by Equation 1.20. Kresheck et al.[226] assumed that the rate-determining step is the loss of the first monomer from the micelle. In other words, the micelle reluctantly parts with one monomer molecule and then explodes. The role of equilibria involving the associations of intermediate species such that

$$A_j + A_k \underset{k_{-1}^{Aj,Ak}}{\overset{k_2^{j,k}}{\rightleftharpoons}} A_{j+k} \quad \text{where } j \text{ and } k > 1 \text{ and} < n \quad (1.21)$$

is also assumed to be negligible. Equation 1.21 represents the so-called micellar fusion–fission reaction mechanism where A_j, A_k, and A_{j+k} represent aggregates made up of j, k, and j + k surfactant monomers. The main appeal of these assumptions, which have been partially criticized in detail by Muller,[227] is the mathematical feasibility of deriving a simple relationship between relaxation time (τ) and rate constants as expressed by Equation 1.22

$$1/\tau = -k_{-1}^{A1,An-1}(n-1) + (nk_{-1}^{A1,An-1}C_D/[A_1]_e) \qquad (1.22)$$

where n is the aggregation number of micelle, C_D is the total concentration of micelle-forming surfactant ($C_D = [A_1]_e + n[A_n]_e$ where subscript e stands for equilibrium state), and $[A_1]_e$ is the equilibrium monomer concentration. The rate constant $k_{-1}^{A1,An-1}$ can be calculated from Equation 1.22 provided n is known.

An alternative assumption put forward by Muller[227] is that micelle formation and breakup may be expressed by Equation 1.19 with equal forward and backward rate constants for each step. This leads to Equation 1.23, which fails to account for the observed surfactant concentration dependence of τ.

$$1/\tau = 2k_{-1}^{A1,An-1}/n^2 \qquad (1.23)$$

Sams et al.[228] have offered an alternative two-state model for micellization, i.e., an associated state of micelles of various sizes and a nonassociated state of monomers, which predicts a single relaxation time (τ) and a relation between τ and the total concentration of surfactant (C_D) as revealed by Equation 1.24

$$1/\tau = k_f C_D - k_d \qquad (1.24)$$

where k_f and k_d represent rate constants for micelle formation and micelle dissociation, respectively.

Early results of the study of the rate of formation of micelles produced the relaxation times which, depending on the methods used, fell into two time domains: one occurring in the 10^{-6} to 10^{-9} sec (τ_1) and the other in the 1 to 10^{-3} sec (τ_2) range. In the late 1970s, it was realized that these relaxation times represent two different processes of the dynamics of micellization. It is still generally believed that the faster process, represented by τ_1, represents the dissociation of micelles as expressed by the equilibrium process

$$A_1 + A_{n-1} \underset{k^-}{\overset{k^+}{\rightleftharpoons}} A_n \qquad (1.25)$$

where k^+ and k^- represent second-order and first-order rate constants for the association of monomer A_1 with aggregate A_{n-1} and exit of A_1 from A_n, respectively. The slower process, represented by τ_2, is assigned to micelle formation to breakup as shown in Figure 1.3.[229] The fast relaxation time τ_1 can be attributed to a change in surfactant monomer concentration at constant concentration of micelles, whereas the slow relaxation time τ_2 involves the redistribution of the number density of micelles and can be attributed to changes in the concentration of the micelles, which remain effectively in equilibrium with the monomers throughout the slow process.

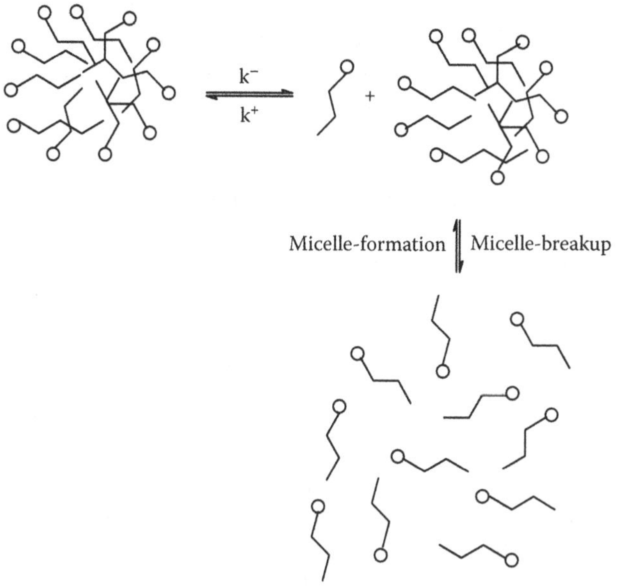

FIGURE 1.3 Intermicellar exchange of a surfactant monomer and of the micelle formation to breakup.

Aniansson and Wall (A–W) appear to be the first to develop a relatively more accurate and convincing kinetic model for micellization in conjunction with the multiple-equilibrium reaction scheme as shown by Equation 1.20.[230,231] The superiority of the A–W model over the others is that it predicts the presence of two discrete relaxation times (τ_1 and τ_2) during the course of micelle formation in the aqueous solutions of a single surfactant above CMC — a fact revealed by many experimental observations in related studies.[232] Although this model successfully predicts the presence of two discrete relaxation times, it is not fully tested in terms of (1) reproducibility of kinetic parameters derived from this model by using various chemical relaxation methods, and (2) kinetic parameters obtained from both relaxation times τ_1 and τ_2 have reasonably acceptable values.

The A–W theoretical model for the kinetics of micelle formation contains the following assumptions: (1) The concentration of monomers A_1 is always much higher than that of multimers (A_j with $1 < j < n$) so that bimolecular equilibria involving only aggregates (Equation 1.21) do not have to be considered; and (2) a direct contribution to the sonic spectrum from oligomers, A_j, is excluded by the presumed shape of the size distribution in thermal equilibrium of n-mers. To derive relations between parameters of the micelle solutions and parameters of ultrasonic spectra, which are suitable for verification of A–W theory, Teubner and Kahlweit considered the A–W model assuming small disturbances from thermal equilibrium due to harmonically alternating sonic fields.[233,234] Two relaxation processes with discrete relaxation times were identified. Broadband (0.7 MHz to 3 GHz) ultrasonic absorption studies of aqueous solutions of anionic, cationic, zwitterionic, and nonionic micelles, however, reveal more complicated spectra.[235,236] All micelle systems exhibit a relaxation term with relaxation time τ_0 between 0.1 and 0.3 nsec (25°C). In addition, the relaxation process with relaxation time τ_1 appears not to be one with discrete relaxation time.[235] Ultrasonic absorption spectra between 100 kHz and 2 GHz of aqueous solutions of *n*-heptylammonium chloride at 25°C reveal two relaxation regions, one at frequencies around a few megahertz and the other at around 1 GHz. But, in contrast to the theoretical predictions from the accepted models of the kinetics of micelle formation, the relaxation region at low frequency is subject to a relaxation time distribution.[237] The width of the distribution function is particularly large near the CMC. Close to the CMC, in addition, the relaxation time and relaxation amplitude do not follow the theoretically predicted dependencies upon concentration. The high-frequency relaxation seems to be partly due to the mechanism of chain (rotational) isomerization in the micellar core. The shear viscosity data support the idea that, at surfactant concentrations in the CMC region and slightly above the CMC, a variety of geometrically less defined supramolecular structures exist in the solutions of surfactant molecules rather than globular micelles that follow a comparatively narrow size distribution.[237]

Triblock copolymers of the type poly(ethylene oxide)-poly(propylene oxide)-poly(ethylene oxide) of general formula $EO_nPO_mEO_n$ are high-molecular-mass nonionic surfactants, which form micelles above CMC and CMT (critical micelle temperature) in aqueous solutions. The kinetics of micelle formation of EO_n-PO_mEO_n have been studied extensively by using ultrasonic relaxation,[232g,238,239] TJ,[234,232g,238,239] and SF[240] methods. In all these studies, the micellar system is characterized by two well-defined relaxation times as predicted by the A–W model. However, kinetics of micellization of block copolymers ($EO_nPO_mEO_n$), studied by the TJ method revealed only one relaxation time (τ) and the kinetic data analysis in terms of A–W model gave rate constants, k^+, for association of A_1 with A_{n-1}, in the range 0.2×10^7 to 2×10^7 M^{-1} sec^{-1}, which are not diffusion controlled.[241] On the other hand, three different relaxation processes have been observed in the kinetics of micellization of block copolymers, studied by iodine laser TJ technique.[242–244] The first and fastest relaxation (τ_1) occurred in the millisecond time range and is associated with unimer insertion into micelles. The second and third relaxations (τ_2 and τ_3) occurred in the time range 1 to 100 msec and are associated with micellar size distribution and micellar clustering, respectively. The values of $1/\tau_1$ and $1/\tau_2$ increased, whereas the values of $1/\tau_3$ decreased with increase in temperature. Waton et al.,[245] however, provided evidence that does not support the suggestion of a change from micelle formation–breakup to micellar clustering with change in temperature.

In a thorough and critical examination of the work on the kinetics of micellization of nonionic triblock copolymers published during 1996–2001, Thurn et al.[246] noticed the claim that the fast relaxation time associated with the single monomer/micelle exchange exists in two different time domains, namely, 10^{-4} to 10^{-5} sec (TJ) and 10^{-7} sec (UAS) for the same triblock copolymers. No satisfactory explanations could be found in the literature. In order to resolve this serious problem, Thurn et al.[246] carried out careful and detailed systematic studies, including ultrasonic relaxation measurements, on a number of triblock copolymers in both H_2O and D_2O as solvents. The data analysis of these studies led the authors[246] to conclude that the ultrasonic relaxation observed in micellar solutions of triblock copolymers is not associated with monomer/micelle exchange as defined in the A–W model. This conclusion is based on the following: (1) the unreasonable values of the resulting rate constants for association ($k^+ \sim 2.4 \times 10^{11}$ M^{-1} sec^{-1}) and (2) the amplitude values of the relaxation process being incompatible with those predicted for monomer/micelle exchange. A few alternative molecular mechanisms have been suggested for observed unusually short ultrasonic relaxation times.

1.1.6.2 Kinetics of Micellar Solubilization

Experimental observations based on diverse experimental techniques show that just as micelles are in rapid equilibrium with monomers (A_1), micellized solubilizate molecules are also in rapid equilibrium with solubilizate molecules in aqueous pseudophase as represented by Equation 1.26

$$S + A_n \underset{k_{sol}^-}{\overset{k_{sol}^+}{\rightleftharpoons}} A_{n+S} \qquad (1.26)$$

where S and A_{n+S} represent nonmicellized and micellized solubilizate, respectively. The second-order rate constant k_{sol}^+ stands for association of S with micelle (A_n) and k_{sol}^- is the first-order rate constant for exit of a solubilizate molecule from the micelle. It is interesting and surprising to note that the two rapid equilibrium processes, represented by Equation 1.25 and Equation 1.26, are found to be independent of each other. Thus, it is obvious that the dynamic behavior of micellar solubilization must be very similar to that of micellization. This speculation is supported by the observed values of k_{sol}^+ for various types of solubilizate molecules that fall within the range 10^9 to 10^{10} M^{-1} sec^{-1}, whereas the values of k_{sol}^- depend very much on the molecular characteristics of solubilizates, and they are in the range 10^4 to 10^6 sec^{-1} for a large number of nonionic solubilizates.[229] The variation in values of k_{sol}^- with the structural features of nonionic solubilizates appear to be reasonable.

It is apparent from the definitions of k_{sol}^+ and k_{sol}^- that the ratio k_{sol}^+/k_{sol}^- represents the micellar binding constant (K) of a solubilizate (S), which is related to apparent micellar binding constant $K_S = [S_M]/([S][Suf]_T-CMC)$ (where $[S_M]$ and $[Suf]_T$ represent concentration of micellized solubilizate and total concentration of surfactant, respectively) by the relationship: $K = n K_S$ with n representing the aggregation number of the micelle. The value of K_S of a solubilizate can be determined directly by various experimental techniques, and such a value should be comparable with K/n obtained from rate constants k_{sol}^+ and k_{sol}^- and aggregation number n. The values of k_{sol}^+, k_{sol}^-, and K_S (determined directly by the use

TABLE 1.10
Values of k^+_{sol}, k^-_{sol} and K_S for Selected Solubilizates in Micelles

Solubilizate	$k^+_{sol}/$ $(M^{-1}\,s^{-1})^a$	$k^-_{sol}/(s^{-1})^a$	$K/(M^{-1})^b$	$K_S/(M^{-1})^c$	K/K_S
Acetophenone	$\cong 2 \times 10^{10}$	7.8×10^6 [d]	2564	$17–48^d$, $18–26^e$, 19^f	151–53
Propiophenone	$\cong 2 \times 10^{10}$	3×10^6 [d]	6667	$27–81^d$, 49^e	247–82
Benzophenone	$\cong 2 \times 10^{10}$	2×10^6 [d]	10000	450^d	22
Acetone	$> 10^{10}$	$1–4 \times 10^8$ [d,e]	$> 25–100$	0.9^d	$> 28–111$
Ethyliodide	2×10^{10}	5×10^6 [d]	4000	22^d	182
1-Pentanol	7×10^9	8.4×10^6 [e]	833	$8–21^d$, $7.4–10^f$	104–40
1-Hexanol	1.5×10^9	4×10^6 [e]	375	$7–55^d$, 26^e, $19–42^f$	54–7
1-Bromonaphthalene	4×10^{10}	2.5×10^4 [d]	1.6×10^6		
		3×10^4 [d,g]	2×10^5 [g]		
Naphthalene		2×10^5 [d,g]	2×10^4	$352–400^d$, 1500^e	57–50
1-Methylnaphthalene				$1150–1300^d$	
Anthracene		2×10^4 [d,g]	4×10^5	$5340–6500^d$	75–31
				43000^e	
Pyrene		4×10^3 [d,g]	2×10^6 [g]		
Benzene		4×10^6 [d,g]	2×10^3 [g]	$19–33^d$, $36–46^e$,	105–61
				$18–35^f$	
Bromobenzene				$167–228^e$	
Toluene				$50–61^d$, $83–146^e$	
Ethylbenzene				274^d, $154–600^e$	
Oxygen	1.2×10^9 [i]	5×10^7 [d,g]	3×10^2 [g]	0.7^d, 0.72^e	429
Dodecylpyridinium ion [h]			4625^i		

[a] Unless otherwise noted all the values are obtained from Reference 229.

[b] $K = k^+_{sol}/k^-_{sol}$.

[c] Unless otherwise noted all the values are obtained from Reference 247.

[d] Values in SDS micelles.

[e] Values in CTABr micelles.

[f] Values in dodecyltrimethylammonium bromide (DTABr) micelles.

[g] Values obtained from Reference 248.

[h] Values obtained from Reference 249.

[i] Values in CTACl micelles.

of various experimental techniques) for a few selected solubilizates are summarized in Table 1.10. The values of K/K_S (= n) for SDS micelles vary from 7 to 429 (Table 1.10). Such a large variation in n with variation in the nature of solubilizates may be attributed to the weakness in both mechanisms or models and techniques used to determine k^+_{sol}, k^-_{sol}, and K_S.

Chemical relaxation methods such as SF, TJ, PJ, concentration jump, and ultrasonic relaxation techniques have been used to study the kinetics of the rate of micellar solubilization.[250,251] The residence time of a micelle at a given aggregation number is of the order of 0.2 to 10 μsec, whereas the mean lifetime of a micelle belongs to the millisecond time scale. As the time window usually applied in time-resolved fluorescence-quenching experiments does not exceed a few microseconds, micelles can be assumed to be frozen aggregates within the time domain of a few microseconds. The presence of certain additives, such as short-chain alcohols, can decrease the mean lifetime of a micelle enough to allow probe and quencher to be exchanged between micelles during the fluorescence decay. From the measurement of these exchange rates by fluorescence-quenching methods, the dynamic properties of micelles and micellar solubilization could be investigated. Thus, fluorescence quenching of a micellized fluorescent probe molecule by fluorescence quenchers (solubilizates) has been a most valuable tool to study the rate of micellization of fluorescence quencher molecules.[249,252,253] The kinetics of quenching of the fluorescence of micelle-solubilized probes by various quenchers both organic[254–259] and inorganic[260–262] have been the subject of extensive investigations.

1.1.6.3 Other Mechanisms of Intermicellar Exchange

Apart from well-accepted mechanisms for monomer/micelle exchange as expressed by Equation 1.25, which is the basis of the A–W model, there are two other less commonly suggested mechanisms for the fast rate of exchange of a monomer between micelles: (1) monomer exchange between two surfactant aggregates, A_n and A_{n-1}, occurring through transient intersurfactant molecular aggregate complexes, $(A_{n-1} \cdot A_n)$, $(A_{n-1} \cdot A_1 \cdot A_{n-1})$ and $(A_n \cdot A_{n-1})$ (Equation 1.27), rather than through intermicellar medium, i.e., aqueous pseudophase (Equation 1.25) and (2) exchange through fusion–fission

$$A_{n-1} + A_n \rightleftarrows (A_{n-1} \cdot A_n) \rightleftarrows (A_{n-1} \cdot A_1 \cdot A_{n-1}) \rightleftarrows (A_n \cdot A_{n-1}) \rightleftarrows A_n + A_{n-1} \quad (1.27)$$

reaction involving surfactant aggregates as expressed by Equation 1.21. Processes (1) and (2) occur less frequently than the exchange via the intermicellar phase.[263]

Various chemical relaxation methods have been used to determine the rates of various relaxation processes. However, it is difficult, with these data alone, to establish the mechanism associated with each rate. For instance, although the A–W model explains satisfactorily the fast relaxation process of micelle dynamics, it fails to give a satisfactory explanation for the slow relaxation process under certain specific conditions. In the extensive PJ and TJ measurements on both nonionic[264] and ionic[265] micelles, it was found that the surfactant concentration dependence of the slow relaxation rate was inconsistent with the A–W model (mechanism) at relatively high surfactant concentrations. These observations led to the proposal of the occurrence of additional mechanisms as expressed by Equation 1.21. A reasonably good fit of the experimentally observed data to a derived rate law based on a proposed reaction mechanism is generally considered a necessary but not sufficient proof or evidence for the validity of the proposed

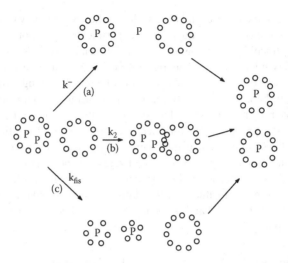

FIGURE 1.4 (a) Exchange via water mechanism, (b) Collision–exchange–separation mechanism, (c) Fission–growth mechanism.

mechanism. Thus, additional experimental proof for the mechanisms expressed by Equation 1.21, Equation 1.25, and Equation 1.26 are desirable.

There appear to be no techniques that could allow one to observe and distinguish between fusion and fission processes of normal micelles. Winnik et al.[75,266–271] have recently used water-insoluble or extremely low water-soluble fluorescent probes to study the kinetics of slow relaxation processes in micelles. The basic hypothesis that these authors considered in the development of this kinetic measurement method is that the intermicellar exchange of water-insoluble solutes involves micelle fusion or fission processes. The complete and most probable mechanisms for the solubilizate exchange among micelles are shown in Figure 1.4. Hilczer et al.[272] developed a detailed theoretical model to describe the kinetics of the exchange of solubilizate among micelles by the mechanisms shown in Figure 1.4. Pyrene derivatives have been used as fluorescent probes to study the solute-exchange kinetics that occur through micellar fusion and fission, for micelles of nonionic[266–269] and ionic[75,270,271] surfactants. Solutions are prepared in which some micelles contain more than one molecule of the fluorescent probe. The fluorescence spectra of these solutions are characterized by a prominent excimer emission. When one of these solutions is mixed under stopped flow conditions with an excess of empty micelles, exchange takes place. One can follow this process in a time-scan experiment by monitoring either the growth in intensity of the blue "monomer" emission I_M or the decrease in intensity of the green "excimer" emission I_E.

The magnitude of I_E is proportional to the fraction of micelles $P(t)$ bearing a pair of fluorescent probe molecules and the decrease in $P(t)$ due to the exchange probe molecule among micelles is given by Equation 1.28

$$I_E \propto P(t) = P(0) \exp(-k_{obs}t) \qquad (1.28)$$

where k_{obs} is pseudo-first-order rate constant for the exchange processes. Thus, Figure 1.4 and Equation 1.28 can lead to

$$k_{obs} = k^- + k_{fis} + k_2 \text{ [empty micelles]} \qquad (1.29)$$

where k^- represents first-order rate constant for the exit of probe (P) from the micelles, k_{fis} represents the first-order rate constant for the fission of micelles monitored through the decay of I_E, and k_2 is the second-order rate constant for the collision–fusion–fission process of the micelles where rate of fusion of micelles is assumed to be the rate-determining step. For a given micelle containing two probe molecules, a single event of fission or fusion–fission could have two outcomes. Both probes may remain in one micelle, leaving the other empty; or each probe could occupy one of the micelles. Only the second process can be observed in the experiment designed to study the kinetics of the solute exchange among micelles.[269] Thus, in the fusion–fission process, the true rate constant for this process is twice the experimental value of k_2. The fission can occur in many different ways, and fragments that are too small will be unable to solubilize a probe molecule. Thus, k_{fis} is a lower bound to the overall fragmentation rate.

If one carries out exchange measurements with a probe that is completely insoluble in water, the exit process is almost stopped; i.e., $k^- \approx 0$ and, under such conditions, Equation 1.29 reduces to Equation 1.30. The values of k_{fis} and k_2 for different probes and micelles have been obtained from Equation 1.30

$$k_{obs} = k_{fis} + k_2 \text{ [empty micelles]} \qquad (1.30)$$

and these results are summarized in Table 1.11.

Py-C$_8$, R = n-octyl

Py-C$_{12}$, R = n-dodecyl

Pyrene, R = H

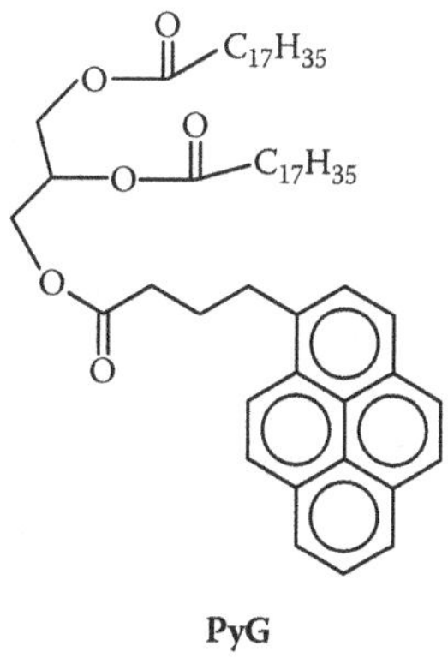

PyG

TABLE 1.11
Values of Rate Constants k_{fis} and k_2 for Solute Exchange in Micelles

Micelle	Solute	Temp/°C	k_{fis}/s^{-1}	$E_a{}^a$	$k_2/M^{-1} s^{-1}$	$E_a{}^a$	Reference
Triton X-100	PyG	24.6	10	110	1.5×10^6	160	266
$A_7{}^b$	PyG	24.2	0.85		8.7×10^4		269
	Py-C12	24.2	1.9		9.4×10^4		269
	Py-C8	24.2	3.4		8.5×10^4		269
Triton X-100	PyG	24.2	14		1.5×10^6		269
	Py-C12	24.2	40		1.8×10^6		269
	Py-C8	24.2	47		2.0×10^6		269
SDS	PyG	23	3.7×10^2		38		75
Triton X-100	Pyrene	25			5×10^8		273
			k_{obs}/s^{-1} ^{c}				
SDS	none		260–1600				274
LABA^d	none		0.2		$[Suf]_T{}^e = 0.4$ mM		275
			4		$[Suf]_T = 0.8$ mM		
			100		$[Suf]_T = 1.5$ mM		
			200		$[Suf]_T = 2.7$ mM		
C_{12}-maltose^f	none		0.1		$[Suf]_T{}^e = 0.4$ mM		275
			1		$[Suf]_T = 0.8$ mM		
			20		$[Suf]_T = 1.5$ mM		
			150		$[Suf]_T = 2.7$ mM		
TEDAd^g	none		0.3		$[Suf]_T{}^e = 0.8$ Mm		275
			15		$[Suf]_T = 1.4$ mM		
			200		$[Suf]_T = 2.4$ mM		
			900		$[Suf]_T = 3.7$ mM		

^{a} Energy of activation with units kJ/mol.
^{b} $A_7 = CH_3(CH_2)_8CH_2(OCH_2CH_2)_{6.5}OH$.
^{c} The rate constant k_{obs} was obtained from dynamic surface tension data.[274]
^{d} LABA = N-dodecyllactobionamide.
^{e} Total concentration of surfactant.
^{f} C_{12}-maltose = maltose 6′-O-dodecanoate.
^{g} TEDAd = *tetra*(ethylene oxide) dodecyl amide.

The bimolecular process [k_2-step] in Equation 1.30 for the exchange of solubilizates between micelles is believed to occur through a stepwise mechanism involving a reactive intermediate. The nature of this intermediate is unknown. Rharbi et al.[266] suggest two possible mechanisms for the k_2-step of Equation 1.30. One involves complete fusion of two micelles to form a super-micelle with a hydrophobic core in which solubilizates move freely. The other one involves adhesion (a sticky collision) of two micelles for a short time synchronized with exchange of solubilizate by permeation through the layer consisting of micellar head groups. These proposed mechanisms for k_2-step are shown in Figure 1.5. These authors[266] proposed $k_1{}^2$-step and $k_4{}^2$-step as the rate-determining step in

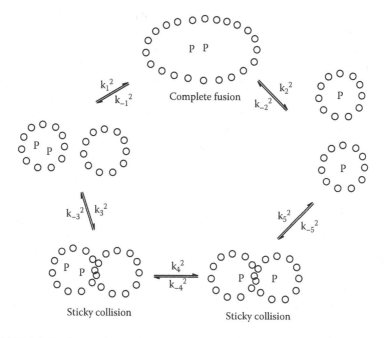

FIGURE 1.5 Exchange via complete fusion–fission mechanism (k_1^2-path), and via sticky collision–fission mechanism (k_3^2-path).

these alternative mechanisms. Based on the plausible assumptions that rate constants k_1^2 should be insensitive and k_4^2 should be sensitive to the structural features of solubilizates, the almost equal values of experimentally determined rate constants, k_2, for solubilizates PyG, Py-C_{12}, and Py-C_8 in A_7 micelles led Rharbi et al.[269] to conclude that exchange follows a mechanism that involves the formation of a completely fused super-micelle intermediate as the rate-limiting step. However, the values of second-order rate constants for solubilization of neutral solubilizates of diverse structural features such as acetone and pyrene are nearly diffusion controlled (Table 1.10). Thus, the equal values of k_2 for PyG, Py-C_{12}, and Py-C_8 in A_7 micelles may not be considered strong evidence for the preference of complete fusion mechanism over the sticky-collision mechanism.

The rate constants of k_{fis} and k_2, obtained for the exchange of PyG in Triton X-100 micelles,[266] have been used to calculate entropy of activation (ΔS^*) and enthalpy of activation (ΔH^*) by using the Eyring equation

$$k = (k_B T/h) \exp((\Delta S^*/R) \exp(-\Delta H^*/RT) \qquad (1.31)$$

where all the symbols have their usual meaning. The nonlinear least-squares calculated values of (ΔS^*) and (ΔH^*) are 4.2 ± 7.9 calK^{-1}mol^{-1} and 17.14 ± 2.34 kcal mol^{-1} for k_{fis} and 109 ± 5 cal K^{-1} mol^{-1} and 41.45 ± 1.67 kcal mol^{-1} for k_2. Theoretically, a unimolecular disintegration chemical process should involve the positive entropy of activation. Although the calculated value of ΔS^* (4.2 calK^{-1}mol^{-1}) for k_{fis} is associated with a standard deviation of more than 100%, a slightly positive value of ΔS^* is consistent with the proposal that this exchange is due to a fission-growth process. Significantly large positive value of ΔS^* (= 109 calK^{-1}mol^{-1}) for rate constant k_2 reveals that the k_1^2-step (Figure 1.5) cannot be the rate-limiting step. Thus, if the exchange of solubilizate among micelle follows the complete fusion mechanism, then the k_2^2-step should be the rate-limiting step, which is plausible because the rate of fission of a moderately big micelle is expected to be slower than the rate of association of two small micelles to form a big micelle. On the other hand, if the exchange follows the sticky-collision mechanism (Figure 1.5), then the k_4^2-step or k_5^2-step is most likely the rate-determining step.

Pseudo-first-order rate constants, k_{obs}, for exchange of PyG between SDS micelles are highly sensitive to the counterion concentration at a constant [SDS].[75] The exchange rate is not very sensitive to the concentration of SDS micelles, which indicates that the kinetics are dominated by a first-order process. This process is attributed to a fission-growth mechanism in which the fission rate is rate limiting. The energy barrier to fission is proposed as the creation of surface instabilities, which are enhanced in the proximity of the micelle-to-rod transition.[75] The significant second-order contribution to the exchange rate has been observed at a considerably higher salt concentration.[270]

1.1.7 Water Penetration into Micellar Environment

The extent of water distribution in the micellar pseudophase remains the topic of conflicting conclusions based on experimental data obtained to understand this extremely important aspect of micellar pseudophase. Most physicochemical properties of micellized solute molecules (i.e., solubilizates) are governed by the concentration and structural features of water molecules in the micellar microenvironment of solubilizates. However, in view of the forces involved in the formation of highly dynamic micelles, common sense suggests that not all of the hydrocarbon tail of a surfactant molecule or ion is removed from the contact with water in the micellar pseudophase, and there should be a continuous decrease in water concentration with increase in the distance from the exterior to the centre (core) of the micelle. The surface area per head group is larger than the cross-sectional area of the hydrocarbon chain (i.e., tail) for both ionic and nonionic surfactants,[276] which might allow water to penetrate between the chains, provided the radius of the spherical micelle is not significantly smaller than the extended hydrocarbon chain length of micelle-forming surfactant molecules. The presence of water molecules between adjacent hydrocarbon chains is considered to be energetically unfavorable, for the reason that it decreases the attractive van der Waals interaction involving dispersion forces between hydrocarbon chains separated partially or fully by water molecules.

Water penetration into micellar pseudophase is a controversial topic. In an excellent account,[90] Menger has summed up a bewildering array of opinions on this topic in a few words as follows: "Almost any conceivable hydration picture finds support somewhere in the literature. At one extreme lies the "reef" model: water does not penetrate beyond the ionic head group. At the other extreme, one finds the "fjord" model in which water percolates nearly to the center of the micelle."[277-282] Experimental observations of a diverse nature on the physicochemical behavior of micelles, which indirectly revealed water penetration much deeper than merely the head group (i.e., Stern layer/palisade layer) of micelles, led Menger to a micellar model named *porous cluster model* of micelle.[90] The basic assumptions involved in this model are described as follows:

1. The great majority of chains have six or more water-exposed carbons outside the nucleus.
2. The nucleus is defined as the region of direct chain contact which contains virtually no water and possesses a dielectric constant similar to that of pure hydrocarbon.
3. The nucleus comprises only a fraction of total micelle volume (perhaps 15 to 20% depending on where one locates the borders).
4. With a few of the chains, water molecules could reach well beyond the first six carbons.
5. Water concentration decreases as the distance increases from micellar surface or exterior region toward nucleus.
6. Stern layer should not really be called a "layer"; a better name would be "Stern region."
7. Stern regions penetrate effectively into micelles and occupy a large proportion of the micellar volume.

Although the porous cluster micellar model may not be considered completely different from the two-state micellar model of Hartley in view of the fact that the so-called Stern layer and micellar core of Hartley micelles have never been clearly and precisely defined, a constructive scientific debate has started among the proponent and opponent scientific groups of porous cluster model of micelle. Proponents[283-291] and opponents[292-299] of porous cluster micellar structure have published a series of papers over more than a decade since Menger proposed this model in 1979, but, as usual, the findings described in these papers always have almost equally convincing alternative explanations and thus, the real issue on water penetration in micelles remains unclear.

Fourier-transform infrared (FT-IR) spectra of the fluorescence probes n-(anthroyloxy) stearic acids (where n, the position of the anthroyloxy moiety, = 2, 6, 9, and 12), which presumably locate at different depth within the mixed Triton X-100-SDS micelles, indicated a very hydrophobic core, with SDS being in a much more open structure than Triton X-100.[298] Melo et al.[292] used n-(9-anthroyloxy) stearic acids (n = 2, 6, 9, and 12) to probe the water content in the vicinity of the anthroate chromophore in micelles of SDS, dodecytrimethylammonium chloride (DTC), and Triton X-100. The 12-(9-anthroyloxy)stearic acid probe revealed water content for DTC micelles as 19 M (from λ_f data) or 1 to 6

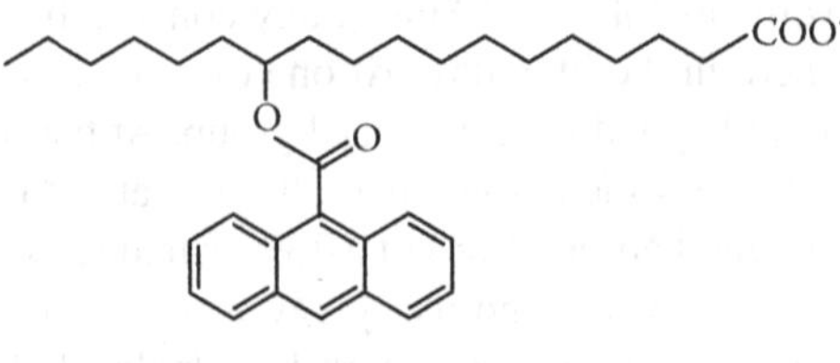

12-(9-anthroyloxy)stearic acid

M (from quenching data), and for SDS micelles as 2.5 M (from λ_f data) or 17 to 22 M (from quenching data). The obvious conclusion from these data, namely that micelle interiors are wet, was summarily dismissed in favor of other possibilities (including one in which the anthroyloxy group perturbs the local environment and allows the entry of water).[292] Menger and Mounier[289] synthesized cationic surfactants (CS), which, upon micellization in water solvent, buried a carboxylic substituent in the micelle interior. CMCs (determined tensiometrically) and pK_a (determined spectrophotometrically) of CS–8, CS–12, and CS–16 reveal that

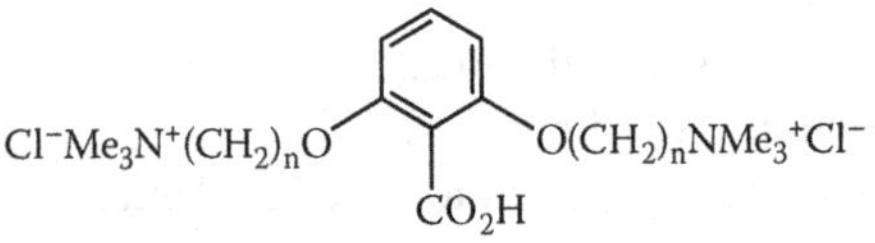

CS-8, n = 8
CS-12, n = 12
CS-16, n = 16

ionization has little effect upon the CMC, and micellization has little effect upon pK_a. A likely explanation for these facts is that micelles are loose molecular "brush heaps" with the interstices occupied by water. The carboxyl groups reside in the aqueous interstices. The alternative possibility that both chains simultaneously distort themselves so as to place all three polar groups onto the surface of normal-sized micelles is ruled out for the reason that it will create an extremely large chain disorder. Micellar environmental effects on vibronic band intensities in pyrene can be interpreted as a measure of the compactness of the head group structures and the extent of surface charge.[300] An alternative explanation can also be sought in terms of the extent of water penetration into the micellar systems. In micelles with compact head group such as SDS and lauryl ammonium chloride, the vibronic band intensities in pyrene monomer fluorescence indicate a smaller water penetration in these micelles compared to micelles with larger head groups such as CTABr and Brij35.[300] Certain solubilizates such as urea has been found to induce deeper water penetration in Triton X-100 micelles.[301]

Infrared spectral studies of pure sodium 7-oxooctane (7-oxo-Na-C$_8$, CMC ≈ 0.25 M) and mixed 7-oxo-Na-C$_8$ + sodium octanoate (Na-C$_8$) as well as 5-nonanone + Na-C$_8$ reveal the coexistence of at least two regions. One is devoid of water, whereas the other contains water.[194] This conclusion assumes that individual molecules of 7-oxo-Na-C_8 are able to adopt all conformations adopted by those of Na-C_8. The possibility that the 7-oxo-Na-C_8 molecules come out of the micelles, engage in hydrogen bonding to the water solvent, and then "drag" the water molecules inside, thereby modifying the micelle, is an energetically unfavorable process and hence may not occur. However, because of the polarity of CO group in 7-oxo-Na-C_8, the conformational arrangement of hydrophobic chains in 7-oxo-Na-C_8 micelles may not be similar to that in Na-C_8 micelles. The hydrophobic chains in 7-oxo-Na-C_8 micelles are energetically forced to attain a conformational arrangement in which 7-CO group is close to head groups where water concentration is significantly large. This possibility may be experimentally verified by studying the size and aggregation number of both 7-oxo-Na-C_8 and Na-C_8 micelles where the diameter of the spherical micelle of 7-oxo-Na-C_8 should be smaller than that of Na-C_8. The extent of water penetration into these micelles as perceived by the existence of hydrogen bonding to the carbonyl group of 7-oxo-Na-C_8 monomers remained essentially unchanged with the increase in the external pressure up to < ~ 18 kbar.[302] In contrast, 5-nonanone dissolved in Na-C_8 micelles show that relatively low external pressures (0 to 5 kbar) favor solubilization of 5-nonanone into dry micellar core, whereas higher pressures (> 5 kbar) make 5-nonanone to come in contact with water, which may be caused by translocation of 5-nonanone, inducing a fixed immobilized bent ∩ conformation in the lamellar phase.

The molecular dynamics calculations on sodium octanoate micelle in water reveal that the inner core of the micelle is devoid of water molecules, while there is considerable water penetration into the outer region of the core.[303] Dynamic simulation of hydration of spherical micelles of $C_{12}E_6$ in aqueous solution reveals that the micellar interface is separated in an inner part composed of water and hydrophobic and hydrophilic moieties and an outer part with hydrophilic moiety and water only.[304] Temperature dehydration occurs in the inner region only and is related to the presence of water molecules directly in contact with the hydrophobic core at low temperature.[304]

1.1.8 INTERNAL VISCOSITY (MICROVISCOSITY) OF MICELLES

Just as the extent of water penetration into micelles still remains a controversial topic, micellar internal viscosity (microviscosity) also suffers from a similar fate.[90] As the determination of both micellar internal viscosity and the extent of water penetration in micelles involves an almost similar experimental approach, that is, the use of some kind of probe, it is not surprising to see that both topics still remain controversial in nature.

Progress in achieving a clear and thorough understanding of the physicochemical properties of micelles has been limited[305] because of the lack of a deep understanding of both the structural and dynamic properties of micelles. There does not exist a single technique capable of yielding information on both structural and dynamic behavior of micelles unambiguously. Thus, generally two or more than two different techniques are used to increase more information on physicochemical properties of micelles and thereby decrease the degree of uncertainties about these properties. Microviscosity is one of the important micellar internal properties and, hence, considerable effort has been directed toward its measurement during the last few decades.[306–322] The techniques normally used to determine microviscosity of micelles include fluorescence,[306–322] ESR,[323–325] and NMR.[326]

The most commonly used fluorescence techniques involve the introduction (in the medium under investigation) of fluorescent probes having fluorescence properties — such as the intensity of emission, or the ratio of the intensities of two emission bands, or the wavelength at emission maximum — that are sensitive to the viscosity of the probe environment. However, reported values of microviscosity (η_m) of micelles based on fluorescence method show a significantly large variation. For instance, microviscosities of alkyltrimethylammonium bromide micelles, determined by using intermolecular excimer formation of fluorescent probe pyrene,[308] are an order of magnitude higher than those determined by Shinitzky et al.[307] (η_m range 17–50 cP) who employed the fluorescence depolarization method. The intermolecular excimer formation method relies on the translational diffusion of the pyrene molecules to form the excited dimmer, whereas fluorescence depolarization method is said to rely on rotational diffusion of the pyrene molecules. Rodgers and Wheeler[311] have pointed out that viscosity (which characterizes translational motion) is much higher than that which describes the rotational diffusion and, consequently, the intermolecular excimer-formation method compared to the fluorescence depolarization method senses higher microviscosities. Subsequent studies,[313,314] which used intramolecular excimer-forming probes, reported microviscosities that are comparable to those reported by Shinitzky et al.[307] Menger and Jerkunica,[326] by using ^{13}C-NMR technique, have reported the value of η_m of 8.3 cP. (By way of comparison, the viscosities of water, dodecane, and 1-octanol are 1.0, 1.3, and 8.9 cP, respectively.)[90]

Micellar microviscosities determined with two different probes using the same technique are often not the same.[312] Such inconsistencies in the reported values of η_m are sometimes rationalized qualitatively in terms of various factors such as nonisotropic microviscosity of micelle, significant contribution from the rotation of the micelle to the measured steady-state anisotropy with certain fluorescent probes, different micellar locations of probe molecules, different structural features of probe molecules, and perturbation of the micellar environment in which probe molecules are embedded.

Despite this somewhat jaundiced view of the reported values of microviscosity of micelles, it should be noted that they remain a great source of help for our understanding about the complex structural and dynamic properties of micelles and mechanistic aspects of micellar-mediated reactions. Although the fine details of structural features and dynamic behavior of micelle are still far from being fully understood, microviscosity studies remain meaningful for investigations involving a series of homologous surfactants.

With the intention of addressing the question, "are the experimentally determined microviscosities of the micelles probe dependent," Dutt[210] determined microviscosities of SDS, CTABr, 2-phenyldodecanesulfonate (2-PDS), 3-PDS,

4-PDS, and 6-PDS using the time-resolved fluorescence depolarization method involving two dissimilar probes, 2,5-dimethyl-1,4-dioxo-3,6-diphenylpyrrolo[3,4-c]pyrrole (DMDPP) and coumarin 6 (C6). The decay of anisotropy for both probes in all the six micelles has been rationalized on the basis of a two-step model consisting of fast-restricted rotation of the probe and slow lateral diffusion of the probe in the micelle that are coupled to the overall rotation of the micelle. On the basis of the assumption that the fast and slow motions are separable, the experimentally obtained slow and fast reorientation times (τ_{slow} and τ_{fast}) are related to the time constants for lateral diffusion (τ_L), wobbling motion (τ_W), and rotation of the micelle as a whole (τ_M) by the following relationships:

$$1/\tau_{slow} = 1/\tau_L + 1/\tau_M \qquad (1.32)$$

$$1/\tau_{fast} = 1/\tau_W + 1/\tau_{slow} \qquad (1.33)$$

In view of the fact that the rotation of the micelle also contributes to the depolarization of the fluorescence of the probe, microviscosity should be calculated by using τ_L rather than τ_{slow}. Dutt[210] calculated η_m values of SDS, CTABr, 2-PDS, 3-PDS, 4-PDS, and 6-PDS micelles by using the calculated values of τ_L, which show that both probes sense almost identical microviscosity for a given micelle. Both probes, DMDPP and C6, are concluded to reside in the Stern layer of all six micelles.

The values of η_m of SDS, determined by using the fluorescent tetracene probe in mixed water–propanol solvent, vary from 8 to 4 cP as the propanol concentration increases from 0 to 10% v/v.[327] The value of η_m (= 8 cP) in pure water is similar to the one determined by ^{13}C NMR[326] but nearly 1.5-fold smaller than the one determined by Dutt.[210] The increase in the aggregation number of dodecyltrimethylammonium bromide and chloride shows a modest increase in microviscosity of the polar shells of these cationic micelles as deduced from the rotational correlation time of the nitroxide moiety of the spin probe.[325] Ordinarily, one uses different probes to determine different microproperties of micelles such as micropolarity and microviscosity. Pistolis and Malliaris[319] have found fluorescent probes, which could be used to determine simultaneously both micropolarity and microviscosity of a micelle.

The fluorescent probe methods were used to determine microviscosity (η_m) of potassium N-acylalaninates and potassium N-acylvalinates micelles.[318] The results obtained show that η_m on the micellar surface is larger in N-acylalaninates than in N-acylvalinates, whereas η_m in the micellar core remains same in the micelles of the two surfactant series. The values of η_m of SDS, lithium dodecyl sulfate (LDS), CTACl, and $C_{12}E_6$, determined at 15°C by monomer/excimer intensity ratio and excimer lifetime of dipyrenylpropane dissolved in micelles, are 19, 19, 39, and 57 cP, respectively.[317] The fluorescent probe technique was used to determine η_m values of SDS, CTABr, and CTACl micelles at different applied pressure.[315] The derived η_m values of SDS, CTABr, and CTACl micelles at 25°C and atmospheric pressure are 12, 47, and 27 cP, respectively. The addition of NaCl, Na_2SO_4, and ethanol to the micellar solution generally decrease the η_m values. The increase in SDS micellar size has been found to cause a linear increase in η_m.[323] The η_m of the polar shell of mixed micelles of SDS and a nonionic sugar-based surfactant (dodecylmalono-bis-N-methylglucamide, DBNMG) at 45°C, as deduced from the rotational correlation time of the spin probe, varies from 2.79 cP for pure SDS to 13.1 cP for pure DBNMG departing only slightly from a linear dependence on the mole fraction of DBNMG.[324] The η_m value of pure DBNMG micelles decreases from 32 cP at 21°C to 13.1 cP at 45°C. The fluorescent probes, C6 and merocyanine 540 (MC540), have been used to determine the values of η_m of TX-100 micelles at 25°C and at different contents of ethylene glycol (EG) in mixed aqueous solvents.[322] The increase in the content of EG from 0 to 60% w/w increases η_m of TX-100 micelles from 26.2 to 35.2 cP for C6 and from 20.1 to 23.7 cP for MC450. The effects of various parameters, such as surfactant concentration, temperature, surfactant chain length, nature of the head group, surfactant chemical structure, and nature of the counterion on η_m of micelles of 60 surfactants of diverse structural features have been studied by fluorescence probing technique.[321] The fluorescence depolarization dynamics of organic fluorescent dye probes were studied in cationic, anionic, and nonionic micelles by picosecond time-resolved single-photon-counting technique. The concept of microviscosity in the micelles has been critically discussed in light of the rotational and translational diffusion coefficients and their temperature dependance.[328]

1.2 HEMIMICELLES

Nonionic and ionic surfactant molecules adsorb on solid surface at water-insoluble solid–water interface such as silica gel–water and alumina–water interface through van der Waals, hydrophobic, hydrogen bonding, and polar/ionic interactions at low concentrations of surfactant. Then, at a relatively higher and specific concentration known as *critical hemimicelle concentration* (CHMC), the adsorption increases dramatically as hemimicelles form on the adsorbent involving forces that characterize hydrophobic, van der Waals, and polar/ionic forces. Hemimicelles are two-dimensional molecular aggregates whose structural and physicochemical behavior have not been studied as extensively as those of normal micelles and, consequently, they are not understood even at a very rudimentary level.

In 1956, Fuerstenau[329] coined the term *hemimicelle* to describe the two-dimensional aggregates that he proposed were responsible for the rise through several orders of magnitude in the observed absorption of surfactant molecules on certain mineral oxide surfaces from aqueous solutions. The study on coadsorption of pinacyanol chloride (cyanine dye) with sodium p-(1-propylnonyl)-benzenesulfonate (dissolved in water) on Al_2O_3 surface reveals that surfactant adsorption is independent of the concentration of pinacyanol chloride, and the surfactant solution remains clear at total surfactant concentration ([Surf]$_T$) below CMC, whereas the solid surface becomes dyed blue, the color characteristic of pinacyanol solubilized in micelles. These observations are attributed as visual evidence for the presence of hemimicelles in the aqueous solution of mixed surfactant, pinacyanol chloride, and Al_2O_3 at $[Surf]_T < CMC$.[330]

The effects of added inorganic salts (NaCl and NaBr), alcohols (ethanol and n-butanol), and urea on the adsorption of alkylpyridinium chlorides from aqueous solution on silica gel show that all the adsorption isotherms exhibit two plateau either with or without various additives.[331] The first plateau is ascribed to the adsorption of individual surface-active cations on the negatively charged silica gel surface through both ionic attraction and specific adsorption. Then, at a particular concentration known as the CHMC, the adsorption increases dramatically, which has been attributed to hemimicelle formation on the adsorbent[329] through association of tails of the adsorbed surfactant molecules. The adsorption reaches the second plateau at $[Surf]_T \geq CMC$ of the surfactant. Based on the assumption that each adsorbed surface-active cation in the first plateau is an active site for surface aggregation, the minimum average aggregation number of a hemimicelle (nhm) is equal to the ratio between the adsorption amounts of two plateau.[331] Following the thermodynamic formulations of adsorption of surface-active ions, Gu et al.[332] have derived equations to calculate the nhm, equilibrium constant (K), and standard free energy ($\Delta G°$) for hemimicellization. In order to explain S-type adsorption isotherm, obtained for adsorption of nonionic surfactants at the silica gel–water interface, Gu and Zhu[333] used Equation 1.34

$$\Gamma = (\Gamma_\infty Kc^n)/(1 + Kc^n) \qquad (1.34)$$

where Γ is the amount of surfactant adsorbed at c (= total concentration of nonionic surfactant), Γ_∞ the amount adsorbed in the limiting adsorption at high concentrations, n ($\equiv$ nhm) the aggregation number of a hemimicelle, and K the equilibrium constant for the equilibrium process, Equation 1.35.

$$\text{solid (adsorbent)} + n \text{ monomer (adsorbate)} \xrightleftharpoons{K} \text{hemi-micelle} \qquad (1.35)$$

The apparent flaw in Equation 1.34 originates from Equation 1.35 where n is considered the average aggregation number of hemimicelle. In actuality, n should be represented as the number of surfactant monomers that remain in monomeric form at equilibrium, that is, $n = n_t - n_{hm}$ where n_t is the total number of surfactant molecules and n_{hm} is the number of surfactant molecules, used up in the formation of hemimicelles. The relationship $n = n_t - n_{hm}$ is true only if n remains independent of total concentration of surfactant, which is normally true and one of the basic assumptions of the micellization process.

Pyrene and dinaphthylpropane fluorescent probes have been used to investigate the structural details of adsorbed layer of SDS at the alumina–water interface.[334] This study supports the basic concepts of hemimicellization. Adsorption isotherm turns out to be S-type. The values of microviscosity of adsorbed layer and normal micelle are 90 to 120 and 8 cP, respectively. The mobility of the probe

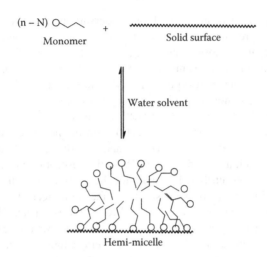

FIGURE 1.6 Hemimicelle formation on solid surface, where n is the total number of monomers (surfactant molecules), and N is the aggregation number of the hemimicelle.

is more restricted in the adsorbed layer than in the normal micelle. Fluorescence decay methods enabled the determination of the size (aggregation number) of the aggregates at the solid–liquid interface as a function of surface coverage.

In view of the mass-action model, which has been applied in normal micellization process of surfactants, hemimicelles are considered to be formed in a single step, as shown in Figure 1.6, in which solid surface (adsorbent) shows strong attractive interaction with surfactant head groups. The display of structural features of hemimicelle in Figure 1.6 remains highly speculative within the domain of our present limited knowledge about the dynamics and physicochemical behavior of hemimicelles. Although hemimicellar structure is not yet fully understood, it seems that the nature of both adsorbate and adsorbent greatly influence the structural features of hemimicelle. For ionic surfactant adsorption on hydrophobic surfaces (such as graphite), there is general agreement that the lower-density adsorbed layer (below the CMC) is oriented with surfactant tails parallel to the adsorbent surface plane (Figure 1.7a). However, there have been two schools of thought regarding the higher-density adsorbate structure. Some authors[335] propose a gradual change from a parallel to a perpendicular adsorption scheme (Figure 1.7) as the concentration approaches the CMC. The others[336] propose hemimicelle formation as being responsible for the increased adsorption near the CMC. As these conflicting models for hydrophobic adsorbents predict similar overall adsorbate density and degree of surface ionization, it is difficult to resolve this issue using surface techniques. In an attempt to resolve this issue, Manne et al.[337] used an atomic force microscope (AFM) to image the adsorbate structure of an ionic surfactant (CTABr) adsorbed from aqueous solution onto hydrophobic surfaces (graphite). The results obtained indicate the formation of hemimicelles above the CMC. This technique provided the first direct imaging

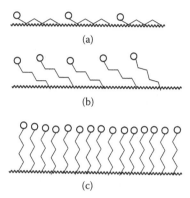

FIGURE 1.7 An adsorption model for ionic surfactant molecules at the graphite–aqueous solution interface. This model attempts to explain an observed increase in adsorbate density with increasing surfactant concentration in solution; (a) at low concentration, molecules adsorb with their alkane chains extended on the graphite substrate plane; (b) as the concentration is increased, the chains gradually desorb so that a portion of the adsorbate molecule is oriented perpendicular to the graphite substrate plane; (c) at concentrations near the critical hemimicelle concentration (CHMC) the adsorbates are oriented perpendicular to the graphite substrate plane, with the hydrophilic head groups completely shielding the hydrophobic substrate from solution.

of hemimicelles with a hemicylindrical morphology whose cross-sectional view is shown in Fig. 1.8. These authors postulate that the monolayer structure of adsorbate at the surface at low surfactant concentration serves as a template for hemimicelle formation as the concentration is increased. Thus, hemimicellar shape need not be related to bulk micellar shape; rather, it can be determined primarily by the monolayer structure.

Direct images of ionic surfactant (TTABr, CTAOH, and SDS) that aggregate at a gold surface in aqueous solutions have been obtained by the use of AFM.[338] These images indicate linear aggregates (cylindrical or half-cylindrical) lying on the gold surface. For the surfactants CTAOH and SDS, in which counterions do not have adsorption affinity to the gold surface, AFM images have been proposed to represent half-cylindrical aggregates on the gold surface whose cross-sectional models are similar to the one shown in Figure 1.8. But for TTABr, in which

FIGURE 1.8 Schematic representation of a perpendicular cross section through two neighboring hemimicelles (i.e., the cylindrical axis is into the plane). The number of molecules shown in each hemimicelle is somewhat arbitrary and is not meant to suggest an actual aggregation number.

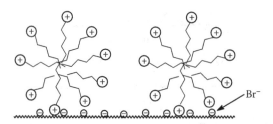

FIGURE 1.9 Cross-sectional model of a plausible aggregation morphology consistent with observed AFM image of $C_{14}TAB$ aggregates on gold substrate. Counterions (Br) of $C_{14}TAB$ adsorb perpendicularly to the gold surface, creating a negatively charged layer. Surfactant adsorption is driven primarily by head-group-surface electrostatic forces, leading to full cylindrical aggregates. The surface interacts with only a small portion of the cylindrical aggregate, leading to weak surface control and flexible stripes.

chemisorption of bromide ions (i.e., counterions) occurs spontaneously on the gold surface in aqueous solution,[339] AFM images have been suggested to represent cylindrical aggregates lying on the gold surface. The cross-sectional models of two such cylindrical aggregates are shown in Figure 1.9.[338] Images of adsorbed surfactant aggregates, obtained from AFM technique, show that these aggregates bear remarkable similarities to aqueous solution self-assembly structures. These images have been interpreted in terms of the proposal that the surfactant-adsorbed layer consists of full spheres,[340–342] flexible cylinders,[340–342] and bilayers[340,341] on hydrophilic surfaces such as mica and rigid hemicylinder on graphite.[337,340,343] The adsorption of SDS on graphite-solution interface in the presence of Mg^{2+}, Mn^{2+}, and Ca^{2+} was imaged by AFM, which showed long parallel hemicylindrical surface aggregates under most conditions.[344] Electrochemical measurements, AFM, and scanning tunneling microscopy (STM) have been combined to present the first direct images of the potential-controlled phase transition between hemimicellar and condensed monolayer of an SDS-adsorbed film at the Au(111) electrode surface.[345] The hemimicellar aggregates may be re-formed by decreasing the charge density at the electrode surface. A theoretical model has been developed for the transition from an adsorption monolayer to hemicylindrical surface micelles on the basis of one-dimensional nucleation-growth-collision.[346] The role of geometric constraints in self-assembly is discussed in terms of the surface packing parameter, G. This parameter should fulfill the condition $1/3 < G < 0.5$ for the formation of hemicylindrical surface aggregates.

1.3   REVERSED MICELLES

Amphipathic nature of the surfactant molecules causes them to undergo rapid self-organization when they, along with or without a limited number of water molecules, are mixed with a nonpolar solvent such as liquid hydrocarbon. Under such a condition, the rapid self-organization of surfactant and water molecules occur in a specific manner that minimizes repulsive and maximizes attractive molecular interactions between surfactant–surfactant, water–water, surfactant–nonpolar solvent, water–nonpolar solvent, and surfactant–water molecules. Such energetic requirement, above a critical concentration of surfactant (known as CMC or critical aggregate concentration, CAC), forces the surfactant and water molecules to attain a molecular aggregate structure, which is called *reversed micelle* (Figure 1.10), in which water and polar/ionic head groups of surfactant molecules protect themselves from being in contact with hydrophobic solvent molecules. Such surfactant molecular aggregates are in rapid equilibrium with monomeric surfactant molecules (Figure 1.10). At a very low concentration of surfactant, the reversed micelles are very close to spherical in which water molecules occupy the central part of the sphere, thus forming a so-called micro water-pool, and these water molecules are in contact with head groups of reversed-micelle-forming surfactant molecules. The tails of these surfactant molecules are extended toward bulk nonpolar solvent phase.

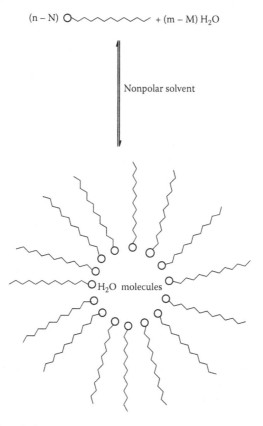

FIGURE 1.10 Speculative representation of reversed-micelle formation, where n is the total number of monomers (surfactant molecules); N, the aggregation number of reversed micelle; m, the total number of water molecules; and M, the total number of water molecules present inside the reversed micelles.

The distinction between reversed micelles (RMs) and reversed microemulsions is often ill-defined. Water molecules added to an RM are not distributed evenly throughout the hydrocarbon continuum but are found associated with the surfactant head groups. This is also described as *swollen reversed micelle*. The volume of water that can be taken up and be stabilized in the swollen reversed micelle is limited. The swollen reversed micelles are also often called *reversed microemulsion*. In RMs, the amount of solubilized water is less than or equal to the amount necessary to hydrate the surfactant head groups. Solubilization of water over and above this threshold results in the formation of an isotropic and thermodynamically stable water-in-oil microemulsion.[347,348]

RMs have been investigated for decades as microreactors whose size can be changed by increasing or decreasing the amount of solubilized water in the inner core of the micellar aggregates.[349] It is almost universally accepted that water molecules are solubilized in the polar core of RMs. However, there is now evidence that some surfactants do not allow this idealized compartmentalization of solubilized water and that other organizational variations are possible for the nanostructure of hydrated RMs.[350] ^{1}H NMR measurements of the chemical shift of water protons indicate that nickel(II) bis(2-ethylhexyl)phosphate reversed micellar solubilized water molecules are situated in a different environment compared with the water molecules in classical sodium bis(2-ethylhexyl) sulfosuccinate (Aerosol-OT, AOT) reversed micelles. The molecular modeling shows that water molecules can be localized in channels within the surface of some rodlike micellar aggregates, thereby confirming the "open water-channel (OWC) model" of RMs.[350,351] Various investigators have reported that the divalent transition-metal, such as Ni(II) and Co(II) salts of bis(2-ethylhexyl)phosphoric acid (HDEHP), forms polymeric species in nonpolar solvents.[352–355] However, it is equally plausible to interpret this in terms of rodlike RMs.[356] Neuman et al.[357] presented evidence obtained by using a variety of scattering, spectroscopic, and analytical techniques for the conclusion that the aggregates that form in the nonpolar phase by the nickel salt of HDEHP are cylindrical-type RMs containing between five and six water molecules per HDEHP monomer. Steytler et al.[358] also concluded (based on SANS measurements) that divalent metal salts of HDEHP in cyclohexane solvent form short rodlike RMs. The length of the rodlike RMs formed is strongly dependent upon the nature of the counterion (M^{n+}).

Although the very existence of CMC[359] of surfactant dissolved in nonpolar has been questioned,[360] the UV absorption and fluorescence emission measurements of AOT-solubilized solubilizates, such as *tert*-butylphenol, pyrene, and hemi-Mg salt of 8-anilino-1-naphthalenesulfonic acid, in isooctane provided evidence for the existence of CMC at different values of R = [H_2O]/[AOT] (with R < 3) and CMC value increases with increase in R.[361] The values of CMC, obtained at different temperatures, were used to calculate standard enthalpy of micellization (ΔH_m^0) from Equation 1.13 and standard entropy of micellization (ΔS_m^0) from Equation 1.14, and these respective values vary from 38.07 to 44.81 kJ mol^{-1} and 184.60 to 202.60 JK^{-1} mol^{-1} with increase in temperature from 20 to 45°C.[361] The positive ΔH_m^0 values arise mainly from the dismantling of hydrated ions in the quasi-lattice of AOT interior, whereas positive values of ΔS_m^0 may be due to the release of free isooctane molecules at the onset of micellization to some extent. The microviscosity of AOT micellar interior is concluded to decrease with increasing water content. The polar solvent media as well as dioxane and benzene show positive enthalpies, whereas the alkanes, except decane, show negative enthalpies of micellization of AOT.[362] The entropies of micellization of these two categories of solvents are also of opposite signs.

The effects of two trihydroxy bile salts, sodium taurocholate and 3-[(3-chol-amidylpropyl)dimethylammonio]-1-propane sulfonate, on the size, shape, and percolation temperature of RMs formed by AOT in isooctane show that (1) the percolation temperature of RMs decreases upon inclusion of bile salts, indicating increased water uptake; (2) there is consistent enlargement of RMs upon addition of the bile salts; and (3) the inclusion of the enzyme yeast alcohol dehydrogenase increases the percolation temperature and distorts the spherical structure of the AOT RMs. But the spherical structure is restored upon addition of bile salts.[363] The distribution constants of vitamin E partitioned between apolar organic phase and water-containing RMs of AOT, didodecyldimethylammonium bromide, soybean phosphatidylcholine (lecithin), and tetraethylene glycol monododecyl ether ($C_{12}E_4$) have been evaluated by a spectrophotometric method, and the results suggest that vitamin E is partitioned between the micellar palisade layer and the organic solvent.[364] The binding strength of vitamin E to RMs depends mainly on specific interactions between the head group of vitamin E and that of the surfactant. The dependence of the binding constants upon the molar ratio R (R = [H$_2$O]/[Surfactant]) indicates a competition between water and vitamin E for the binding sites at the water/surfactant interface.[364] The interactions between poly(vinylpyrrolidone) (PVP) and the RMs composed of water, AOT, and n-heptane have been studied by the use of various photo-physical techniques in which the FTIR results suggest that the contents of SO$_3$-bound water and Na$^+$-bound water both decrease, whereas the content of the bulklike water increases with added PVP.[365] However, the trapped water in the hydrophobic chain of the surfactant is nearly unaffected by PVP.

The states and structure of the solubilized water in RMs of AOT and sodium bis(2-ethylhexyl) phosphate (NaDEHP) in heptane were characterized by FTIR and NMR spectroscopic

AOT

NaDEHP

Lecithin

parameters.[366] According to the four-component hydrational model, the free, anion-bound, bulklike, and cation-bound water are present in RMs of AOT. The ^{1}H NMR measurements show the presence of in-core anion-bound, bulklike, and cation-bound water molecules. The effects of electrolytes on the solubilized states of water by RMs of AOT have been examined using ^{1}H NMR and near-IR spectroscopic techniques, and the results obtained show that two types of solubilized water molecules, i.e., water bound directly to the ionic head groups of the surfactant in RMs and water interacting with the hydrated head groups in swollen micelles through hydrogen bonds, are almost unaffected by any electrolytes, whereas much greater effects have been observed for bulklike water in W/O microemulsions.[367] The presence of four principal microenvironments (an inner, free water-pool, a bound-water region, the interface, and the surrounding hydrocarbon continuum) is described in a review article on fluorescence probes for structural and distance effect studies in micelle, RMs, and microemulsions.[368] Calorimetric investigations of solutions of RMs are reviewed under the topics (1) reversed micelles as nanocontainers, (2) intermicellar interaction and percolation, (3) water-containing reversed micelles as nanosolvents and the solubilization of nonionic solutes, and (4) reversed micelles as nanoreactors.[369]

The classical and perhaps the most widely accepted picture of RMs seems to be that of Zensli who describes the water in the interior water-pool of microemulsions using a two-state model.[370] A very viscous water, close to the interface, would be in equilibrium with that in the center of the pool, which exhibits properties similar to bulk water. FTIR and ^{1}H NMR studies of the AOT reversed micellar-solubilization of pure and aqueous 1,2-ethanediol (ED and ED-W)[371] as well as 1,2,3-propanetriol (PT and PT-W)[372] show that the reversed-micellar-solubilized ED or ED-W as well as PT or PT-W molecules do not seem to coexist in "layers" of different structures as suggested by the multistate water solubilization micellar model.

The effects of additives on the structure and behavior of wet RMs (i.e., RMs with the number of water molecules less or sufficient to hydrate polar head groups of reversed-micelle-forming surfactant molecules) and highly polar liquid additives on dry RMs (i.e., RMs with $[H_2O] = 0$) have been studied extensively.[373–377] In contrast, confinement of finite amounts of strongly polar solid additives into reversed-micellar cores has not received significant attention.[378] The solubilization of solid urea by lecithin, NaDEHP, and AOT RMs has been studied by FTIR spectroscopy.[379,380] Analysis of FTIR spectra leads to the hypothesis that urea is confined within the polar reversed-micellar core. The encapsulation of urea involves some changes of the urea NH stretching band with respect to that of the pure solid urea, which is attributed to the specific interactions between the urea NH_2 and anionic head groups of RMs. For the urea/AOT/n-heptane system, the SANS results are consistent with the hypothesis that urea is encapsulated as small-sized ellipsoidal hydrogen-bonded clusters within the polar core of AOT RMs.[380] Lecithin and AOT RMs containing cyanamide (H_2NCN) have been investigated by small-angle x-ray scattering, FTIR, and ^{1}H NMR spectroscopy, and these data are consistent with a model of cyanamide molecules confined in RMs in which cyanamide molecules are distributed uniformly among the surfactant polar head groups without being segregated to form a separate micellar interior core.[381] FTIR and ^{1}H NMR data suggest strong interactions between solubilized cyanamide molecules and surfactant polar head groups.

Polarity of AOT reversed-micellar interfaces has been evaluated in terms of effective dielectric constants by the use of the preferential salvation concepts in reversed-micellar systems containing polarity-sensitive probe molecules.[382] The effective dielectric constant at the interface increases from 2.3 ([H_2O]/[AOT] = R = 0) to approximately 9.0 before reaching a plateau at R = 12. This is explained by the solvation of the polar head groups of the surfactant.[382]

An adsorption isotherm equation is derived for adsorption of surfactant molecules on solids from nonpolar solvents by considering a two-step adsorption model (i.e., in the fist step, the surfactant molecules are adsorbed through van der Waals attractive interaction between surfactant molecules and solid surface, then in the second step, the so-called reversed hemimicelles form through hydrogen bonding and/or polar interactions between the adsorbed surfactant molecules) and the mass-action law.[383] The result is supported quantitatively by adsorption data of 1-decanol from heptane at the solution–graphite interface.[384]

REFERENCES

1. McBain, J.W. Solutions of soaps and detergents as colloidal electrolytes, *Colloid Chem.* **1944**, *5*, 102–120.
2. Hartley, G.S. *Aqueous solutions of Paraffin-Chain Salts*, Hermann and Cia, Paris, 1936.
3. Essquena, J., Solans, C. Influence of the HLB parameter of nonionic surfactants on normal and reversed-phase thin layer chromatography. *Colloids Surf. A* **2001**, *189*(1–3), 85–92.
4. Sahoo, L. Organization of Nonionic Amphiphiles in Solution and at Solid-Liquid Interface. Thesis submitted to Sambalpur University, India, for Ph. D. in Science **2002**, and the reference cited therein.
5. Clunie, J.S., Goodman, J.F., Symons, P.C. Electrical conductance of sodium n-alkyl sulfonates in aqueous solution. *Trans. Faraday Soc.* **1967**, *63*(3), 754–758.
6. Sahoo, L., Sarangi, J., Misra, P.K. Organization of amphiphiles, Part 1: evidence in favor of pre-micellar aggregates through fluorescence spectroscopy. *Bull. Chem. Soc. Jpn.* **2002**, *75*, 859–865.

7. (a) Duynstee, E.F.J., Grunwald, E. Organic reactions occurring in or on micelles. I. Reaction rate studies of the alkaline fading of triphenylmethane dyes and sulfonephthalein indicators in the presence of detergent salts. *J. Am. Chem. Soc.* **1959**, *81*, 4540–4542. (b) Duynstee, E.F.J., Grunwald, E. Organic reactions occurring in or on micelles. II. Kinetic and thermodynamic analysis of the alkaline fading of triphenylmethane dyes in the presence of detergent salts. *J. Am. Chem. Soc.* **1959**, *81*, 4542–4548. (c) Menger, F.M., Portnoy, C.E. On the chemistry of reactions proceeding inside molecular aggregates. *J. Am. Chem. Soc.* **1967**, *89*(18), 4698–4703. (d) Duynstee, E.F.J., Grunwald, E. Kinetic neutral-salt effects of organic salts on the alkaline hydrolysis of methyl 1-naphthoate. *Tetrahedron* **1965**, *21*(9), 2401–2412. (e) Romsted, L.R., Cordes, E.H. Secondary valence force catalysis. VII. Catalysis of hydrolysis of p-nitrophenyl hexanoate by micelle-forming cationic detergents. *J. Am. Chem. Soc.* **1968**, *90*(16), 4404–4409. (f) Bunton, C.A., Robinson, L. Micellar effects upon nucleophilic aromatic and aliphatic substitution. *J. Am. Chem. Soc.* **1968**, *90*(22), 5972–5979. (g) Bunton, C.A., Robinson, L. Electrolyte and micellar effects upon the reaction of 2,4-dinitrofluorobenzene with hydroxide ion. *J. Org. Chem.* **1969**, *34*(4), 780–785.

8. (a) Zana, R., Yiv, S., Kale, K.M. Chemical relaxation and equilibrium studies of association in aqueous solutions of bolaform detergenets. *J. Colloid Inteface Sci.* **1980**, *77*(2), 456–465. (b) Bunton, C.A., Savelli, G. Organic reactivity in aqueous micelles and similar assemblies. *Adv. Phys. Org. Chem.* **1986**, *22*, 213–309 and references cited therein.

9. Domingo, X. In *Amphoteric Surfactants*. Lomax, E.G., Ed., Surfactant science series, Vol. 59, Marcel Dekker, New York, **1996**, chap. 3.

10. Tsujii, K. *Surfactant Activity: Principles, Phenomena and Applications*. Academic Press, New York, **1998**.

11. (a) Elworthy, P.H., Florence, A.T., Macfarlane, C.B. *Solubilization of Surface Active Agents and Its Applications in Chemistry and the Biological Sciences*. Chapman and Hall, London, **1968**. (b) Shinoda, K., Nakagawa, T., Tamamushi, B., Isemura, T. *Colloidal Surfactants: Some Physico-Chemical Properties*. Academic Press, New York, **1963**. (c) Mukerjee, P., Mysels, K.J. Critical micelle concentrations of aqueous surfactant systems. NSRDS-NBS-36, Superintendent of Documents, Washington, D.C., **1971**.

12. Patist, A. Determining critical micelle concentration. *Handbook of Applied Surface and Colloid Chemistry*. Vol. 2, Wiley, Chichester, **2002**, pp. 239–249.

13. Senz, A., Gsponer, H.E. Micellar binding of phenoxide ions to cetyltrimethylammonium chloride. *J. Colloid Interface Sci.* **1994**, *16*, 60–65.

14. Yiv, S., Zana, R. Chemical relaxation and equilibrium studies of association in aqueous solutions of Bolaform detergents. *J. Colloid Interface Sci.* **1980**, *77*(2), 449–455.

15. Mukerjee, P., Mysels, K.J. Re-evaluation of the spectral-change method of determining critical micelle concentration. *J. Am. Chem. Soc.* **1955**, *77*, 2937–2943.

16. Toullec, J., Couderc, S. Aqueous hexadecyltrimethylammonium acetate solutions: pH and critical micelle concentration evidence for dependence of the degree of micelle ionic dissociation on acetate ion concentration. *Langmuir* **1997**, *13*(7), 1918–1924.

17. Kalyanasundaram, K., Thomas, J.K. Environmental effects on vibronic band intensities in pyerene monomer fluorescence and their application in studies of micellar systems. *J. Am. Chem. Soc.* **1977**, *99*(7), 2039–2044.

18. Kalyanasundaram, K. Photochemistry in microheterogeneous systems. Academic Press, New York, **1987**.

19. (a) Aguiar, J., Carpena, P., Molina-Bolivar, J.A., Ruiz, C.C. On the determination of the critical micelle concentration by the pyrene 1:3 ratio method. *J. Colloid Interface Sci.* **2003**, *258*, 116–122. (b) Frindi, M., Michels, B., Zana, R. Ultrasonic absorption studies of surfactant exchange between micelles and bulk phase in aqueous micellar solutions of nonionic surfactants with a short alkyl chain. 3. Surfactants with a sugar head group. *J. Phys. Chem.* **1992**, *96*(20), 8137–8141.

20. Turro, N.J., Kuo, P.-L., Somasundaram, P., Wong, K. Surface and bulk interactions of ionic and nonionic surfactants. *J. Phys. Chem.* **1986**, *90*(2), 288–291.

21. Lei, X.-G., Tang, X.-D., Liu, Y.-C., Turro, N.J. Microenvironmental control of photochemical reactions. 3. Additive effects on micellar structure and properties of TX-100. *Langmuir* **1991**, *7*(12), 2872–2876.

22. Zana, R., Levy, H., Danino, D., Talmon, Y., Kwetkat, K. Mixed micellization of cetyltrimethylammonium bromide and an anionic dimeric (Gemini) Surfactant in aqueous solution. *Langmuir* **1997**, *13*(3), 402–408.

23. Zana, R., Levy, H., Kwetkat, K. Mixed micellization of dimeric (Gemini) Surfactants and conventional surfactants. I. Mixtures of an anionic dimeric surfactant and of the nonionic surfactants $C_{12}E_5$ and $C_{12}E_8$. *J. Colloid Interface Sci.* **1998**, *197*(2), 370–376.

24. Ruiz, C.C., Aguiar, A. Mixed micelles of Triton X 100: interaction, composition, stability and micro-environmental properties of the aggregates. *Mol. Phys.* **1999**, *97*(10), 1095–1103.

25. Ruiz, C.C., Aguiar, A. Mixed micelles of Triton X 100: interaction, composition, stability and micro-environmental properties of the aggregates. [Erratum to document cited in CA132:80062]. *Mol. Phys.* **2000**, *98*(10), 699.

26. Ruiz, C.C., Aguiar, A. Interaction, stability and micro-environmental properties of mixed micelles of Triton X 100 and N-alkytrimethylammonium bromides: influence of alkyl chain length. *Langmuir* **2000**, *16*(21), 7946–7953.

27. Oremusova, J., Greksakova, O. Determination of the critical micelle concentration, hydrodynamic micelle radius and experimental partition coefficient of (-)-N-dodecyl-N-methylephedrinium bromide. *Tenside Surfactants Deterg.* **2003**, *40*(2), 90–95.

28. Akhter, M.S., Alawi, S.M. A comparison of micelle formation of ionic surfactants in formamide, in N-methylformamide and in N,N-dimethylformamide. *Colloids Surf. A* **2003**, *219*(1–3), 281–290.

29. Oremusova, J., Greksakova, O. Micellization parameters of cationic surfactants consisting of the [1-(ethoxycarbonyl)-pentadecyl]-trimethylammonium ion and various anions. I. Effect of temperature and type of anion on the critical micelle concentration and thermodynamic functions of micellization. *Tenside Surfactants Deterg.* **2003**, *40*(1), 35–39.

30. El-Badry, M., Schubert, R. Determination of critical micelle concentration (CMC) of different pluronics with two different methods. *Bull. Pharm. Sci.* **2002**, *25*(2), 201–205.

31. Anghel, D.F., Galatanu, A.-N. Spectrophotometric determination of critical micelle concentration of ethoxylated alkylphenols. *Analele Universitatii Bucuresti, Fizica* **1999**, *48*, 71–75.

32. Murphy, A., Taggart, G. A comparison of predicted and experimental critical micelle concentration values of cationic and anionic ternary surfactant mixtures using molecular-thermodynamic theory and pseudophase-separation theory. *Colloids Surf. A* **2002**, *205*(3), 237–248.

33. Priev, A., Zalipsky, S., Cohen, R., Barenholz, Y. Determination of critical micelle concentration of lipopolymers and other amphiphiles: comparison of sound velocity and fluorescent measurements. *Langmuir* **2002**, *18*(3), 612–617.

34. Hait, S.K., Moulik, S.P. Determination of critical micelle concentration (CMC) of nonionic surfactants by donor-acceptor interaction with iodine and correlation of CMC with hydrophile-lipophile balance and other parameters of the surfactants. *J. Surfactants Deterg.* **2001**, *4*(3), 303–309.

35. (a) Amato, M.E., Caponetti, E., Chillura Martino, D., Pedone, L. ^{1}H and ^{19}F NMR Investigation on mixed hydrocarbon-fluorocarbon micelles. *J. Phys. Chem. B* **2003**, *107*(37), 10048–10056. (b) Perez-Rodriguez, M., Prieto, G., Rega, C., Varela, L.M., Sarmiento, F., Mosquera, V. A comparative study of the determination of the critical micelle concentration by conductivity and dielectric constant measurements. *Langmuir* **1998**, *14*(16), 4422–4426.

36. Garcia-Anton, J., Guinon, J.L. Determination of Hyamine 2389 critical micelle concentration (CMC) by means of conductometric, spectrophotometric and polarographic methods. *Colloids Surf.* **1991**, *61*, 137–145.

37. Ghosh, S., Krishnan, A., Das, P. K., Ramakrishnan, S. Determination of critical micelle concentration by hyper-Rayleigh scattering. *J. Am. Chem. Soc.* **2003**, *125*(6), 1602–1606.

38. Dwivedi, K.K., Ghosh, S., Singh, S., Sinha, D., Srivastava, A., Bhat, S.N. Determination of critical micelle concentration (CMC) of surfactants by a nuclear track microfilter. *J. Surface Sci. Technol.* **1997**, *13*(2–4), 53–60.

39. Ysambertt, F., Vejar, F., Paredes, J., Salager, J.-L. The absorbance deviation method: a spectrophotometric estimation of the critical micelle concentration (CMC) of ethoxylated alkyphenol surfactants. *Colloids Surf. A* **1998**, *137*(1–3), 189–196.

40. Huettenrauch, R., Fricke, S., Koehler, M. Experimental confirmation of the change of water structure in the critical range of micelle formation: a new method of critical micelle concentration (CMC) determination. *Pharm. Res.* **1988**, *5*(11), 726–728.

41. (a) Broxton, T.J., Christie, J.R., Chung, R.P.-T. Micellar catalysis of organic reactions. 23. Effect of micellar orientation of the substrate on the magnitude of micellar catalysis. *J. Org. Chem.* **1988**, *53*(13), 3081–3084. (b) Broxton, T. J., Christie, J.R., Dole, A.J. Micellar catalysis of organic reactions. Part 35. Kinetic determination of the critical micelle concentration of cationic micelles in the presence of additives. *J. Phys. Org. Chem.* **1994**, *7*(8), 437–441.

42. Khan, M.N., Arifin, Z. Kinetics and mechanism of intramolecular general base-catalyzed methanolysis of ionized phenyl salicylate in the presence of cationic micelles. *Langmuir* **1996**, *12*(2), 261–268.

43. Berndt, D.C., Pamment, M.G., Fernando, A.C.M., Zhang, X., Horton, W.R. Micellar kinetics in aqueous acetonitrile. Part A. Sodium-hydrogen ion-exchange constant for 1-dodecanesulfonate surfactant. Part B. Substrate structural effects for micellar catalysis by perfluorooctanoic acid as reactive counterion surfactant. *Int. J. Chem. Kinet.* **1997**, *29*(10), 729–735.

44. Khan, M.N. Effects of salts and mixed CH_3CN-H_2O solvents on alkaline hydrolysis of phenyl benzoate in the presence of ionic micelles. *Colloids Surf. A* **1998**, *139*, 63–74.

45. Rico, I., Halvorsen, K., Dubrule, C., Lattes, A. Effect of micelles on cyclization reactions: the use of N-hexadecyl-2-chloropyridinium iodide as an amphiphilic carboxyl-activating agent in lactonization and lactamization. *J. Org. Chem.* **1994**, *59*(2), 415–420.

46. Khan, M.N., Arifin, Z. Effects of mixed CH_3CN-H_2O solvents on intramolecular general base-catalyzed methanolysis of ionized phenyl salicylate in the presence of cationic micelles. *Colloids Surf. A* **1997**, *125*, 149–154.

47. Bunton, C.A., Fendler, E.J., Sepulveda, L., Yang, K.-U. Micellar-catalyzed hydrolysis of nitrophenyl phosphates. *J. Am. Chem. Soc.* **1968**, *90*(20), 5512–5518.

48. Khan, M.N., Arifin, Z., Wahab, I.B., Ali, S.F.M., Ismail, E. Effects of cationic micelles on rate of intramolecular general base-catalyzed reaction of ionized phenyl salicylate with tris-(hydroxymethyl)aminomethane (Tris). *Colloids Surf. A* **2000**, *163*, 271–281.

49. Khan, M.N., Arifin, Z. Effects of cationic micelles on the intramolecular general base-catalyzed hydrolysis of ionized phenyl salicylate at different temperatures. *J. Chem. Res. (S)* **1995**, 132–133.

50. Wang, Z.-w., Huang, D.-y., Gong, S.-p., Li, G.-z. Prediction on critical micelle concentration of nonionic surfactants in aqueous solution: quantitative structure-property relationship approach. *Chin. J. Chem.* **2003**, *21*(12), 1573–1579.

51. Yuan, S., Cai, Z., Xu, G., Jiang, Y. Quantitative structure-property relationships of surfactants: prediction of the critical micelle concentration of nonionic surfactants. *Colloid Polym. Sci.* **2002**, *280*(7), 630–636.

52. Wang, Z., Li, G., Zhang, X., Wang, R., Lou, A. A quantitative structure-property relationship study for the prediction of critical micelle concentration of nonionic surfactants. *Colloids Surf. A* **2002**, *197*(1–3), 37–45.

53. Jalali-Heravi, M., Konouz, E. Multiple linear regression modeling of the critical micelle concentration of alkyltrimethylammonium and alkylpyridinium salts. *J. Surfactants Deterg.* **2003**, *6*(1), 25–30.

54. Yuan, S., Cai, Z., Xu, G., Jiang, Y. Quantitative structure-property relationships of surfactants: critical micelle concentration of anionic surfactants. *J. Dispersion Sci. Technol.* **2002**, *23*(4), 465–472.

55. Anoune, N., Nouiri, M., Berrah, Y., Gauvrit, J.-Y., Lanteri, P. Critical micelle concentrations of different classes of surfactants: a quantitative structure property relationship study. *J. Surfactants Deterg.* **2002**, *5*(1), 45–53.

56. Burns, J.A. In *Thermodynamic Behaviour of Electrolytes in Mixed Solvents*. Furter, W.F., Ed., Advances in Chemistry Series 155, American Chemical Society, Washington D.C., **1976**.

57. Pandya, K., Lad, K., Bahadur, P. Effect of additives on the clouding behavior of an ethylene oxide-propylene oxide block copolymer in aqueous solution. *J. Macromol. Sci. — Pure Appl. Chem.* **1993**, *A30*(1), 1–18.

58. Bahadur, P., Pandya, K., Almgren, M., Li, P., Stilbs, P. Effect of inorganic salts on the micellar behavior of ethylene oxide-propylene oxide block copolymer in aqueous solution. *Colloid Polym. Sci.* **1993**, *271*(7), 657–667.

59. Bahadur, P., Li, P., Almgren, M., Brown, W. Effect of potassium fluoride on the micellar behavior of pluronic F-68 in aqueous solution. *Langmuir* **1992**, *8*(8), 1903–1907.

60. Penders, M.H.G.M., Nilsson, S., Picuiell, L., Lindman, B. Clouding and diffusion of a poly(ethylene oxide)-poly(propylene oxide)-poly(ethylene oxide) block copolymer in agarose gels and solutions. *J. Phys. Chem.* **1994**, *98*(21), 5508–5513.

61. Alexandridis, P., Athanassiou, V., Hatton, T.A. Pluronic-P105 PEO-PPO-PEO block copolymer in aqueous urea solutions: micelle formation, structure and microenvironment. *Langmuir* **1995**, *11*(7), 2442–1450.

62. Alexandridis, P., Holzwarth, J.F. Differential scanning calorimetry investigation of the effect of salts on aqueous solution properties of an amphiphilic block copolymer (poloxamer). *Langmuir* **1997**, *13*(23), 6074–6082.

63. Jorgensen, E.B., Hvidt, S., Brown, W., Schillen, K. Effects of salts on the micellization and gelation of a triblock copolymer studied by rheology and light scattering. *Macromolecules* **1997**, *30*(8), 2355–2364.

64. Desai, P.R., Jain, N.J., Sharma, R.K., Bahadur, P. Effect of additives on the micellization of PEO/PPO/PEO block copolymer F127 in aqueous solution. *Colloids Surf. A* **2001**, *178*(1–3), 57–69.

65. Alexandridis, P., Yang, L. SANS investigation of polyether block copolymer micelle structure in mixed solvents of water and formamide, ethanol, or glycerol. *Macromolecules* **2000**, *33*(15), 5574–5587.

66. Mukerjee, P. Olubilization in aqueous micellar systems. In *Solution Chemistry of Surfactants*, Vol. 1. Mittal, K.L., Ed., Plenum Press, New York, **1979**, pp. 153–174.

67. Soni, S.S., Sastry, N.V., Joshi, J.V., Seth, E., Goyal, P.S. Study on the effects of nonelectrolyte additives on the phase, thermodynamics, and structural changes in micelles of silicone surfactants in aqueous solutions from surface activity, small angle neutron scattering, and viscosity measurements. *Langmuir* **2003**, *19*(17), 6668–6677.

68. Jiang, B.-y., Du, J., Cheng, S.-q., Wang, Q., Zeng, X.-c. Effects of amine additives on critical micelle concentration of ionic surfactants. *J. Dispersion Sci. Technol.* **2003**, *24*(6), 755–760.

69. Khan, M.N., Arifin, Z. Effects of cationic micelles on rate of intramolecular general base-catalyzed ethanediolysis of ionized phenyl salicylate (PS⁻). *Langmuir* **1997**, *13*(25), 6626–6632.

70. Misra, P.K., Misra, B.K., Behera, G.B. Micellization of ionic surfactants in tetrahydrofuran-water and acetonitrile-water mixed solvent systems. *Colloids Surf.* **1991**, *57*(1–2), 1–10.

71. Rodriguez, A., Graciani, M. del M., Munoz, M., Moya, M.L. Water-ethylene glycol alkyltrimethylammonium bromide micellar solutions as reaction media: study of spontaneous hydrolysis of phenyl chloroformate. *Langmuir* **2003**, *19*(18), 7206–7213.

72. Graciani, M. del M., Rodriguez, A., Munoz, M., Moya, M.L. Water-ethylene glycol alkyltrimethylammonium bromide micellar solutions as reaction media: study of the reaction methyl 4-nitrobenzenesulfonate + Br⁻. *Langmuir* **2003**, *19*(21), 8685–8691.

73. Kumar, S., Sharma, D., Khan, Z.A., Kabir-ud-Din. Salt-induced cloud point in anionic surfactant solutions: role of the head group and additives. *Langmuir* **2002**, *18*(11), 4205–4209.

74. Mukerjee, P., Ray, A. The effect of urea on micelle formation and hydrophobic bonding. *J. Phys. Chem.* **1967**, *67*, 190–193.

75. Rharbi, Y., Winnik, M.A. Salt effects on solute exchange and micelle fission in sodium dodecyl sulfate micelles below the micelle-to-rod transition. *J. Phys. Chem. B* **2003**, *107*(7), 1491–1501 and reference cited therein.

76. Fisher, L.R., Okenfull, D.G. Micelles in aqueous solution. *Chem. Soc. Rev.* **1977**, *6*(1), 25–42.

77. Emerson, M.F., Holtzer, A. The ionic strength dependence of micelle number. *J. Phys. Chem.* **1965**, *69*(11), 3718–3721.

78. Zana, R. Critical micellization concentration of surfactants in aqueous solution and free energy of micellization. *Langmuir* **1996**, *12*(5), 1208–1211.

79. Jolicoeur, C., Philip, P.R. Enthalpy-entropy compensation for micellization and other hydrophobic interactions in aqueous solutions. *Can. J. Chem.* **1974**, *52*, 1834–1839.

80. Sarmiento, F., del Rio, J.M., Prieto, G., Attwood, D., Jones, M.N., Mosquera, V. Thermodynamics of micelle formation of chlorhexidine digluconate. *J. Phys. Chem.* **1995**, *99*(49), 17628–17631.

81. Kim, H.-U., Lim, K.-H. A model on the temperature dependence of critical micelle concentration. *Colloids Surf. A* **2004**, *235*, 121–128.

82. Bunton, C.A. Reaction kinetics in aqueous surfactant solutions. *Catal. Rev. — Sci. Eng.* **1979**, *20*(1), 1–56.

83. Hek, S.R., Chetry, S., Jayasree, V., Bhat, N. Effects of solvents and temperature on the critical micellar concentration of sodium dodecyl sufate and cetylpyridinium chloride. *J. Indian Chem Soc.* **2000**, *77*(10), 478–481.

84. Engberts J.B.F.N. Mixed aqueous solvent effects on kinetics and mechanisms of organic reactions. Frannks, F., Ed., *Water: A Comprehensive Treatise.* Vol. 6, Plenum Press, New York, **1979**, pp. 139–237, 411–436, chap. 4.

85. Oremusova, J., Greksakova, O., Peterek, K. Thermodynamics of micellization of hexadecylpyridinium bromide in aqueous and alcoholic (C_1–C_4) solutions. *Coll. Czech. Chem. Commun.* **2000**, *65*(9), 1419–1437.

86. Sharma, K.S., Rakshit, A.K. Thermodynamics of micellization and interfacial adsorption of polyoxyethylene (10) lauryl ether ($C_{12}E_{10}$) in water. *Indian J. Chem.*, Section A **2004**, *43A*(2), 265–269.

87. Bijma, K., Blanmer, M.J., Engberts, J.B.F.N. Effect of counterions and headgroup hydrophobicity on properties of micelles formed by alkylpyridinium surfactants. 2. Microcalorimetry. *Langmuir* **1998**, *14*(1), 79–83.

88. Shimizu, S., Pires, P.A.R., Loh, W., El Seoud, O.A. Thermodynamics of micellization of cationic surfactants in aqueous solutions: consequences of the presence of the 2-acylaminoethyl moiety in the surfactant head group. *Colloid Polym. Sci.* **2004**, *282*(9), 1026–1032.

89. Hartley, G. S. The application of the Debye-Huckel theory to colloidal electrolytes. *Trans. Faraday Soc.* **1935**, *31*, 31–50.

90. Menger, F.M. On the structure of micelles. *Acc. Chem. Res.* **1979**, *12*, 111–117.

91. Dill, K.A., Flory, P.J. Interphases of chain molecules: monolayers and lipid bilayer membranes. *Proc. Natl. Acad. Sci. U.S.A.* **1980**, *77*(6), 3115–3119.

92. Dill, K.A., Flory, P.J. Molecular organization in micelles and vesicles. *Proc. Natl. Acad. Sci. U.S.A.* **1981**, *78*(2), 676–680.

93. Dill, K.A., Koppel, D.E., Cantor, R.S., Dill, J.D., Bendedouch, D., Chen, S.-H. Molecular conformations in surfactant micelles. *Nature* **1984**, *309*(5963), 42–45.

94. Fromherz, P. The surfactant-block structure of micelles, synthesis of the droplet and of the bilayer concept. *Ber. Bunsen-Gesellschaft* **1981**, *85*(10), 891–899.

95. Butcher, J.A., Jr., Lamb, G.W. The relationship between domes and foams: application of geodesic mathematics to micelles. *J. Am. Chem. Soc.* **1984**, *106*(5), 1217–1220.

96. Rusanov, A.I., Grinin, A.P., Kuni, F.M., Shchekin, A.K. Nanostructural models of micelles and premicellar aggregates. Russian. *J. General Chem.* (Translation of Zhurnal Obshchei Khimii) **2002**, *72*(4), 607–621.

97. Balasubramanian, S., Pal, S., Bagchi, B. Hydrogen-bond dynamics near a micellar surface: origin of the universal slow relaxation at complex aqueous interface. *Phys. Rev. Lett.* **2002**, *89*(11), 115505/1–115505/4.

98. Khan, M.N. Effects of anionic micelles on the intramolecular general base-catalyzed hydrolysis of phenyl and methyl salicylates. *J. Mol. Catal. A: Chem.* **1995**, *102*, 93–101.

99. Khan, M.N. Spectrophotometric determination of anionic micellar binding constants of ionized and non-ionized phenyl and methyl salicylates. *J. Phys. Org. Chem.* **1996**, *9*, 295–300.

100. Nusselder, J.J.H., Engberts, J.B.F.N. A search for a relation between aggregate morphology and the structure of 1,4-dialkylpyridinium halide surfactants. *J. Org. Chem.* **1991**, *56*(19), 5522–5527.

101. Israelachvili, J.N., Mitchell, J., Ninham, B.W. Theory of self-assembly of hydrocarbon amphiphiles into micelles and bilayers. *J. Chem. Soc. Faraday Trans. II* **1976**, *72*(9), 1525–1568.

102. Porte, G., Poggi, Y., Appel, J., Maret, G. Lage micelles in concentrated solutions: the second critical micellar concentration. *J. Phys. Chem.* 2 **1984**, *88*(23), 5713–5720.

103. Grell, E., Lewitzki, E., von Raumer, M., Hormann, A. Lipid-similar thermal transition of polyethylene glycol alkyl ether detergents. *J. Therm. Anal. Calorimetry* **1999**, *57*(1), 371–375.

104. Grell, E., Lewitzki, E., Schneider, R., Ilgenfritz, G., Grillo, I., von Raumer, M. Phase transitions in non-ionic detergent micelles. *J. Therm. Anal. Calorimetry* **2002**, *68*(2), 469–478.

105. Bernheim-Grosswasser, A., Zana, R., Talmon, Y. Sphere-to-cylinder transition in aqueous micellar solution of dimeric (Gemini) surfactants. *J. Phys. Chem. B* **2000**, *104*(17), 4005–4009.

106. May, S., Ben-Shaul, A. Molecular theory of the sphere-to-rod transition and the second CMC in aqueous micellar solutions. *J. Phys. Chem. B* **2001**, *105*(3), 630–640.

107. Weber, V., Narayanan, T., Mendes, E., Schosseler, F. Micellar growth in salt-free aqueous solutions of a Gemini cationic surfactant: evidence for a multimodal population of aggregates. *Langmuir* **2003**, *19*(4), 992–1000.

108. Kim, H.-U., Lim, K.-H. Sizes and structures of micelles of cationic octadecyltrimethylammonium chloride and anionic ammonium dodecyl sulfate surfactants in aqueous solutions. *Bull. Korean Chem. Soc.* **2004**, *25*(3), 382–388.

109. Young, C.Y., Missel, P.J., Mazer, N.A., Benedek, G.B., Carey, M.C. Deduction of micellar shape from angular dissymmetry measurements of light scattered from aqueous sodium dodecyl sulfate solutions at high sodium chloride concentrations. *J. Phys. Chem.* **1978**, *82*(12), 1375–1378.

110. Porte, G., Appel, J. Growth and size distributions of cetylpyridinium bromide micelles in high ionic strength aqueous solutions. *J. Phys. Chem.* **1981**, *85*(17), 2511–2519.

111. Imae, T., Ikeda, S. Sphere-rod transition of micelles of tetradecyltrimethylammonium halides in aqueous sodium halide solutions and flexibility and entanglement of long rodlike micelles. *J. Phys. Chem.* **1986**, *90*(21), 5216–5223.

112. Backlund, S., Hoeiland, H., Kvammen, O.J., Ljosland, E. An ultrasonic study of sphere to rod transitions in aqueous solutions of hexadecyltrimethylammonium bromide. *Acta Chem. Scand. A* **1982**, *A36*(8), 698–800.

113. Lindblom, G., Lindman, B., Mandell, L. Effect of micellar shape and solubilization on counterion binding studied by bromine-81 NMR. *J. Colloid Interface Sci.* **1973**, *42*(2), 400–409.

114. Reiss-Husson, F., Luzzati, V. The structure of the micellar solutions of some amphiphilic compounds in pure water as determined by absolute small-angle-x-ray scattering technique. *J. Phys. Chem.* **1964**, *68*(12), 3504–3511.

115. Roux-Desgranges, G., Roux, A.H., Grolier, J.P.E., Viallard, A. Role of alcohol in microemulsions as determined from volume and heat capacity data for the water-sodium dodecylsulfate-n-butanol system at 25°C. *J. Solution Chem.* **1982**, *11*(5), 357–375.

116. Perron, G., DeLisi, R., Davidson, I., Genereux, S., Desnoyers, J.E. On the use of thermodynamic transfer functions for the study of the effect of additives on micellization: volumes and heat capacities of solutions of sodium octanoate systems. *J. Colloid Interface Sci.* **1981**, *79*(2), 432–442.

117. DeLisi, R., Milioto, S., Castagnolo, M., Inglese, A. Standard partial molar volumes of alcohols in aqueous dodecyltrimethylammonium bromide solutions. *J. Solution Chem.* **1990**, *19*(8), 767–791.

118. DeLisi, R., Milioto, S., Triolo, R. Heat capacities, volumes and solubilities of pentanol in aqueous alkyltrimethylammonium bromides. *J. Solution Chem.* **1988**, *17*(7), 673–696.

119. Treiner, C., Chattopadhyay, A.K., Bury, R. Heat and solution of various alcohols in aqueous micellar solutions of hexadecyltrimethylammonium bromide as a function surfactant concentration: the preferential solvation phenomenon. *J. Colloid Interface Sci.* **1985**, *104*(2), 569–578.

120. Quirion, F., Desnoyers, J.E. A thermodynamic study of the postmicellar transition of cetyltrimethylammonium bromide in water. *J. Colloid Interface Sci.* **1986**, *112*(2), 565–572.

121. Ekwall, P., Mandell, L., Solyom, P. Aqueous cetyltrimethylammonium bromide solutions. *J. Colloid Interface Sci.* **1971**, *35*(4), 519–528.

122. Attwood, D., Elworthy, P.H., Kayne, S.B. Membrane osmometry of aqueous micellar solutions of pure nonionic and ionic surfactants. *J. Phys. Chem.* **1970**, *74*(19), 3529–3534.

123. Brown, W., Johnsen, R., Stilbs, P., Lindman, B. Size and shape of nonionic amphiphile ($C_{12}E_6$) micelles in dilute aqueous solutions as derived from quasielastic and intensity light scattering, sedimentation, and pulsed-field-gradient nuclear magnetic resonance self-diffusion data. *J. Phys. Chem.* **1983**, *87*(22), 4548–4553.

124. Richtering, W.H., Burchard, W., Jahns, E., Finkelmann, H. Light scattering from aqueous solutions of a nonionic surfactant ($_{14}E_8$) in a wide concentration range. *J. Phys. Chem.* **1988**, *92*(21), 6032–6040.

125. Menge, U., Lang, P., Findenegg, G.H. From oil-swollen wormlike micelles to microemulsion droplets: a static light scattering study of the L1 phase of the system water + $C_{12}E_5$ + decane. *J. Phys. Chem.* **1999**, *103*(28), 5768–5774.

126. Glatter, O., Fritz, G., Lindner, H., Brunner-Popela, J., Mittelbach, R., Strey, R., Egelhaaf, S.U. Nonionic micelles near the critical point: micellar growth and attractive interactions. *Langmuir* **2000**, *16*(23), 8692–8701.

127. Drye, T.J., Cates, M.E. Living networks: the role of cross-links in entangled surfactant solutions. *J. Phys. Chem.* **1992**, *96*(2), 1367–1375.

128. Bohbot, Y., Ben-Shaul, A., Granek, R., Gelbart, W.M. Monte Carlo and mean-field studies of phase evolution in concentrated surfactant solutions. *J. Chem. Phys.* **1995**, *103*(19), 8764–8782.

129. Bernheim-Grosswasser, A., Tlusty, T., Safran, S.A., Talmon, Y. Direct observation of phase separation in microemulsion networks. *Langmuir* **1999**, *15*(17), 5448–5453.

130. Danino, D., Talmon, Y., Levy, H., Beinert, G., Zana, R. Branched threadlike micelles in an aqueous solution of a trimeric surfactant. *Science* **1995**, *269*(5229), 1420–1421.

131. Oda, R., Panizza, P., Schmutz, M., Lequeux, F. Direct evidence of the shear-induced structure of wormlike micelles: gemini surfactant 12-2-12. *Langmuir* **1997**, *13*(24), 6407–6412.

132. Lin, Z. Branched worm-like micelles and their networks. *Langmuir* **1996**, *12*(7), 1729–1737.

133. Bernheim-Groswasser, A., Wachtel, E., Talmon, Y. Micellar growth, network formation, and criticality in aqueous solutions of the nonionic surfactant $C_{12}E_5$. *Langmuir* **2000**, *16*(9), 4130–4140.

134. Tlusty, T., Safran, S.A., Strey, R. Topology, phase instabilities, and wetting of microemulsion networks. *Phys. Rev. Lett.* **2000**, *84*(6), 1244–1247.

135. Mukerjee, P. In *Physical Chemistry: Enriching Topics from Colloidal and Surface Science.* van Olphen, H., Mysels, K. Eds., Iupac Theorex, La Jolla, CA, **1975**, chap. 9.

136. Ilgenfritz, G., Schneider, R., Grell, E., Lewitzki, E., Ruf, H. Thermodynamic and kinetic study of the sphere-to-rod transition in nonionic micelles: aggregation and stress relaxation in $C_{14}E_8$ and $C_{16}E_8/H_2O$ systems. *Langmuir* **2004**, *20*(6), 1620–1630.

137. Israelachvili, J.N. In *Aggregation of Amphiphilic Molecules into Micelles.* Academic Press, London, 1985, chap. 16.

138. (a) Imae, T., Abe, A., Ikeda, S. Viscosity behavior of semiflexible rodlike micelles of alkyltrimethylammonium halides in dilute and semidilute solutions. *J. Phys. Chem.* **1988**, *92*(6), 1548–1553. (b) Mancini, G., Schiavo, C., Cerichelli, G. Trapping of counterions and water on the surface of cationic micelles. *Langmuir* **1996**, *12*(15), 3567–3573. (c) Al-Lohedan, H.A. Quantitative treatment of micellar effects upon nucleophilic substitution. *J. Chem. Soc. Perkin Trans.* 2 **1995**, 1707–1713 and references cited therein.

139. Helndl, A., Strnad, J., Kohler, H.-H. Effect of aromatic solubilisates on the shape of CTABr micelles. *J. Phys. Chem.* **1993**, *97*(3), 742–746.

140. Prud'homme, R.K., Warr, G.G. Elongational flow of solutions of rodlike micelles. *Langmuir* **1994**, *10*(10), 3419–3426.

141. Lindemuth, P.M., Bertrand, G.L. Calorimetric observations of the transition of spherical to rodlike micelles with solubilized organic additives. *J. Phys. Chem.* **1993**, *97*(29), 7769–7773.

142. (a) Kumar, S., Aswal, V.K., Goyal, P.S., Kabir-ud-Din. Micellar growth in the presence of quaternary ammonium salts: a SANS study. *J. Chem. Soc., Faraday Trans.* **1998**, *94*(6), 761–764. (b) Kabir-ud-Din., David, S.L., Kumar, S. Effect of counterion size on the viscosity behavior of sodium dodecyl sulfate micellar solutions. *J. Mol. Liq.* **1998**, *75*(1), 25–32.

143. Schurtenberger, P., Jerke, G., Cavaco, C. Cross-section structure of cylindrical and polymer-like micelles from small-angle scattering data. 2. Experimental results. *Langmuir* **1996**, *12*(10), 2433–2440 and references cited therein.

144. Kumar, S., Aswal, V.K., Naqvi, A.Z., Goyal, P.S., Kabir-ud-Din. Cloud point phenomenon in ionic micellar solutions: a SANS study. *Langmuir* **2001**, *17*(9), 2549–2551.

145. Mchedlov-Petrossyan, N.O., Vodolazkaya, N.A., Reichardt, C. Unusual findings on studying surfactant solutions: displacing solvatochromic pyridinium N-phenolate towards outlying areas of rodlike micelles. *Colloids Surf. A* **2002**, *205*(3), 215–229.

146. Kumar, S., Kirti, Kumari, K., Kabir-ud-Din. Role of alkanols in micellar growth: a viscometric study. *J. Am. Oil Chem. Soc.* **1995**, *72*(7), 817–821.

147. Kabir-ud-Din, Kumar, S., Aswal, V.K., Goyal, P.S. Effect of the addition of n-alkylamines on the growth of sodium decyl sulfate micelles. *J. Chem. Soc., Faraday Trans.* **1996**, *92*(13), 2413–2415.

148. Kumar, S., Aswal, V.K., Singh, H.N., Goyal, P.S., Kabir-ud-Din. Growth of sodium dodecyl sulfate micelles in the presence of n-octylamine. *Langmuir* **1994**, *10*(11), 4069–4072.

149. Hoffmann, H., Ebert, G. Surfactants, micelles and fascinating phenomena. *Angew. Chem., Int. Ed. Engl.* **1988**, *27*(7), 902–912.

150. Kabir-ud-Din, Kumar, S., Kirti, Goyal, P.S. Micellar growth in presence of alcohols and amines: a viscometric study. *Langmuir* **1996**, *12*(6), 1490–1494.

151. Kabir-ud-Din, Bansal, D., Kumar, S. Synergistic effect of salts and organic additives on the micellar association of cetylpyridinium chloride. *Langmuir* **1997**, *13*(19), 5071–5076.

152. (a) Kumar, S., David, S.Z., Aswal, V.K., Goyal, P.S., Kabir-ud-Din, Growth of sodium dodecyl sulfate micelles in aqueous ammonium salts. *Langmuir* **1997**, *13*(24), 6461–6464. (b) David, S.L., Kumar, S., Kabir-ud-Din. Viscosities of sodium dodecyl sulfate solutions in aqueous ammonium salts. *J. Chem. Eng. Data* **1997**, *42*(6), 1224–1226.

153. Kumar, S., Bansal, D., Kabir-ud-Din. Micellar growth in the presence of salts and aromatic hydrocarbons: influence of the nature of the salt. *Langmuir* **1999**, *15*(15), 4960–4965.

154. Kumar, S., Naqvi, A.Z., Kabir-ud-Din. Micellar morphology in the presence of salts and organic additives. *Langmuir* **2000**, *16*(12), 5252–5256.

155. Kumar, S., Sharma, D., Kabir-ud-Din. Cloud point phenomenon in anionic surfactant + quaternary bromide systems and its variation with additives. *Langmuir* **2000**, *16*(17), 6821–6824.

156. Kumar, S., Naqvi, A.Z., Kabir-ud-Din. Solubilization-site-dependent micellar morphology: effect of organic additives and quaternary ammonium bromides. *Langmuir* **2001**, *17*(16), 4787–4792.

157. Yue, Y., Wang, J., Dai, M. Volumetric and fluorescence studies of aqueous solutions containing n-octylamine, cetyltrimethylammonium bromide, and salt. *Langmuir* **2000**, *16*(15), 6114–6117.

158. Goyal, P.S., Menon, S.V.G. Role of van der Waaals forces on small angle neutron scattering from ionic micellar solutions. *Chem. Phys. Lett.* **1993**, *211*(6), 559–563.

159. Christov, N.C., Denkov, N.D., Kralchevsky, P.A., Ananthapadmanabhan, K.P., Lips, A. Synergistic sphere-to-rod micelle transition in mixed solutions of sodium dodecyl sulfate and cocoamidopropyl betaine. *Langmuir* **2004**, *20*(3), 565–571.

160. Warr, G.G., Zemb, T.N., Drifford, M. Liquid-liquid phase separation in cationic micellar solutions. *J. Phys. Chem.* **1990**, *94*(7), 3086–3092.

161. Almgren, M., Swarup, S. Size of sodium dodecyl sulfate micelles in the presence of additives. 3. Multivalent and hydrophobic counterions, cationic and nonionic surfactants. *J. Phys. Chem.* **1983**, *87*(5), 876–881.

162. (a) Stangret, J., Gampe, T. Hydration sphere of tetrabutylammonium cation. FTIR studies of HDO spectra. *J. Phys. Chem. B.* **1999**, *103*(18), 3778–3783. (b) Koga, Y., Nishikawa, K., Westh, P. Icebergs: or no "icebergs" in aqueous alcohols?: composition-dependent mixing schemes. *J. Phys. Chem. A* **2004**, *108*(17), 3873–3877.

163. (a) Cerichelli, G., Mancini, G. Role of counterions in the solubilization of benzene by cetytrimethylammonium aggregates: a multinuclear NMR investigation. *Langmuir* **2000**, *16*(1), 182–187. (b) Soldi, V., Keiper, J., Romsted, L.S., Cuccovia, I.M., Chaimovich, H. Arenediazonium salts: new probes of the interfacial compositions of association colloids. 6. Relationships between interfacial counterion and water concentrations and surfactant headgroup size, sphere-to-rod transitions, and chemical reactivity in cationic micelles. *Langmuir* **2000**, *16*(1), 59–71. (c) Haverd, V.E., Warr, G.G. Cation selectivity at airanionic surfactant solution interface. *Langmuir* **2000**, *16*(1), 157–160.

164. Ulmius, J., Wennerstroem, H., Johansson, L.B.-A., Lindblom, G., Gravsholt, S. Viscoelasticity in surfactant solutions. Characteristics of the micellar aggregates and the formation of periodic colloidal structures. *J. Phys. Chem.* **1979**, *83*(17), 2232–2236.

165. Anet, F.A.L. Novel spin-spin splitting and relaxation effects in the proton NMR spectra of sodium salicylate in viscoelastic micelles. *J. Am. Chem. Soc.* **1986**, *108*(22), 7102–7103.

166. Rao, U.R.K., Manohar, C., Valaulikar, B.S., Iyer, R.M. Micellar chain model for the origin of the viscoelasticity in dilute surfactant solutions. *J. Phys. Chem.* **1987**, *91*(12), 3286–3291.

167. Bachofer, S.J., Turbitt, R.M. The orientational binding of substituted benzoate anions at the cetyltrimethylammonium bromide interface. *J. Colloid Interface Sci.* **1990**, *135*(2), 325–334.

168. Imai, S.-i., Shikata, T. Viscoelastic behavior of surfactant threadlike micellar solutions: effect of additives 3. *J. Colloid Interface Sci.* **2001**, *244*(2), 399–404.

169. Olsson, U., Soderman, O., Guering, P. Characterization of micellar aggregates in viscoelastic surfactant solutions: a nuclear magnetic resonance and light scattering study. *J. Phys. Chem.* **1986**, *90*(21), 5223–5232.

170. Clausen, T.W., Vinson, P.K., Minter, J.R., Davis, H.T., Talmon, Y., Miller, W.G. Viscoelastic micellar solutions: microscopy and rheology. *J. Phys. Chem.* **1992**, *96*(1), 474–484.

171. (a) Underwood, A.L., Anacker, E.W. Organic counterions and micellar parameters: substituent effects in a series of benzoates. *J. Phys. Chem.* **1984**, *88*(11), 2390–2393. (b) Underwood, A.L., Anacker, E.W. Organic counterions and micellar parameters: unusual effects of hydroxyl- and chlorobenzoates. *J. Colloid Interface Sci.* **1985** *106*(1), 86–93.

172. Imae, T., Kohsaka, T. Size and electrophoretic mobility of tetradecyltrimethylammonium salicylate (C_{14}TASal) micelles in aqueous media. *J. Phys. Chem.* **1992**, *96*(24), 10030–10035.

173. Lin, M.Y., Hanley, H.J.M., Sinha, S.K., Straty, G.C., Peiffer, D.G., Kim, M.W. Shear-induced behavior in a solution of cylindrical micelles. *Phys. Rev. E* **1996**, *53*(5-A), R4302–R4305.

174. Berret, J.-F., Appel, J., Porte, G. Linear rheology of entangled wormlike micelles. *Langmuir* **1993**, *9*(11), 2851–2854.

175. Mishra, B.K., Samant, S.D., Prafhan, P., Mishra, S.B., Manohar, C. A new strongly flow birefringent surfactant system. *Langmuir* **1993**, *9*(4), 894–898.

176. (a) Bijma, K., Rank, E., Engberts, J.B.F.N. Effect of counterion structure on micellar growth of alkylpyridinium surfactants in aqueous solution. *J. Colloid Interface Sci.* **1998**, *205*(2), 245–256. (b) Bijma, K., Engberts, J.B.F.N. Effect of counterions on properties of micelles formed by alkylpyridinium surfactants. 1. Conductometry and ^{1}H NMR chemical shifts. *Langmuir* **1997**, *13*(18), 4843–4849.

177. (a) Carver, M., Smith, T.L., Gee, J.C., Delichere, A., Caponetti, E., Magid, L.J. Tuning of micellar structure and dynamics in aqueous salt-free solutions of cetyltrimethylammoinium mono- and dichlorobenzoates. *Langmuir* **1996**, *12*(3), 691–698. (b) Kreke, P.J., Magid, L.J., Gee, J.C. ^{1}H and ^{13}C NMR studies of mixed counterion, cetyltrimethylammonium bromide/cetyltrimethylammonium dichlorobenzoate, surfactant solutions: the interaction of aromatic counterions. *Langmuir* **1996**, *12*(3), 699–705.

178. Soltero, J.F.A., Puig, J.E., Manero, O. Rheology of the cetytrimethylammonium tosilate-water system. 2. Linear viscoelastic regime. *Langmuir* **1996**, *12*(11), 2654–2662.

179. Kaler, E.W., Herrington, K.L., Murthy, A.K., Zasadzinski, J.A.N. Phase behavior and structures of mixtures of anionic and cationic surfactants. *J. Phys. Chem.* **1992**, *96*(16), 6698–6707.

180. Magid, L.J., Han, Z., Warr, G.G., Casidy, M.A., Butler, P.W., Hamilton, W.A. Effect of counterion competition on micellar growth horizons for cetltrimethylammonium micellar surfaces: electrostatics and specific binding. *J. Phys. Chem. B* **1997**, *101*(40), 7919–7927.

181. Brown, W., Johansson, K., Almgren, M. Threadlike micelles from cetyltrimethylammonium bromide in aqueous sodium naphthalenesulfonate solutions studied by static and dynamic light scattering. *J. Phys. Chem.* **1989**, *93*(15), 5888–5894.

182. Salkar, R.A., Hassan, P.A., Samant, S.D., Valaulikar, B.S., Kumar, V.V., Kern, F., Candu, S.J., Manohar, C. A thermally reversible vesicle to micelle transition driven by a surface solid-fluid transition. *J. Chem. Soc. Chem. Commun.* **1996**, 1223–1224.

183. Hassan, P.A., Valaulikar, B.S., Manohar C., Kern, F., Bourdieu, L., Candu, S.J. Vesicle to micelle transition: rheological investigations. *Langmuir* **1996** *12*(18), 4350–4357.

184. Schubert, B.A., Wagner, N.J., Kaler, E.W., Raghavan, S.R. Shear-induced phase-separation in solutions of wormlike micelles. *Langmuir* **2004**, *20*(9), 3564–3573.

185. Kern, F., Zana, R., Candau, S.J. Rheological properties of semidilute and concentrated aqueous solutions of cetytrimethylammonium chloride in the presence of sodium salicylate and sodium chloride. *Langmuir* **1991**, *7*(7), 1344–1351.

186. Rehage, H., Hoffmann, H. Rheological properties of viscoelastic surfactant systems. *J. Phys. Chem.* **1988**, *92*(16), 4712–4719.

187. Acharya, D.P., Khalid, H.M., Jin, F.S., Takaya, K.H. Phase and rheological behaviour of viscoelastic wormlike micellar solutions formed in mixed nonionic surfactant systems. *Phys. Chem. Chem. Phys.* **2004**, *6*(7), 1627–1631.

188. Sommer, C., Pedersen, J.S., Egelhaaf, S.U., Cannavacciuolo, L., Kohlbrecher, J., Schurtenberger, P. Wormlike micelles as "Equilibrium Polyelectrolytes": light and neutron scattering experiments. *Langmuir* **2002**, *18*(7), 2495–2505.

189. Lu, B., Li, X., Scriven, L.E., Davis, H.T., Talmon, Y., Zakin, J.L. Effect of chemical structure on viscoelasticity and extensional viscosity of drag-reducing cationic surfactant solutions. *Langmuir* **1998**, *14*(1), 8–16.

190. Almgren, M., Gimel, J.C., Wang, K., Karisson, G., Edwards, K., Brown, W., Mortensen, K. SDS micelles at high ionic strength: a light scattering, neutron scattering, Fluorescence quenching, and cryo TEM investigation. *J. Colloid Interface Sci.* **1998**, *202*(2), 222–231.

191. Manohar, C., Rao, U.R.K., Valaulikar, B.S., Iyer, R.M. On the origin of viscoelasticity in micellar solutions of cetyltrimethylammonium bromide and sodium salicylate. *J. Chem. Soc., Chem. Commun.* **1986**, 379–381.

192. Zilman, A., Safran, S.A., Sottmann, T., Strey, R. Temperature dependence of the thermodynamics and kinetics of micellar solutions. *Langmuir* **2004**, *20*(6), 2199–2207.

193. Menger, F.M., Boyer, B.J. Water penetration into micelles as determined by optical rotary dispersion. *J. Am. Chem. Soc.* **1980**, *102*(18), 5936–5938.

194. Casal, H.L. The water content of micelles: infrared spectroscopic studies. *J. Am. Chem. Soc.* **1988**, *110*(15), 5203–5205.

195. Wennerstroem, H., Lindman, B. Water penetration into surfactant micelles. *J. Phys. Chem.* **1979**, *83*(22), 2931–2932.

196. Aniansson, G.E.A. Dynamics and structure of micelles and other ampiphile structures *J. Phys. Chem.* **1978**, *82*(26), 2805–2808.

197. (a) Malliaris, A., Le Moigne, J., Sturm, J., Zana, R. Temperature dependence of the micelle aggregation number and rate of intramicellar excimer formation in aqueous surfactant solutions. *J. Phys. Chem.* **1985**, *89*(12), 2709–2713. (b) Lianos, P., Viriot, M.L., Zana, R., Study of the solubilization of aromatic hydrocarbons by aqueous micellar solutions. *J. Phys. Chem.* **1984**, *88*(6), 1098–1101. (c) Hirose, C., Sepulveda, L. Transfer free energies of p-alkyl-substituted benzene derivatives, benzene, and toluene from water to cationic and anionic micelles and to n-heptane. *J. Phys. Chem.* **1981**, *85*(24), 3689–3694.

198. Sabate, R., Gallardo, M., Estelrich, J. Location of pinacyanol in micellar solutions of N-alkyl trimethylammonium bromide surfactants. *J. Colloid Interface Sci.* **2001**, *233*(2), 205–210.

199. Berr, S.S., Coleman, M.J., Jones, R.R.M., Johnson, J.S., Jr. Small-angle neutron scattering study of the structural effects of substitution of tetramethylammonium for sodium as the counterion in dodecyl sulfate micelles. *J. Phys. Chem.* **1986**, *90*(24), 6492–6499.

200. Berr, S.S., Caponetti, E., Johnson, J.S., Jr., Jones, R.R.M., Magid, L.J. Small-angle neutron scattering from hexadecyltrimethylammonium bromide micelles in aqueous solutions. *J. Phys. Chem.* **1986**, *90*(22), 5766–5770.

201. Berr, S.S. Solvent isotope effects on alkyltrimethylammonium bromide micelles as a function of alkyl chain length. *J. Phys. Chem.* **1987**, *91*(18), 4760–4765.

202. DelaCruz, J.L., Blanchard, G.J. Understanding the balance between ionic and dispersion interactions in aqueous micellar media. *J. Phys. Chem. B* **2003**, *107*(29), 7102–7108 and references cited therein.

203. Mishra, A., Behera, R.K., Behera, P.K., Mishra, B.K., Behera, G.B. Cyanines during 1990s: a review. *Chem. Rev.* **2000**, *100*(6), 1973–2011.

204. Mishra, A., Patel, S., Behera, R.K., Mishra, B.K., Behera, G.B. Dye-surfactant interaction: role of an alkyl chain in the localization of styrylpyridinium dyes in a hydrophobic force field of a cationic surfactant (CTAB). *Bull. Chem. Soc. Jpn.* **1997**, *70*(12), 2913–2918.

205. Laschewsky, A. Polymerized micelles with compartments. *Curr. Opin. Colloid Interface Sci.* **2003**, *8*(3), 274–281.

206. Itoh, H., Ishido, S., Nomura, M., Hayakawa, T., Mitaku, S. Estimation of the hydrophobicity in microenvironments by pyrene fluorescence measurements. *J. Phys. Chem.* **1996**, *100*(21), 9047–9053.

207. Graetzel, M., Kalyanasundaram, K., Thomas, J. K. Proton nuclear magnetic resonance and laser photolysis studies of pyrene derivatives in aqueous and micellar solutions. *J. Am. Chem. Soc.* **1974**, *96*(26), 7869–7874.

208. Okano, L.T., Quina, F.H., El Seoud, O.A. Fluorescence and light-scattering studies of the aggregation of cationic surfactants in aqueous solution: effects of headgroup structure. *Langmuir* **2000**, *16*(7), 3119–3123.

209. Kim, J.-H., Domach, M.M., Tiltoni, R.D. Pyrene solubilization capacity in octaethylene glycol monododecyl ether ($C_{12}E_8$) micelles. *Colloids Surf. A* **1999**, *150*(1–3), 55–68.

210. Dutt, G.B. Are the experimentally determined microviscosities of the micelles probe dependent? *J. Phys. Chem. B* **2004**, *108*(11), 3651–3657.

211. Bachofer, S.J., Simonis, U., Nowicki, T.A. Orientational binding of substituted naphthoate counterions to the tetradecyltrimethylammonium bromide micellar interface. *J. Phys. Chem.* **1991**, *95*(1), 480–488.

212. Suratkar, V., Mahapatra, S. Solubilization of organic perfume molecules in sodium dodecyl sulfate micelles: new insights from proton NMR studies. *J. Colloid Interface Sci.* **2000**, *225*(1), 32–38.

213. Cerichelli, G., Mancini, G. Role of counterions in the solubilization of benzene by cetyltrimethylammonium aggregates: a multinuclear NMR investigation. *Langmuir* **2000**, *16*(1), 182–187.

214. Eriksson, J.C., Gillberg, G. NMR-studies of the solubilization of aromatic compounds in cetyltrimethylammonium bromide solution. II. *Acta Chem. Scand.* **1966**, *20*(8), 2019–2017.

215. Fendler, J.H., Patterson, L.K. Solubilization of benzene in aqueous cetyltrimethylammonium bromide measured by differential spectroscopy. Comment. *J. Phys. Chem.* **1971**, *75*(25), 3907; *J. Phys. Chem.* **1970**, *74*(26), 4608–4609.

216. Rehfeld, S.J. Solubilization of benzene in aqueous cetyltrimethylammonium bromide measured by differential spectroscopy. *J. Phys. Chem.* **1971**, *75*(25), 3905–3906.

217. Cardinal, J.R., Mukerjee, P. Solvent effects on the ultraviolet spectra of benzene derivatives and naphthalene: identification of polarity sensitive spectral characteristics. *J. Phys. Chem.* **1978**, *82*(14), 1614–1620.

218. Mukerjee, P., Cardinal, J.R. Benzene derivatives and naphthalene solubilized in micelles: polarity of microenvironments, location and distribution in micelles, and correlation with surface activity in hydrocarbon-water systems. *J. Phys. Chem.* **1978**, *82*(14), 1620–1627.

219. Nagarajan, R., Chaiko, M.A., Ruckenstein, E. Locus of solubilization of benzene in surfactant micelles. *J. Phys. Chem.* **1984**, *88*(13), 2916–2922.

220. Kandori, K., McGreevy, R.J., Schechter, R.S. Solubilization of phenol and benzene in cationic micelles: binding sites and effect on temperature. *J. Phys. Chem.* **1989**, *93*(4), 1506–1510.

221. (a) Dutt, G.B. Rotational diffusion of hydrophobic probes in Brij-35 micelles: effect of temperature on micellar internal environment. *J. Phys. Chem. B* **2003**, *107*(38), 10546–10551. (b) Duschl, J., Michl, M., Kunz, W. A porphyrin dye with monoexponential fluorescence intensity and anisotropy decay behavior in spherical micelles. *Angew. Chem. Int. Ed. Engl.* **2004**, *43*(5), 634–636.

222. Patist, A., Kanicky, J.R., Shukla, P.K., Shah, D.O. Importance of micellar kinetics in relation to technological processes. *J. Colloid Interface Sci.* **2002**, *245*(1), 1–15.

223. Fendler, J.H., Fendler, E.J. *Catalysis in Micellar and Macromolecular Systems*, Academic Press, New York, **1975**, chap. 2.

224. Lang, J., Zana, R. Chemical relaxation methods. *Surfactant Solutions*. Vol. 22, Surfactant Science Series, **1987**, pp. 405–452.

225. Goodeve, C.F. General discussion on "soap micelles." *Trans. Faraday Soc.* **1935**, *31*, 197–198.

226. Kresheck, G.C., Hamori, E., Davenport, G., Scheraga, H.A. Determination of the dissociation rate of dodecylpyridinium iodide micelles by a temperature-jump technique. *J. Am. Chem. Soc.* **1966**, *88*(2), 246–253.

227. Muller, N. Kinetics of micelle dissociation by temperature-jump techniques: a reinterpretation. *J. Phys. Chem.* **1972**, *76*(21), 3017–3020.

228. Sams, P.J., Wyn-Jones, E., Rassing, J. New model describing the kinetics of micelle formation from chemical relaxation studies. *Chem. Phys. Lett.* **1972**, *13*(3), 233–236.

229. Zana, R. Dynamics of micellar systems. *Encyclopedia of Surface and Colloid Science*. Marcel Dekker, New York, **2002**, pp. 1515–1528.

230. Aniansson, E.A.G., Wall, S.N. On the kinetics of step-wise micelle association. *J. Phys. Chem.* **1974**, *78*(10), 1024–1030.

231. Aniansson, E.A.G., Wall, S.N. A correction and improvement of "on the kinetics of step-wise micelle association." *J. Phys. Chem.* **1975**, *79*(8), 857–858.

232. (a) Aniansson, E.A.G., Wall, S.N., Almgren, M., Hoffmann, I., Ulbricht, W., Zana, R., Lang, J., Tondre, C. Theory of the kinetics of micellar equilibria and quantitative interpretation of chemical relaxation studies of micellar solutions of ionic surfactants. *J. Phys. Chem.* **1976**, *80*(9), 905–922. (b) Frindi, M., Michels, B., Zana, R. Ultrasonic absorption studies of surfactant exchange between micelles and bulk phase in aqueous micellar solutions of amphoteric surfactants. *J. Phys. Chem.* **1994**, *98*(26), 6607–6611. (c) Lang, J., Zana, R. Effects of alcohols and oils on the kinetics of micelle formation: breakdown in aqueous solutions of ionic surfactants. *J. Phys. Chem.* **1986**, *90*(21), 5258–5265. (d) Jobe, D.J., Verrall, R.E., Skalski, B., Aicart, E. Ultrasonic relaxation studies of mixed micelles formed from alcohol-decyltrimethylammonium bromide-water. *J. Phys. Chem.* **1992**, *96*(16), 6811–6817. (e) Aicart, E., Jobe, D.J., Skalski, B., Verrall, R.E. Ultrasonic relaxation studies of mixed micelles formed from propanol-decyltrimethylammonium bromide-water. *J. Phys. Chem.* **1992**, *96*(5), 2348–2355. (f) Goldmints, I., Holzwarth, J.F., Smith, K.A., Hatton, T.A. Micellar dynamics in aqueous solutions of PEO-PPO-PEO block copolymers. *Langmuir* **1997**, *13*(23), 6130–6134. (g) Michels, B., Waton, G., Zana, R. Dynamics of micelles of poly(ethylene oxide)-poly(propylene oxide)- poly(ethylene oxide) block copolymers in aqueous solutions. *Langmuir* **1997**, *13*(12), 3111–3118. (h) Frindi, M., Michels, B., Zana, R. Ultrasonic absorption studies of surfactant exchange between micelles and bulk phase in aqueous micellar solutions of nonionic surfactants with a short alkyl chain. 3. Surfactants with a sugar head group. *J. Phys. Chem.* **1992**, *96*(20), 8137–8141. (i) Frindi, M., Michels, B., Zana, R. Ultrasonic absorption studies of surfactant exchange between micelles and bulk phase in aqueous micellar solutions of nonionic surfactants with a short alkyl chain. 2. C_6E_3, C_6E_5, C_8E_4, and C_8E_8. *J. Phys. Chem.* **1992**, *96*(14), 6095–6102. (j) Frindi, M., Michels, B., Levy, H., Zana, R. Alkanediyl-α,ω-bis(dimethylalkylammonium bromide) surfactants. 4. Ultrasonic absorption studies of amphiphile exchange between micelles and bulk phase in aqueous micellar solution. *Langmuir* **1994**, *10*(4), 1140–1145.

233. Teubner, M. Theory of ultrasonic absorption in micellar solutions. *J. Phys. Chem.* **1979**, *83*(22), 2917–2920.

234. Kahlweit, M., Teubner, M. On the kinetics of micellization in aqueous solutions. *Adv. Colloid Interface Sci.* **1980**, *13*(1–2), 1–64.

235. Kaatze, U., Lautscham, K., Berger, W. Ultrasonic and hypersonic absorption and molecular relaxation in aqueous solutions of anionic and cationic micelles. *Z. Phys. Chem.* (Muenchen, Germany) **1988**, *159*(2), 161–174.

236. Kaatze, U., Berger, W., Lautscham, K. High frequency ultrasonic absorption spectroscopy on aqueous solutions of zwitterionic and nonionic micelles. *Ber. Bunsen-Ges. Phys. Chem.* **1988**, *92*(8), 872–877.

237. (a) Telgmann, T., Kaatze, U. On the kinetics of the formation of small micelles. 1. Broadband ultrasonic absorption spectrometry. *J. Phys. Chem. B* **1997**, *101*(39), 7758–7765. (b) Telgmann, T., Kaatze, U. Monomer exchange and concentration fluctuations in poly(ethylene glycol) monoalky ether/water mixtures: toward a uniform description of acoustical spectra. *Langmuir* **2002**, *18*(8), 3068–3075.

238. Waton, G., Michels, B., Zana, R. Dynamics of micelles of polyethylene oxide-polypropylene oxide-polyethylene oxide block copolymers in aqueous solutions. *J. Colloid Interface Sci.* **1999**, *212*(2), 593–596.

239. Waton, G., Michels, B., Zana, R. Dynamics of micelles of poly(ethylene oxide)-poly(propylene oxide)-poly(ethylene oxide) block copolymers in aqueous solutions. *Langmuir* **1997**, *13*(12), 3111–3118.

240. Kositza, M.J., Bohne, C., Hatton, T.A., Holzwarth, J.F. Micellization dynamics of poly(ethylene oxide)-poly(propylene oxide)-poly(ethylene oxide) block copolymers measured by stopped flow. *Prog. Colloid Polym. Sci. (Trends in colloid and interface science XIII)*, **1999**, *112*, 146–151.

241. Hecht, E., Hoffmann, H. Kinetic and calorimetric investigations on micelle formation of block copolymers of the poloxamer type. *Colloids Surf. A* **1995**, *96*, 181–197.

242. Kositza, M.J., Bohne, C., Alexandridis, P., Hatton, T.A., Holzwarth, J.F. Dynamics of micro- and macrophase separation of amphiphilic block-copolymers in aqueous solution. *Macromolecules* **1999**, *32*(17), 5539–5551.

243. Kositza, M.J., Bohne, C., Alexandridis, P., Hatton, T.A., Holzwarth, J.F. Micellization dynamics and impurity solubilization of the block-copolymer L64 in an aqueous solution. *Langmuir* **1999**, *15*(2), 322–325.

244. Kositza, M.J., Rees, G.D., Holzwarth, A., Holzwarth, J.F. Aggregation dynamics of the block-copolymer L64 in aqueous solution: copolymer-sodium dodecyl sulfate interactions studied by laser T-jump. *Langmuir* **2000**, *16*(23), 9035–9041.

245. Waton, G., Michels, B., Zana, R. Dynamics of block copolymer micelles in aqueous solution. *Macromolecules* **2001**, *34*(4), 907–910.

246. Thurn, T., Couderc-Azouani, S., Bloor, D.M., Holzwarth, J.F., Wyn-Jones, E. Ultrasonic relaxation in micellar solutions of nonionic triblock copolymers. *Langmuir* **2003**, *19*(10), 4363–4370.

247. Quina, F.H., Alonso, E.O., Farah, J.P. Incorporation of nonionic solutes into aqueous micelles: a linear salvation free energy relationship analysis. *J. Phys. Chem.* **1995**, *99*(30), 11708–11714.

248. Turro, N.J., Graetzel, M., Braun, A.M. Photophysical and photochemical processes in micellar systems. *Angew. Chem.* **1980**, *92*(9), 712–734.

249. Malliaris, A., Lang, J., Zana, R. Dynamic behavior of fluorescence quenchers in cetyltrimethylammonium chloride micelles. *J. Chem. Soc., Faraday Trans. 1*, **1986**, *82*, 109–118.

250. (a) Amgren, M. Migration and partitioning of pyrene and perylene between lipid vesicles in aqueous solutions studied with a fluorescence stopped-flow technique. *J. Am. Chem. Soc.* **1980**, *102*(27), 7882–7887. (b) Lang, J., Zana, R. Effects of alcohols and oils on the kinetics of micelle formation: breakdown in aqueous solutions of ionic surfactants. *J. Phys. Chem.* **1986**, *90*(21), 5258–5265. (c) Baumueller, W., Hoffmann, H., Ulbricht, W., Tondre, C., Zana, R. Chemical relaxation and equilibrium studies of aqueous solutions of laurylsulfate micelles in the presence of divalent metal ions. *J. Colloid Interface Sci.* **1978**, *64*(3), 430–449. (d) Tashiro, R., Inoue, T., Shimozawa, R. Pressure-jump apparatus with optical detection and its application to tetradecylpyridinium iodide micellar solutions. *Fukuoka Daigaku Rigaku Shuho* **1981**, *11*(1), 31–38.

251. Wan-Badhi, W., Bloor, D.M., Wyn-Jones, E. The partition of n-hexanol into micelles of cetyltrimethylammonium bromide in aqueous solution: ultrasonic relaxation and head-space analysis measurements. *Langmuir* **1994**, *10*(7), 2219–2222.

252. (a) Gehlen, M.H., De Schryver, F.C. Time-resolved fluorescence quenching in micellar assemblies. *Chem. Rev.* **1993**, *93*(1), 199–221. (b) Alonso, E.O., Quina, F.H. Dynamics of counterion exchange in aqueous micellar solution: salt effects on the counterion exit rate. *Langmuir*, **1995**, *11*(7), 2459–2463. (c)

253. Yoshida, N., Moroi, Y., Humphry-Baker, R., Graetzel, M. Dynamics for solubilization of naphthalene and pyerene into n-decyltrimethylammonium perfluorocarboxylate micelles. *J. Phys. Chem. A* **2002**, *106*(16), 3991–3997.

254. Loefroth, J.E., Almgren, M. Quenching of pyrene fluorescence of alkyl iodides in sodium dodecyl sulfate micelles. *J. Phys. Chem.* **1982**, *86*(9), 1636–1641.

255. Croonen, Y., Gelade, E., Van der Zegel, M., Van der Auweraer, M., Vandendriessche, H., De Schryver, F.C., Almgren, M. Influence of salt, detergent concentration, and temperature on the fluorescence quenching of 1-methylpyrene in sodium dodecyl sulfate with m-dicyanobenzene. *J. Phys. Chem.* **1983**, *87*(8), 1426–1431.

256. Atik, S.S., Thomas, J.K. Photoprocesses in cationic microemulsion systems. *J. Am. Chem. Soc.* **1981**, *103*(15), 4367–4371.

257. Atik, S.S., Thomas, J.K. Photoinduced electron transfer in organized assemblies. *J. Am. Chem. Soc.* **1981**, *103*(12), 3550–3555.

258. Almgren, M., Loefroth, J.E. Determination of micelle aggregation numbers and micelle fluidities from time-resolved fluorescence quenching studies. *J. Colloid Interface Sci.* **1981**, *81*(2), 486–499.

259. Van der Auweraer, M., Dederen, C., Palmans-Windels, C., De Schryver, F.C. Fluorescence quenching by neutral molecules in sodium dodecyl sulfate micelles. *J. Am. Chem. Soc.* **1982**, *104*(7), 1800–1804.

260. Dederen, J.C., Van der Auweraer, M., De Schryver, F.C. Fluorescence quenching of solubilized pyrene and pyrene derivatives by metal ions in SDS micelles. *J. Phys. Chem.* **1981**, *85*(9), 1198–1202.

261. Grieser, F. The dynamic behavior of iodide ion in aqueous dodecyltrimethylammonium chloride solutions: a model for counter-ion movement in ionic micellar systems. *Chem. Phys. Lett.* **1981**, *83*(1), 59–64.

262. Grieser, F., Tausch-Treml, R. Quenching of pyrene fluorescence by single and multivalent metal ions in micellar solutions. *J. Am. Chem. Soc.* **1980**, *102*(24), 7258–7264.

263. (a) Zana, R. Dynamics of organized assemblies of amphiphiles in solution. In *Dynamic Properties of Interfaces and Association Structures.* Pillai, V., Shah, D.O., Eds., AOCS Press, Champaign, IL, **1996**, p. 142, chap. 6. (b) Kahlweit, M. Kinetics of formation of association colloids. *J. Colloid Interface Sci.* **1982**, *90*(1), 92–99.

264. Hermann, C.U., Kahlweit, M. Kinetics of micellization of Triton X-100 in aqueous solutions. *J. Phys. Chem.* **1980**, *84*(12), 1536–1540.

265. Lessner, E., Teubner, M., Kahlweit, M. Relaxation experiments in aqueous solutions of ionic micelles. 2. Experiments on the system water-sodium dodecyl sulfate-sodium perchlorate and their theoretical interpretation. *J. Phys. Chem.* **1981**, *85*(21), 3167–3175.

266. Rharbi, Y., Li, M., Winnik, M.A., Hahn, K.G., Jr. Temperature dependence of fusion and fragmentation kinetics of Triton X-100 micelles. *J. Am. Chem. Soc.* **2000**, *122*(26), 6242–6251; **2001**, *123*(5), 1016. [Erratum to document cited in CA133:110343.]

267. Rharbi, Y., Winnik, M.A., Hahn, K.G., Jr. Kinetics of fusion and fragmentation nonionic micelles: triton X-100. *Langmuir* **1999**, *15*(14), 4697–4700.

268. Rharbi, Y., Winnik, M.A. Solute exchange between surfactant micelles by micelle fragmentation and fusion. *Adv. Colloid Interface Sci.* **2001**, *89–90*, 25–46.

269. Rharbi, Y., Bechthold, N., Landfester, K., Salzman, A., Winnik, M.A. Solute exchange in synperonic surfactant micelles. *Langmuir* **2003**, *19*(1), 10–17.

270. Rharbi, Y., Chen, L., Winnik, M.A. Exchange mechanisms for sodium dodecyl sulfate micelles: high salt concentration. *J. Am. Chem. Soc.* **2004**, *126*(19), 6025–6034.

271. Rharbi, Y., Winnik, M.A. Salt effects on solute exchange in sodium dodecyl sulfate micelles. *J. Am. Chem. Soc.* **2002**, *124*(10), 2082–2083.

272. Hilczer, M., Brzykin, A.V., Tachiya, M. Theory of the stopped-flow method for studying micelle exchange kinetics. *Langmuir* **2001**, *17*(14), 4196–4201.

273. Zana, R., Weill, C. Effect of temperature on the aggregation behavior of nonionic surfactants in aqueous solutions. *J. Phys. Lett.* **1985**, *46*(20), 953–960.

274. Frese, C., Ruppert, S., Schmidt-Lewerkuhne, H., Witten, K.P., Eggers, R., Fainerman, V.B., Miller, R. Adsorption dynamics of micellar solutions of a mixed anionic-cationic surfactant system. *Colloids Surf. A* **2004**, *239*(1–3), 33–40.

275. Kjellin, U.R.M., Reimer, J., Hansson, P. An investigation of dynamic surface tension, critical micelle concentration, and aggregation number of three nonionic surfactants using NMR, time-resolved fluorescence quenching, and maximum bubble pressure tensiometry. *J. Colloid Interface Sci.* **2003**, *262*, 506–515.

276. Tanford, C. Thermodynamics of micelle formation: prediction of micelle size and size distribution. *Proc. Natl. Acad. Sci. U.S.A.* **1974**, *71*(5), 1811–1815.

277. Stigter, D. Micelle formation by ionic surfactants. II. Specificity of head groups, micelle structure. *J. Phys. Chem.* **1974**, *78*(24), 2480–2484.

278. Muller, N., Birkhahn, R.H. Investigation of micelle structure by fluorine magnetic resonance. I. Sodium 10,10,10-trifluorocaprate and related compounds. *J. Phys. Chem.* **1967**, *71*(4), 857–962.

279. Corkill, J.M., Goodman, J.F., Walker, T. Partial molar volumes of surface-active agents in aqueous solution. *Trans. Faraday Soc.* **1967**, *63*(3), 768–772.

280. Kurz, J.L. Effects of micellization on the kinetics of the hydrolysis of monoalkyl sulfates. *J. Phys. Chem.* **1962**, *66*, 2239–2245.

281. Svens, B., Rosenholm, B. Investigation of the size and structure of the micelles in sodium octanoate solutions by small-angle x-ray scattering. *J. Colloid Interface Sci.* **1973**, *44*(3), 495–504.

282. Menger, F.M., Jerkunica, J.M., Johnston, J.C. The water content of a micelle interior: the fjord vs. reef models. *J. Am. Chem. Soc.* **1978**, *100*(15), 4676–4678.

283. Menger, F.M., Venkatasubban, K.S., Das, A.R. Solvolysis of a carbonate and a bemzhydryl chloride inside micelles: evidence for a porous cluster micelle. *J. Org. Chem.* **1981**, *46*(2), 415–419.

284. Varadaraj, R., Bock, J., Valint, P., Jr., Brons, N. Micropolarity and water penetration in micellar aggregates of linear and branched hydrocarbon surfactants. *Langmuir* **1990**, *6*(8), 1376–1378.

285. Menger, F.M., Bonicamp, J.M. Experimental test for micelle porosity. *J. Am. Chem. Soc.* **1981**, *103*(8), 2140–2141.

286. Menger, F.M., Chow, J.F. Testing theoretical models of micelles: the acetylenic probe. *J. Am. Chem. Soc.* **1983**, *105*(16), 5501–5502.

287. Menger, F.M., Doll, D.W. On the structure of micelles. *J. Am. Chem. Soc.* **1984**, *106*(4), 1109–1113.

288. Menger, F.M. Molecular conformations in surfactant micelles: comments. *Nature* (London) **1985**, *313*(6003), 603.

289. Menger, F.M., Mounier, C.E. A micelle that is insensitive to its ionization state: relevance to the micelle wetness problem. *J. Am. Chem. Soc.* **1993**, *115*(25), 12222–12223.

290. Al-Lohedan, H., Bunton, C.A., Mhala, M.M. Micellar effects upon spontaneous hydrolyses and their relation to mechanism. *J. Am. Chem. Soc.* **1982**, *104*(24), 6654–6660.

291. Khan, M.N., Naaliya, J., Dahiru, M. Effect of anionic micelles on intramolecular general base-catalyzed hydrolyses of salicylate esters: evidence for a porous cluster micelle. *J. Chem. Res.* (S) **1988**, 116–117; *J. Chem. Res.* (M) **1988**, 1168–1176.

292. Melo, E.C.C., Costa, S.M.B., Macanita, A.L., Santos, H. The use of the n-(9-anthroyloxy) stearic acids to probe the water content of sodium dodecyl sulfate, dodecyltrimethylammonium chloride, and Triton X-100 micelles. *J. Colloid Interface Sci.* **1991**, *141*(2), 439–453.

293. Dill, K.A., Koppel, D.E., Cantor, R.S., Dill, J.D., Bendedouch, D., Chen, S.H. Molecular conformations in surfactant micelles. *Nature* (London) **1984**, *309*(5963), 42–45.

294. Umemura, J., Mantsch, H.H., Cameron, D.G. Micelle formation in aqueous n-alkanoate solutions: a Fourier transform infrared study. *J. Colloid Interface Sci.* **1981**, *83*(2), 558–568.

295. Bendedouch, D., Chen, S.H., Koefler, W.C. Structure of ionic micelles from small angle neutron scattering. *J. Phys. Chem.* **1983**, *87*(1), 153–159.

296. Cabane, B., Zemb, T. Water in the hydrocarbon core of micelles: comments. *Nature* (London) **1985**, *314*(6009), 385.

297. Ulmius, J., Lindman, B. Fluorine-19 NMR relaxation and water penetration in surfactant micelles. *J. Phys. Chem.* **1981**, *85*(26), 4131–4135.

298. Prieto, M.J.E., Villalain, J., Gomez-Fernandez, J.C. Water penetration in micelles by FT-IR spectroscopy: a quantitative approach. Alix, A.J.P., Bernard, L., Manfait, M., Eds., *Sectrosc. Biol. Mol., Proc. 1st Eur. Conf.*, **1985**, 296–298.

299. Villalain, J., Gomez-Fernandez, J.C., Prieto, M.J.E. Structural information on probe solubilization in micelles by FT-IR spectroscopy. *J. Colloid Interface Sci.* **1988**, *124*(1), 233–237.

300. Kalyanasundaram, K., Thomas, J.K. Environmental effects on vibronic band intensities in pyrene monomer fluorescence and their application in studies of micellar systems. *J. Am. Chem. Soc.* **1977**, *99*(7), 2039–2044.

301. Raghuraman, H., Pradhan, S.K., Chattopadhyay, A. Effect of urea on the organization and dynamics of Triton X-100 micelles: a fluorescence approach. *J. Phys. Chem. B* **2004**, *108*(7), 2489–2496.

302. Casal, H.L., Wong, P.T.T. A Fourier-transform infrared spectroscopy study of the effect of external pressure on water penetration in micelles. *J. Phys. Chem.* **1990**, *94*(2), 777–780.

303. Watanabe, K., Ferrario, M., Klein, M.L. Molecular dynamics study of a sodium octanoate micelle in aqueous solution. *J. Phys. Chem.* **1988**, *92*(3), 619–821.

304. Sterpone, F., Pierleoni, C., Briganti, G., Marchi, M. Molecular dynamics study of temperature dehydration of a $C_{12}E_6$ spherical micelle. *Langmuir* **2004**, *20*(11), 4311–4314.

305. Grieser, F., Drummond, C.J. The physicochemical properties of self-assembled surfactant aggregates as determined by some molecular spectroscopic probe techniques. *J. Phys. Chem.* **1988**, *92*(20), 5580–5593.

306. Dorrance, R.C., Hunter, T.F. Absorption and emission studies of solubilization in micelles. 1. Pyrene in long-chain cationic micelles. *J. Chem. Soc., Faraday Trans. I* **1972**, *68*(Pt. 7), 1312–1321.

307. Shinitzky, M., Dianoux, A.-C., Weber, G. Microviscosity and order in the hydrocarbon region of micelles and membranes determined with fluorescent probes. I. Synthetic micelles. *Biochemistry* **1971**, *10*(11), 2106–2113.

308. Pownall, H.J., Smith, L.C. Viscosity of the hydrocarbon region of micelles: measurement by excimer fluorescence. *J. Am. Chem. Soc.* **1973**, *95*(10), 3136–3140.

309. Chen, M., Gratzel, M., Thomas, J.K. Kinetic studies in bile acid micelles. *J. Am. Chem. Soc.* **1975**, *97*(8), 2052–2057.

310. Kalyanasundaram, K., Gratzel, M., Thomas, J.K. Electrolyte-induced phase transitions in micellar systems: proton and carbon-13 nuclear magnetic resonance relaxation and photochemical study. *J. Am. Chem. Soc.* **1975**, *97*(14), 3915–3922.

311. Rodgers, M.A.J., Da Silva, E., Wheeler, M.E. Fluorescence from pyrene solubilized in aqueous micelles: a model for quenching by inorganic ions. *Chem. Phys. Lett.* **1976**, *43*(3), 587–591.

312. Zachariasse, K.A. Intramolecular excimer formation with diarylalkanes as a microfluidity probe for sodium dodecyl sulfate micelles. *Chem. Phys. Lett.* **1978**, *57*(3), 429–432.

313. Emert, J., Behrens, C., Goldenberg, M. Intramolecular excimer-forming probes of aqueous micelles. *J. Am. Chem. Soc.* **1979**, *101*(3), 771–772.

314. Turro, N.J., Aikawa, M., Yekta, A. A comparison of intermolecular and intramolecular excimer formation in detergent solutions: temperature effects and microviscosity measurements. *J. Am. Chem. Soc.* **1979**, *101*(3), 772–774.

315. Turro, N.J., Okubo, T. Micellar microviscosity of ionic surfactants under high pressure. *J. Am. Chem. Soc.* **1981**, *103*(24), 7224–7228.

316. Zinsli, P.E. Inhomogeneous interior of Aerosol OT microemulsions probed by fluorescence and polarization decay. *J. Phys. Chem.* **1979**, *83*(25), 3223–3231.

317. Turley, W.D., Offen, H.W. Micellar microfluidities at high pressures. *J. Phys. Chem.* **1985**, *89*(13), 2933–2937.

318. Miyagishi, S., Suzuki, H., Asakawa, T. Microviscosity and aggregation number of potassium N-acylalaninate micelles in potassium chloride solution. *Langmuir* **1996**, *12*(12), 2900–2905.

319. Pistolis, G., Malliaris, A. Simultaneous double probing of the microenvironment in colloidal systems and molecular assemblies by DPH derivatives. *Langmuir* **1997**, *13*(6), 1457–1462.

320. Zana, R., In, M., Levy, H., Duportail, G. Alkanediyl-α,ω-bis(dimethylalkylammonium bromide). 7. Fluorescence probing studies of micelle micropolarity and microviscosity. *Langmuir* **1997**, *13*(21), 5552–5557.

321. Zana, R. Microviscosity of aqueous surfactant micelles: effect of various parameters. *J. Phys. Chem. B* **1999**, *103*(43), 9117–9125.

322. Ruiz, C.C., Molina-Bolivar, J.A., Aguiar, J., MacIsaac, G., Moroze, S., Palepu, R. Thermodynamic and structural studies of Triton X-100 micelles in ethylene glycol-water mixed solvents. *Langmuir* **2001**, *17*(22), 6831–6840.

323. Bales, B.L., Stenland, C. Statistical distributions and collision rates of additive molecules in compartmentalized liquids studied by EPR spectroscopy. 1. Sodium dodecyl sulfate micelles, 5-doxylstearic acid ester, and cobalt(II). *J. Phys. Chem.* **1993**, *97*(13), 5418–5433.

324. Bales, B.L., Ranganathan, R., Griffiths, P.C. Characterization of mixed micelles of SDS and a sugar-based nonionic surfactant as a variable reaction medium. *J. Phys. Chem. B* **2001**, *105*(31), 7465–7473.

325. Bales, B.L., Zana, R. Characterization of micelles of quaternary ammonium surfactants as reaction media I: dodecyltrimethylammonium bromide and chloride. *J. Phys. Chem. B* **2002**, *106*(8), 1926–1939.

326. Menger, F.M., Jerkunica, J.M. Anistropic motion inside a micelle. *J. Am. Chem. Soc.* **1978**, *100*(3), 688–691.

327. Wirth, M.J., Chou, S.H., Piasecki, D.A. Frequency-domain spectroscopic study of the effect of n-propanol on the internal viscosity of sodium dodecyl sulfate micelles. *Anal. Chem.* **1991**, *63*(2), 146–151.

328. Maiti, N.C., Krishna, M.M.G., Britto, P.J., Periasamy, N. Fluorescence dynamics of dye probes in micelles. *J. Phys. Chem. B* **1997**, *101*(51), 11051–11060.

329. Fuerstenau, D.W. Streaming-potential studies on quartz in solutions of aminium acetates in reaction to the formation of hemi-micelle at the quartz-solution interface. *J. Phys. Chem.* **1956**, *60*, 981–985.

330. Nunn, C.C., Schechter, R.S., Wade, W.H. Visual evidence regarding the nature of hemimicelles through surface solubilization of pinacyanol chloride. *J. Phys. Chem.* **1982**, *86*(16), 3271–3272.

331. Gao, Y., Du, J., Gu, T. Hemimicelle formation of cationic surfactants at silica gel-water interface. *J. Chem. Soc., Faraday Trans. I* **1987**, *83*(8), 2671–2679.

332. Gu, T., Gao, Y., He, L. Hemimicelle formation of cationic surfactants at silica gel-water interface. *J. Chem. Soc., Faraday Trans. I* **1988**, *84*(12), 4471–4473.

333. Gu, T., Zhu, B.-Y. The S-type isotherm equation for adsorption of nonionic surfactants at the silica gel-water interface. *Colloids Surf.* **1990**, *44*, 81–87.

334. Chandar, P., Somasundaran, P., Turro, N.J. Fluorescence probe studies on the structure of the adsorbed layer of dodecyl sulfate at the alumina-water interface. *J. Colloid Interface Sci.* **1987**, *117*(1), 31–46.

335. Zettlemoyer, A.C. Hydrophobic surfaces. *J. Colloid Interface Sci.* **1968**, *28*(3–4), 343–369.

336. Koganovskii, A.M. Influence of electrolytes on the micelle formation of humic and apocrenic acids, and on their adsorption from aqueous solutions. *Kolloidnyi Zhurnal* **1962**, *24*(1), 34–41.

337. Manne, S., Cleveland, J.P., Gaub, H.E., Stucky, G.D., Hansma, P.K. Direct visualization of surfactant hemimicelles by force microscopy of the electrical double layer. *Langmuir* **1994**, *10*(12), 4409–4413.

338. Jaschke, M., Butt, H.-J., Gaub, H.E., Manne, S. Surfactant aggregates at a metal surface. *Langmuir* **1997**, *13*(6), 1381–1384.

339. (a) Tao, N.J., Lindsay, S.M. In situ scanning tunneling microscopy study of iodine and bromine adsorption on gold(111) under potential control. *J. Phys. Chem.* **1992**, *96*(13), 5213–5217. (b) Wandlowski, Th., Wang, J.X., Magnussen, O.M., Ocko, B.M. Structural and kinetic aspects of bromide adsorption on Au(111). *J. Phys. Chem.* **1996**, *100*(24), 10277–10287. (c) Bockris, J.O'M., Paik, W.-K., Genshaw, M.A. Adsorption of anions at the solid-solution interface: ellipsometric study. *J. Phys. Chem.* **1970**, *74*(24), 4266–4275.

340. Manne, S., Gaub, H.E. Molecular organization of surfactants at solid-liquid interfaces. *Science* (Washington, D.C.) **1995**, *270*(5241), 1480–1482.

341. Manne, S., Schaffer, T.E., Huo, Q., Hansma, P.K., Morse, D.E., Stucky, G.D., Aksay, I.A. Gemini surfactants at solid-liquid interfaces: control of interfacial aggregate geometry. *Langmuir* **1997**, *13*(24), 6382–6387.

342. Patrick, H.N., Warr, G.G., Manne, S., Aksay, I.A. Surface micellization parameters of quaternary ammonium surfactants on mica. *Langmuir* **1999**, *15*(5), 1685–1692.

343. Wanless, E.J., Ducker, W.A. Organization of sodium dodecyl sulfate at the graphite-solution interface. *J. Phys. Chem.* **1996**, *100*(8), 3207–3214.

344. Wanless, E.J., Ducker, W.A. Weak influence of divalent ions on anionic surfactant surface-aggregation. *Langmuir* **1997**, *13*(6), 1463–1474.

345. Burgess, I., Jeffrey, C.A., Cai, X., Szymanski, G., Galus, Z., Lipkowski, J. Direct visualization of the potential-controlled transformation of hemimicellar aggregates of dodecyl sulfate into a condensed monolayer at the Au(111) electrode surface. *Langmuir* **1999**, *15*(8), 2607–2616.

346. Retter, U. One-dimensional nucleation-growth-collision in the formation of surface hemimicelles of amphiphiles. *Langmuir* **2000**, *16*(20), 7752–7756.

347. Silber, J.J., Biasutti, A., Abuin, E., Lissi, E. Interactions of small molecules with reverse micelles. *Adv. Colloid Interface Sci.* **1999**, *82*(1–3), 189–252.

348. El Seoud, O.A. Reversed micelles and water-in-oil microemulsions: formation and some relevant properties. In *Organized Assemblies in Chemical Analysis*, Vol. 1, Hinze, W.L., Ed. JAI Press, Greenwich, **1994**, pp. 1–36.

349. Eicke, H.F., Kvita, P. In *Reverse Micelles*. Luisi, P.L., Straub, B.E., Eds., Plenum Press, New York, **1984**, pp. 21–35.

350. Neuman, R.D., Ibrahim, T.H. Novel structural model of reversed micelles: the open water-channel model. *Langmuir* **1999**, *15*(1), 10–12.

351. Ibrahim, T.H., Neuman, R.D. Nanostructure of open water-channel reversed micelles. I. ^{1}H NMR spectroscopy and molecular modeling. *Langmuir* **2004**, *20*(8), 3114–3122.

352. Brisk, M.L., McManamey, W.J. Liquid extraction of metals from sulfate solutions by alkyl phosphates. 1. Equilibrium distributions of copper, cobalt, and nickel with bis(2-ethylhexyl) hydrogen phosphate. *J. Appl. Chem.* **1969**, *19*(4), 103–108.

353. Kolarik, Z., Grimm, R. Acidic organophosphorus extractants. XXIV. The polymerization behavior of copper(II), cadmium(II), zinc(II), and cobalt(II) complexes of di(2-ethylhexyl) phosphoric acid in fully loaded organic phases. *J. Inorg. Nucl. Chem.* **1976**, *38*(9), 1721–1727.

354. Thiyagarajan, P., Diamond, H., Danesi, P.R., Horwitz, E.P. Small-angle neutron-scattering studies of cobalt(II) organophosphorus polymers in deuteriobenzene. *Inorg. Chem.* **1987**, *26*(25), 4209–4212.

355. Kunzmann, M.W., Kolarik, Z. Extraction of mono and polynuclear complexes of zinc(II) with di(2-ethylhexyl) phosphoric acid. In *Solvent Extraction 1990*, Sekine, T., Ed., Part(A). Elsevier, New York, **1990**, pp. 207–211.

356. Yu, Z.-J., Ibrahim, T.H., Neuman, R.D. Aggregation behavior of cobalt(II), nickel(II), and copper(II) bis(2-ethylhexyl) phosphate complexes in n-heptane. *Solvent Extr. Ion Exch.* **1998**, *16*(6), 1437–1463.

357. Neuman, R.D., Park, S.J. Characterization of association microstructures in hydrometallurgical nickel extraction by di(2-ethylhexyl) phosphoric acid. *J. Colloid Interface Sci.* **1992**, *152*(1), 41–53.

358. Steytler, D.C., Jenta, T.R., Robinson, B.H., Eastoe, J., Heenan, R.K. Structure of reversed micelles formed by metal salts of bis(ethylhexyl) phosphoric acid. *Langmuir* **1996**, *12*(6), 1483–1489.

359. Fendler, J.H., Fendler, E.J. *Catalysis in Micellar and Macromolecular Systems*, Academic Press, New York, **1975**, pp. 320–325, chap. 10.

360. Ruckenstein, E., Nagarajan, R. Aggregation of amphiphiles in nonaqueous media. *J. Phys. Chem.* **1980**, *84*(11), 1349–1358.

361. Manoj, K.M., Jayakumar, R., Rakshit, S.K. Physicochemical studies on reverse micelles of sodium bis(2-ethylhexyl) sulfosuccinate at low water content. *Langmuir* **1996**, *12*(17), 4068–4072.

362. Mukherjee, K., Moulik, S.P., Mukherjee, D.C. Thermodynamics of micellization of Aerosol OT in polar and nonpolar solvents: a calorimetric study. *Langmuir* **1993**, *9*(7), 1727–1730.

363. Yang, H., Erford, K., Kiserow, D.J., McGown, L.B. Effects of bile salts on percolation and size of AOT reversed micelles. *J. Colloid Interface Sci.* **2003**, *262*(2), 531–535.

364. Avellone, G., Bongiorno, D., Ceraulo, L., Ferrugia, M., Turco, L.V. Spectrophotometric investigation of the binding of vitamin E to water-containing reversed micelles. *Int. J. Pharm.* **2002**, *234*(1–2), 249–255.

365. Laun, Y., Xu, G., Dai, G., Sun, Z., Liang, H. The interaction between poly(vinylpyrrolidone) and reversed micelles of water/AOT/n-heptane. *Colloid Polym. Sci.* **2003**, *282*(2), 110–118.

366. Zhou, N., Li, Q., Wu, J., Chen, J., Weng, S., Xu, G. Spectroscopic characterization of solubilized water in reversed micelles and microemulsions: sodium bis(2-ethylhexyl) sulfosuccinate and sodium bis(2-ethylhexyl) phosphate in *n*-heptane. *Langmuir* **2001**, *17*(15), 4505–4509.

367. Hamada, K., Ikeda, T., Kawai, T., Kon-No, K. Ionic strength effects of electrolytes on solubilized states of water in AOT reversed micelles. *J. Colloid Interface Sci.* **2001**, *233*(2), 166–170.

368. Behera, G.B., Mishra, B.K., Behera, P.K., Panda, M. Fluorescent probes for structural and distance effect studies in micelles, reversed micelles and microemulsions. *Adv. Colloid Interface Sci.* **1999**, *82*(1–3), 1–42.

369. Liveri, V.T., Calorimetric investigations of solutions of reversed micelles. *Surfactant Sci. Ser.* **2001**, *93*, 1–22.

370. Zinsli, P.E. Inhomogeneous interior of Aerosol OT microemulsions probed by fluorescence and polarization decay. *J. Phys. Chem.* **1979**, *83*(25), 3223–3231.

371. Novaki, L.P., Correa, N.M., Silber, J.J., El Seoud, O.A. FTIR and ^{1}H NMR studies of the solubilization of pure and aqueous 1,2-ethanediol in the reversed aggregates of Aerosol-OT. *Langmuir* **2000**, *16*(13), 5573–5578.

372. El Seoud, O.A., Correa, N.M., Novaki, L.P. Solubilization of of pure and aqueous 1,2,3-propanetriol by the reversed aggregates of Aerosol-OT in isooctane probed by FTIR and ^{1}H NMR spectroscopy. *Langmuir* **2001**, *17*(6), 1847–1852.

373. Hayes, D.G., Gulari, E. Ethylene glycol and fatty acid have a profound impact on the behavior of water-in-oil microemulsions formed by the surfactant Aerosol-OT. *Langmuir* **1995**, *11*(12), 4695–4702 and references cited therein.

374. Luisi, P.L., Magid, L.J. Solubilization of enzymes and nucleic acids in hydrocarbon micellar solutions. *CRC Crit. Rev. Biochim.* **1986**, *20*(4), 409–474.

375. Aveyard, R., Binks, B.P., Fletcher, P.D.I., Kirk, A.J., Swansbury, P. Adsorption and aggregation of sodium bis(2-ethylhexyl) sulfosuccinate in systems containing toluene plus water, 1,2-ethanediol, or 1,2,3-propanetriol. *Langmuir* **1993**, *9*(2), 523–530.

376. D'Aprano, A., Lizzio, A., Liveri, V.T. Enthalpies of solution and volumes of water in reversed AOT micelles. *J. Phys. Chem.* **1987**, *91*(18), 4749–4751.

377. Cavallaro, G., La Manna, G., Liveri, V.T., Aliotta, F., Fontanella, M.E. Structural investigation of water/lecithin/cyclohexane microemulsions by FT-IR spectroscopy. *J. Colloid Interface Sci.* **1995**, *176*(2), 281–285.

378. (a) Ruggirello, A., Turco Liveri, V. FT-IR investigation of the acetamide state in AOT reversed micelles. *Colloid Polym. Sci.* **2003**, *281*(11), 1062–1068. (b) Calvaruso, G., Minore, A., Liveri, V.T. FT-IR investigation of the UREA state in AOT reversed micelles. *J. Colloid Interface Sci.* **2001**, *243*(1), 227–232. (c) Calvaruso, G., Ruggirello, A., Turco Liveri, V. FT-IR investigation of the *n*-methylurea state in AOT reversed micelles. *J. Nanoparticle Res.* **2002**, *4*(3), 239–246.

379. (a) Ruggirello, A., Liverri, V.T. FT-IR investigation of the urea state in lecithin and sodium bis(2-ethylhexyl)phosphate reversed micelles. *J. Colloid Interface Sci.* **2003**, *258*(1), 123–129. (b) Ceraulo, L., Dormond, E., Mele, A., Turco, L.V. FT-IR and nuclear overhauser enhancement study of the state of urea confined in AOT-reversed micelles. *Colloids Surf. A* **2003**, *218*(1–3), 255–264.

380. Caponetti, E., Chillura-Martino, D., Ferrante, F., Pendone, L., Ruggirello, A., Liveri, V.T. Structure of urea clusters confined in AOT reversed micelles. *Langmuir* **2003**, *19*(12), 4913–4922.

381. Calandra, P., Longo, A., Ruggirello, A., Liveri, V.T. Physico-chemical investigation of the state of cyanamide confined in AOT and lecithin reversed micelles. *J. Phys. Chem. B* **2004**, *108*(24), 8260–8268.

382. Belletete, M., Lachapelle, M., Durocher, G. Polarity of AOT micellar interfaces: use of the preferential salvation concepts in the evaluation of the effective dielectric constants. *J. Phys. Chem.* **1990**, *94*(13), 5337–5341 and references cited therein.

383. Zhu, B.Y., Gu, T. Reverse hemimicelle formation of 1-decanol from heptane at the solution/graphite interface. *Colloids Surf.* **1990**, *86*(2–4), 339–345.

384. Findenegg, G.H., Koch, C., Liphard, M. Adsorption of decan-1-ol from heptane at the solution/graphite interface. In *Adsorption from Solution, Symposium*. Ottewill, R.H., Rochester, C.H., Smith, A.L., Eds., Academic Press, London, **1983**, pp. 87–97.

2 Catalysis in Chemical Reactions: General Theory of Catalysis

2.1 INTRODUCTION

Chemical catalysis is an area of tremendous interest because of its occurrence in reactions important to both biochemical, biotechnological, and industrial processes. The word catalysis (*katalyse*) was coined by Berzelius in 1835. In his words, "*Catalysts* are substances, which by their mere presence evoke chemical reactions that would not otherwise take place."

In a 14th-century Arabian manuscript, Al Alfani described the "xerion, alaksir, noble stone, magisterium, that heals the sick and turns base metals into gold, without in itself undergoing the least change."[1] A small amount of catalyst suffices to bring about great changes without itself being consumed.

Wilhelm Ostwald was the first to emphasize the effects of a catalyst on the rate of a chemical reaction and his famous definition of catalyst was: "a catalyst is a substance that changes the velocity of a chemical reaction without itself appearing in the end products." Based on the essence of the first law of thermodynamics, Ostwald showed that a catalyst cannot change the equilibrium. Because equilibrium $K = k_f/k_b$ (where k_f and k_b represent rate constants for forward and backward reactions in an equilibrium chemical process, respectively), the catalyst must change both k_f and k_b in the same proportion and in the same direction. This statement, although true in its essence, has been vaguely interpreted in some popular books of physical chemistry. For instance, "any claim that an equilibrium can be shifted by an enzyme or a catalyst (which are the materials that change the rate of reaction without suffering any net change) can be dismissed by appealing to the laws of thermodynamics."[2] "Thus catalysts that accelerate the hydrolysis of esters must also accelerate the esterification of alcohols; enzymes like pepsin and papain that catalyze the splitting of peptides must also catalyze their synthesis from the amino acids."[1]

The statement that a catalyst cannot change the magnitude of an equilibrium constant is correct only if (1) the forward and backward reactions of an equilibrium process follow strictly the same mechanism, i.e., forward and backward reactions must involve the same transition states (if there is one or more than one) and same reactive intermediates (if any) on the reaction paths, (2) the catalyst should have either no effect or same effect on ground states of both reactants and products, and (3) products of a catalyzed reaction do not undergo an irreversible chemical change with or without the participation of catalyst. In other words, a catalyst cannot change equilibrium constant if the equilibrium process follows the principle of microscopic reversibility. These points can be easily understood in terms of mechanisms of a few equilibrium reactions, which are discussed in the following subsections.

2.1.1 TAUTOMERIZATION OF KETONES

Ketones containing α and α hydrogen undergo keto-enol tautomerization. The net reaction may be shown as

$$\underset{\alpha}{RCH_2} - CO - \underset{\alpha}{CH_2R_1} \underset{k_b}{\overset{k_f}{\rightleftharpoons}} RCH=C(OH)CH_2R_1 \qquad (2.1)$$

The rate of interconversion of keto and enol form is catalyzed by both specific acid (H^+ or H_3O^+) and specific base (HO) catalysts. The formation of enol from ketone and ketone from enol involve the same intermediates and transition states as shown in Scheme 2.1 and Scheme 2.2, respectively. The transition states involved in the interconversion of Ke and I_1, I_1 and En, Ke and I_2, as well as I_3 and En are TS_1, TS_2, TS_3, and TS_4, respectively. Thus, in these reactions, the

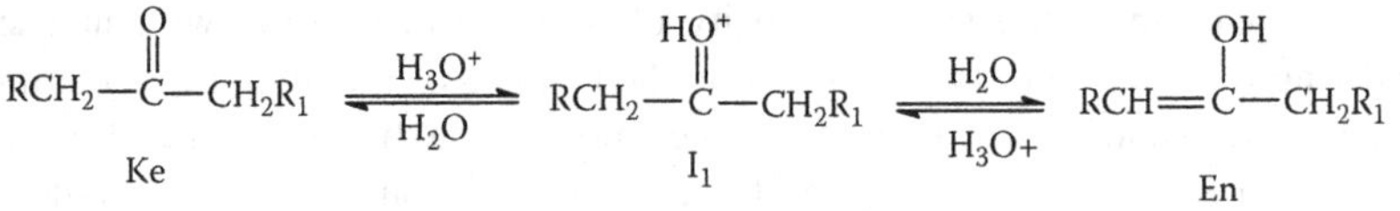

SCHEME 2.1

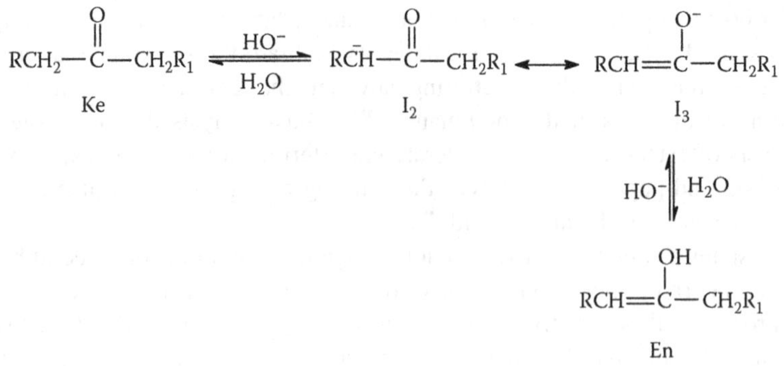

SCHEME 2.2

catalyst (H_3O^+ or HO^-) increases the magnitudes of both rate constants, k_f and k_b, equally, because the ground state stabilities of keto and enol are unaffected by the presence of these catalysts. The presence of either H_3O^+ or HO^- increases the rate of tautomerization (i.e., increases the values of k_f and k_b) without affecting the equilibrium concentrations of keto and enol forms (i.e., without affecting the values of k_f/k_b).

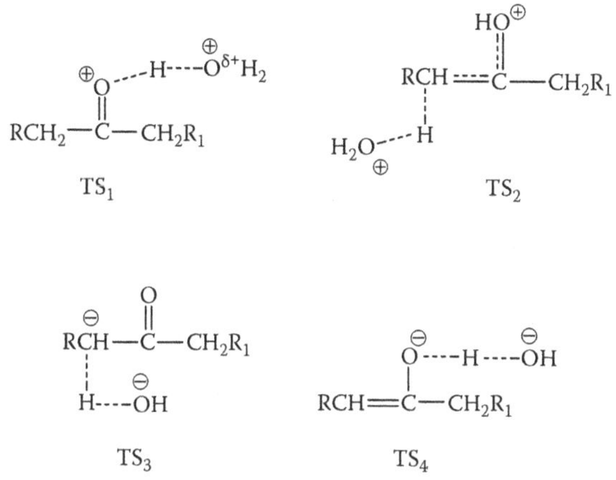

TS$_1$ TS$_2$

TS$_3$ TS$_4$

2.1.2 HYDROLYSIS OF ESTERS

The overall reaction for ester hydrolysis may be described as

$$RCOOR_1 + H_2O \underset{k_b}{\overset{k_f}{\rightleftarrows}} RCOOH + R_1OH \qquad (2.2)$$

The rates of hydrolysis of esters and related compounds are catalyzed by specific acid and specific base catalysts. Fine details of general mechanisms of specific acid and specific base catalyzed hydrolysis of esters are shown in Scheme 2.3 and Scheme 2.4, respectively. It is apparent from Scheme 2.3 that the presence of catalyst (H_3O^+) should equally increase the rate hydrolysis of ester (i.e., k_f value) and the rate of alkanolysis of acid (i.e., k_b value) without affecting the ratio k_f/k_b (i.e., equilibrium concentrations of ester and acid). However, Scheme 2.4 shows that in the presence of specific base catalyst (HO^-), the net reaction shown by Equation 2.2 loses its reversibility. The hydrolysis product, RCOOH is a stronger acid than the conjugate acid (H_2O) of the specific base catalyst and, consequently, HO^-/R_1O^- reacts irreversibly with product RCOOH to produce a more stable product, $RCOO^-$, under such conditions. Thus, it is obvious to say that a specific base cannot catalyze the rate of reverse reaction, i.e., rate of reaction between RCOOH and R_1OH of Equation 2.2.

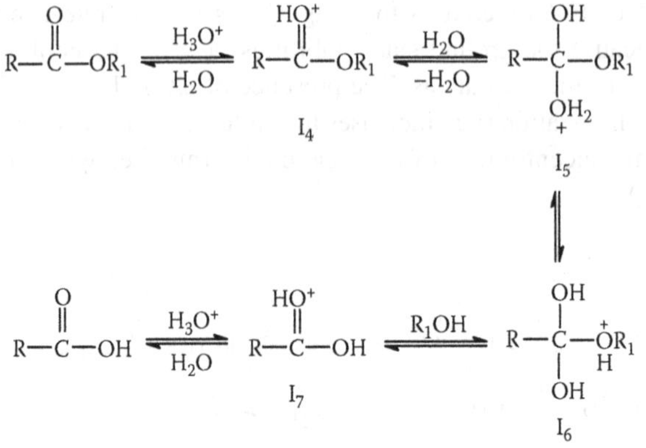

SCHEME 2.3

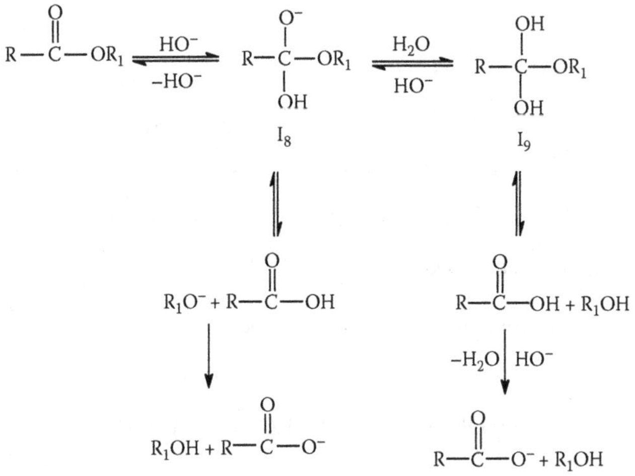

SCHEME 2.4

2.1.3 CLEAVAGE OF PHTHALAMIDE UNDER MILD ALKALINE pH

Product characterization studies on the cleavage of phthalamide and related compounds under mild alkaline pH ($\approx$ 9) reveal the net reaction, which is expressed by Equation 2.3.

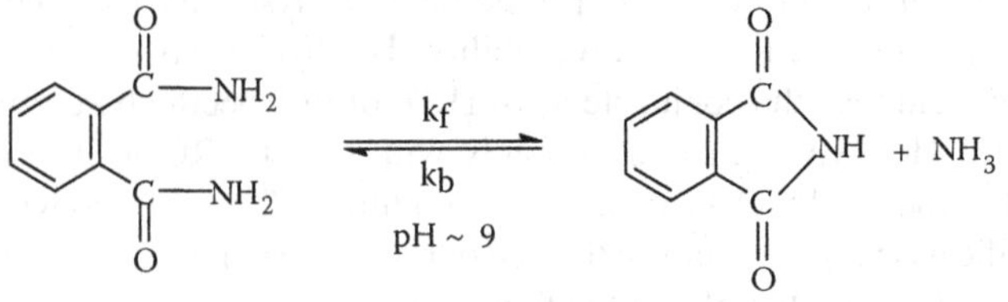

(2.3)

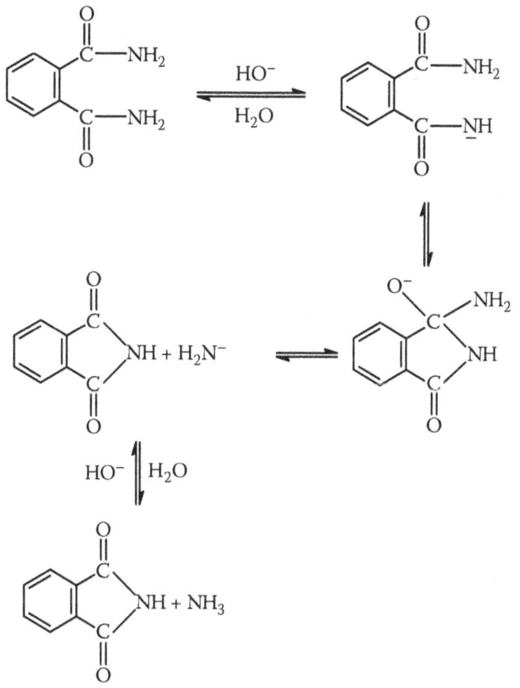

SCHEME 2.5

The rate of noncatalyzed cleavage of phthalamide is negligible compared to hydroxide ion-catalyzed cleavage of phthalamide in an aqueous solution of pH ≈ 9. The cleavage of phthalamide involves hydroxide ion-catalyzed intramolecular nucleophilic attack by o-$CONH_2$ group at the carbonyl carbon of other adjacent $CONH_2$ group as shown in Scheme 2.5. The increase in pH or [HO$^-$] increases the rate of cleavage of phthalamide and decreases the equilibrium concentration of reactant (phthalamide) because of the irreversible conversion of immediate product phthalimide to ionized phthalamic acid as shown in Equation 2.4. Thus, the increase in pH increases the value of k_f and

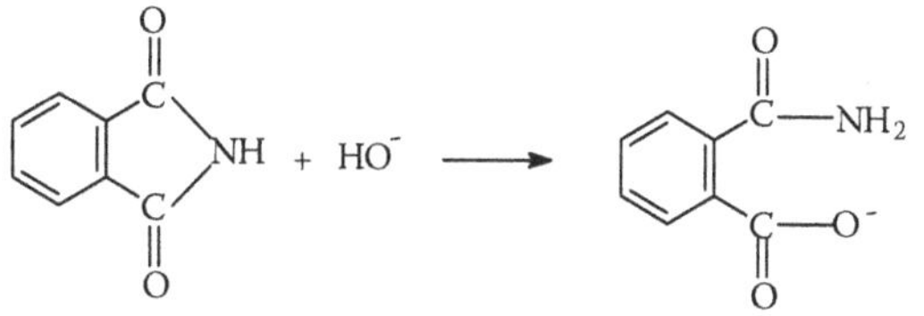

$$(2.4)$$

apparently decreases the value of k_b because the catalyst, HO$^-$, reacts with the product, phthalimide, and produces irreversibly a more stable product, ionized phthalamic acid. It is apparent that the reversibility of the reaction shown by

Equation 2.3 is a function of pH of the reaction medium. The mechanistic details[3,4] of the reaction expressed by Equation 2.3 show that the reverse reaction, (k_b-step), may not be catalyzed by a specific base.

2.1.4 ENZYME-CATALYZED HYDROLYSIS OF PEPTIDES (AMIDE BONDS)

Serine proteases, which consist of trypsin, chymotrypsin, and elastase, catalyze the hydrolysis of protein peptide bonds. The net reaction for the enzyme (Enz)-catalyzed hydrolysis of peptide bond (i.e., amide bond) may be expressed as

$$\underset{\text{XCH C}-\text{NHC HY}}{\overset{\text{R   O}\quad\ \ \text{R}_1}{|\quad||\quad\ |}} + H_2O \underset{k_b}{\overset{k_f}{\rightleftharpoons}} \underset{\text{XC HC OOH}}{\overset{\text{R}}{|}} + \underset{\text{NH}_2\text{CHY}}{\overset{\text{R}_1}{|}} \quad (2.5)$$

If the enzyme-catalyzed hydrolysis of peptide bond involves a simple reversible reaction as shown by Equation 2.5 then, indeed, the enzyme must catalyze the rate of formation of peptide bond from amino acids (i.e., k_b-step), provided the amino acids do not react irreversibly with the enzyme. Incidentally, if the function of serine proteases is to catalyze both the rate of hydrolytic cleavage and the rate of formation of protein peptide bond, then, probably, these enzymes cannot digest the proteins that we eat and, consequently, the results would have been disastrous for all protein-eating creatures — which certainly Nature will never allow. Although the mechanisms of most of the enzyme-catalyzed reactions are unknown, even at a very rudimentary level, the mechanism of α-chymotrypsin-catalyzed hydrolysis of peptide bond has been relatively well understood. The reaction has been almost ascertained to involve acylation and deacylation of enzyme as shown by Equation 2.6. Widely accepted mechanisms for acylation and deacylation steps are shown in Scheme 2.6 and Scheme 2.7.[5,6]

$$\underset{\text{XCHC}-\text{NHCHY}}{\overset{\text{R O}\quad\ \text{R}_1}{|\ \ ||\quad\ |}} + \text{Enz} \xrightarrow[\underset{\underset{\text{R}_1}{|}}{-\ \text{YCHNH}_2}]{\text{acylation}} \underset{\underset{\text{R}}{|}}{\overset{\text{O}}{\overset{||}{\text{XCHC Enz}}}} \xrightarrow[H_2O]{\text{deacylation}} \underset{\underset{\text{R}}{|}}{\text{XCHCOOH}} + \text{Enz} \quad (2.6)$$

The products, XCH(R)COOH and NH$_2$CH(R$_1$)Y, of enzyme-catalyzed cleavage of peptide bond are expected to react irreversibly to form the ammonium salt, XCH(R)COO$^+$NH$_3$CH(R$_1$)Y. This extremely fast irreversible reaction does not allow the enzyme to catalyze the formation of peptide bond from amino acids.

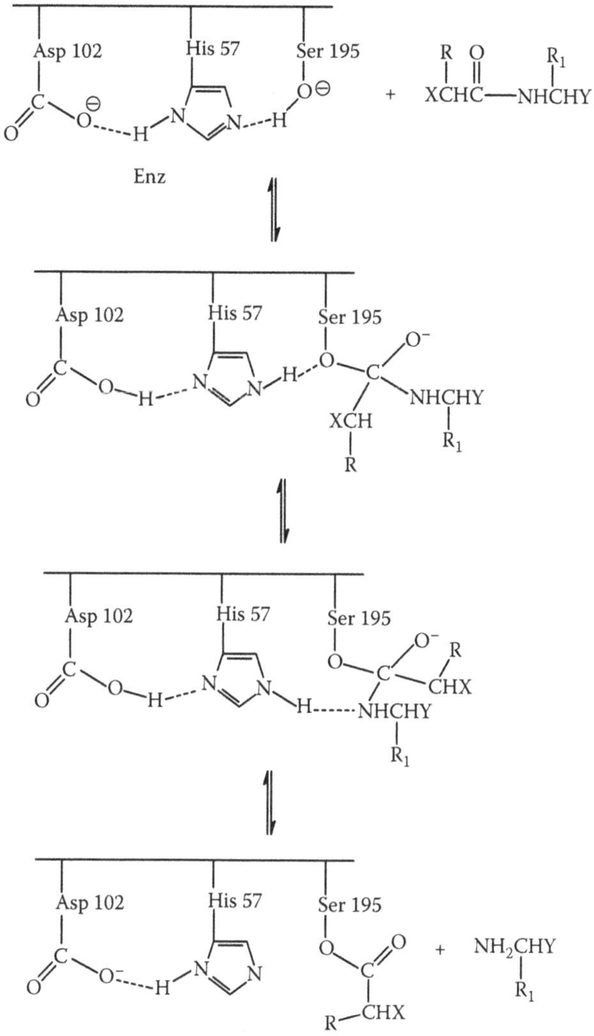

Asp 102   His 57   Ser 195

Enz

$+$ XCHC—NHCHY (with R, O, R_1 groups)

Asp 102   His 57   Ser 195

Asp 102   His 57   Ser 195

Asp 102   His 57   Ser 195

$+$ NH_2CHY (R_1)

SCHEME 2.6

2.2 CATALYSIS AND FREE-ENERGY REACTION COORDINATE DIAGRAM

In a classical sense, a catalyst is regarded as a chemical species that increases the rate of a reaction (compared to the rate of the same reaction in the absence of catalyst under exactly similar experimental conditions) without being consumed during the course of the reaction. But now there is a general perception that a chemical species or a molecular aggregate can either increase or decrease

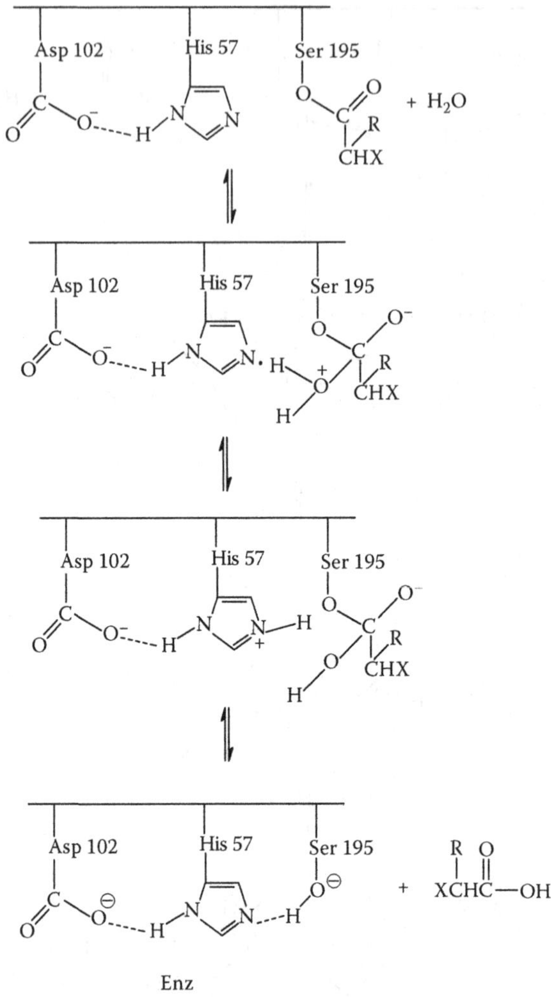

SCHEME 2.7

the rate of a reaction without being consumed during the course of the reaction. Both aspects (rate acceleration and rate deceleration/retardation) of such chemical species or molecular aggregates are equally important and fascinating chemical events, which should be understood mechanistically in terms of fine details of molecular interaction and recognition. Thus, it is tempting to define a catalyst as a chemical species (such as an atom or a group of atoms, i.e., a molecule), or a molecular segment, or a molecular aggregate that changes the rate of a reaction without being consumed during the course of the reaction and, thus, is capable of being reused in changing the rate of the same reaction until the reaction is completely ceased or over. A catalyst may be a positive catalyst/accelerator (when it increases the rate of reaction) or a negative catalyst/inhibitor* (when it decreases the rate of a reaction). For the sake of mere clarity, the terms *positive catalyst* and *negative catalyst* are used in this book to refer to chemical species or molecular aggregate that increases and decreases the rate of a reaction, respectively, without being consumed during the course of the reaction.

The essence of positive and negative catalysis may best be explained in terms of a free-energy-reaction coordinate diagram. A positive catalyst may increase the rate of a reaction by either destabilizing ground state (GS) more strongly than the transition state (TS) or by stabilizing TS more strongly than GS in a simple one-step reaction as shown in Figure 2.1. On the other hand, a negative catalyst may decrease the rate of a simple one-step reaction by either stabilizing GS more strongly than TS or destabilizing TS more strongly than GS as shown in Figure 2.2. In a multistep reaction, a catalyst changes the overall activation barrier for the whole reaction process by affecting the energy barriers of individual reaction steps in a complex manner, which could be understood only in terms of a correct reaction mechanism. Say for example, it is possible that the catalyst changes the rate of a multistep reaction by the combinations of preferential stabilization/destabilization of GS of the first step followed by the preferential stabilization/destabilization of TS of the second step of the reaction.

Interaction between a catalyst molecule or molecular aggregate and a reactant molecule may involve one or more than one of the following molecular or atomic interactions:

1. Covalent bonding
2. Electrovalent (ionic) bonding
3. Resonance
4. Electrostatic
   a. Ion–ion
   b. Ion–dipole
   c. Dipole–dipole
   d. Hydrogen bonding
5. Hydrophobic
6. van der Waals'/steric interactions

2.2.1 COVALENT BONDING INTERACTION

Covalent bonding interaction between reaction sites of substrate and catalyst may cause different amounts of stabilization or destabilization of GS and TS and,

---

* The term *inhibitor* has a slightly different meaning in different classes of reactions. For example, inhibitor in an enzyme-mediated reaction is a chemical species, which inactivates the catalytic efficiency of an enzyme by an irreversible chemical reaction between inhibitor and enzyme — a chemical process, which probably consumes the inhibitor during the course of the reaction. The inhibitor in a chain reaction involves the reaction between chain carrier and product (acts as an inhibitor) — a chemical process, which consumes the inhibitor during the course of reaction.

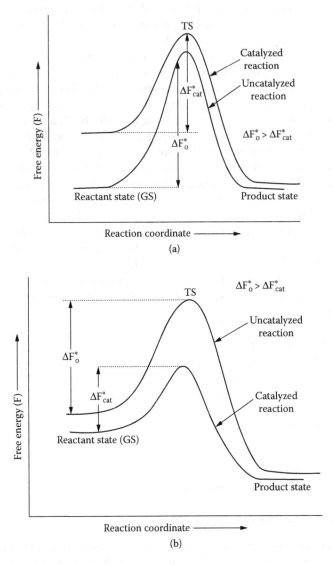

FIGURE 2.1 Free energy (F) — reaction coordinate diagrams for the catalyzed and uncatalyzed (reference) reaction systems in which (a) catalytic effect destabilizes ground state (GS) more strongly than the transition state (TS) and (b) catalytic effect stabilizes TS more strongly than GS.

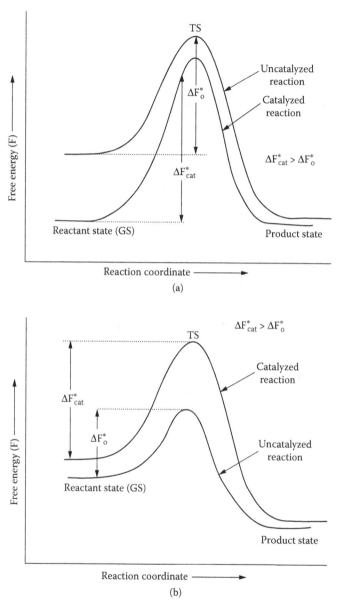

FIGURE 2.2 Free energy (F) — reaction coordinate diagrams for the catalyzed and uncatalyzed (reference) reaction systems in which (a) catalytic effect stabilizes ground state (GS) more strongly than the transition state (TS) and (b) catalytic effect destabilizes TS more strongly than GS.

consequently, may affect the rate of the reaction. This interaction, perhaps, constitutes the major source of catalytic effects in various catalyzed reactions. In
view of the basic definition of a catalyst, covalent bonding interaction, which
causes the formation of a strong covalent bond between the substrate and catalyst
must produce an intermediate, which should be more reactive toward final product
formation than substrate under the reaction conditions of catalytic process.

To illustrate this catalytic process, let us consider the cleavage of phthalimide
in the presence of buffers of trimethylamine, 1,4-diazabicyclo[2.2.2]octane
(DABCO), and carbonate dianion.[7] The values of second-order rate constants
(k_b) for the reaction of nonionized phthalimide with trimethylamine and DABCO
are 0.0062 and 0.0027 M^{-1} sec^{-1}, respectively, whereas triethylamine, triethanolamine, and carbonate dianion did not show detectable reactivity toward phthalimide (i.e., $k_b \approx 0$ for these nucleophiles). The end product of the reaction
mixtures is the hydrolysis product phthalamic acid. Tertiary-amine-buffer-depen-
dent cleavage of phthalimide may be attributed to either nucleophilic catalysis
or kinetically equivalent general base catalysis of hydrolysis. However, the
possibility of the occurrence of general base catalysis of hydrolysis through TS$_5$
has been ruled out for the following reason: significant reactivity of trimethylamine (pK_a 9.95) and DABCO (pK_2 9.10) and insignificant reactivity of carbonate dianion (pK_2 9.68) toward nonionized phthalimide show that the transition state TS$_5$ is unlikely to exist on the reaction path, because the carbonate
dianion appeared to show the normal expected reactivity in a few related
reactions[8] in which general base-catalysis of hydrolysis has been unequivocally
established. It is also known[9] that in comparison with nucleophilic catalysis,
general base-catalyzed water attack at carbonyl carbon is less subject to the
steric requirements of tertiary amines. Thus, the lack of detectable reactivity of
triethylamine and triethanolamine toward nonionized phthalimide may be attributed to the absence of TS$_5$.

TS$_5$

The absence of catalytic cleavage of nonionized phthalimide in the presence
of carbonate buffer solution cannot be attributed to the steric requirement of
carbonate dianion, because trimethylamine and DABCO may be considered to
have the same or even more steric requirements than carbonate dianion. The most

plausible explanation for the presence and absence of reactivity of nonionized phthalimide toward trimethylamine or DABCO and carbonate, respectively, may be attributed to the relative stabilities of the addition intermediates formed during stepwise nucleophilic catalysis. Based purely on electrostatic interactions, the zwitterionic addition intermediate ($T^\pm$) formed by the nucleophilic attack of a tertiary amine at the carbonyl carbon of nonionized phthalimide is apparently more stable than the dianionic addition intermediate (T^{2-}) formed by the nucleophilic attack of CO_3^{2-} at the carbonyl carbon of nonionized phthalimide. Thus, the higher stability of $T^\pm$ compared with that of T^{2-} makes the overall energy barrier much higher for the reaction of nonionized phthalimide with CO_3^{2-} than with tertiary amines of comparable basicity.

$$T^\pm \qquad\qquad T^{2-}$$

2.2.2 ELECTROVALENT BONDING INTERACTION

A catalyst ($CatE^+$) with electron-deficient atom/centre (E^+) can accept a pair of electrons from an electron-rich atom/center (N) of a substrate (NXSub) and thus forms a stable electrovalent bond. Such electrovalent bonding interaction between the catalyst and substrate will increase the electrophilicity of the atom X bonded covalently to the electron-rich atom (N) of the substrate as exhibited by Equation 2.7

$$\text{Cat—E}^+ + \text{N—X—Sub} \rightarrow \text{Cat—E—N}^+\text{—X—Sub} \qquad (2.7)$$

Such an interaction will increase the rate of nucleophilic reaction by (1) stabilizing the TS if nucleophilic attack at X and bond formation between E^+ and N occur simultaneously and (2) by destabilizing the GS if the nucleophilic attack at X occurs after the equilibrium formation of Cat–E–N$^+$–X–Sub.

Brown and Neverov[10] provided clear evidence for complex formation of anionic phosphate diesters to one as well as two lanthanide metal ions (La^{3+}). The La^{3+}-catalyzed methanolysis of anionic methyl 4-nitrophenyl phosphate revealed the rate acceleration of 3.3×10^5-fold with one La^+ and 1.7×10^9-fold with two La^{3+} at pH 8.5. Large rate acceleration due to lanthanide catalysis has been explained in terms of plausible transition states TS_6 and TS_7, which show the occurrence of bifunctional catalysis (Lewis acid and nucleophilic).

TS$_6$ TS$_7$

2.2.3 RESONANCE INTERACTION

Unlike electrostatic interaction, which may be energetically either favorable or unfavorable, resonance interaction is always energetically favorable and, hence, it causes an energetically stabilizing/favorable effect. If the covalent bonding interaction between a catalyst and a substrate results in the loss of resonance in the catalyst, or in the substrate, or in both catalyst and substrate, then such interaction will decrease the stability of the addition adduct formed between the substrate and catalyst by the amount of free energy that is lost owing to the loss of resonance in the formation of the addition adduct. Thus, the catalyst is expected to decrease the rate of reaction if the formation of addition adduct is the rate-determining step because, under such a situation, the transition state is destabilized due to partial loss of resonance compared to the reactant state. However, if the breakdown of the addition adduct is the rate-determining step, then the catalyst participation will increase the rate of reaction. It may be worth noting that if the formation of the addition adduct involves the loss of a large amount of resonance, then it may be too unstable to exist on the reaction path, and this may lead to the observation of nondetectable catalysis. The absence of nucleophilic catalysis in the cleavage of phthalimide in the aqueous buffer solutions of carbonate may be attributed partly to the loss of resonance of carbonate dianion when it formed a nucleophilic addition adduct or intermediate (T^{2-}) with nonionized phthalimide.

2.2.4 ELECTROSTATIC INTERACTION

Predominant energetically favorable and unfavorable electrostatic interactions between the catalyst as well as the TS and GS of the reactant, respectively, in the rate-determining step will increase the rate of a reaction, whereas negative catalysis will be exhibited by respective predominant energetically unfavorable and favorable electrostatic interactions between the catalyst as well as the TS and GS of the reactant in the rate-determining step.

2.2.4.1 Ion–Ion Interaction

Strong electrostatic interaction between an ionic catalyst and an ionic substrate may drastically change the rate of a reaction depending on the nature of electrostatic interaction, and its predominant effects on TS or GS of the reactant in the rate-determining step. Energy of interaction (E) between two point charges may be expressed as

$$E = \frac{e_1 e_2}{Dr} \qquad (2.8)$$

where e_1 and e_2 are the magnitudes of the charges, r is the distance between them, and D is the dielectric constant of the surrounding medium. Whether E is favorable (i.e., negative) or unfavorable (i.e., positive) depends on the algebraic signs of the point charges.

Pseudo-first-order rate constants (k_{obs}) for the reaction of anionic N-hydroxyphthalimide (NHP) with HO^- increased by ~ > 3-fold and 15-fold in the presence of inert monocations (Li^+, Na^+, K^+, and Cs^+) and dication (Ba^{2+}), respectively, in aqueous solvent containing 2% v/v acetonitrile.[11] Catalytic effects of these cations increased with the increase in the contents of acetonitrile in mixed aqueous solvents.[12] The presence of anions such as Cl^- and CO_3^{2-} did not show a kinetically detectable effect on k_{obs} for the alkaline hydrolysis of NHP^-. The catalytic effects have been explained quantitatively in terms of an ion-pair mechanism in which cations produced a predominantly stabilizing effect on TS rather than on GS. The overall catalytic effect of inert cations is apparently the combined effect of ion-pair formation between cation and anionic reactants, which causes the increase in electrophilicity of carbonyl carbon of NHP^- for nucleophilic attack and decrease in the nucleophilicity of nucleophile (HO^-).

2.2.4.2 Ion–Dipole Interaction

Ion–dipole interaction is a weaker electrostatic interaction compared to ion–ion interaction. Energy of such interaction varies in accordance to Equation 2.9[13]

$$E = \frac{e_1(\delta+)l\cos\theta}{Dr^2} \qquad (2.9)$$

where θ is the angle between the line (r) joining the point charge (e_1, a positive charge) and the middle of the dipole and the line (l) joining the middle of the dipole and that end of the dipole ($\delta+$) having a charge of the same algebraic sign as the interacting point charge (e_1), and l is the distance between the two equal and opposite charges ($\delta+$) and ($\delta-$) of dipole. Because the dipole moment μ of the dipole is equal to the product of the absolute charge of dipole (i.e., $|(\delta+)|$, which is equal to $|(\delta-)|$) and distance l separating the two charged poles, Equation 2.9 further reduces to Equation 2.10.

$$E = \frac{e_1 \mu \cos\theta}{Dr^2} \qquad (2.10)$$

It is evident from Equation 2.10 that whether E is positive or negative depends on the magnitude of θ. In view of the set criteria based on which Equation 2.9 has been derived, the positive pole of the dipole will be nearer to the positive point charge e_1 when $\theta < 90°$ and $\theta > 270°$, and the negative pole of the dipole will be nearer to the positive point charge e_1 when $\theta > 90°$ and $\theta < 270°$.

In order to illustrate the effects of ion–dipole interaction on reaction rates, let us consider the effects of buffer solutions of methylamine, dimethylamine, and trimethylamine on the rate of cleavage of ionized phthalimide. The values of second-order rate constants (k_2) for the reactions of ionized phthalimide with methylamine[14] (pK$_a$ 10.85) and dimethylamine[7] (pK$_a$ 11.05) are 0.323 M^{-1} sec^{-1} and 0.316 M^{-1} sec^{-1}, respectively. But, no detectable reactivities of trimethylamine (pK$_a$ 9.95) and DABCO (pK$_2$ 9.10) toward ionized phthalimide were observed.[7] These observations were explained in terms of nucleophilic addition–elimination mechanism, because the immediate products of the reactions of phthalimide with primary and secondary amines were corresponding N-substituted phthalamides. Significantly high nucleophilic reactivity of methylamine and dimethylamine toward ionized phthalimide was attributed to the presence of highly reactive addition intermediate ($T^{\neq}$), which is stabilized by intramolecular ion–dipole inter-action/internal hydrogen bonding interaction (because in ion–dipole interaction,

$T^{\neq}$

if the interacting dipole end is the hydrogen atom, then such an interaction is also called *hydrogen bonding interaction*) between anionic nitrogen of phthalim-ide moiety and slightly positively charged hydrogen attached to nitrogen of nucleophile moiety of $T^{\neq}$. In case of tertiary amine as nucleophile, there is no hydrogen atom attached to nucleophilic nitrogen, which could stabilize $T^{\neq}$ through intramolecular ion–dipole interaction and, consequently, addition of intermediate formed from the nucleophilic addition of tertiary amine and ionized phthalimide becomes too unstable to exist as an intermediate on the reaction path.*[15]

---

* An intermediate on the reaction path is defined as the chemical framework/entity that could survive for a period > 10^{13} sec (time period for a critical molecular bond vibration is ~ 10^{-13} sec).

2.2.4.3 Dipole–Dipole Interaction

Criteria similar to the ones used to derive Equation 2.10 can be used to express the energy of interaction (E) between two dipoles separated by the distance r between the middle points of two dipoles as a function of their dipole moments μ_1 and μ_2, angles θ and ϕ between the line r and line joining the middle as well as the end of the dipole having a charge of the same algebraic sign as the interacting charges, and distance r.

$$E = \frac{2\mu_1\mu_2 \cos\theta \cos\phi}{Dr^3} \tag{2.11}$$

Pseudo-first-order rate constants (k_{obs}) for intramolecular carboxylic-acid-catalyzed conversion of N-substituted phthalamic and related acids to phthalic anhydride (PAn) gave negative ρ (Hammett reaction constant)[16] and ρ^* (Taft reaction constant).[17] But the value of k_{obs} for the formation of PAn from *N*-methoxyphthalamic acid[18] is about 10-fold smaller than the corresponding k_{obs} for the formation of PAn from *N*-hydroxyphthalamic acid[19] under similar conditions. The nearly 10-fold larger value of k_{obs} for *N*-hydroxyphthalamic acid than that for *N*-methoxyphthalamic acid is attributed to the occurrence of intramolecular dipole–dipole interaction through transition state TS_8 in the catalyzed cleavage of *N*-hydroxyphthalamic acid.

TS_8

2.2.4.4 Hydrogen Bonding Interaction

Hydrogen bonds may be classified as conventional and nonconventional hydrogen bonds.[20] In an energetically favorable ion–dipole or dipole–dipole interaction, if the positive interacting pole of the dipole is hydrogen atom, then this interaction is also called *conventional hydrogen bonding interaction*, and such hydrogen bonds are classified as protic hydrogen bonds (A---H⁺---B). In essence, conventional hydrogen bonding interaction is always electrostatically attractive in nature. Nonconventional hydrogen bonds are classified as hydric hydrogen bonds (A---H⁻---B) and protic-hydric hydrogen bonds (A---H⁺---H⁻---B).[20]

2.2.5 HYDROPHOBIC INTERACTION

Hydrophobic effects and hydrophobic–hydrophobic interactions have been the topics of intense scrutiny for the last few decades for the simple reason that these interactions appear to play key roles in membrane and micelle formation, protein binding, ligand–protein and protein–protein binding, possibly nucleic acid interactions, and partitioning of drugs, metabolites, and toxins throughout the environment and living systems. However, precise mechanisms of hydrophobic and hydrophobic–hydrophobic interactions in aqueous medium is still not well understood.[21] But fine details of the mechanisms of such complex molecular interactions are now emerging at a relatively faster rate.[22] Excellent and detailed descriptions of the nature and origin of hydrophobic interactions can be found in a few papers,[22,23] reviews,[24] and books.[25] Unlike electrostatic interaction, which could be both attractive and repulsive in nature, hydrophobic–hydrophobic interaction is always attractive and hydrophobic–hydrophilic interaction is always repulsive. More precisely, energetically unfavorable hydrophobic–hydrophilic interaction causes the energetically favorable hydrophobic–hydrophobic interaction. It seems that hydrophobic–hydrophobic interaction originates owing to (1) attractive van der Waals–London dispersion forces and (2) the mutual attraction of the hydrophilic solvent molecules, which dissolve hydrophobic molecules. Thus, unlike electrostatic interaction, hydrophobic–hydrophobic interaction is very weak molecular interaction. It is apparent that in a solution containing both hydrophilic and hydrophobic molecules, if an additive decreases the energy of interaction between hydrophilic molecules, then it will also indirectly decrease the energy of interaction between hydrophobic molecules.

There are a few examples of both rate deceleration and acceleration, which are caused by factors including apparently hydrophobic–hydrophobic interactions. The effects of the concentrations of caffeine (**1**) on pseudo-first-order rate constants (k_{obs}) for alkaline hydrolysis of ethyl 4-aminobenzoate (benzocaine) in aqueous solvents are probably clear manifestations of the effects of hydrophobic interactions on reaction rates.[26] The contribution due to $k_w[H_2O][benzocaine]$ should be insignificant compared to $k_{OH}[HO^-][benzocaine]$ because of the fact that the value of k_w for phenyl acetate is 3×10^{-10} M^{-1} sec^{-1} at 25°C.[27] The most plausible mechanism for alkaline hydrolysis of benzocaine at pH $\geq$ 7 is shown in Scheme 2.8 where k_2-step is the rate-determining step.

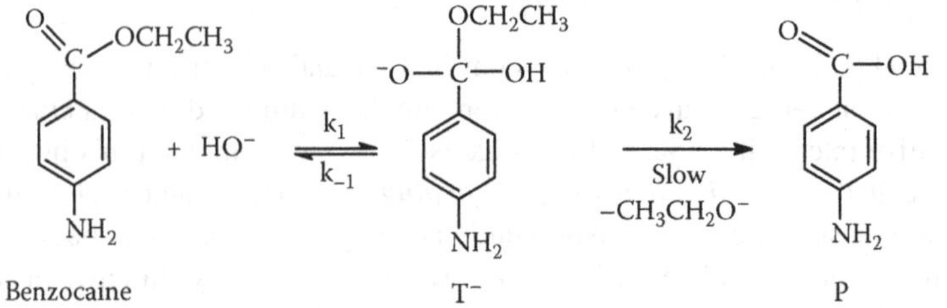

SCHEME 2.8

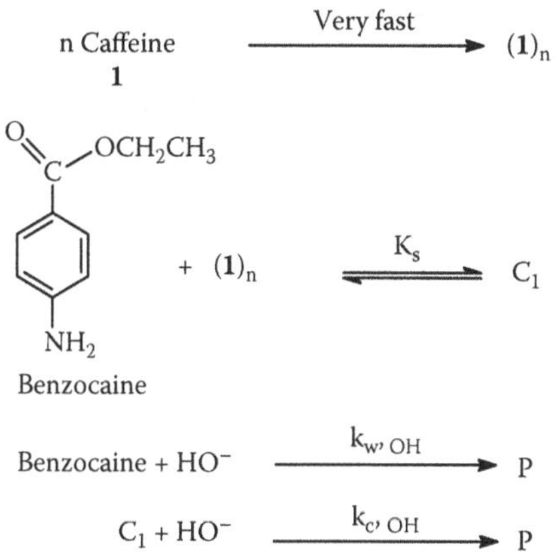

$$\text{n Caffeine} \xrightarrow{\text{Very fast}} (1)_n$$
$$\mathbf{1}$$

Benzocaine

$$\text{Benzocaine} + (1)_n \underset{}{\overset{K_s}{\rightleftharpoons}} C_1$$

$$\text{Benzocaine} + HO^- \xrightarrow{k_{w, \, OH}} P$$

$$C_1 + HO^- \xrightarrow{k_{c, \, OH}} P$$

SCHEME 2.9

The most obvious explanation for rate deceleration (negative catalysis) is the formation of a nonproductive complex between benzocaine and **1** through hydrophobic interaction. The reaction scheme in terms of such complex formation is shown in Scheme 2.9, where K_S represents binding constant between benzocaine and aggregate $(1)_n$, $k_{w,OH}$ and $k_{c,OH}$ are second-order rate constants for the reaction of HO^- with benzocaine and C_1, respectively. The observed rate law, rate $= k_{obs}[Im]_T$ (where $[Im]_T = [\text{benzocaine}]_T = [\text{benzocaine}] + [C_1]$), and Scheme 2.9 can lead to Equation 2.12

$$k_{obs} = \frac{k_w + k_c K_S[\mathbf{1}]}{1 + K_S[\mathbf{1}]} \qquad (2.12)$$

where $[\mathbf{1}] = [(\mathbf{1})_n]$, $k_w = k_{w,OH}[HO^-]$, and $k_c = k_{c,OH}[HO^-]$.

The binding constant (K_S) of benzocaine with **1** is 60 M^{-1}, and the rate of hydrolysis of the complex is negligible compared with that of the free ester, i.e., $k_c \, K_S \, [\mathbf{1}] << k_w$ in Equation 2.12.[26] The rate of alkaline hydrolysis of a few so-called activated esters (phenyl benzoates) in the presence of **1** and mixed aqueous solvent containing 31% CH_3CN also follow the reaction mechanism shown by Scheme 2.9 with replacement benzocaine by phenyl benzoates. The values of k_w/k_c vary from 1.7 to 2.4, whereas the values of K_S vary from 25 to 5 M^{-1} with change in ester from least reactive 3-methylphenyl benzoate to most reactive 4-nitrophenyl benzoate.[28] Significantly lower values of K_S for phenyl benzoates compared to that for benzocaine may be attributed to the presence of 31% CH_3CN in aqueous solution, because phenyl benzoates are apparently more hydrophobic than benzocaine. Acetonitrile is an additive that produces water-structure-breaking effect. The effects of the concentration of theophylline (**2**) on the rate of alkaline hydrolysis of phenyl benzoates show a more effective retarding rate on the alkaline hydrolysis of complex formed from **2** and phenyl benzoate esters.

TABLE 2.1

Pseudo-First-Order Rate Constants (k_{obs}) for Hydrolysis of 3 in the Presence of $(CH_3CH_2CH_2CH_2)_4Br$ and 0.001-M HCl at 25°C[a]

$[(CH_3CH_2CH_2CH_2)_4Br]/M$	$10^4 k_{obs}/sec^{-1}$	$10^4 k_{calcd}/sec^{-1}$ [b]
0.05	36.1	35.4
0.10	36.7	37.1
0.20	42.9	43.5
0.30	53.4	53.8
0.50	87.0	86.2
0.80	164	164

[a] In water solvent; data from Menninga, L., Engberts, J.B.F.N. Hydrolysis of arylsulfonylmethyl perchlorates: a simple model to explain electrolyte effects on a water-catalyzed deprotonation reaction. *J. Am. Chem. Soc.* **1976**, *98*(24), 7652–7657.
[b] Calculated from Equation 2.13 with $10^4 k_w = 34.6$ sec[1], $10^4 k_{s1} K_{S1} = 4.9$ M^{-1} sec[-1], and $10^4 k_{s2} K_{S2} = 196$ M^{-2} sec[-1].

The values of k_w/k_c and K_S vary from 6.1 to 5.3 and 25 to 5 M^{-1}, respectively, with the change from least reactive 3-methylphenyl benzoate to most reactive 4-nitrophenyl benzoate.[28] It is interesting to note that the values of K_S remain nearly the same for various phenyl benzoates with both **1** and **2**, but the values of k_c are 2- to 4-fold smaller in the presence of **2** than the corresponding k_c values in the presence of **1**. Significantly lower values of k_c with **2** compared to those with **1** may be attributed to the destabilization of anionic transition by the ionized carboxylic group in **2**.

Rate acceleration due to hydrophobic interaction between substrate and catalyst with large segment of hydrophobic moiety may be seen in a few examples of simple reactions. The increase in $[(CH_3CH_2CH_2CH_2)_4NBr]$ from 0.0 to 0.8 M increases pseudo-first-order rate constants (k_{obs}) for hydrolysis of 4-nitrophenylsulfonylmethyl perchlorate (**3**) at 0.001-M HCl in aqueous solvent from 32.5 × 10^4 to 164 × 10^{-4} sec^{-1} (Table 2.1). Although such remarkable positive catalysis (~ 5-fold) has been explained qualitatively in terms of electrostatic ion–water interactions and electrolyte-induced changes in the diffusionally average water structure,[29] an alternative explanation of these data may be given in terms of addition complex formation between **3** and $(CH_3CH_2CH_2CH_2)_4NBr$ through hydrophobic interaction in Scheme 2.9 with replacement of benzocaine by **3**. If the complexation has 1:1 and 1:2 stoichiometry, then the values of k_{obs} would follow Equation 2.13 derived from Scheme 2.9, provided 1 >> ($K_{S1}[R_4NBr]$ + $K_{S2}[R_4NBr]^2$).

$$k_{obs} = k_w + k_{s1} K_{S1}[R_4NBr] + k_{s2} K_{S2}[R_4NBr]^2 \qquad (2.13)$$

In Equation 2.13, K_{S1} and K_{S2} are association/binding constants for 1:1 and 1:2 stoichiometric addition complex formation, respectively, whereas k_{s1} and k_{s2} represent respective rate constants for hydrolysis of complex with 1:1 and 1:2 stoichiometry. Observed data (Table 2.1) fit satisfactorily to Equation 2.13 as evident from calculated values of rate constants (k_{calcd}) as shown in Table 2.1. The least-squares calculated values of k_w, $k_{s1}K_{S1}$, and $k_{s2}K_{S2}$ are $(34.6 \pm 0.79) \times 10^{-4}$ sec^{-1}, $(4.9 \pm 5.1) \times 10^{-4}$ M^{-1} sec^{-1}, and $(196 \pm 6) \times 10^{-4}$ M^{-2} sec^{-1}, respectively. The value of $k_{s1}K_{S1}$ with standard deviation of slightly more than 100% diminishes its importance compared with other terms of Equation 2.13. The maximum contribution of $k_{s1}K_{S1}[R_4NBr]$ when compared with $k_w + k_{s2}K_{S2}[R_4NBr]^2$, obtained at maximum value of $[R_4NBr]$ (= 0.8 M), is only 2.4%. The inequality 1 >> $(K_{S1}[R_4NBr] + K_{S2}[R_4NBr]^2)$ implies that the values of K_{S1} and K_{S2} should be less than 1, indicating a weak binding affinity of **3** with $(CH_3CH_2CH_2CH_2)_4NBr$, which is plausible because **3** contains a large moderately hydrophilic moiety.

2.2.6 VAN DER WAALS ATTRACTIVE AND REPULSIVE (STERIC) INTERACTIONS

Molecules with no permanent poles (positive and negative) such as noble gases and other nonpolar molecules do aggregate to form liquids and solids depending upon the temperature. The nature of this attraction is governed by London forces.[13] van der Waals attractive interaction is caused by London forces, apparently, when the interacting molecular sites remain at the *van der Waals distance* — the distance at which the atoms in interacting molecular sites are at an energy minimum. Half the van der Waals distance is called *van der Waals radius.* van der Waals repulsive interaction, which is also known as *steric hindrance/steric strain*, results when the distance between two atoms of molecular segments becomes less than the sum of their van der Waals radii.

Weak molecular bonding interactions are defined as those whose bond energies are comparable to thermal energies within the temperature range 0 to 100°C.[30] Most of these weak interactions are highly short range. For example, the energy of interaction for the weakest molecular interaction caused by London forces varies inversely with the power of six of the distance between interacting sites (i.e., van der Waals distance) of molecular segments. van der Waals attractive forces cause perhaps the weakest molecular interaction. This and other weak attractive molecular interactions are the basis for the stability of various biomolecular structural networks and consequently for various biomolecular activities.

Rate acceleration and deceleration of catalyst-mediated reactions occur due to predominant effects of such energetically favorable interactions (between reaction site of substrate and catalyst) on transition state and reactant state, respectively. However, the van der Waals' repulsive interactions (between reaction site of substrate and catalyst) can also cause rate acceleration and deceleration by predominant destabilization effects on reactant state and transition state, respectively. Because the van der Waals attractive interactions are responsible for hydrophobic–hydrophobic interactions between nonpolar molecules in aqueous solution, van der Waals attractive and hydrophobic interactions may be described as two facets of essentially the same molecular interactions. van der Waals repulsive interaction, which is more commonly known as *steric hindrance* or *steric strain*, not only affects the rates of catalyst-mediated reactions, but also provides effective chemical tools to optimize the stereoselectivity or enantiomeric excess in stereoselective reactions. A catalyst, by creating predominant strain on GS and TS can exhibit positive and negative catalysis, respectively. Sometimes the steric hindrance is so large that it almost stops the reaction.

An example of catalysis almost ceasing owing to steric hindrance destabilization of TS is the tertiary-amine-buffer-catalyzed hydrolysis of maleimide.[31] The brief reaction scheme for the aqueous cleavage of maleimide (MI) in the buffer solutions of DABCO, trimethylamine, and triethylamine is expressed by Equation 2.14

$$MI + H_2O \xrightarrow{\ R_1R_2R_3N\ } MA \tag{2.14}$$

where MA represents maleamic acid. Pseudo-first-order rate constants for the reaction, obtained at different tertiary amine concentrations and pH follow Equation 2.15

$$k_{obs} = k_w + k_{OH}[HO^-] + k_b[R_1R_2R_3N] + k_{ga}[R_1R_2R_3N][R_1R_2R_3NH^+] \tag{2.15}$$

where k_w is a pseudo-first-order rate constant for the reaction of H_2O with MI, k_{OH} and k_b represent second-order rate constants for the reaction of MI with HO^- and $R_1R_2R_3N$, respectively, and k_{ga} is third-order rate constant for general acid ($R_1R_2R_3NH^+$)-catalyzed reaction of MI with $R_1R_2R_3N$. The value of k_w for the hydrolysis of phthalimide (PT) is 9×10^{-6} sec^{-1} at 100°C [32], and the respective values of k_{OH} at 30°C for MI and PT are 72 M^{-1} sec^{-1} and 26 M^{-1} sec^{-1}. These results show that the value of k_w is negligible compared with $k_{OH}[HO^-]$ even at pH 6 to 7. The values of k_b for the reaction of nonionized MI with DABCO (pK$_a$ 9.10), trimethylamine (pK$_a$ 9.95), and triethylamine (pK$_a$ 10.63) are 3.34×10^{-2} M^{-1} sec^{-1}, 5.15×10^{-2} M^{-1} sec^{-1}, and 0.0 M^{-1} sec^{-1}, respectively. The presence of significant reactivity of trimethylamine and absence of such reactivity of triethylamine toward MI have been explained in terms of the mechanism shown by Scheme 2.10, in which the steric requirements of triethylamine could not allow the formation of reactive intermediate $T_1^\pm$ on the reaction path. It is worthy to note that the half-lives ($t_{1/2}$) for the reaction expressed by Equation 2.14 are ~ 30 to 150 h and ~ 70 sec in the absence and presence of 0.2 M trimethylamine (free base).

In terms of the Eyring equation, the rate changes (i.e., k_0/k_c, where k_0 and k_c represent respective rate constants in the absence and presence of catalyst) of 4- and 20-fold, brought about by the presence of a catalyst, require the changes in free energy of activation of only 0.835 and 1.804 kcal/mol respectively. Such small changes in free energy of activation can be easily achieved if the catalyst creates steric hindrance to the approach of the two reacting sites in the TS. Steric

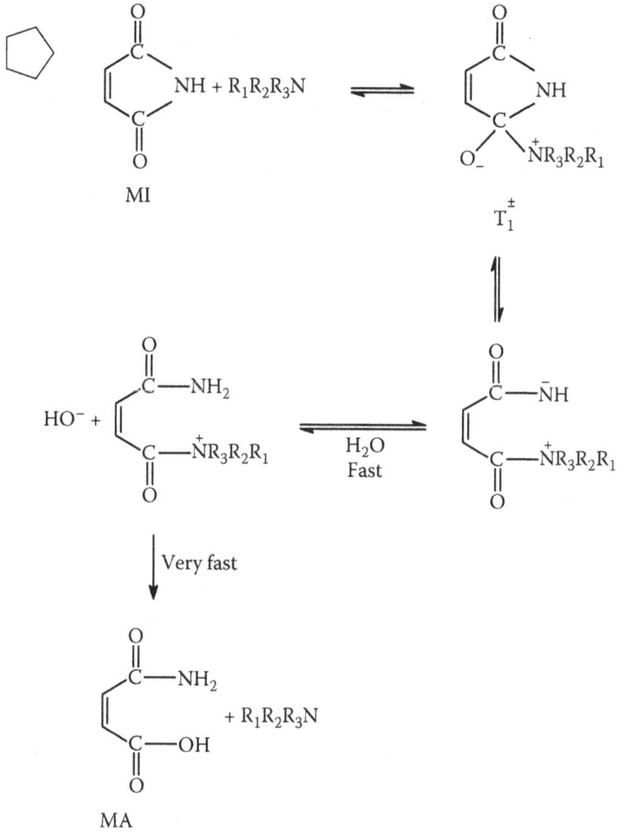

SCHEME 2.10

hindrance can be manipulated by the structural variations in the substrate (reactant), catalyst, or both, as well as solvent molecules. Selective steric hindrance or shielding of one of the two opposite sides or faces of the reaction site from the approach of the other reaction site can lead to so-called regioselective, stereoselective, or enantioselective reactions depending upon the structural features of the appropriate substrate or reactant. To demonstrate the enantioselectivity of the reactions caused primarily by the structural features of catalyst, let us consider a few chemical examples.

Enantioselective imidation of alkyl aryl sulfides with N-alkoxycarbonyl azides as a nitrene precursor is effected by using (OC)Ru(salen) complex as catalyst. The steric and electronic nature of the N-alkoxycarbonyl group strongly affect the enantioselectivity and the reaction rate.[33] In a systematic and well-executed study of ligand effects on Lewis-acid-catalyzed Diels–Alder reaction, it has been shown that the attachment of aromatic α-amino acid ligands to copper(II) ions leads to an increase in the overall rate of the Diels–Alder reaction between 3-phenyl-1-(2-pyridyl)-2-propene-1-one (Din) and cyclopentadiene

(Die) in water.[34] Such ligand-accelerated catalysis is observed for several aromatic α-amino acid ligands, which is attributed to the arene–arene interaction between the aromatic ring of the α-amino acid ligand and the pyridine ring of the dienophile (Din). The arene–arene interaction shields one face of the dienophile from attack by the diene, inducing up to 70% enantioselectivity in the reaction. Effects of solvents on enantioselectivity in the reaction show that it is significantly enhanced by water compared to organic solvents, which is partly attributed to hydrophobic interactions.

Din   Die

Study of catalytic enantioselective additions of dimethylzinc to carbonyl compounds shows that well-validated reaction conditions in terms of the relationship of 2,6-disubstituted benzyl aldehyde substrates, catalyst (dimethylzinc), solvents, and their ratios promoted enantioselectivity from 25% ee to more than 80% ee.[35] The reaction of 4,4-disubstituted 2,3-allenamides and organic iodides in toluene containing 1 mol% $Pd(PPh_3)_4$ catalyst and K_2CO_3-TBAB as the base afforded stereospecifically iminolactones in > 90% yields. But a similar reaction with 4-monosubstituted 3,3-allenamides gave γ-hydroxy-γ-lactams in relatively lower yields. This N/O-attack selectivity is attributed to the steric effect at the 4-position of 2,3-allenamides.[36] Steric hindrance has also been used to design new inhibitors of class C β-lactamases.[37] β-Lactam antibiotics that form an acyl-intermediate with the enzyme but subsequently are hindered from forming a catalytically competent conformation seem to be inhibitors of β-lactamases. Using steric hindrance of deacylation as a design guide, penicillin and carbacephem substrates have been converted into effective β-lactamase inhibitors, and antiresistance antibiotics.[37] Macromolecular crowding accelerates the rate of cleavage of DNA catalyzed by DNA nucleases. The reason for the acceleration is attributed to the possible condensation of the reactants caused by macromolecular crowding.[38]

2.3 MECHANISTIC CHANGE IN CATALYZED REACTIONS

Sometimes, a catalyzed reaction follows a different reaction mechanism compared to that of its uncatalyzed counterpart, and by changing the reaction mechanism, a catalyst changes the overall activation energy or overall energy barrier of the reaction through predominant effects on either GS or TS.

When 0.2 M of 2-mercaptoethanol, (2-ME), was allowed to react with 4.2 × 10⁻⁵ M of 9-anilinoacridine, (9-ANA), at pH 8.8 and 37°C for about 19 h, the products obtained were aniline and 9-acridiny 2-mercaptoethyl ether (**4**). The reaction scheme, based on these products' characterization, is given by Equation 2.16

$$9\text{-ANA} + \underset{\text{(2-ME)}}{\text{HSCH}_2\text{CH}_2\text{OH}} \longrightarrow \mathbf{4} + \underset{\text{(aniline)}}{\text{C}_6\text{H}_5\text{NH}_2} \qquad (2.16)$$

Detailed kinetic studies on the rates of hydrolysis[39,40], aminolysis[41], and thiolysis[40,42] of 9-aminoacridine and its N-substituted derivatives revealed that the rate of thiolysis is much faster than those of aminolysis and hydrolysis under similar experimental conditions. This difference in the reactivity of thiols and amines as well as HO⁻/H₂O toward 9-aminoacridine and its N-substituted derivatives is ascribed partly to the fact that HO⁻/H₂O, RO⁻/ROH, and amines are hard, whereas thiols are soft nucleophiles,[43] and as the reactive centers in 9-aminoacridine and its N-substituted derivatives are "soft electrophilic" in nature, a soft nucleophile should react much faster than a hard nucleophile of comparable basicity.

The value of the pseudo-first-order rate constant (k_{obs}) for hydrolysis of 9-ANA at pH 6.45, 0.6-M phosphate buffer and 30°C is 4.6×10^{-5} min⁻¹.[41] If the value of k_{obs}, under such conditions, is assumed to be due to uncatalyzed reaction between H₂O and 9-ANA only, then the second-order rate constant (k_2^h) for hydrolysis is 8.4×10^{-7} M^{-1} min⁻¹. This value of k_2^h might have been augmented owing to possible occurrence of general acid (phosphate monoanion) catalysis because thiolysis of 9-ANA is highly sensitive to general acid catalysis. Thus, the value of k_2^h for uncatalyzed hydrolysis of 9-ANA may be even smaller than 8.4×10^{-7} M^{-1} min⁻¹. As both H₂O and ethanol are hard nucleophiles, the value of uncatalyzed second-order rate constant (k_2^{et}) for the reaction of ethanol with 9-ANA at pH 8.8 and 0.2-M CH₃CH₂OH may not be significantly different from k_2^h, because [H₂O]/[CH₃CH₂OH] = 275 at 0.2-M ethanol. The value of pK_a of 9-ANAH⁺ is 7.93 at 30°C,[44] and nucleophilic reactivity of a nucleophile is about 300-fold larger with 9-ANAH⁺ than with 9-ANA.[40] Thus, the value of k_{obs} for uncatalyzed ethanolysis of 9-ANA at 0.2-M ethanol, pH 8.8, and 30°C would be $\ll 5 \times 10^{-5}$ min⁻¹. The value of k_{obs} for thiolysis of 9-ANA at pH 8.8, 0.2 ME and 30°C is 0.024 min⁻¹.[42]

In a typical kinetic run at 37°C, the UV-visible spectra of the reaction mixture containing 0.2-M ME buffer of pH 8.8 and 4.2×10^{-5} M 9-ANA were scanned at different reaction time ranging from 90 sec to 19 h. The appearance and disappearance of the intermediate product, 9-acridinyl 2-hydroxyethy thioether (IP), was confirmed owing to increase in the absorbance value of a characteristic absorption peak at 363 nm because of IP in the initial phase of the reaction (reaction time, t ranges from 90 sec to 44 min) followed by the decrease in its absorbance value in the final phase of the reaction (t ranges from 44 min to 19 h). The UV-visible spectrum of the reaction mixture scanned at t = 44 min was similar to that of 9-acridinyl n-propyl thioether.[40] The UV-visible spectrum of the reaction mixture scanned at t = 19 h was similar to a typical absorption spectrum

OCH$_2$CH$_2$SH   HN—⟨ ⟩   SCH$_2$CH$_2$OH

4   9-ANA   IP

of 9-methoxyacridine. Thus, IP converted to **4** with a rate slightly slower than the rate of 2-hydroxyethanethiolysis of 9-ANA at pH 8.8. Although the half-life period for ethanolysis of 9-ANA at pH 8.8 is expected to be > 230 h, the half-life period for the formation of **4** from IP is slightly more than 30 min.

More than 500-fold larger rate of formation of **4** from IP than the rate of ethanolysis of 9-ANA at pH 8.8 is because (1) the rate of formation of **4** from IP involves intramolecular nucleophilic substitution reaction at C$_9$ of acridine moiety, which is also known as *Smiles rearrangement*,[45] and the rate of an intramolecular reaction is larger by a factor of 10 to 10^{10} compared to the rate of analogous intermolecular reaction[46] and (2) the leaving group, RS$^-$, in the conversion of IP to **4** is a much better leaving group (pK$_a$ of SH group in 2-ME is 9.45) than the leaving group, C$_6$H$_5$NH$^-$ (pK$_a$ of aniline is ~ 27 [47]), in the ethanolysis of 9-ANA. It is apparent that SH (mercaptan) group of 2-ME acts as a positive catalyst, which accelerates the conversion of 9-ANA to **4** through a mechanism that is different in some respects from the one expected for the reaction of RO$^-$/ROH with 9-ANA.

2.4 PROXIMITY AND SHIELDING EFFECTS OF CATALYST

A positive catalyst may affect the rate of a bimolecular reaction by forcing reactant molecules to remain in close proximity, thereby increasing the rate of reaction at the expense of the substantial loss of freedom of translational motions of the reactant molecules. This effect has been given several names, including *proximity effect*. Similarly, the rate of a bimolecular reaction is affected by a negative catalyst, which forces the reactant molecules away from each other. This effect may be called *shielding effect* because a negative catalyst protects one of the reactants from a close approach by the other reactants.

Crown ethers (macrocyclic ethers) are known to catalyze reactions involving an anion as one of the reactants by increasing (1) the solubility of the salt of anion and (2) the intrinsic reactivity of anion owing to tight binding between the counterion (i.e., cation) and crown ether.[48] This differential/specific binding of anions with crown ethers results in a moderate rate retardation in metal-ion-catalyzed reactions.[49] Crown ethers have also been found to catalyze reactions involving S$_N$^{1} and S$_N$^{2} mechanisms merely by acting as phase transfer agents.[50] Remarkable specific metal ion catalysis has been observed in S$_N$^{2} and E$_2$ reactions occurring in proximity to a crown ether ring.[51] The specificity of the metal ion (strontium ion) catalysis in these reactions is attributed to merely proximity effect, which requires the metal to be located asymmetrically on one side of the crown ether ring surface. It is interesting to note that in the case of monovalent cations, such as sodium and potassium ions, which might fit inside the cavity of 15-crown-5 cavity (0.92 Å), the catalytic effects are less pronounced or even absent when one of the reactants (the electrophile) is covalently bonded to the ring of the crown ether.[51]

2.5 CLASSIFICATION OF CATALYSIS

The general classification of catalysis is shown in Figure 2.3.

Although we confine our attention to mainly micellar catalysis (microheterogeneous catalysis) in this book, a brief mention of other kinds of catalysis is also included.

2.5.1 HOMOGENEOUS CATALYSIS

Homogeneous catalysts and the reaction components of these catalyst-catalyzed reactions remain in the same phase. Thus, the homogeneous catalysis requires both the catalyst and the reactants or the reaction components to be present in the same phase. For example, the net reaction between acetone and bromine in an acidic aqueous medium is expressed as the following:

$$CH_3COCH_3 + Br_2 \xrightarrow{k/H_3O^+} CH_3COCH_2Br + HBr \qquad (2.17)$$

for which the observed rate law, rate = $k[CH_3COCH_3][H_3O^+]$, does not contain $[Br_2]$ but the net reaction (Equation 2.17) does show Br_2 as one of the reactants. On the other hand, H_3O^+ does not appear in the net reaction (a basic requirement for a chemical species to be called a catalyst), whereas the observed rate of reaction is proportional to $[H_3O^+]$. These observed facts may be explained in terms of a plausible reaction mechanism as shown in Scheme 2.11.

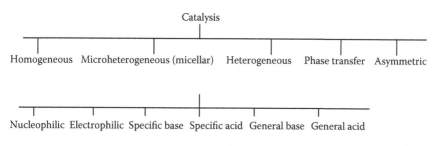

FIGURE 2.3 Representation of various types of catalysis and subcatalysis.

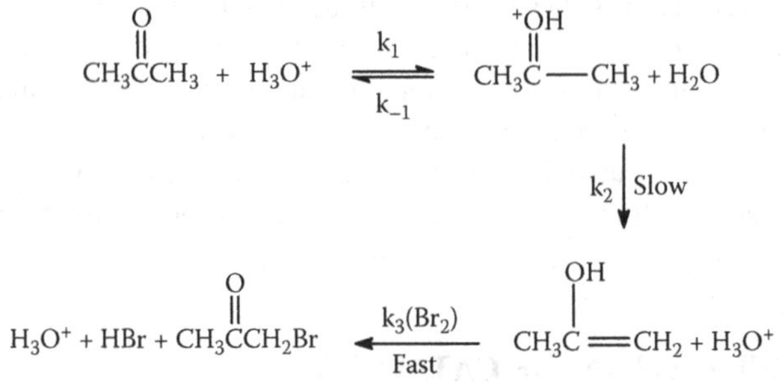

$$\underset{\text{CH}_3\overset{\text{O}}{\overset{\|}{\text{C}}}\text{CH}_3 \;+\; \text{H}_3\text{O}^+}{} \underset{k_{-1}}{\overset{k_1}{\rightleftharpoons}} \underset{\text{CH}_3\overset{^+\text{OH}}{\overset{\|}{\text{C}}}-\text{CH}_3 \;+\; \text{H}_2\text{O}}{}$$

$$k_2 \downarrow \text{Slow}$$

$$\underset{\text{H}_3\text{O}^+ + \text{HBr} + \text{CH}_3\overset{\text{O}}{\overset{\|}{\text{C}}}\text{CH}_2\text{Br}}{} \underset{\text{Fast}}{\overset{k_3(\text{Br}_2)}{\longleftarrow}} \underset{\text{CH}_3\overset{\text{OH}}{\overset{|}{\text{C}}}=\text{CH}_2 + \text{H}_3\text{O}^+}{}$$

SCHEME 2.11

The observed rate law for the reaction shows that the k_3-step cannot be the rate-determining step. The rate of bromination of acetone shows substantial amount of deuterium primary kinetic isotope effect ($k_H/k_D \approx 7$ where k_H and k_D represent rate constants for the reaction of Br_2 with CH_3COCH_3 and CD_3COCD_3, respectively), which implies k_2-step as the rate-determining step.

2.5.2 HETEROGENEOUS CATALYSIS

Heterogeneous catalyst and reactants of a catalyzed reaction remain in two different phases. Heterogeneous catalysis affects the rate of a reaction, which occurs at the interface of two distinct phases — for example, the hydrogenation of alkenes in the presence of solid catalysts such as Pt/C, Pd/C, and Ni. The details of the mechanisms of catalytic hydrogenation and related reaction are not yet well understood.

$$CH_3CH = CHCH_3 \xrightarrow[\text{H}_2]{\text{Pt/C}} CH_3CH_2CH_2CH_3 \tag{2.18}$$

2.5.3 MICROHETEROGENEOUS (MICELLAR) CATALYSIS

Aqueous micellar solutions are transparent to UV-visible electromagnetic radiation. But the existence of invisible micelles and even much larger aggregates, vesicles, in aqueous solution containing micelle- or vesicle-forming surfactant molecules has been ascertained experimentally beyond any doubt. Thus, the effect of such molecular aggregates on reaction rates is referred to as *microheterogeneous catalysis* although not all of such reactions are necessarily taking place strictly at the surface of the aggregate.

Direct or indirect predominant effect of hydrophobic interaction between nonionic micelle and either reactants or TS appears to be the source for micellar catalysis. Micelles also affect the rate of the reactions by causing proximity or shielding effect. However, if both the micelle and reactants are ionic, then the predominant combined effect of both hydrophobic and electrostatic interactions on either GS or TS will be the cause for micellar catalysis. Thus, in a broad sense, micellar catalysis and hydrophobic or hydrophobic-electrostatic catalysis are the same. However, micelles, with a large average aggregation number, are dynamic yet very stable and have a well-defined structure as well as an environment of varying medium characteristics such as polarity, water concentration, and ionic strength (in the case of ionic micelles). These micelles affect the rate of a reaction by various factors including medium effect. The details of these factors are discussed in Chapter 4 describing micellar catalysis.

2.5.4 NUCLEOPHILIC CATALYSIS

Nucleophilic catalysis is defined as the catalytic process in which a catalyst or a catalytic group forms a covalent bond with the reaction center of the substrate in the initial phase of the reaction followed by the recovery of the catalyst by the cleavage of the covalent bond between the reaction center of the substrate and the catalyst in the latter phase of the reaction. Such catalysis is also known as *covalent catalysis*. Nucleophilic catalysis may be either an intermolecular (if the substrate and the catalyst are not covalently attached to each other in the reactant state) or a semiintramolecular (if the catalyst becomes a part of the structural network of the substrate during the course of the reaction) molecular event.

2.5.4.1 Intermolecular Nucleophilic Catalysis

To understand exactly how nucleophilic catalysis works, let us see an example that involves the cleavage of 4-nitrophenyl acetate (4-NPA) in the presence of buffers of a few tertiary amines ($R_1R_2R_3N$). The overall reaction may be expressed by Equation 2.19

$$4\text{-}NO_2C_6H_4OOCCH_3 + H_2O \xrightarrow{\ R_1R_2R_3N\ } 4\text{-}NO_2C_6H_4OH + CH_3COOH \quad (2.19)$$

It may be noted that the half-lives ($t_{1/2}$) of the reaction are ~ 350 h and 50 sec in the absence and presence of 0.07 M $CH_3CH_2CH_2CH_2N(CH_3)_2$, respectively.[52] Pseudo-first-order rate constants (k_{obs}) for the reaction were found to depend on the concentration of free amine base ($[R_1R_2R_3N]$) as expressed by Equation 2.20

$$k_{obs} = k_w + k_{OH}[HO^-] + k_b[R_1R_2R_3N] \quad (2.20)$$

where k_w is pseudo-first-order rate constant for the reaction of H_2O with 4-NPA, whereas k_{OH} and k_b represent second-order rate constants for the reaction of 4-NPA with HO^- and $R_1R_2R_3N$, respectively. The respective values of k_w and k_{OH} are 5.5×10^{-7} sec^{-1} and 12 M^{-1} sec^{-1}.[52] The values of k_b for different tertiary amines yielded a Bronsted plot of slope (β) of 0.70 and intercept (c) of 8.45 M^{-1} sec^{-1}. In order to account for the participation of $R_1R_2R_3N$ molecule in the cleavage of 4-NPA, the suggested and most plausible mechanism for the reaction is shown in Scheme 2.12. Most of the aliphatic tertiary amines are strong bases

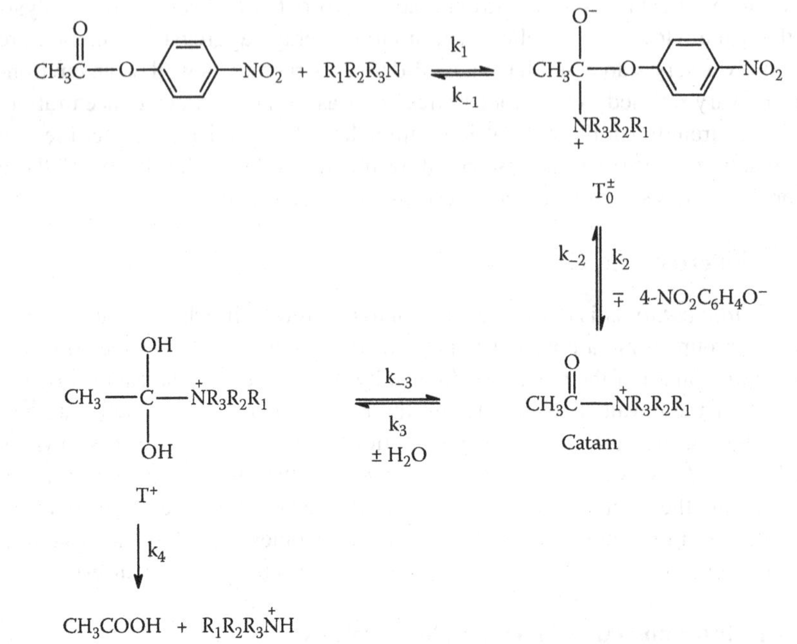

SCHEME 2.12

($pK_a \geq 10$), and that is why these amines, although sterically hindered, act as efficient nucleophiles. The intermediates $T_0^{\pm}$ and T_1^{+} are so unstable in aqueous solution that they cannot be directly detected by any usual spectral technique. However, indirect evidence for their existence on the reaction path come from various types of studies as elegantly described by Jencks.[25]

Cationic amides such as Catam (Scheme 2.12) are far less stable and, hence, more reactive toward nucleophiles compared to neutral amide partly because of the loss of a substantial amount of resonance in cationic amides as expressed in Equation 2.21. The cationic amide such as Catam becomes

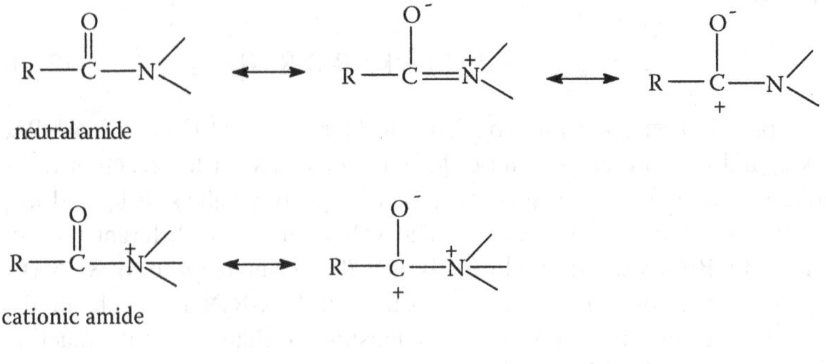

(2.21)

too unstable in water solvent to be detected directly by any usual experimental technique. Although water is a much weaker nucleophile compared to 4-$NO_2C_6H_4O^-$, ~ > 10^6-fold larger concentration of water ($[H_2O] \approx 55$ M in the solvent containing 99% v/v H_2O) compared to that of 4-$NO_2C_6H_4O^-$ ([4-$NO_2C_6H_4O^-$] = < 2×10^{-5} M when the initial concentration of 4-NPA is 2×10^{-5} M) makes k_3 [H_2O] > k_{-2}[4-$NO_2C_6H_4O^-$] (Scheme 2.12). Considerably large value of Bronsted slope (β) of 0.7 may be compared with reported values of β of 0.9[5] and 1.05[53] derived from the Bronsted plots, which included various primary, secondary, and tertiary amines in the aminolysis of acetate esters. The value of β of 0.9 is attributed to the expulsion of the leaving groups (aryl ions, ArO^-) from the zwitterionic tetrahedral addition intermediate (such as $T_0^{\pm}$) as the rate-determining step in the nucleophilic reactions of primary, secondary, and tertiary amines with phenyl and 4-nitrophenyl acetates.[5] The value of β of 0.7 may be attributed to the k_2-step as the rate-determining step (Scheme 2.12).

In the absence of direct detection of Catam, a skeptic might argue and suggest that the presence of positive catalysis due to tertiary amines could be because of the occurrence of an alternative reaction mechanism as shown by Scheme 2.13, which involves intermolecular general base catalysis instead of nucleophilic catalysis. The formation of reactive intermediate T^- occurs through the transition state TS_9. Although the proposal for the existence of transition states similar to TS_9 is not uncommon in the vast literature on the related

TS_9

reaction, caution must be exercised for such a proposal because of the fact that even in solution phase reactions, the probability of the effective termolecular collision with proper stereoelectronic requirements for bond formation and bond cleavage with large reactant molecules is extremely low. An alternative stepwise mechanism for the formation of T^- through a zwitterionic tetrahedral intermediate $T^{\pm}$ may be ruled out because $T^{\pm}$ is so unstable (and consequently the energy barrier for its formation is so large) that it may never exist on the reaction path as an intermediate.[54]

However, the occurrence of nucleophilic catalysis (Scheme 2.12) is supported by the experimental evidence as described in the following text.

1. The reaction mechanism shown in Scheme 2.12 and Equation 2.20 can lead to Equation 2.22

$$k_b = \frac{k_1' k_2'}{k' + k_{-1}'[ArO^-]}$$
(2.22)

where $k_1' = k_1 k_2$, $k_2' = k_3$, and $k_{-1}' = k_{-1} k_{-2}$ provided $k_3[H_2O] \ll k_{-2}$ $[ArO^-]$. In Equation 2.22, $k_b = (k_{obs} - k_w - k_{OH} [HO^-])/[R_1R_2R_3N]$ and $ArO^- = 4\text{-}NO_2C_6H_4O^-$. It is evident from Equation 2.22 that the values of k_b should decrease nonlinearly with the increase in $[ArO^-]$. The values of k_b, obtained at different values of $[ArO^-]$ (by externally added aryloxide ions) for the cleavages of aryl acetates[55] and carbonates[56] in the buffer solutions of tertiary amines, were found to fit to Equation 2.22. These observations cannot be explained in terms of reaction mechanism shown by Scheme 2.13. Werber and Shalitin[57] also found the inhibitory effects of added leaving-group ions (owing to the so-called mass law effect or common ion/common leaving-group effect) in the cleavages of some of the activated esters in the buffers of tertiary amines. Similar inhibitory effect of the added imidazole on k_b for the reaction of acetylimidazolium ion with N-methylaimidazole was observed at pH 6.8.[58] But no such inhibition of added imidazole was found in the reaction of quinuclidine with acetylimidazole at pH 10.6.[59] However, this negative result may be the consequence of the fact that the immediate nucleophilic substitution product is imidazole anion, which may quickly and irreversibly convert to neutral imidazole at pH 10.6 (pK_a of imidazole is 14.2).

2. Second-order rate constants (k_b) for the reactions of 4-NPA with various primary, secondary, and tertiary amines yielded the same Bronsted plot with slope β of 0.9 ± 0.1.[5] The reaction products for the reactions of 4-NPA with primary and secondary amines were corresponding amides (i.e., nucleophilic products). It is highly

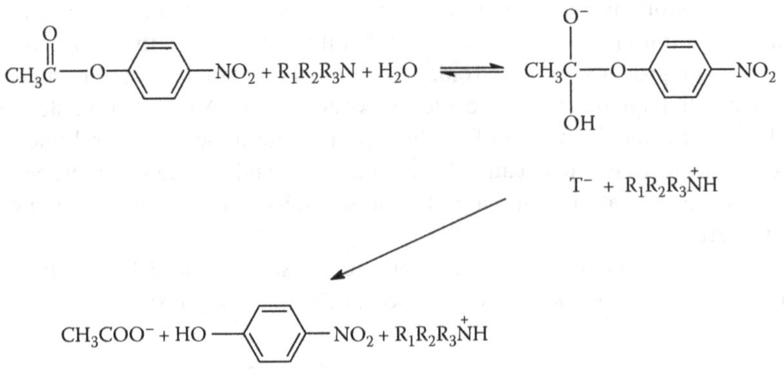

SCHEME 2.13

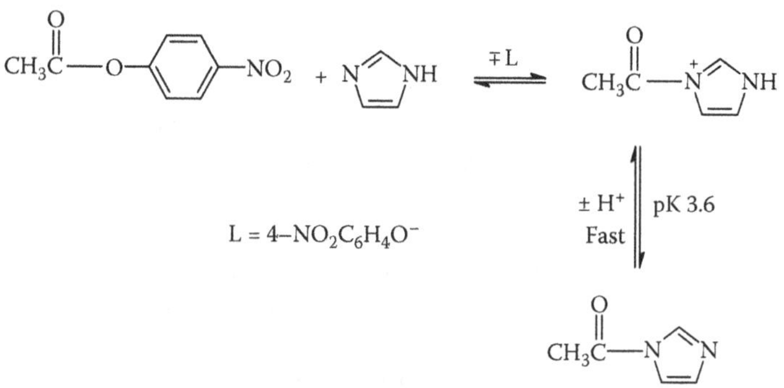

$$L = 4\text{-}NO_2C_6H_4O^-$$

SCHEME 2.14

unlikely that k_b for all various primary, secondary, and tertiary amines would give Bronsted plots of the same slope and intercept if the reactions of 4-NPA with tertiary amines followed the reaction mechanism shown by Scheme 2.13.

3. Imidazole contains two types of nitrogen — a tertiary nitrogen and a secondary nitrogen. Resonance interaction between the lone pairs of electrons of nitrogen atom and pi–electrons of the ring predict that tertiary nitrogen is more basic than secondary nitrogen. Thus, the nucleophilicity of imidazole is governed by their tertiary nitrogen and, consequently, imidazole behaves as a tertiary amine in its reactions with electrophiles. The cleavage of 4-NPA in the presence of imidazole buffers revealed the intermediate formation of acetylimidazole, which could be recognized by its characteristic absorption at 245 nm. The immediate product acetylimidazolium ion (similar to Catam in Scheme 2.12) is quickly converted to thermodynamically more stable acetyl-imidazole by a fast deprotonation process at pH > 5 as shown in Scheme 2.14.[60] Similarly, Fersht and Jencks[61] observed spectrophotometrically the formation and hydrolysis of the acetylpyridinium ion intermediate in the pyridine-catalyzed hydrolysis of acetic anhydride at 280 nm.

The kinetic study on tertiary-amine-catalyzed ester formation from benzoyl chloride (BC) and phenol in dichloromethane at 0°C showed the following: (1) the lack of phenoxide in the UV-visible spectral study, which probably rules out specific base catalysis, (2) the presence of benzoylammonium salt (AAS) observed in low-temperature NMR experiments, (3) the presence of small or negligible deuterium isotope effect (i.e., $k_H/k_D \approx 1$), which rules out the general base catalysis, and (4) the rates of the reactions of BC with quinuclidine and triethyl amine (the most basic amines in pure aqueous solvent) differed by at least 3 orders of magnitude.[62] These observations suggest the occurrence of nucleophilic catalysis for phenyl benzoate formation in the reaction of BC with phenol under the presence of tertiary amines. However, butyl benzoate formation from the reaction of BC with butanol in the presence of tertiary amine resulted in a huge amount of deuterium isotope effect ($k_H/k_D > 100$). Theoretically, the maximum value of k_H/k_D should be about 7 at 25°C and in pure water solvent. This unusual observation appears to be inconsistent with nucleophilic catalysis.

The occurrence of nucleophilic catalysis has been reported in the reactions of dimethyl carbonate with carboxylic acids[63] and indoles[64] under the presence of tertiary amines such as 1,8-diazabi*cyclo*[5.4.0]undec-7-ene (DBU) and DABCO. However, it remained unclear why tertiary amines such as *N*-methyl-morpholine and tertiary butylamine revealed almost zero catalytic efficiency.

2.5.4.2   Rate Relationship between Intra- and Intermolecular Reactions with Identical Reaction Sites in Terms of Energetics

When two reaction sites (A and B), attached to each other covalently, react to give product P_1 as shown in Equation 2.23, then such a reaction is referred to as an *intramolecular reaction*. In order to assess the effect of intramolecularity of such a reaction, the first-order rate constant (k_1) for intramolecular reaction such as the one shown by Equation 2.23 is generally compared with second-order rate constant (k_2) for a bimolecular reaction involving molecules with exactly the same reaction sites (A and B) and the same reaction conditions as shown by Equation 2.24. The products P_1 and P_2 may not be necessarily same.

$$\begin{matrix} A \\ \\ B \end{matrix} \quad \xrightarrow{k_1} \quad P_1 \qquad\qquad (2.23)$$

$$A \ + \ B \quad \xrightarrow{k_2} \quad P_2 \qquad\qquad (2.24)$$

The efficiency of the intramolecular reaction (Equation 2.23) over its inter-molecular counterpart (Equation 2.24) is measured by the ratio k_1/k_2, which has units of molarity (M). Despite the accumulation of large amounts of kinetic data on intramolecular reactions during the last (nearly) four decades, which show the values of k_1/k_2 ranging from 0.5 [65] to ~10^{15} M[66], the attempts to explain these values of k_1/k_2 created nearly 24 seemingly different suggestions[65,67] as summarized in Table 2.2, and resulted in an uncomfortable debate among some of the most productive researchers in this area.[65,67, 68a] Apparently, some of the suggestions describe essentially same molecular interactions but with different, attractive, and fanciful nomenclature.

The rate constant for a one-step reaction is basically governed by the free energy of activation. Because of the covalent bonding between reactive sites A and B in the intramolecular reaction, there is a significantly large amount of loss of translational and rotational freedom (i.e., entropy) of reactants of reaction (of

TABLE 2.2
Various Explanations for the Efficiency of Intramolecular Reactions

Explanation	References
The 55 figure	Jencks[25]
Anchimeric assistance	Winstein et al.[69]
Approximation, propinquity, proximity	Jencks[25]; Bruice and Benkovic[70]
Entropy	Page and Jencks[71]
Togetherness	Jencks and Page[72]
Rotamer distribution	Bruice[73]
Distance distribution function	De Lisi and Crothers[74]
Orbital steering	Storm and Koshland[75a]; Dafforn and Koshland[75b]; Storm and Koshland[75c,d]
Stereopopulation control	Milstein and Cohen[76]
Substrate anchoring	Reuben[77]
Vibrational activation	Firestone and Christensen[78]
Vibrational activation entropy	Cook and McKenna[79]
Orbital perturbation theory	Ferreira and Gomes[80]
Group transfer hydration	Low and Somero[81]
Electrostatic stabilization	Warshel[82]
Electric field effect	Hol, van Duijnen, and Berendsen[83a]; van Duijnen, Thole, and Hol[83b]
Catalytic configurations	Henderson and Wang[84]
Direct proton transfer	Wang[85]
Coupling between conformational fluctuations	Olavarria[86]
Gas phase analogy	Dewar and Storch[87]
Torsional strain	Mock[88]
Circe effect	Jencks[89]
Freezing at the reactive centers of enzymes	Nowak and Mildvan[90]
Spatiotemporal postulate	Menger[68a]; Menger and Venkataram[68b]

Equation 2.23) compared to that of the other reaction (of Equation 2.24). Because the translational degrees of freedom of transition states of both reactions remain essentially the same, there may be little change in the entropy of TS of reaction (of Equation 2.23) compared to that of the other reaction. Under such circumstances, a large portion of rate enhancement due to intramolecularity (i.e., the value of k_1/k_2) may be accounted for by the larger value of entropy of activation for reaction {of Equation 2.23} than that for the other reaction {i.e., of Equation 2.24}. However, in the assessment of entropy factor, it is also essential to consider the degree of proximity/propinquity/spatiotemporal of the reaction sites A and B in the reactions. For instance, in the aqueous acidic cleavages of phthalamic acid (**5**) and terephthalamic acid (**6**), **5** reacts nearly 500 times faster than **6** under identical reaction conditions.[16b] Such a large reactivity of **5** compared to **6** is attributed to the intramolecular carboxylic group participation in the cleavage of amide bond in **5**. Although the same carboxylic group is also covalently attached

to benzene ring in **6**, the position of attachment is such that it cannot participate intramolecularly in the cleavage of amide bond (i.e., degree of proximity in this case is zero). Apart from entropy and degree of proximity factors, effects of aqueous solvents and geometrical strain on both reactant and transition states of both reactions may also contribute significantly to the values of k_1/k_2.

$$\begin{array}{cc} 5 & 6 \end{array}$$

2.5.4.2.1 Intramolecular Induced Nucleophilic Reaction

In an intermolecular nucleophilic reaction, if the reactant with an electrophilic center is activated by the intramolecular reaction, then such a reaction may be called *intramolecular induced nucleophilic reaction*. For instance, *N*-benzyloxycarbonyl derivatives of α-aminophosphonochloridates (**7**, **8**, and **9**) react nearly 10^3 to 10^4 times faster than $ClCH_2P(O)(OCH_3)Cl$ (**10**) with isopropanol (Pr^{i}OH) to give substitution products as shown in Equation 2.25 and Equation 2.26.[91] Significantly high reactivity of **7**, **8**, and **9** compared to **10** is attributed to the occurrence of reaction

$$C_6H_5CH_2OCONRC(R_1)_2P(O)(OCH_3)Cl + Pr^iOH \longrightarrow$$
$$C_6H_5CH_2OCONRC(R_1)_2P(O)(OCH_3)OPr^i + HCl \qquad (2.25)$$

7, R = R$_1$ = H; **8**, R = H, R$_1$ = CH$_3$; **9**, R = CH$_3$, R$_1$ = H

$$ClCH_2P(O)(OCH_3)Cl + Pr^iOH \longrightarrow ClCH_2P(O)(OCH_3)OPr^i + HCl \quad (2.26)$$

mechanism as shown by Scheme 2.15, which involves the initial formation of the reactive cationic cyclic intermediate through intramolecular nucleophilic substitution reaction followed by intermolecular nucleophilic substitution reaction between Pr^{i}OH and cationic cyclic intermediate. The reaction between **10** and Pr^{i}OH involves only intermolecular substitution reaction.

2.5.4.2.2 Induced Intramolecular Nucleophilic Reaction

Such a reaction involves the formation of an addition covalent adduct between electrophilic and nucleophilic reactants in the initial step followed by intramolecular reaction in the subsequent step of the reaction. An example of such a reaction is the methanolysis of 4-methoxyphenyl 2-formylbenzenesulfonate in basic media.[92] The reaction of methanol with 4-methoxyphenyl 2-formylbenzenesulfonate (**11**) in the presence of anhydrous potassium carbonate gives the dimethyl acetal of 2-formylbenzenesulfonic acid in excellent yield. But the reac-

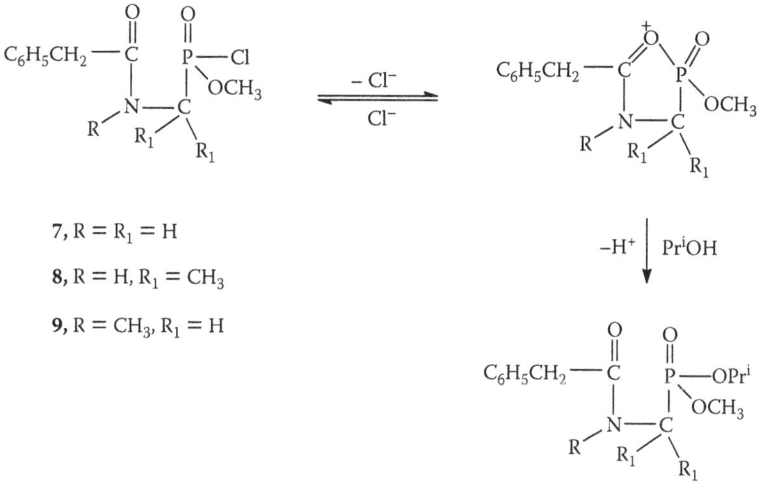

SCHEME 2.15

tion of methanol with 4-methoxyphenyl 4-formylbenzenesulfonate (**12**) under identical reaction conditions remains unaffected. These results provide evidence for the pseudocatalytic involvement of the neighboring aldehyde carbonyl group and, consequently, the suggested mechanism is shown in Scheme 2.16. The structural features of **12** do not allow the occurrence of reaction mechanism shown by Scheme 2.16 in its reaction with CH_3O^-.

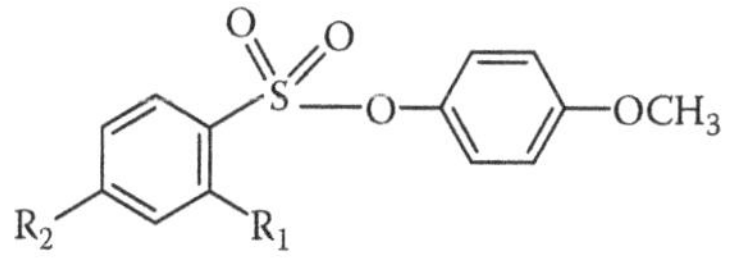

11, $R_1 = CHO$, $R_2 = H$

12, $R_1 = H$, $R_2 = CHO$

2.5.4.3 Induced Intramolecular and Intramolecular Induced Nucleophilic Catalysis

Induced intramolecular nucleophilic catalysis involves the formation of a molecular addition complex (Ad) between the reactant/substrate (R–A) and catalyst (W–B) followed by intramolecular nucleophilic substitution reaction between two reaction sites A and B giving products P_3 and P_4, where one of the products, say P_4, is unstable, i.e., more reactive, and, hence, P_4 undergoes further reaction to reproduce catalyst (W–B) and give the final stable product P_5. This whole chemical process may be shown by Equation 2.27. Intramolecular induced nucleophilic catalysis involves the intermolecular nucleophilic substitution reaction between R–A and catalyst, (C–W–B), where nucleophilicity of the nucleophile is enhanced owing to intramolecular interaction between molecular sites B and C of catalyst,

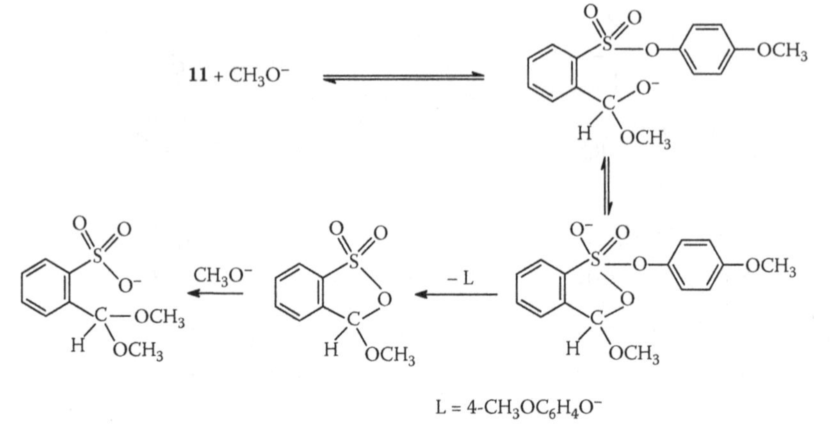
11 + CH₃O⁻
CH₃O⁻
– L
L = 4-CH₃OC₆H₄O⁻

SCHEME 2.16

Cat, producing products P_3 and P_4. One of these products, say P_4, is more reactive and therefore it undergoes further reaction to yield more stable product P_5 and catalyst (Cat) as shown in Equation 2.28. The intermolecular nucleophilic catalyzed reaction is shown by Equation 2.29 where B is catalyst.

$$\mathrm{R{-}A} + \mathrm{W{-}B} \xrightarrow{k_{ad}} \underset{\mathrm{Ad}}{\begin{matrix}\mathrm{R{-}A}\\ \mathrm{W{-}B}\end{matrix}} \xrightarrow[-\,P_3]{k_1} P_4 \xrightarrow{k_2} P_5 + \mathrm{W{-}B} \tag{2.27}$$

$$\mathrm{R{-}A} + \underset{\mathrm{Cat}}{\mathrm{W}\begin{matrix}\mathrm{B}\\ \mathrm{C}\end{matrix}} \xrightarrow[-\,P_3]{k_1} P_4 \xrightarrow{k_2} P_5 + \underset{\mathrm{Cat}}{\mathrm{W}\begin{matrix}\mathrm{B}\\ \mathrm{C}\end{matrix}} \tag{2.28}$$

$$\mathrm{R{-}A} + \mathrm{B} \xrightarrow{k_2} P_3 + P_5 + \mathrm{B} \tag{2.29}$$

The fine details of the structural features of most of the enzymes, which are prerequisite to know the detailed mechanistic aspects of enzyme catalysis, are not well understood. α-Chymotrypsin is probably one of only a few enzymes whose active site has been the subject of detailed structural study through X-ray diffraction and finer details of the structural features of this enzyme have been reported in 1969.[93] This report becomes the catalyst for the surge of interest in the mechanistic details of the simple organic reactions, which are considered to be, at least, partial models to the catalysis exhibited by α-chymotrypsin and related enzymes in aqueous medium by assuming that the structural features of active sites of these enzymes remain unchanged with the change in the reaction medium/phase from solid to aqueous. Though no single enzyme-catalyzed reac-

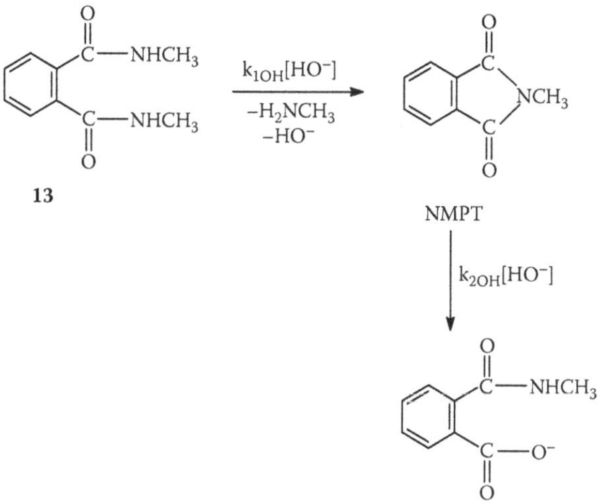

SCHEME 2.17

tion is as well understood mechanistically as almost any carefully studied simple organic reaction, catalytic important reaction steps in α-chymotrypsin-catalyzed reactions are believed to be similar to those shown by Equation 2.27. Perhaps it is a common perception among, at least, physical organic chemists that the major (if not entire) part of enzyme (α-chymotrypsin) catalytic efficiency results is due to the occurrence of intramolecular reactions that occur in k_1- and k_2-steps in the reaction of Equation 2.27.

Reports on the rate enhancements in intramolecular nucleophilic reactions compared to analogous intermolecular nucleophilic reactions are plenty. But such reports on induced intramolecular or intramolecular induced nucleophilic catalysis are rare. One might argue that the induced intramolecular/intramolecular induced nucleophilic reactions also represent the corresponding nucleophilic catalysis. But this statement is not totally correct in view of the definition of a catalyst. For example, hydroxide-ion-catalyzed second-order rate constant (k_{OH}) for the cleavage of one of the two amide bonds in N,N-dimethylphthalamide (**13**) at 25.3°C is 7.6 M^{-1} sec^{-1}.[94] The value of k_{OH} for the cleavage of amide bond in N-methylbanzamide (**14**) at 100.4°C is 7.2 $\times$ 10^{-4} M^{-1} sec^{-1}, which gives k_{OH} as 5 $\times$ 10^{-8} M^{-1} sec^{-1} at 25°C.[95] Nearly 10^7-fold larger reactivity of **13** than that of **14** is attributed to the occurrence of intramolecular induced nucleophilic reaction as shown in Scheme 2.17, because the polar effect of 2-CONHCH$_3$ substituent can cause only ~ 2- to 3-fold increase in k_{OH} for **14** and pK$_a$ values of **14** and

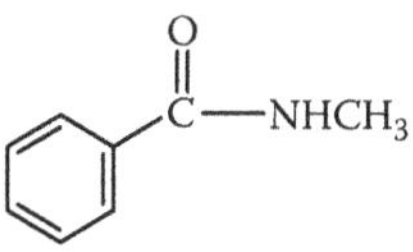

14

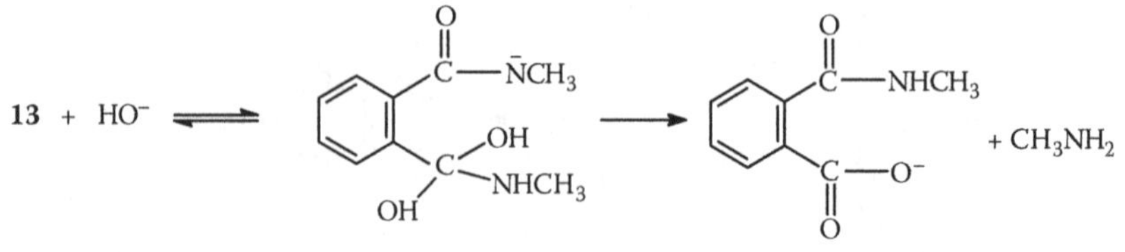

SCHEME 2.18

H_2O are nearly the same. The value of k_{2OH}/k_{1OH} is ~ 3, which shows that the intermediate NMPT is more reactive than **13** and, consequently, NMPT is quickly converted to a much more stable product, *N*-methylphthalamate, under alkaline aqueous medium. It is evident from Scheme 2.17 that 2-CONHCH$_3$ group in **13** is not acting as an intramolecular catalyst, because it is not regenerated in the last reaction step in a molecular shape, which could allow it to again catalyze the cleavage of another amide bond through an intramolecular substitution reaction process.

An alternative reaction mechanism involving intramolecular electrophilic-assisted hydroxide-ion-nucleophilic attack at carbonyl carbon (TS$_{10}$), shown in Scheme 2.18, may be ruled out because of the fact that *N*-methylphthalimide has been detected spectrophotometrically as an intermediate during the course of the reaction.

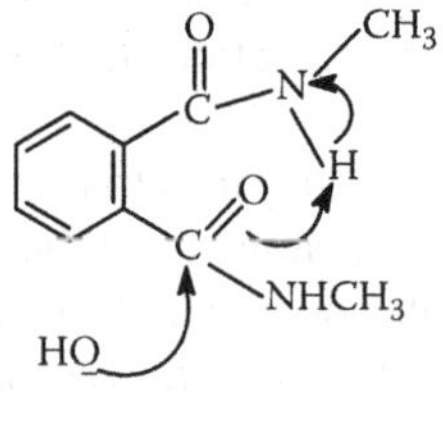

TS$_{10}$

If both reactions and so-called catalytic sites are the parts of the same molecular skeleton and, at the end of the reaction process, the reaction site or both the reaction and so-called catalytic sites are transformed into other molecular forms (i.e., products), then such a reaction should be called an *intramolecular reaction* and not intramolecular catalysis. Even if the structural features of the so-called catalytic site remain unchanged during the conversion of reaction site into product, it cannot further catalyze reaction intramolecularly, because the reaction site is no longer part of the same molecule. Unfortunately, the literature is not completely free from the ambiguous usage of intramolecular reaction and intramolecular catalysis.

Since the discovery of catalytic RNA molecules in the early 1980s, the mechanisms of the enzymatic and nonenzymatic hydrolysis of RNA have been of great interest.[96] Despite extensive efforts by many research groups around the

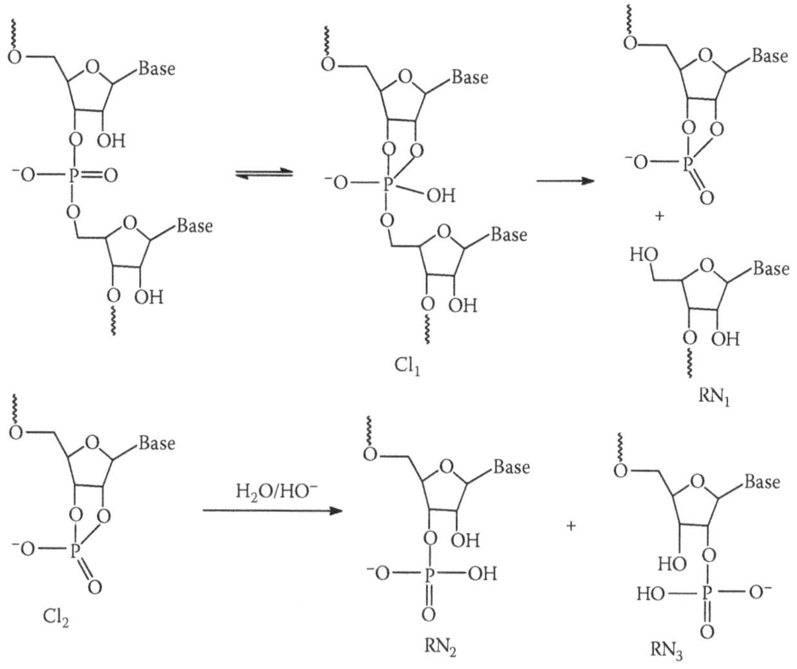

SCHEME 2.19

world over nearly two decades, the mechanistic details of even much simpler nonenzymatic hydrolysis of RNA is not yet fully understood.[97] However, perhaps the salient features of the mechanism of nonenzymatic hydrolysis of RNA is shown in Scheme 2.19 (for simplicity, the reaction steps involving external catalysts are not included in the reaction scheme). Although highly reactive intermediate CI_1 is plausible within the domain of our knowledge about related model reactions, it has not been detected experimentally under both acidic and alkaline pH. The indirect proof for its existence, only in acidic pH, comes from the identification of the migration product with 2′,5′-linkage RNA to 3′,5′-linkage RNA. The rate enhancement due to 2′-OH participation in the expulsion of the 5-oxygen of the next nucleotide to produce cyclic phosphodiester, CI_2, and cleaved RNA, RN_1, is difficult to quantify because of the lack of precise data (especially kinetic data) for such purpose. The intermediate cyclic phosphodiester, CI_2, is hydrolyzed in the second step to yield products RN_2 and RN_3 (Scheme 2.19). It is evident from Scheme 2.19 that the role of 2′-OH group in the hydrolysis of RNA cannot be considered that of an intramolecular nucleophilic catalyst, because the regenerated 2′-OH group in both RN_2 and RN_3 may not be able to accelerate the rate of expulsion of 5′-oxygen of the next nucleotide in the same or different RNA molecule. Thus, Scheme 2.19 shows that the enhanced hydrolysis of RNA due to participation of 2′-OH group is merely an intramolecular nucleophilic reaction.

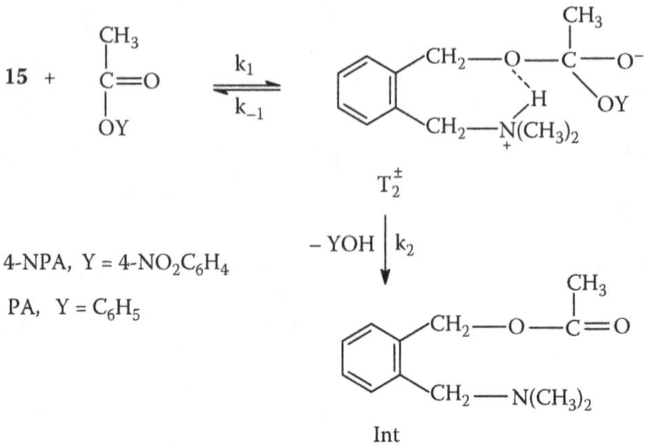

SCHEME 2.20

2.5.4.3.1 Intramolecular Induced Nucleophilic Catalysis

An example of intramolecular induced nucleophilic catalysis is found in the aqueous cleavages of activated esters such as 4-nitrophenyl acetate (4-NPA) and phenyl acetate (PA) in buffer solutions of 2-(N,N-dimethylaminomethyl)benzyl alcohol (15).[52] Pseudo-first-order rate constants (k_{obs}), obtained for the formation of 4-nitrophenoxide ion from 4-NPA or phenoxide ion from PA at different pH and concentrations of free amine base, fit to Equation 2.20. The calculated values of k_b for 4-NPA and PA are 0.057 M^{-1} sec^{-1} and 0.072 M^{-1} sec^{-1}, respectively. But the values of k_b for 2-(N,N-dimethylaminomethyl)benzyl methyl ether (16) and N,N-dimethylbenzylamine are not different from zero, which is ascribed to the presence of steric hindrance in the nucleophilic attack by tertiary amine nitrogen at carbonyl carbon of esters. These observations suggest that significantly large buffer catalysis with 15 is not due to amino group of 15 acting as nucleophile. Studies on pK_a determination of 15 and 16 as well as product characterization of reaction mixture reveal the presence of extensive internal hydrogen bonded form of 15 (17) in aqueous solution and hydroxyl group of 15 acting as nucleophile. The mechanism for 15-catalyzed cleavages of 4-NPA and PA is shown in Scheme 2.20.

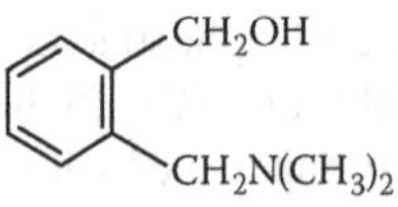

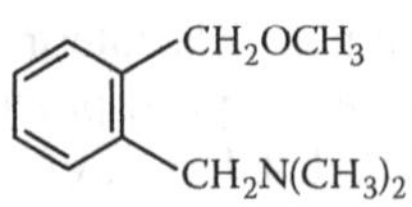

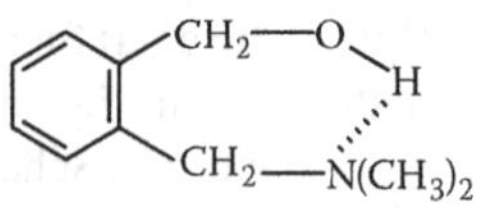

The deuterium oxide solvent isotope effect of 1.18 for 4-NPA reveals that significant proton transfer is not involved in the rate-determining step.[52] This observation suggests the k_2-step as the rate-determining step (Scheme 2.20), because the proton transfer catalysis occurs in both k_1- and k_{-1}-steps and there is no demand for proton transfer catalysis in k_2-step owing to the low pK_a (~ 7) of the conjugate acid of 4-nitrophenoxide ion. It is evident from Scheme 2.20 that the inhibitory mass law effect inhibition on k_{obs}, which is found in reactions that obey the reaction mechanism shown by Scheme 2.12, cannot be expected to occur in reactions that follow the reaction mechanism shown by Scheme 2.20. The study intended to check the inhibitory mass law effect inhibition on k_{obs} for **15**-catalyzed cleavage of 4-NPA reveals the absence of such an effect. It is perhaps noteworthy that the internally hydrogen-bonded amine **17** could have its hydrogen-bonding proton in a series of potential dips. The proton has an energetically restricted choice of lying near the nitrogen (with essentially very near to complete unit charge development on oxygen and nitrogen, **18**). The conformers **17** and **18** constitute the two extremes of the spectrum of the internal hydrogen bonding in **15**. Actually, **18** is the internally hydrogen-bonded form of zwitterions, and it exists in equilibrium with internally nonhydrogen-bonded form of zwitterions, **19**. The equilibrium concentrations of various conformers vary in the order **17** > --- **18** > --- **19**, whereas the nucleophilicity of hydroxylic oxygen varies in the reverse order. The observed nucleophilic second-order rate constant, k_b, is actually the sum of the contributions made by all conformers including **17**, **18**, and **19**. The contribution to k_b owing to **19** is only about 2%. It may be noted that the presence of various possible conformers (**17** through **19**) constitutes a kind of process that can take place with serine esterases such as α-chymotrypsin.

The intermediate, Int, in Scheme 2.20 is a so-called nonactivated ester and, hence, it should be stable enough to be detected experimentally. However, the same structural features that make the activation energy for going from reactants to $T_2^{\ddagger}$ particularly low must lower the activation energy for the reverse process by exactly the same amount. IR spectrophotometric technique reveals the formation and decay of Int with the progress of the reaction as shown in Figure 2.4.[98] The brief reaction

18   **19**

Scheme for this chemical process is given by Equation 2.30

$$\text{4-NPA} + \textbf{15} \xrightarrow[\text{-4-NO}_2\text{C}_6\text{H}_4\text{OH}]{k_{1obs}} \text{Int} \xrightarrow{k_{2obs}} \text{CH}_3\text{COOH} + \textbf{15} \qquad (2.30)$$

where k_{1obs} and k_{2obs} represent pseudo-first-order rate constants for the formation and decay of Int, respectively. The values of nucleophilic second-order rate constants, k_n, at 25°C, calculated by dividing $k_{1obs} - k_0$ (where k_0 is the pseudo-

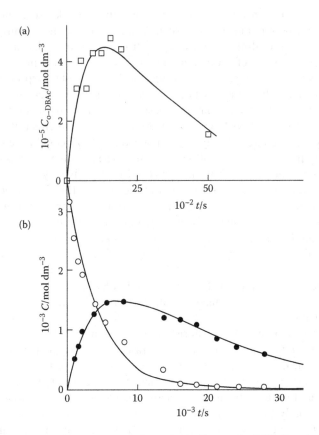

FIGURE 2.4 Plots showing the variation of (a) $C_{o\text{-DBAc}}$ {o-(N,N-dimethylaminome-thyl)benzyl acetate} with time for a kinetic run using the acetohydroxamate method ($\square$) and (b) C_{PNPA} (p-nitrophenyl acetate} (O) and $C_{o\text{-DBAc}}$ ($\bullet$) with time for a kinetic run using the I.R. method.[98] (Reproduced with permission of the Royal Society of Chemistry.)

first-order rate constant for hydrolysis of 4-NPA) by the concentrations of free amine base, are about 0.04 to 0.05 M^{-1} sec^{-1}, which may be compared to the k_n value (= 0.05 M^{-1} sec^{-1}) obtained by the more reliable UV-visible spectrophoto-metric technique.

The values of k_{2obs} at pH 8.68 – 8.33 and 9.09 – 8.70 give the values of k_{OH} (second-order rate constant for the reaction of HO$^-$ with Int) of ~ 10 -18 M^{-1} sec^{-1}. These values are based on the assumption that the pseudo-first-rate constant k_w (for the reaction of H$_2$O with Int) is negligible compared with k_{OH}[HO$^-$]. The values of k_{OH} are nearly 100 times larger than the corresponding value of 7.5 × 10^{-2} M^{-1} sec^{-1} for the cleavage of ethyl acetate under similar experimental con-ditions.[53] The polar effect of 2-CH$_2$N(CH$_3$)$_2$ group is concluded to have almost no effect on the reactivity of Int towards HO$^-$. The most likely explanation for ~ 100-fold larger reactivity of Int than that of ethyl acetate towards HO$^-$ may be attributed to the occurrence of intramolecular electrophilic induced nucleophilic reaction involving transition state TS_{11} (if the expulsion of leaving group is the rate-determining step) or TS_{12} (if the nucleophilic attack is the rate-determining step). Transition state TS_{11} is kinetically indistinguishable from TS_{13}, which involves intramolecular general base induced nucleophilic reaction.

Buffer solutions of *2-endo*-[(dimethylamino)methyl]-*3-endo*-(hydroxyme-thyl)bicyclo[2.2.1]hept-5-ene (**20**) show catalytic efficiency in acylation [i.e., k_1-step in Equation 2.30] and deacylation [i.e., k_2-step in Equation 2.30] in the presence of 4-NPA, which is almost similar to that of **15**.[99] Thus, appropriately positioned amino and hydroxyl groups in amino alcohols (**15** and **20**) exhibit significant catalytic efficiency in both acylation and deacylation processes in the presence of activated esters. But these amino alcohols fail to show detectable catalytic effects in acylation with nonactivated esters such as methyl formate and acetylcholine chloride, and the reason for such observations is given in terms of a plausible mechanism for the acylation process.[100]

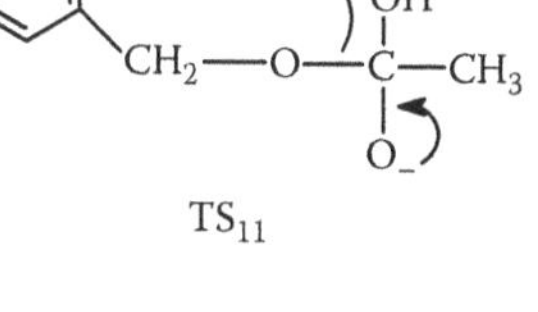

TS_{11} 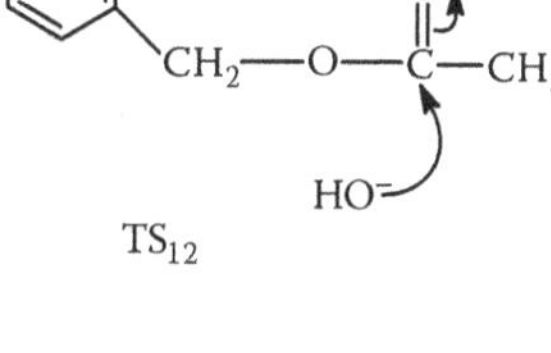 TS_{12}

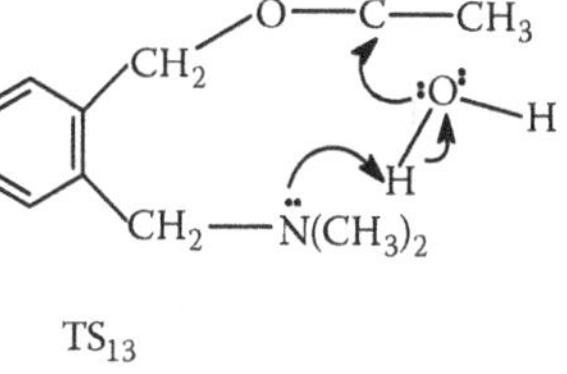

TS_{13} 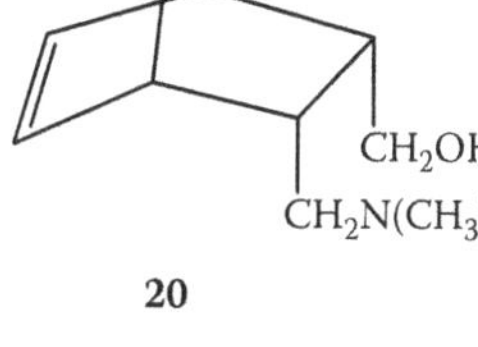   **20**

2.5.4.3.2   Induced Intramolecular Nucleophilic Catalysis

Nature's most efficient catalysts often bring about the maximum possible reduction in the overall free energy of activation for both simple and extremely complex reactions. The molecular machinery involved in reactions of these catalysts is fascinating and awesome, and gives rise to the idea of designing synthetic catalysts that could mimic some, if not all, the characteristic molecular features of Nature's catalysts. Perhaps, this is the fastest way to come up with an efficient catalyst for a specific purpose. In order to synthesize such a cost-effective catalyst, it is essential to know the fine details of the mechanism of the reaction catalyzed by Nature's catalyst. This aspect of such effort is indeed extremely difficult, if not impossible, to understand. Some motivated and enthusiastic researchers or groups of researchers are trying to develop such catalysts. However, these catalysts, having complex molecular structures, utilize several modes of catalytic processes (Figure 2.3) in a concerted or stepwise manner, including induced intramolecular nucleophilic catalysis.

Antibody-catalyzed cleavage of an amide bond has been studied by incorporating to antibodies, some selective catalytic modes such as selective binding of TS analogs,[101,102] proximity effects supplied by a metal cofactor,[103] intramolecular rearrangements,[104,105] and complementary acid–base catalysis.[106] In an interesting report on antibody-catalyzed cleavage of an amide bond, a few antibodies with external nucleophilic cofactor (phenol) catalyze the rate of cleavage of propionyl 4-nitroanilide (**21**) at pH 8.0 by a factor of nearly 100 compared to the rate of uncatalyzed hydrolysis of **21** at pH 8.0.[107] It is also found that the rate constants for uncatalyzed hydrolysis of **21** at pH 8.0 remain unchanged with the change in phenol concentration, indicating that phenol does not participate in the uncatalyzed hydrolysis of amide **21**. Though the exact mechanism for this catalytic process is difficult to know, the suggested and, probably, the most plausible mechanism involves hydrophobically induced intramolecular nucleophilic attack by phenolic oxygen at carbonyl carbon of **21** as shown in Scheme 2.21.

It is interesting to note that the aqueous pK_a values of conjugate acids of nucleophile phenol (C_6H_5OH) and leaving group $C_6H_5NH^-$ are nearly −7 and 25, respectively,[108] and replacement of 4-H by 4-NO_2 group in $C_6H_5NH^-$ cannot reduce its pK_a very much. Thus, the effects of catalytic antibodies on TS_{14} is such that the apparent or effective pK_a of conjugate acid of nucleophile (phenol) become larger than that of the leaving group (4-$NO_2C_6H_4NH^-$). How the catalytic antibodies cause the apparent change in the relative pK_a of nucleophile and leaving group from nearly 32 to less than zero is something fascinating and essential to understand.

The lack of product inhibition on rate of catalyzed cleavage of amide bond is attributed to probable free diffusion of product **22** out of the binding pocket of antibody. The value of the pseudo-first-order rate constant (k) for hydrolysis of **22** in the buffer employed (k = 2.0×10^{-4} min^{-1} at pH 8.0) is nearly 10 times larger than the pseudo-first-order rate constant (k_{cat}) for catalyzed reaction of phenol with **21**. Thus, the present antibodies with external phenol cofactor catalyzes the acylation step (i.e., reaction between phenol and **21**). But these catalysts have no effect on the deacylation step (i.e., hydrolysis of **22**), simply because **22** probably has no productive binding affinity with antibody catalysts.

2.5.5 ELECTROPHILIC CATALYSIS

If the electrophilic catalyst is a Lewis acid or proton, then this catalysis is also called *Lewis acid catalysis* or *proton catalysis*. Large amount of work on electrophilic catalysis involves metal ions as Lewis acids, and under such specific conditions, this catalysis is also called *metal ion catalysis*. Similarly, all the protic acids are called *Bronsted acids* and, hence, catalysis involving protic acids as

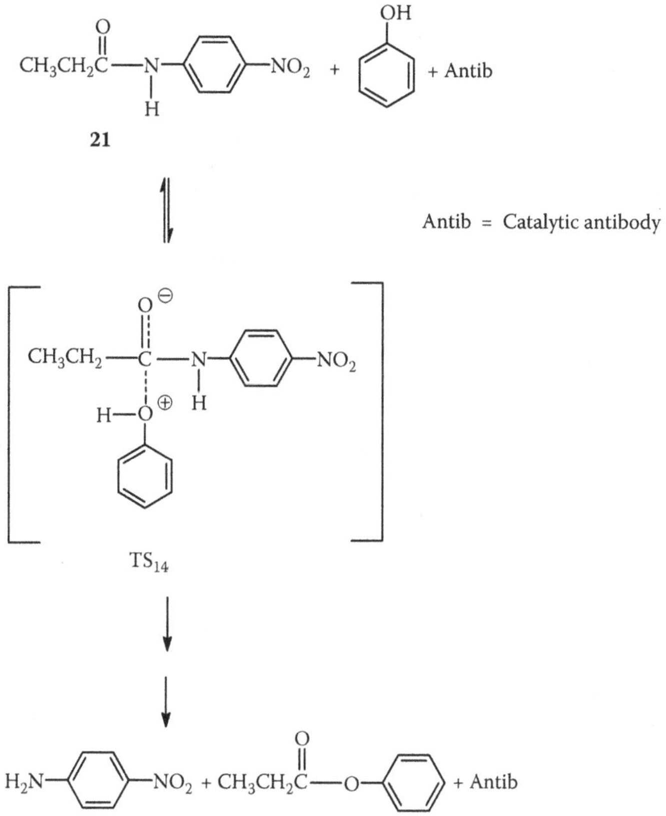

SCHEME 2.21

catalysts may be called *Bronsted acid catalysis*. A look at the vast literature on this topic reveals that all the Bronsted acids except water are named as *general acids (GA)* and proton as *specific acid (SA)*, whereas all the conjugate bases of Bronsted acids are named as *general bases (GB) or Lewis bases* and hydroxide ion as *specific base (SB)*. The SB–SA and GB–GA catalyses are described in Subsections 2.5.6 and 2.5.7 and Subsections 2.5.8 and 2.5.9, respectively, whereas electrophilic catalysis that includes metal ion catalysis and neutral Lewis acid catalysis is described in this section.

In such catalysis, the reaction or the catalytic site in a catalyst is always electron-deficient and, consequently, it accepts a pair of electrons from an electron donor. In the catalytic reduction of overall free energy of activation for positive catalysis, the predominant destabilization of GS or predominant stabilization of TS must involve the pair of electron transfers (partially or fully) from the reaction site in the reactant (substrate) to the reaction site in the catalyst (neutral/cationic atom/molecule). The interaction between catalyst and reactant should involve weak, ionic, or semiionic interactions (if the reaction medium is water) so that the catalyst could be easily regenerated at the end of the reaction. Ionic or semiionic bonds are generally less stable compared to the corresponding neutral covalent bond in a polar solvent such as water. The reaction between the reactant with neutral reaction site and electrophilic catalyst yields intermediate product/molecular complex, which might be less stable than the reactant because of the following probable reasons: (1) development of charge separation or loss of resonance in reactant or catalyst or in both, and (2) high charge-stabilizing power of the water solvent makes the polar bond less stable compared to the nonpolar covalent bond in the water solvent.

2.5.5.1 Intermolecular Electrophilic Catalysis

2.5.5.1.1 Metal Ion Catalysis

Metal ions are known to play an important role in many catalytic biological and nonbiological reactions.[109] The mechanistic essence of metal ion catalysis may be described as follows. Let us consider a simple metal ion (M^+)-catalyzed nucleophilic substitution reaction (Equation 2.31)

$$: Y - Sub - X : + : W - H \xrightarrow{M^+} : Y - Sub - W : + : X - H \qquad (2.31)$$

where electrophilic reactant :Y – Sub – X: reacts with nucleophile: W – H to produce substitution products :Y – Sub – W: and :X – H. The electrophilic centre in :Y – Sub – X: lies in segment Sub and Y, X, and W represent atoms or group of atoms in reactants, which contain a lone pair of electrons. One of the possible reaction mechanisms by which metal ions catalyze the reaction is shown in Scheme 2.22, where $\delta+$ and $\delta-$ represent less than unit + and – charge, respectively.

In Scheme 2.22, it is arbitrarily assumed that the metal ion, by virtue of its electrovalent bonding interactions with one or both reactants, destabilizes the ground states of reactants (i.e., $K_{as1} < 1$, $K_{as2} < 1$, and $K_{as3}K_{as3}^a < 1$) more strongly than the corresponding transition states. It is quite possible that the same molecular interaction, which destabilizes the GS, might also stabilize the TS of the reaction, thus causing more significant positive catalysis. It is also possible that there is no significant metal ion effects on the GS (i.e., $K_{as1} = K_{as2} = K_{as3}K_{as3}^a \approx 0$), and metal ion catalysis results owing to predominant stabilization or destabilization of the TS by metal ions. The ionized metal ion–nucleophile complex, $W^\delta–M^{\delta+}$, is an efficient bifunctional catalyst that provides semiintramolecular electrophilic catalytic assistance to nucleophilic attack at the electrophilic center. However, metal ion catalysis also requires that the products in the catalyzed reaction should have either weaker or no nucleophilic affinity compared to reactants and the TS toward electrophilic metal ion catalyst so that the catalyst cannot be consumed by the products.

The catalytic effects of the alkali-metal ions in the ethanolysis of 4-nitrophenyl diphenyl phosphate (**23**) with 4-nitrophenoxide as leaving group reveals the

$$:Y - Sub - X: \; + \; M^+ \; \underset{}{\overset{K_{as1}}{\rightleftharpoons}} \; :Y - Sub - X^{\delta+} - M^{\delta+}$$

$$:Y - Sub - X: \; + \; M^+ \; \underset{}{\overset{K_{as2}}{\rightleftharpoons}} \; :X - Sub - Y^{\delta+} - M^{\delta+}$$

$$H - W: \; + \; M^+ \; \underset{}{\overset{K_{as3}}{\rightleftharpoons}} \; H - W^{\delta+} - M^{\delta+}$$

$$H - W^{\delta+} - M^{\delta+} + H_2O \; \underset{}{\overset{K_{as3}{}^a}{\rightleftharpoons}} \; W^{\delta-} - M^{\delta+} \; + \; H_3O^+$$

$$:Y - Sub - X: \; + \; W^{\delta-} - M^{\delta+} \; \overset{k_1}{\longrightarrow} \; :Y - Sub - W: + X^- + M^+$$

$$:Y - Sub - X^{\delta+} - M^{\delta+} + W^{\delta-} - M^{\delta+} \; \overset{k_2}{\longrightarrow} \; :Y - Sub - W: + X^- + 2\,M^+$$

$$M^{\delta+} - Y^{\delta+} - Sub - X: + W^{\delta-} - M^{\delta+} \; \overset{k_3}{\longrightarrow} \; :Y - Sub - W: + X^- + 2\,M^+$$

SCHEME 2.22

following order of catalytic reactivity: $Li^+ > Na^+ > K^+ > Cs^+$.[110] Pseudo-first-order rate constants (k_{obs}) for ethanolysis of **23** follow Equation 2.32, where $EtO^- = CH_3CH_2O^-$. Equation 2.32 can be derived from the reaction mechanism

$$k_{obs} = k_{EtO^-}[EtO^-] + k_{MOEt}K_{as}[EtO^-]^2 \qquad (2.32)$$

shown in Scheme 2.23.

The respective values of k_{MOEt} for $M^+ = Li^+$, Na^+, K^+, and Cs^+ are nearly 14-, 11-, 9-, and 4-fold larger than that of k_{EtO^-} (= $9.4 \times 10^{-2}\ M^{-1}\ sec^{-1}$). These results are amazing in the sense that the nucleophilicity of EtO^- in ion-pair MOEt must be significantly lower than free EtO^- even in water solvent, because the reported values of K_{as} are 212 M^{-1} for LiOEt, 102 M^{-1} for NaOEt, 90 M^{-1} for KOEt, and 121 M^{-1} for CsOEt.[110] Significantly larger values of k_{MOEt} compared to that of k_{EtO^-} have been explained by suggesting that MOEt ion-pair is one in which the metal cation coordinates to the phosphoryl oxygen (i.e., acting as an electrophilic

$$EtO^- \; + \; M^+ \; \underset{}{\overset{K_{as}}{\rightleftharpoons}} \; MOEt$$

$$23 \; + \; EtO^- \; \overset{k_{EtO^-}}{\longrightarrow} \; Products$$

$$23 \; + \; MOEt \; \overset{k_{MOEt}}{\longrightarrow} \; Products$$

SCHEME 2.23

catalyst) concertedly with nucleophilic attack by ethoxide in a 4-membered transition state TS_{15}.

The catalysis by alkali-metal ions with the selectivity order $Li^+ > Na^+ > K^+ > Cs^+$ is also found in the ethanolysis of 4-nitrophenyl diphenylphosphinate (**24**).[111] But, in the ethanolysis of 4-nitrophenyl benzenesulfonate (**25**), Li^+ brought about an inhibition of rate, and the catalytic order of the other alkali metals was $K^+ > Cs^+ > Na^+$.[112] These observations have been explained in terms of calculated free energy difference between the free energy of TS and free energy of metal ion stabilization of ethoxide ion.[110] Nucleophilic demethylation of tri-CH_3 (**26**) and di-CH_3 2-pyridylmethyl (**27**) phosphates by iodide is catalyzed by alkali-metal ions and the effect increasing in the order $K^+ < Na^+ < Li^+$.[113] The catalytic effect is more pronounced for **27**. The crown ethers reduce, but do not eliminate the catalysis, indicating strong interactions between the reaction TS and the metal ions. Similar effects of crown ethers are observed in the ethanolysis of **23** in the presence of alkali metals.[110]

The apparent second-order rate constant for hydroxide ion attack on trimethyl phosphate (**26**) is increased 400-fold when the phosphoryl moiety is coordinated to an iridium(III) center.[114] Wadsworth has presented evidence that $ZnCl_2$-mediated methanolysis of phosphate triesters involves metal interaction with phosphoryl oxygen and the leaving-group oxygen.[115] The divalent metal ions, Mg^{2+} and Ca^{2+}, catalyze the reaction of 4-nitrophenyl phosphate dianion (**28**) with substituted pyridines but do not increase the association character of the TS, as indicated by the values of the Bronsted coefficient, β_{nuc}, in the range 0.17 to 0.21 for **28** and for its metal ion complex.[116] But the reactions of phosphorylated morpholinopyridine (**29**) are inhibited ~ 2-fold by Mg^{2+} and Ca^{2+}. The kinetic data reveal the formation of 1:1 complexes between divalent metal ions and substrates (**28 and 29**) with association constant K_{as} (**28**·Mg) = 14.8 M^{-1} and K_{as} (**28**·Ca) = 7.5 M^{-1} at 39.2°C; K_{as} (**29**·Mg) = 4.4 M^{-1} and 9.1 M^{-1} at 25.1 and 52.3°C, respectively; and K_{as} (**30**·Mg) = 6.2 M^{-1} and K_{as} (**30**·Ca) = 6.0 M^{-1} at 39.1°C. The three- to sixfold positive catalysis by Mg^{2+} and Ca^{2+} of the reaction of **28** with pyridines is in contrast with ~ twofold negative catalysis of reactions with uncharged leaving groups, suggesting that the bound metal ion interacts with the phenoxide ion leaving-group as well as the phosphoryl oxygen atoms in the TS.

2,4-Dinitrophenyl phosphate dianion (**30**·Ca) reacts with pyridines sevenfold faster than free **30**, and a constant value of β_{nuc} has been found for the reactions of **30** and **30**·Ca with two substituted pyridines.[116] But the reactions of **30**·Mg with pyridines are ~ 8-fold slower than the reactions of free **30** because of an unfavorable interaction of Mg^{2+} and the ortho nitro group of **30** in the TS. The reasons for characteristic different effects of Mg^{2+} and Ca^{2+} on the rates of reactions of **30** with pyridines are unclear. The second-order rate constants of the Mg^{2+} complexes of phosphorylated pyridine monoanions (**31**) with $Mg(OH)^+$ were 10^4- to 10^6-fold larger than the second-order rate constants for their reaction with water.[117] Of the 10^6-fold rate enhancement with the phosphorylated 4-morpholinopyridine-Mg^{2+} complex, approximately 10^4-fold is attributed to the greater nucleophilicity of $Mg(OH)^+$ compared to water. The remaining catalysis of approximately 10^2-fold is attributed to induced intramolecularity from positioning of HO^- and the phodphoryl group by Mg^{2+}.

28, X = H

30, X = NO_2

29, X = N

31, X = H

Lanthanide(III) and other elements in lanthanide series are very effective catalysts for the hydrolysis of the phosphodiester linkages in RNA, whereas nonlanthanide metal ions are virtually inactive.[118] The pseudo-first-order rate constant for the hydrolysis of adenylyl-(3'→5')adenosine (ApA) by $LuCl_3$ (5 mM) at pH 7.2 and 30°C is 0.19 min^{-1}, which gives ~ 10^8-fold rate acceleration compared to rate of hydrolysis of ApA under the same reaction conditions but in the absence of $LuCl_3$. The product is an equimolar mixture of adenosine and its 2′- or 3′-monophosphate without any by-products.[118] The rates of iodination of 2-acetylimidazole (**32**) are independent of iodine concentration and involve SA and GA–GB catalysis. In SA and GA catalysis, the enolization occurs with N-protonation rather than O-protonation. The complex formation between **32** and Zn^{2+} or Cu^{2+} ions increases the rates of the water- and acetate ion-catalyzed cleavage of **32** by the respective factors of 5.3×10^2 and 2.3×10^2 (with Zn^{2+}) and 8.1×10^3 and 1.2×10^3 (with Cu^{2+}) compared to the corresponding rates of reactions of the uncomplexed **32**.[119]

It is well known that the biocatalysts such as enzymes utilize more than one mode of catalysis, and this fact inspires to develop synthetic catalysts with multimode of catalysis.[120–122] In this vein, scientists have developed several catalysts, which are essentially metal ion complexes with diverse nature of metal ions and ligands. Such synthetic catalysts not only provide probable multimode of catalysis, but they also maintain the stability of the electrophilic nature of metal ions under specific conditions in which metal ions cannot retain their electrophilic catalytic characteristic. For instance, catalytic zinc-based hydrolytic agents undergo precipitation of polymeric zinc hydroxide above neutral pH. But Zn(II) complex with **33** retains its catalytic efficiency up to more than pH 10. Although most of these synthetic catalysts are highly efficient toward the cleavage of phosphate diester bond of RNA and RNA model compounds, almost all except a few[123–125] of them lack bond or substrate selectivity in their catalytic actions. One of the several possible reasons for such deficiency of synthetic catalysts is the lack of our critical understanding about the microreaction environment of catalytic RNA cleavage in terms of size/volume, polarity, and water concentration. It may not be correct to assume that polarity and water concentration of the microreaction environment of RNA hydrolysis and bulk water solvent are the same. These medium effects are expected to change effectively the rates of certain reactions by modifying either the reactant state or the TS or both. A recent study reveals that depending on the sequence of DNA, the water activity of the minor groove is, at least, two- to fourfold lower than bulk solution.[126] In a study on medium-controlled intramolecular catalysis in the direct synthesis of 5′-O-protected ribonucleoside 2′- and 3′-dialkylphosphate, it has been found that nonaqueous media change the nature of participation of the *cis*-vicinal hydroxyl group from nucleophilic to electrophilic.[127] This finding is not unusual because the change in the content of organic cosolvent in mixed aqueous solvents causes significant change in the ionization constant of nonisoelectric ionization reactions.[128]

The zinc complex of a tetraaza macrocycle (**33**) catalyzes the hydrolysis of **23** in 50% v/v aqueous CH_3CN.[129] Both Zn(II).**33**(Br)(ClO_4) and (Zn(II).**33**)$_2$(ClO_4)$_3$(OH) do not differ significantly in their catalytic effects toward the hydrolysis of **23**. The values of second-order rate constants k_2^{Zn} for the reactions of **23** with Zn(II).**33**(ClO_4)(OH) and HO⁻ are 2.8×10^{-1} and 2.8×10^{-2} M^{-1} sec⁻¹, respectively. Nearly 10-fold larger reactivity of Zn(II).**33**(ClO_4)(OH) (pK_a of its conjugate acid [Zn(II).**33**(ClO_4)(OH_2)]⁺ is 8.7 in pure aqueous solvent) compared to that of HO⁻ is attributed to the following probable effects: (1) some remarkable polarizability effects required to make Zn(II).**33**(ClO_4)(OH) a better simple nucleophile, (2) bifunctional mechanism in which zinc acts as a Lewis acid, which increases the nucleophilic attack by OH group of Zn(II).**33**(ClO_4)(OH) in a four-membered TS similar to TS_{15}, and (3) **23** forms a complex with Zn that is then attacked by free hydroxide.[129] Tetravalent cations such as Ce(IV), Th(IV), and Zr(IV) are among the most active catalysts for phosphate diester cleavage including RNA and DNA.[130] Macrocyclic tetraamide complex of Th(IV) binds phosphate diesters and catalyzes the cleavages of RNA and **34** with respective pseudo-first-order rate constant (k_{obs}) values of 7.5×10^{-4} and 9.2×10^{-4} sec^{-1} at pH 7.3, 37°C and 1.0 mM Th(IV) complex.[131] These values of k_{obs} are nearly 10^5- and 10^3-fold larger than k_{obs} for uncatalyzed hydrolysis of RNA and **34**, respectively.

Copper(II) complexes of various cyclic amino, bi-, and tripyridyl ligands increase the rate of cleavage of diadenosine 5′,5′-triphosphate (DT) compared to an uncatalyzed one by a factor of more than 2×10^4 under neutral pH, whereas Zn(II).[12]aneN$_3$ complex is clearly the less active catalyst and Mg(II) does not enhance the cleavage of DT to any significant extent.[132] Although the data are not sufficient to suggest a conclusive mechanism, the inability of Mg(II) to catalyze the reaction might be seen in terms of pK_a values of M^{2+}–OH$_2$, pK_a = 12.8 for M^{2+} = Mg^{2+}, pK_a = 7.3 for M^{2+} = Zn^{2+}[12]aneN$_3$, and pK_a = 7.3 for M^{2+} = Cu^{2+}[9]aneN$_3$ (where [9]aneN$_3$ = 1,4,7-triazacy-clononane).[132] In search of catalysts, which could show the bond-selectivity for the hydrolysis of RNA and RNA model compounds, trinuclear Cu(II)com-plex of some typical trinuclear ligands hydrolyzes 2′-5′ and 3′-5′ ribonucle-otides with remarkable substrate selectivity and catalytic effectiveness.[125] The cooperation action of multiple Cu(II) nuclear centers is shown to be essential for effective and bond-selective hydrolysis of ribonucleotides. Remarkable cooperativity between a Zn(II) ion and guanidinium/ammonium groups is attributed to enhanced hydrolysis of the RNA dimer adenylyl phosphoadenine

(ApA) in the presence of specific designed Zn(II) complexes.[124] Cooperativity in these catalyzed reactions is proposed to increase both the electrophilicity of the electrophilic center for nucleophilic attack by 2'-OH group, which is activated by OH group bound to Zn(II), and the leaving group ability of the leaving-group in the cleavage of the P–O bond.

2.5.5.1.2 Neutral Lewis Acid and Rate Activation

Certain compounds such as neutral Lewis acids, $B(OH)_3$ and $AlCl_3$, are needed for the occurrence of certain reactions, but they do not appear to be part of the actual reaction processes. For example, Friedel–Crafts reaction, Equation 2.33, cannot occur in the absence of Lewis acid, such as $AlCl_3$, but the product HCl deactivates $AlCl_3$ by converting it into $AlCl_4^-H^+$. So, $AlCl_3$ does not satisfy the strict definition of a catalyst. But, because without its presence in the reaction mixture, the reaction cannot occur; so it may be called a *rate activator* or *rate promoter*. Although the Friedel–Crafts reaction was discovered in 1877 and since then a huge amount of work has been carried out on this reaction, the fine details of the mechanism of this reaction are not yet fully understood. Friedel–Crafts acylation of alkoxybenzenes is achieved efficiently by reaction with aliphatic acid anhydrides in the presence of catalytic amounts of ferric hydrogen sulfate, $Fe(HSO_4)_3$ in nitromethane.[133]

Occasionally, the presence of boric acid in the reaction mixture increases the rate of reaction because of the change in the nature of reaction from intermolecular reaction to intramolecular reaction owing to complex formation between the reactant and boric acid. Such rate enhancement is sometimes described as *induced catalysis*. This terminology is correct only if boric acid has lower (compared to the reactant) or no binding or reaction affinity with the products. When products react or complex with boric acid with the same or greater affinity as with the reactants, then such a rate enhancement is more correctly described as *boric-acid-induced intramolecular nucleophilic reaction*.

The presence of $B(OH)_3$ increases the rate of hydrolysis of ionized phenyl salicylate (PS^-) by nearly 10^6-fold compared to the rate of hydrolysis of phenyl benzoate under essentially similar conditions (Equation 2.34). However, the hydrolysis of PS^- can also occur with measurable rate in the absence of $B(OH)_3$. Nearly 10^6-fold rate enhancement due to the presence of boric acid is attributed to the boric-acid-induced intramolecular reaction involving transition state TS_{16}.[134] An alternative and kinetically indistinguishable mechanism involving transition state TS_{17}[135,136] has been ruled out on the basis of the absence of enhanced nucleophilic reactivity of tertiary and secondary amines toward phenyl salicylate in the presence of borate buffer.

$$\text{\raisebox{0pt}{\includegraphics}} + CH_3COCl \xrightarrow[\text{benzene}]{AlCl_3} \quad + \quad HCl \tag{2.33}$$

$$PS^-$$

(2.34)

It may be argued that for the reaction of HO⁻ with phenyl esters, the attack is the rate-determining step, and for the reactions of amines with phenyl esters, the leaving-group expulsion is the rate-determining step. Therefore, the possibility that nucleophilic attack is assisted by borate through the TS similar to TS_{17} is not excluded by the absence of a reaction of amines for which attack is not the rate-determining step. Although this argument seems to be reasonable in principle, it can be easily shown that even if breakdown of the tetrahedral intermediate is the rate-determining step, a rate enhancement would still be expected by the presence of borate buffer provided that the partitioning of the tetrahedral addition intermediate does not change. It is apparent that the partitioning of the tetrahedral addition intermediates **35** and **36** would be expected to be the same for a common nucleophile $R_1R_2R_3N$. The probable steric effect for the absence of enhanced reactivity of a few tertiary amines with PS⁻ in the presence of borate buffer has also been ruled out.[134] The rate of reaction of H_2O with ionized 4-nitrophenyl 2,3-dihydroxybenzoate (NDHB)[137] and PS⁻ reveal very large rate enhancement compared to rate hydrolysis of 4-nitrophenyl and phenyl benzoates because of intramolecular GB assistance. Unlike the hydrolysis of PS⁻, boric acid inhibits the rate of hydrolysis of NDHB[137], which could be explained in terms of equilibrium formation of NDHB–boric acid complex (**37**). The geometrical feature of **37** eliminates the possibility of the occurrence of boric acid-induced intramolecular nucleophilic reaction through a transition state similar to TS_{16}.

$$TS_{16} \qquad TS_{17}$$

35   **36**   **37**

2.5.5.2 Intramolecular Electrophilic Catalysis

The rate of a bimolecular nucleophilic or electrophilic addition, substitution, elimination, or addition–elimination reaction may be increased by several orders in the presence of an electrophilic catalyst (E^+), such as metal ion, if E^+ is capable of forming a complex with both reactants (say electrophile:$Y - Sub - X$: and nucleophile:$W - H$) in an appropriate molecular geometry, which allows the reaction to proceed intramolecularly as shown in Equation 2.35.

$$:Y - Sub - X:$$
$$:Y - Sub - X: + :W - H + E^+ \rightleftharpoons | \qquad \rightarrow :Y - Sub - W: + :X - H + E^+$$
$$E^+ - :W - H$$

$$(2.35)$$

In an attempt to clarify some vexing points of a complex mechanism of ribozyme (with Mg^{2+} as cofactor)-catalyzed hydrolysis of RNA and RNA-model compounds, Bruice et al.[138] studied the rate of divalent metal ion (Mg^{2+}, Zn^{2+}, Cu^{2+}, and La^{3+})-promoted-hydrolysis of adenosine 3′-[(8-hydroxyquinolyl)methyl phosphate] (38) and adenosine 3′-[2-(8-hydroxyquinolyl)ethyl phosphate] (39). Pseudo-first-order rate constants (k_{obs}) for M^{n+}-promoted-hydrolysis of 38 and 39 follow kinetic Equation 2.36

$$k_{obs} = (k_0^{OHM} + k_M^{OHM}[M^{n+}])K_m[M^{n+}][HO^-]/(a_H + K_m[M^{n+}]) \qquad (2.36)$$

where $K_m = K_{as}K_a$ with K_{as} and K_a representing metal ion association constant of 38 and 39 and ionization constant of phenolic proton of 38 and 39, respectively, and k_0^{OHM} and k_M^{OHM} are respective second-order and third-order rate constants for uncatalyzed and M^{n+}-catalyzed reactions of (38 or 39)M^{n+} complex with HO^-.

The respective values of k_0^{OHM} for $M^{n+} = Mg^{2+}$, Zn^{2+}, Cu^{2+}, and La^{3+} are $\sim 10^6$-, 10^4-, 10^5-, and 10^9-fold larger compared to the second-order rate constant (k_0^{OH}) for the reaction of HO^- with 38 in the absence of M^{n+}. Almost similar results are obtained for 39. The values of k_M^{OHM} for respective $M^{n+} = Mg^{2+}$, Zn^{2+}, Cu^{2+}, and La^{3+} are 1.73×10^4, 1.28×10^5, ~ 0, and 2.87×10^7 M^{-2} sec^{-1} for 38 and for 39, these values are 7.93×10^3, ~ 0 and 2.29×10^7 M^{-2} sec^{-1} for Zn^{2+}, Mg^{2+}, and Cu^{2+}, respectively. It is interesting to note that only ~ 10-fold larger reactivity is reported for intermolecular reaction between 23 and divalent metal ion bound HO^-.[129]

Nearly 10^4- to 10^6-fold larger values of k_0^{OHM} compared with k_0^{OH} for $M = Zn^{2+}$, Mg^{2+}, and Cu^{2+} are attributed to increase in electrophilicity of phosphoryl phosphorous (P) owing to intramolecular electrovalent bonding interaction between anionic oxygen attached to P and 38-bound metal ion as shown in Scheme 2.24. Nearly 10^9-fold larger value of k_0^{OHM} compared with k_0^{OH} for $M = La^{3+}$ is attributed to greater enhancement of electrophilicity of P because of more extensive coordination between oxygen atoms attached to P and M^{n+} in complexes 38·M^{n+} and 39·M^{n+} when $M^{n+} = La^{3+}$, than when $M^{n+} = Zn^{2+}$, Mg^{2+}, and Cu^{2+} as shown in Scheme 2.24 and Scheme 2.25.

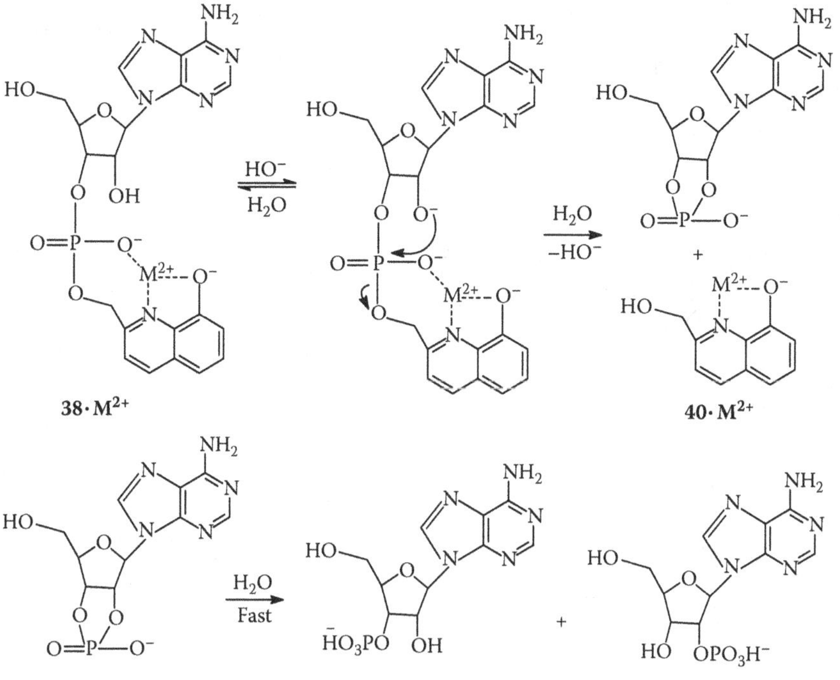

38·M²⁺

40·M²⁺

SCHEME 2.24

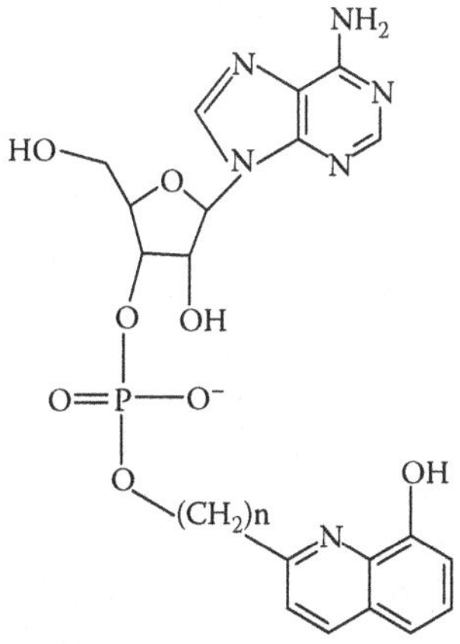

38, n = 1

39, n = 2

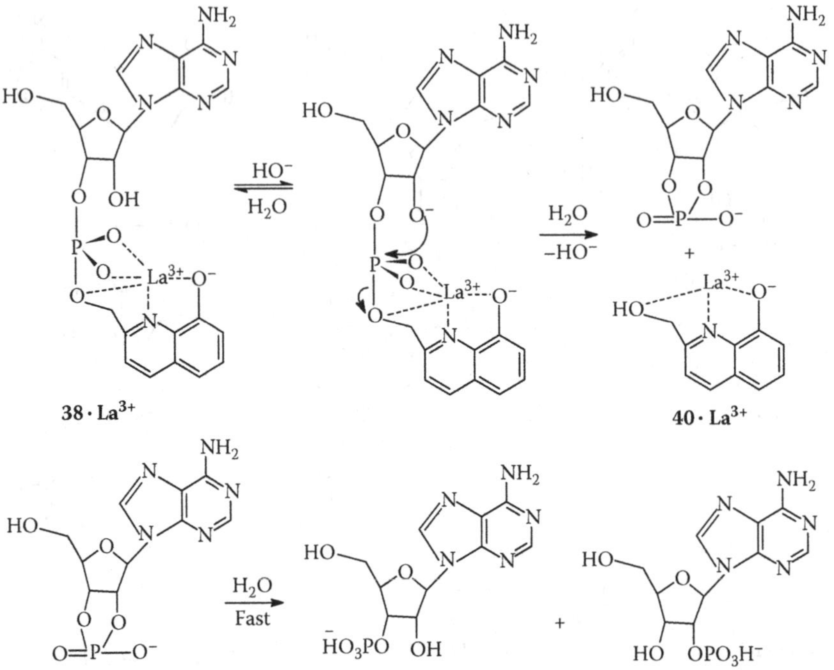

SCHEME 2.25

The values of k_M^{OHM} for hydrolysis of **38·M^{n+}** and **39·M^{n+}** complexes are strongly dependent upon the nature of M^{n+}. The hydrolysis of **38·Cu^{2+}** is not catalyzed by a second Cu^{2+}, whereas the value of k_M^{OHM} for the hydrolysis of **39·Cu^{2+}** is $\sim 10^7 \, M^{-2}$ sec^{-1} indicating some sort of geometrical feature of the substrate is important in determining the facility of the second metal ion catalysis in the instance of Cu^{2+}. The composition of the transition state involving two-metal-ion-promoted hydrolysis of **38** and **39** is $[(\mathbf{38, 39}) \cdot M^{n+}][M^{n+}][HO^-]$, which is kinetically indistinguishable from $[\mathbf{38, 39}) \cdot M^{n+}][M^{(n-1)+}OH]$. The metal-ion-promoted hydrolysis of complexes **38·M^{n+}** and **39·M^{n+}** is suggested to involve GB $M^{(n-1)+}OH$-assisted nucleophilic attack by 2'-OH at P of **38·M^{n+}** and **39·M^{n+}**. The values of metal ion (M^{n+}) association constants (K_{as}) of product **40** are slightly larger than the corresponding K_{as} values of **38** and **39**.[138] These results show that the product **40** is acting as an inhibitor and consequently M^{n+}-promoted hydrolysis is not a perfect catalytic process. However, the magnitude of intramolecular electrophilic catalysis can be correctly assessed from the values of k_0^{OHM}/k_0^{OH}.

The occurrence of intramolecular metal ion catalysis is reported in the hydrolysis of adenylyl(3'-5')adenosine (ApA) under the presence of diaquatetraazacobalt(III) complexes $[Co(N)_4(OH_2)_2]^{3+}$ (N: coordinated nitrogen atom) at pH 7.0.[139] Pseudo-first-order rate constants (k_{obs}) for ApA hydrolysis at 50°C and 0.1 M Co^{III} complexes are $\sim 10^5$ times larger than k_{obs} at 50°C and [Co^{III} complex] = 0. The catalytic activity is only slightly dependent on the structure of amine ligands.

No isomerization of ApA to adenylyl(2'-5')adenosine has been observed. The absence of Co^{III} complex-catalyzed hydrolysis of 2'-deoxyadenylyl(3'-5')2'-deoxyadenosine is considered to be indirect evidence for the intramolecular nucleophilic attack by the 2'-OH of ribose toward the phosphodiester linkage in the hydrolysis of RNA. The proposed mechanism for Co^{III}-complex-catalyzed hydrolysis of ApA as described by Equation 2.37 is similar to Scheme 2.24, except that Scheme 2.24 does not involve the pentacoordinated intermediate similar to PI in Equation 2.37.

$$[Co(N)_4(OH_2)_2]^{3+} + ApA \underset{H_2O}{\overset{K(-H_2O)}{\rightleftharpoons}} [Co(N)_4(OH_2)(ApA)]^{2+} \underset{k_{-1}(H_2O)}{\overset{k_1(HO^-)}{\rightleftharpoons}} \quad (2.37)$$

$$[Co(N)_4(OH_2)(PI)]^{1+} \overset{k_2}{\longrightarrow} Products + [Co(N)_4(OH_2)(OH)]^{2+}$$

Although PI is a plausible intermediate on the reaction path in such reactions, it has not been detected directly by experiment and, hence, its proposed existence remains largely speculative. However, indirect proof for the presence of PI on the reaction path comes from the observed isomerization between 3',5'-linkage and 2',5'-linkage only under acidic pH where PI can exist in the monoanionic form (PIH).[140] Thus, an alternative mechanism (Scheme 2.26) for the metal-ion-assisted hydroxide-ion- and general base (:B)-catalyzed hydrolysis of RNA and RNA-model compounds cannot be completely ruled out if pentacoordinated intermediate (PIH) does exist on the reaction path. Nearly 10^5-fold enhanced catalytic effects of Co^{III} complexes are attributed to the occurrence of intramolecular electrophilic catalysis as shown in the transition state TS_{18}.

The pH-rate constant profile for Co^{III}-complex-mediated hydrolysis of phenyl ester of adenosine 3'-monophosphate (ApΦ) reveals the reaction mechanism (Scheme 2.27), which is different from that described by Equation 2.37. This difference is attributed to the different pK_a of leaving groups in the hydrolysis of ApA and ApΦ.[139] The absence of a kinetic step in Equation 2.37 similar to

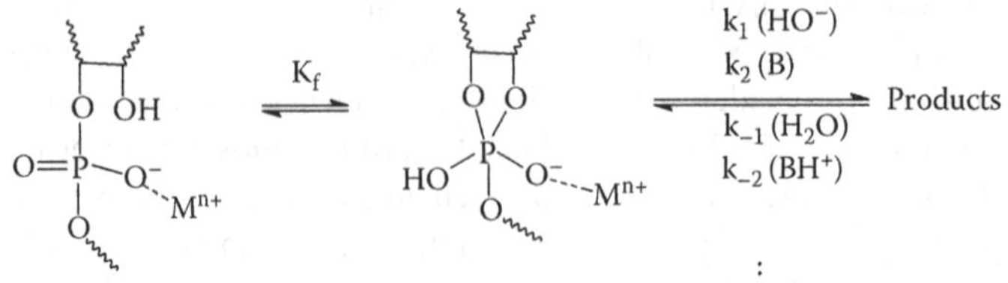

SCHEME 2.26

$$[Co(N)_4(OH_2)_2]^{3+} + Ap\Phi \underset{H_2O}{\overset{K' (-H_2O)}{\rightleftharpoons}} [Co(N)_4(OH_2)(Ap\Phi)]^{2+} \xrightarrow{k_3 (HO^-)}$$

Product(s) + $[Co(N)_4(OH_2)(OH)]^{2+}$

$$[Co(N)_4(OH_2)(OH)]^{2+} + Ap\Phi \underset{H_2O}{\overset{K'' (-H_2O)}{\rightleftharpoons}} [Co(N)_4(OH)(Ap\Phi)]^{1+} \xrightarrow{k_4 (HO^-)}$$

Product(s) + $[Co(N)_4(OH)(OH)]^{1+}$

SCHEME 2.27

the k_4-step of Scheme 2.27 is due to the inability of complex $[Co(N)_4(OH)(ApA)]^{1+}$ to provide intramolecular GA catalysis for the product formation. The value of $k_3/k_4 \approx 30$, which may be explained in terms of greater decrease in electrophilicity of reaction center P in $[Co(N)_4(OH)(Ap\Phi)]^{1+}$ than in $[Co(N)_4(OH_2)(Ap\Phi)]^{2+}$. The pseudo-first-order rate constant for hydrolysis of ApA at pH 7.2 and 30°C is increased by 460 times in the presence of 10-mM La(ClO$_4$)$_3$ and 100-mM H$_2$O$_2$ compared to that in the absence of H$_2$O$_2$ at 10-mM La(ClO$_4$)$_3$.[141] Similar observations are obtained with several other lanthanide ions. Hydrogen peroxide is inactive in the absence of lanthanide ions. The proposed mechanism of the enhanced hydrolysis of ApA by the complex La(ClO$_4$)$_3$-H$_2$O$_2$ involves concerted intramolecular electrophilic and GB catalysis as depicted in the critical transition state TS$_{19}$ in which catalytic species is the trimeric aggregate $[La(O - O)_3La]_3$.

Second-order rate constants (k_2) for the demethylation reactions of methyl diphenylphosphate, (PhO)$_2$P(O)OCH$_3$, by the complexes of polyether ligands with alkali metal iodides, MI (M = Li, Na, and K), increase by a maximum limit of 90-fold compared to k_2 for the reaction of the same ester with tetrahexylammonium iodide in chlorobenzene at 60°C.[142] The catalytic effect is found to depend on both the metal ion charge density and the ligand topology. The values of k_2 increase in the order Li$^+$ > Na$^+$ > K$^+$. The metal ion catalytic effect is rationalized by assuming a transition state (TS$_{20}$) in which the metal ion (M$^+$) complexed by the ligands (M$^+\subset$Lig) can interact with ion-paired anion I$^-$ and, at the same time, with the leaving group (PhO)$_2$PO$_2^-$, i.e., metal ion exerts concerted induced intramolecular nucleophilic attack by I⁻ at sp^3 carbon and induced intramolecular electrophilic assistance (catalysis) for the expulsion of leaving group of ester.

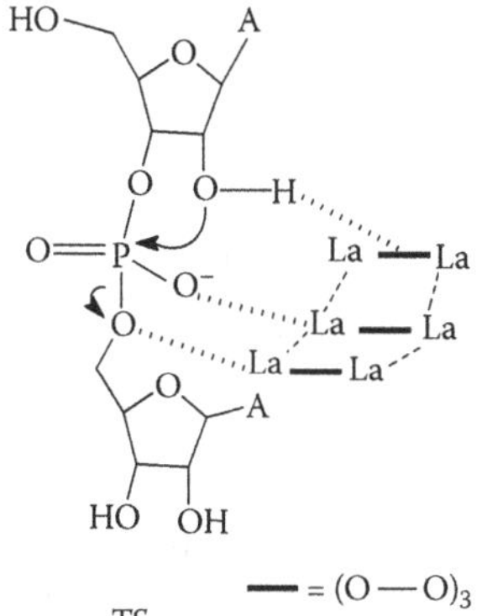

2.5.6 SPECIFIC BASE CATALYSIS

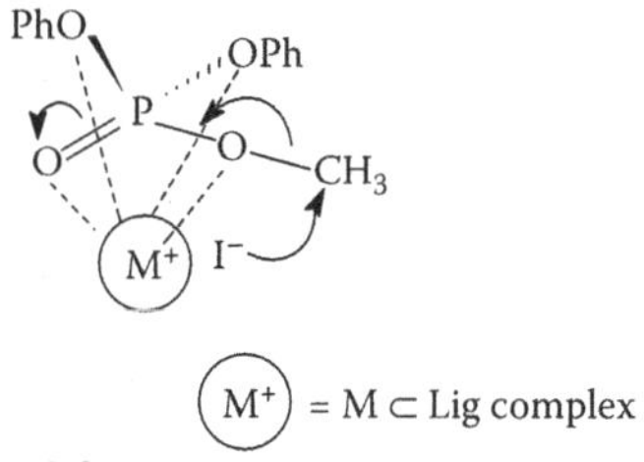

By definition, hydroxide ion is called a *specific base* and, hence, SB catalysis involves hydroxide ion as a catalyst. SB-catalyzed reactions must be only inter-molecular, because in the intramolecular catalytic reaction, the catalyst (hydroxide ion) is required to be covalently or electrovalently attached with the substrate, and under such circumstances, substrate-bound O⁻ or OH group becomes the GB catalyst. There are two possible modes of SB catalysis: (1) hydroxide ion acts as a base in an acid–base reaction of overall catalytic process as shown in Scheme 2.28, and (2) hydroxide ion acts as a nucleophile in the overall catalytic process as shown in Scheme 2.29. It is evident from Scheme 2.28 that Y⁻ must be a stronger base than HO⁻ so that hydroxide ion could not be consumed during the

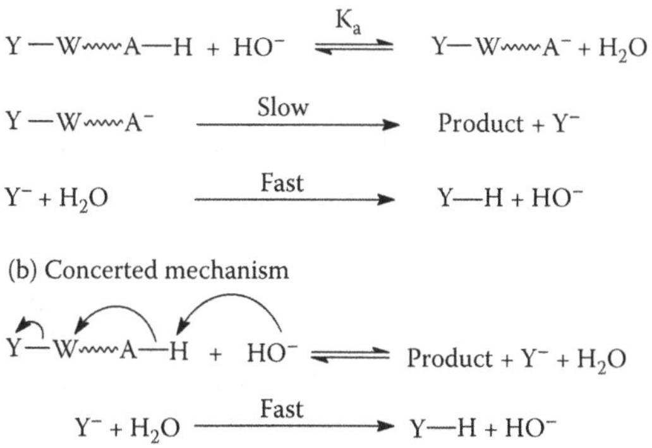

SCHEME 2.28

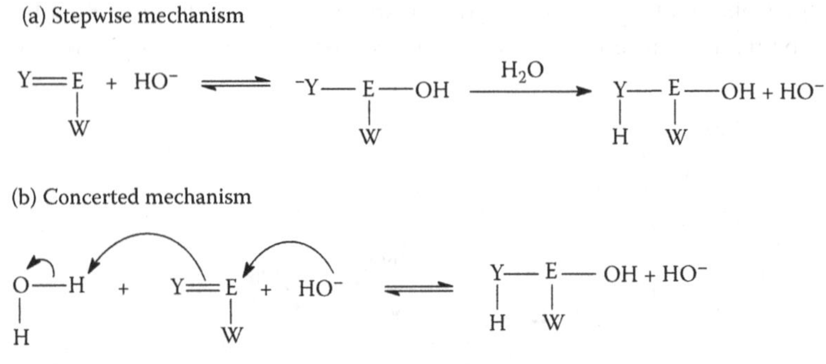

(a) Stepwise mechanism

(b) Concerted mechanism

SCHEME 2.29

course of the reaction. However, Scheme 2.29 predicts that if the hydroxyl group of the product {HOE(W)YH} is a stronger acid than H_2O, then the catalyst hydroxide ion is bound to disappear irreversibly owing to a thermodynamically favorable reaction between HO^- and hydrolysis product {HOE(W)YH}. Under such a situation, hydroxide ion cannot be regarded as a catalyst.

An example to illustrate the catalytic mechanism in Scheme 2.28 is found in the kinetic study on the rate of hydrolysis of 5'-O-pivaloyluridine 3'-dimethylphosphate (**41**) and 5'-O-pivaloyluridine 2'-dimethylphosphate (**42**) at different pH ranging from ~ 0 to ~ 8.5.[143] Pseudo-first-order rate constants (k_{obs}^{int}) for interconversion of **41** and **42** follow the relationship: $k_{obs}^{int} = k_{OH}^{int}[HO^-]$ within the pH range > 2 to < 4. Although the exact mechanisms for this and similar reactions are far from being fully understood, a plausible mechanism, based on existing experimental observations, is shown in Scheme 2.30. The rate of formation of **43** is significantly smaller than its rate of hydrolysis under similar conditions, because the rate of hydrolysis of **43** involves the release of five-membered ring strain, whereas there is no such energetic advantage in the k_{hyd}-step. Nearly 10^6 to 10^8-fold differences in the hydrolytic reactivity between the cyclic and the acyclic phosphates have been ascribed to GS destabilization that arises from ring strain in the cyclic phosphate, which is released in the trigonal-bipyramidal transition state.[144–148] The values of pK_a of the conjugate acids of leaving groups in the k_{int}-step and the k_{hyd}-step are not significantly different from each other. Similar energetic advantage makes k_{int} much larger than k_{hyd}.

The pseudo-first-order rate constants, k_{obs}^{hyd}, for the hydrolysis of **41** and **42** follow the relationship: $k_{obs}^{hyd} = k_{OH}^{hyd}[HO^-]$ at pH > 7, whereas the values of k_{obs}^{hyd} remain unchanged with the change in pH from ~ 2 to ~ 7.[143] The linear increase in k_{obs}^{int} and k_{obs}^{hyd} with the increase in pH at pH > 2 and > 7, respectively, reflects the fact that $k_{OH}^{int} \gg k_{OH}^{hyd}$. The dissimilar pH-rate dependence at pH 2 to 7 for k_{obs}^{int} and k_{obs}^{hyd} is attributed to the leaving-group expulsion in k_{hyd}-step as hydronium ion-catalyzed process, whereas k_{int}-step does not need hydronium ion catalysis, because the sugar hydroxyl groups as oxyanions are 10^5 times better leaving groups than methoxide ion.[143] However, the effects of pH on k_{obs}^{hyd} can also be explained by an alternate plausible mechanism (Scheme 2.30) involving

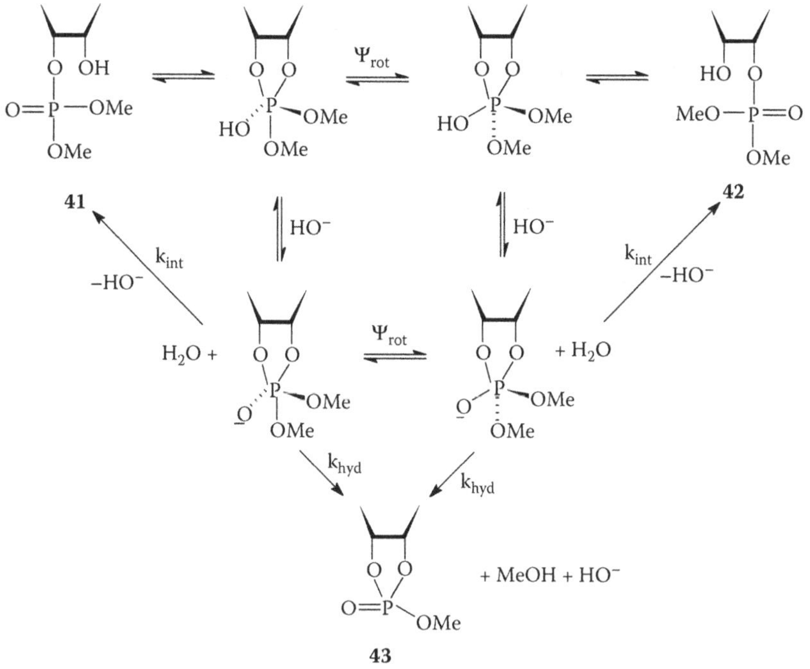

SCHEME 2.30

transition state TS_{21}. The occurrence of a concerted mechanism involving transition state TS_{22} for hydrolysis of **41** or **42** at pH < 8 may be ruled out because such a mechanism cannot explain the observed interconversion of **41** and **42** under such conditions.

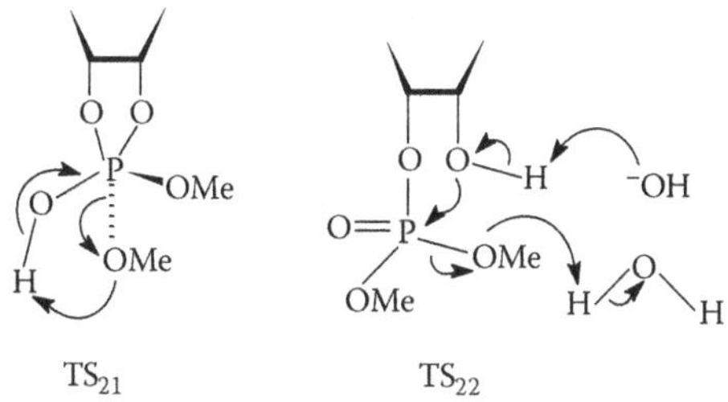

SB catalysis has been detected in many nucleophilic addition–elimination reactions in which the nucleophilic site of a nucleophile is bonded to at least one hydrogen. SB catalyst in these reactions increases the nucleophilicity of the nucleophile if the catalysis occurs through a concerted mechanism (Scheme 2.31) involving transition state TS_{23} in the formation of addition intermediate Int_1. However, the occurrence of such TS suffers from energetically unfavorable

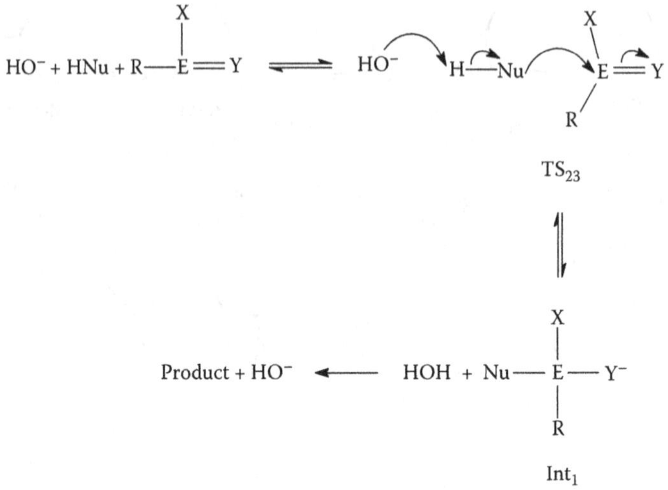

$$HO^- + HNu + R\!-\!\underset{R}{\overset{X}{\underset{|}{E}}}\!\!=\!Y \;\rightleftharpoons\; HO^- \underset{H-Nu}{\frown} \underset{R}{\overset{X}{E}}\!\!=\!Y$$

$$TS_{23}$$

$$\updownarrow$$

$$Product + HO^- \;\longleftarrow\; HOH + Nu\!-\!\underset{R}{\overset{X}{\underset{|}{E}}}\!-\!Y^-$$

$$Int_1$$

SCHEME 2.31

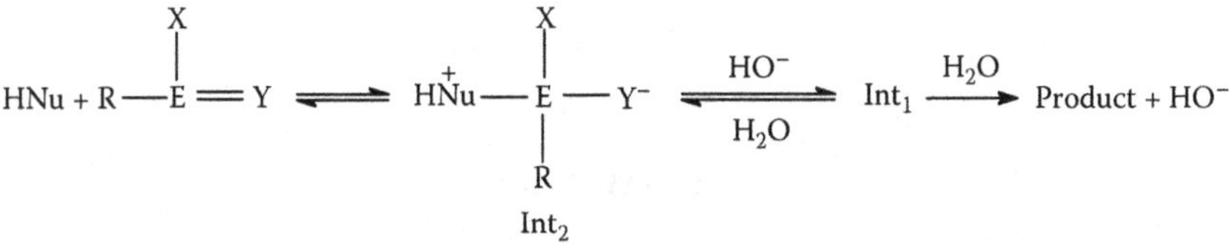

$$HNu + R\!-\!\underset{R}{\overset{X}{\underset{|}{E}}}\!\!=\!Y \;\rightleftharpoons\; H\overset{+}{N}u\!-\!\underset{R}{\overset{X}{\underset{|}{E}}}\!-\!Y^- \;\underset{H_2O}{\overset{HO^-}{\rightleftharpoons}}\; Int_1 \;\xrightarrow{H_2O}\; Product + HO^-$$

$$Int_2$$

SCHEME 2.32

entropic requirement of three interacting species (specific base HO^-, nucleophile HNu, and substrate RE(Y) = X) to be aligned in an appropriate geometrical framework. The probability of effective three-body collision in an even liquid phase is extremely low, especially when the concentrations of colliding bodies are moderately low. The energetically unfavorable entropic barrier for such reactions necessitates following a stepwise mechanism as shown in Scheme 2.32 provided the lifetime of reactive intermediate (Int_2) is larger than the lifetime for an intral rotation (approximately 10^{-12} sec).

The pseudo-first-order rate constants (k_{obs}) for the alkaline hydrolysis of β-sultam (**44**) follow Equation 2.38[149]

$$k_{obs} = k_n^{OH}[HO^-] + k_{sb}^{OH}[HO^-]^2 \qquad (2.38)$$

in which the presence of SB-catalyzed kinetic term $k_{sb}^{OH}[HO]^2$ is explained in terms of stepwise mechanism similar to Scheme 2.32. The calculated values of k_n^{OH} and k_{sb}^{OH} are 8.98×10^{-3} M^{-1} sec^{-1} and 38.7×10^{-3} M^{-2} sec^{-1}, respectively, at 30°C. The kinetic Equation 2.38 has been used in the kinetic studies on alkaline hydrolysis of several amides and imides since this kinetic equation was first used

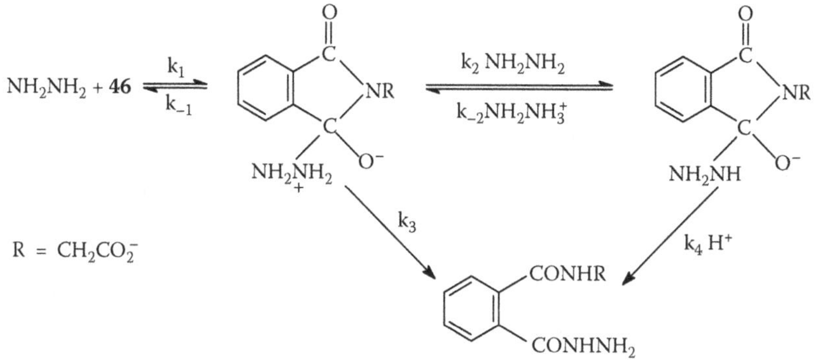

SCHEME 2.33

in the alkaline hydrolysis of acetanilide in 1957.[150] The nucleophilic second-order rate constants (k_n^{am}) for the reactions of cephalosporins (**45**) with primary amines (RNH_2) follow Equation 2.39[151]

$$k_n^{am} = k_{gb}^{am}[RNH_2] + k_{sb}^{am}[HO^-] \qquad (2.39)$$

where k_{gb}^{am} and k_{sb}^{am} represent respective GB- and SB-catalyzed aminolysis of **45**. A stepwise mechanism, similar to Scheme 2.32, is suggested for SB-catalyzed term $k_{sb}^{am}[HO^-][RNH_2][45]$ in the rate law.

A search of the literature on the reactions of primary and secondary amines with esters, amides, and imides reveals that generally SB-catalyzed aminolysis term in the rate law appears only when the rate law contains an additional GB-catalyzed term. It is also quite common that the rate law contains only a GB-catalyzed term. But it is almost rare or nonexistent that a rate law does not contain a GB-catalyzed term but contains an SB-catalyzed term. These general observations can be easily understood in terms of a stepwise mechanism similar to Scheme 2.32 with $HNu = RNH_2$. However, a recent report on the kinetic studies of the cleavage of *N*-phthaloylglysine (**46**) in the buffers of hydrazine shows that pseudo-first-order rate constants (k_{obs}) for the cleavage of **46** follow Equation 2.40[152]

$$k_{obs} - k_0 = (k_n^{am} + k_{sb}^{ap}[HO^-])[NH_2NH_2] \qquad (2.40)$$

where k_0 is the pseudo-first-order rate constant for hydrolysis and k_{sb}^{ap} is the rate constant for the apparent SB-catalyzed hydrazinolysis of **46**. It has been shown in a subsequent paper[153] that Equation 2.40 is kinetically indistinguishable from Equation 2.41, which is derived from the observed rate law coupled with possible reaction mechanism as shown in Scheme 2.33 with condition $k_4[H^+] \ll k_{-2}[NH_2NH_3^+]$. It is apparent that Equation 2.40 is similar to Equation 2.41 with $k_n^{am} =$

$$k_{obs} - k_0 = \frac{k_1 k_3 [NH_2NH_2]}{k_{-1}} + \frac{k_1 k_2 k_4 [NH_2NH_2]^2}{k_{-2}[NH_2NH_3^+]} \qquad (2.41)$$

$k_1 k_3/k_{-1}$ and $k_{sb}^{ap} = k_1 k_2 k_4 K_a/(K_w k_{-1} k_{-2})$ where $K_a = ([NH_2NH_2][H^+])/[NH_2NH_3^+]$ and $K_w = [H^+][HO^-]$. Thus, under the reaction conditions under which $k_4[H^+] \ll k_{-2}[BH^+]$, with BH^+ representing any GA, the SB-catalyzed and GB-catalyzed aminolysis of **46** are kinetically indistinguishable. Similar observations were obtained in the reactions of maleimide with some secondary amines.[154] The kinetic indistinguishability of Equation 2.40 and Equation 2.41 has been qualitatively resolved in the favor of the occurrence of GB catalysis.[153]

44   45   46

2.5.7 SPECIFIC ACID CATALYSIS

Proton, by definition, is called *specific acid*, and if the overall energy barrier (activation energy) of a reaction is reduced in the presence of proton as a catalyst, then the reaction is said to involve *specific acid catalysis*. Generally, catalyst proton reacts with reactant (substrate) in a so-called acid–base reaction process, which, in turn, activates the reaction system (by either the preferential destabilization of reactant state or stabilization of transition state in the rate-determining step) for the product formation. The products do not contain any molecular site, which has enough basicity to trap the proton catalyst irreversibly or even reversibly. Thus, for a detectable SA catalysis, the basicity (measured by the magnitude of basicity constant, K_b) of the basic site of electrophilic reactant, products, and solvent should vary in the order: K_b (for electrophilic reactant) $\geq K_b$ (for solvent) $> K_b$ (for products).

The activation of the reaction system by SA catalyst for the product formation depends on the nature of the reaction system and the structural feature of the reactant (substrate). For instance, in an addition–elimination reaction (Scheme 2.34) with two basic sites, Y and X (as a leaving group), the predominant protonation will occur at the relatively more basic site such as Y when Y = O and X = NR_2, OR, or SR. However, electrophilic sp^2 carbon in both protonated substrates (Sub), SYH$^+$ and SXH$^+$, became more reactive than that in nonprotonated Sub (Scheme 2.34) for the nucleophilic attack. Based on the general principle that an atom or a molecule is more stable in nonionic state than in the ionic state, reactive intermediate Int$_3$ is more stable than Int$_4$. But, the rate of formation of products from Int$_3$ is much slower than that from Int$_4$ because of the following:

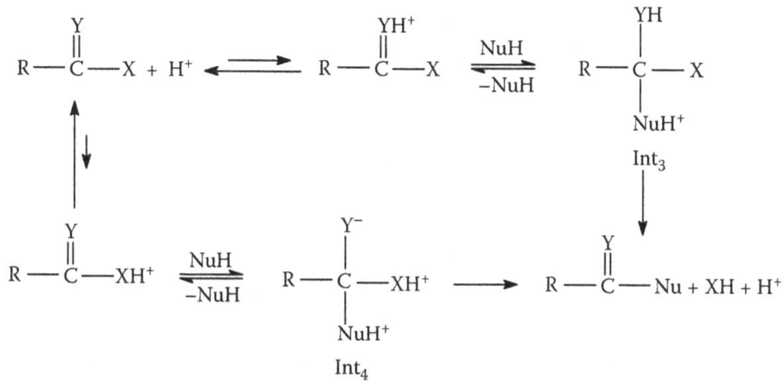

SCHEME 2.34

(1) the electronic push experienced by the leaving-group expulsion from Int_4 is apparently larger than that from Int_3 and (2) fully protonated leaving-group XH^+ is a far better leaving group than nonprotonated or partially protonated leaving-group X. It has been well established that the expulsion of a leaving group from intermediates such as Int_3 or Int_4 does not depend only upon its intrinsic leaving ability (as measured by its proton basicity) but also on the electronic push provided by the other atoms or groups attached to the same atom to which the leaving group is attached.[155]

N-Methyl β-sultam (**47**) undergoes SA-catalyzed hydrolysis in water at 30°C with H^+-catalyzed second-order rate constant, $k_H = 2.79\ M^{-1}\ sec^{-1}$.[156] The value of k_H for SA-catalyzed hydrolysis of acyclic sulfonamide **48** is estimated to be $< 2 \times 10^{-9}\ M^{-1}\ sec^{-1}$ at 30°C. Nearly 10^9-fold larger value of k_H for **47** than that for **48** is attributed to probable release of four-membered ring strain energy in the rate-determining step.[156]

The rate of formation of products via SXH^+ and Int_4 (Scheme 2.34) is governed by two apparent opposing forces. The increase in the basicity of X increases the equilibrium concentration of SXH^+ and decreases the intrinsic leaving ability of the leaving group. These two opposing forces might result in little or no SA catalysis in such reaction systems. An example of this generalization is found in the SA-catalyzed hydrolysis of *N*-substituted β-sultams.[157] The observed pseudo-first-order rate constants, k_{obs}, for the hydrolysis of *N*-(α-carboxybenzyl) β-sultam (**49**) at pH < 5.5 and 30°C fit to Equation 2.42

$$k_{obs} = (k_H^{RCO_2^-} K_a + k_H^{RCO_2H} [H^+])([H^+]/(K_a + [H^+]))\qquad (2.42)$$

where $k_H^{RCO_2^-}$ and $k_H^{RCO_2H}$ represent second-order rate constants for SA-catalyzed hydrolysis of **49** with ionized and nonionized carboxylic acid substituent, respectively, and K_a is the ionization constant of the carboxylic acid group of **49** ($pK_a = 2.62$ at 30 °C). Modeling Equation 2.42 to the experimental data gives $k_H^{RCO_2^-} = 1.54\ M^{-1}\ sec^{-1}$ and $k_H^{RCO_2H} = 0.94\ M^{-1}\ sec^{-1}$. The less than twofold

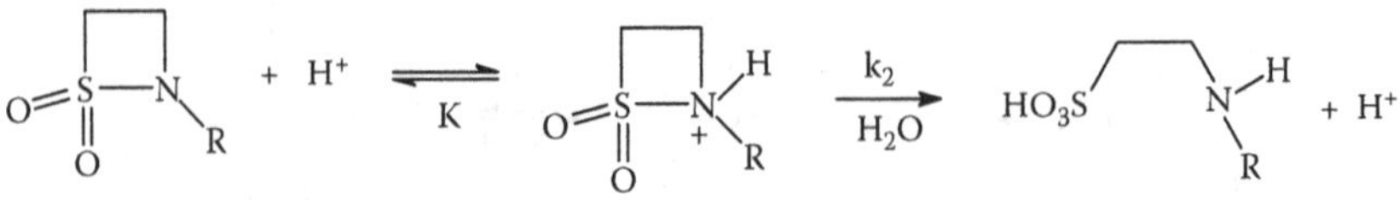

SCHEME 2.35

difference in the values of $k_H^{RCO_2^-}$ and $k_H^{RCO_2H}$ is far too low to be attributable to intramolecular catalysis or any form of neighboring group participation. Furthermore, k_H for SA-catalyzed hydrolysis of N-(α-methoxycarbonylbenzyl) β-sultam (**50**) is 1.02 M^{-1} sec^{-1}, almost identical to that of the nonionized form of **49** and consistent with an inductive effect of the carboxylic acid group rather than any intramolecular reaction.[157] It is perhaps surprising that there is only a small difference between the values of $k_H^{RCO_2^-}$ and $k_H^{RCO_2H}$. The mechanism for the SA-catalyzed hydrolysis of β-sultams involves N-protonation followed by rate-determining S–N bond fission (Scheme 2.35).

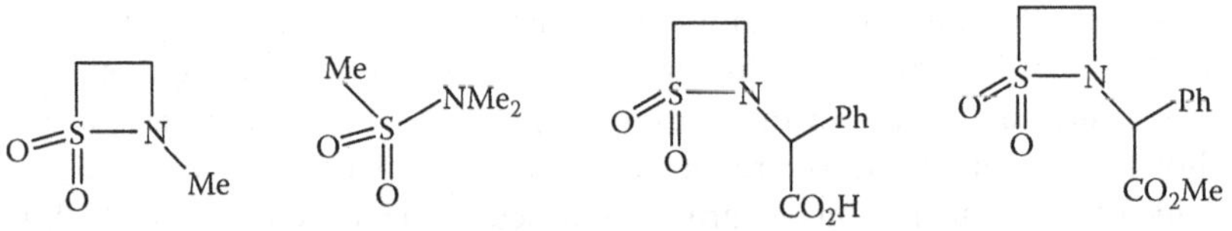

47   **48**   **49**   **50**

The electronic demands for N-protonation and S–N bond cleavage are opposite, that is, the caboxylate anion favors N-protonation and the rate of S–N bond cleavage is faster with the more-electron-withdrawing nonionized carboxylic acid. These opposing factors presumably cancel to give approximately equal values of $k_H^{RCO_2^-}$ and $k_H^{RCO_2H}$. However, these opposing factors may not be always equally effective as is evident from the nearly 50-fold larger value of k_H (= 2.64 M^{-1} sec^{-1}) for **47** than that for N-phenyl β-sultam (k_H = 5.6 × 10^{-2} M^{-1} sec^{-1}).[157]

The term $k_H^{RCO_2^-}$[H$^+$][RCO$_2^-$][H$_2$O] is kinetically equivalent to a term k_0[RCO$_2$H][H$_2$O], which corresponds to the apparent pH-independent reaction of **49** with a nonionized carboxylic acid substituent. The value of k_0 (= $k_H^{RCO_2^-}$ K_a) turns out to be 3.7 × 10^{-3} sec^{-1}. But there is no term (k_0[RCO$_2^-$][H$_2$O]) in the rate law corresponding to the pH-independent hydrolysis of **49** with a carboxylate anion substituent.[157]

Second-order rate constants for SA-catalyzed hydrolysis of RC$_6$H$_4$NHCH$_2$SO$_3^-$ (R = H, p-MeO, p-Me, p-Cl, and m-Cl) yield Hammett reaction constant, $\rho = -2.77$, which is not significantly different from ρ (= −2.30) for uncatalyzed hydrolysis of RC$_6$H$_4$NHCH$_2$SO$_3^-$.[158] The suggested mechanism for uncatalyzed and H$^+$-catalyzed hydrolysis is shown in Scheme 2.36.

It is evident from Scheme 2.36 that the experimentally determined rate constants, k_{ex}^{cat}, for catalyzed hydrolysis, have the relationship: $k_{ex}^{cat} = k_2K_b$, because pseudo-first-order rate constants (k_{obs}) at different [H$^+$] follow the relationship:

$$RC_6H_4NHCH_2SO_3^- + H_3O^+ \xrightleftharpoons{K_b} RC_6H_4NH_2^+CH_2SO_3^- + H_2O$$

$$RC_6H_4NH_2^+CH_2SO_3^- + H_2O \xrightarrow[S_N2]{k_2} RC_6H_4NH_2 + HOCH_2SO_3^- + H^+$$

SCHEME 2.36

$k_{obs} = k_{ex}^0 + k_{ex}^{cat}[H^+]$. The linear plot of k_{obs} vs. $[H^+]$ with definite intercept (= k_{ex}^0) implies that $1 >> K_b[H^+]$ under the experimental conditions attained in the study.[158] Thus, $\rho = \rho_{(for\ k2)} + \rho_{(for\ Kb)}$. It can be easily shown by the use of a free-energy reaction coordinate diagram that the value of $\rho_{(for\ Kb)}$ must be negative and that of $\rho_{(for\ k2)}$ must be positive. Therefore, the observed value of ρ (= −2.77) shows that $|\rho_{(for\ Kb)}| > |\rho_{(for\ k2)}|$.

The kinetic study on the conversion of (Z)-phenylhydrazone of 5-amino-3-benzoyl-1,2,4-oxadiazole (51) into N,5-diphenyl-2H-1,2,3-triazol-4-ylurea (52) reveals for the first time the occurrence of SA catalysis in such monocyclic rearrangement reaction.[159] The plot of pseudo-first-order rate constants (k_{obs}) vs. $[H^+]$ shows that the reactivity tends to a limiting rate constant. This fact suggests an SA-catalyzed mechanism according to Equation 2.43 and Equation 2.44.

$$51 + H_3O^+ \xrightleftharpoons{K_b\ (fast)} 51\text{-}H^+ + H_2O \qquad (2.43)$$

$$51\text{-}H^+ \xrightarrow{k_H\ (slow)} 52 \qquad (2.44)$$

The values of k_{obs} at different $[H_3O^+]$ fit to Equation 2.45 with $k_H = 1.58 \times 10^{-4}$ sec^{-1}, $K_b = 1.23$ M^{-1}, and k_u (uncatalyzed rate constant for the rearrangement) $= 0.03 \times 10^{-6}$ sec^{-1}.

$$k_{obs} = (k_u + k_H K_b[H_3O^+])/(1 + K_b[H_3O^+]) \qquad (2.45)$$

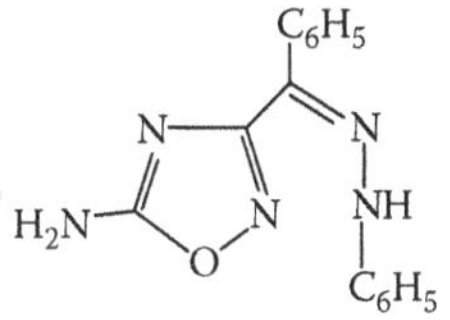

51 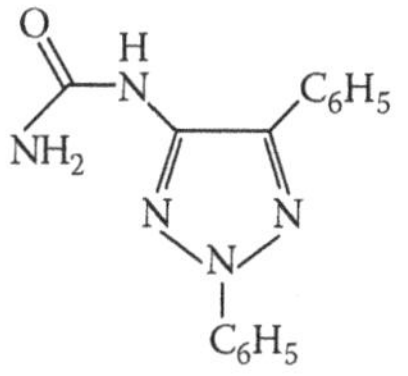   52

2.5.8 General Base Catalysis

All the bases except hydroxide ion (SB) come under the common name *general base*. When the basicity of a GB is defined in terms of proton basicity, then it is also called a *Bronsted base*, whereas when its basicity is defined as a pair of electron donors to an electrophilic center, then it is also called a *Lewis base*. However, a base (whether a GB or SB) exerts its influence upon reactivity of a chemical process through its ability to donate a pair of electrons to the electron-deficient site of the chemical reacting system. A GB catalyst catalyzes the reaction by apparently increasing the nucleophilicity of the nucleophile through the abstraction of proton attached to the nucleophilic site of the nuclephile in or before the rate-determining step. The GB catalyst is recovered through the transfer of proton from the conjugate acid of general base (GBH$^+$) to the leaving group after the rate-determining step.

If a reaction process is highly sensitive to [HO$^-$], the value of [HO$^-$] into the reaction mixture is generally controlled by the use of buffers. A buffer solution contains both the GB and GBH$^+$, and experimentally attainable pH range provided by a particular buffer depends on the pK_a of general acid (GBH$^+$) of buffer components. As both HO$^-$ and GB have the ability to donate a pair of electrons to an electrophilic center, it is logical to believe that if the rate of a reaction is affected by [HO$^-$], then it may be also affected by [GB]. However, if the basicity of GB is much lower than that of hydroxide ion, the effect of [GB] on the reaction rate may not be detected in competition to that of the hydroxide ion. The relative effectiveness of the effects of HO$^-$ and GB on reaction rate depends on both their relative basicity as well as concentration into the reaction mixture.

2.5.8.1 Intermolecular GB Catalysis

Intermolecular GB catalysis, in which the GB catalyst is not an integral part of the substrate molecule, has been extensively studied in addition–elimination reactions involving sp^2 carbon as an electrophilic center including several enzyme model reactions.[160] An interesting account of GB catalysis is found in the literature.[25] Jencks has elegantly argued that the driving force for GB and GA catalyses (Subsection 2.5.9) arises from the large changes in pK_a of reaction sites in the chemical transformation from reactant to product.[161] But even the large amount of work on GB–GA catalysis does not provide the perfect guidelines, which could be safely used to predict the occurrence or nonoccurrence of GB or GA catalysis in even very closely related reactions. For example, GB catalysis was detected in the reaction of hydrazine with phthalimide[162], whereas such catalysis was absent in its reaction with maleimide[163] under similar experimental conditions. Similarly, morpholine exhibited GB catalysis in its reaction with both phthalimide[164] and maleimide[165], whereas such catalysis could not be detected in its reaction with *N*-ethoxycarbonylphthalimide.[166]

Consider a general reaction

$$H-Nu \ + \ R-\overset{\overset{\displaystyle Y}{\|}}{E}-X \ \xrightarrow[\text{H}_2\text{O}]{\text{GB}} \ R-\overset{\overset{\displaystyle Y}{\|}}{E}-Nu \ + \ XH \qquad (2.46)$$

where HNu is a nucleophile with at least one hydrogen attached to its nucleophilic site, X is a leaving group, RE(= Y)X is a substrate with electrophilic site E and leaving group X, and GB is GB catalyst, which may or may not be the same molecule as HNu. In order to detect the occurrence or nonoccurrence of GB catalysis, the rate of the reaction (of Equation 2.46) is generally studied by carrying out kinetic runs at a constant concentration of RE(= Y)X, HNu (if HNu and GB are different molecules), and different total buffer concentrations of GB ($[Buf]_T = [GB] + [GBH^+]$) at a constant pH and temperature. It is preferable to maintain reaction conditions in which $[RE(=Y)X] << [HNu] < [Buf]_T$ so that the rate of reaction obeys the pseudo-first-order rate law. Under such experimental conditions, pseudo-first-order rate constants (k_{obs}) follow Equation 2.47

$$k_{obs} = k_0 + k_n[HNu] + k_{gb1}[HNu]^2 + k_{gb2}[HNu][GB] \qquad (2.47)$$

where k_0 is the pseudo-first-order rate constant for hydrolysis, k_n is nucleophilic second-order rate constant for the nucleophilic cleavage, and k_{gb1} and k_{gb2} are third-order rate constants for catalyzed cleavage of substrate by GB HNu and GB, respectively (HNu is acting both as a nucleophile and as a GB catalyst, whereas GB does not act as a nucleophile but acts as a GB catalyst). As the total concentration of HNu ($[HNu]_T$) is kept constant, the plot of k_{obs} vs. $[Buf]_T$ should be linear with intercept $(c_1) = k_0 + k_n[HNu] + k_{gb1}[HNu]^2$ and slope $(m_1) = k_{gb2}$ f_b $[HNu]$ where $f_b = K_a/([H^+ + K_a)$ with $K_a = ([GB][H^+])/[GBH^+]$ (Figure 2.5, Δ). If the rate of reaction is GB-catalyzed, then the value of m_1 would increase with the increase in the fraction of free base $(= [GB]/[Buf]_T$, i.e., increase in pH) and increase in $[HNu]_T$ at a constant pH.

The mechanism for the GB-catalyzed cleavage of substrate (RE(=Y)X), which gives rise to the kinetic terms $k_{gb1}[HNu]^2$ and $k_{gb2}[HNu][GB]$ in Equation 2.47, may be shown by two alternative reaction mechanisms, a concerted reaction mechanism (Scheme 2.37) and a stepwise reaction mechanism (Scheme 2.38). The occurrence of reaction mechanism Scheme 2.37 is highly unlikely, because the probability of effective termolecular collision is extremely low even in aqueous medium and, consequently, such reaction mechanism requires very high energetically unfavorable entropic barrier for the reaction. Reaction Scheme 2.38 can lead to Equation 2.47 provided $k_3 >> k_{-2}^a[H_2Nu^+]$ and $k_3 >> k_{-2}^b[GBH^+]$ with $k_n = kk_1/k_1$, $k_{gb1} = k_2^a k_1/k_{-1}$, and $k_{gb2} = k_2^b k_1/k_{-1}$. The formation of Int_6 and Int_8 from reactants and catalyst without involving respective intermediate Int_5 and Int_7 may be ruled out for the mere reason that the probability of effective termolecular collision even in aqueous medium is extremely low.

If the values of k_{obs} at a constant pH and temperature turn out to be independent of $[Buf]_T$ (i.e., $m_1 \approx 0$) (Figure 2.5, O), it may be the consequence of at least one of the two possibilities: (1) the rate of reaction is not catalyzed by GB catalyst, that is, $k_{gb2} = 0$, and (2) the rate of reaction is catalyzed by GB catalyst, but under such reaction conditions, $k_{-2}^b[GBH^+] >> k_3$, that is, the k_3-step is the rate-determining step (Scheme 2.38). Under such a condition, Scheme 2.38 can lead to

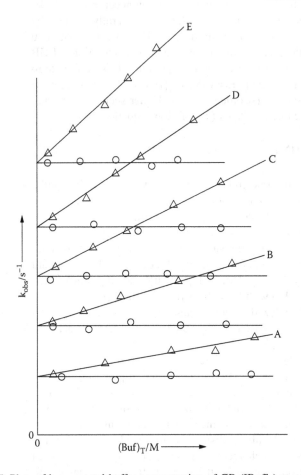

URE 2.5 Plots of k_{obs} vs. total buffer concentration of GB ($[Buf]_T$) at a constant pH different total concentrations of nucleophile ($[HNu]_T = A, B, C, D,$ and E where $A <$ $C < D < E$), which represent the experimental validity of Equation 2.47.

$$R \overset{\overset{Y}{\|}}{\underset{}{E}} X \quad
\begin{cases}
\xrightarrow{k_0(H_2O)} & R \overset{\overset{Y}{\|}}{\underset{}{E}} OH + HX \\[2ex]
\xrightarrow{k_n(HNu)} & R \overset{\overset{Y}{\|}}{\underset{}{E}} Nu + HX \\[2ex]
\xrightarrow{k_{gb1}(HNu)^2} & R \overset{\overset{Y}{\|}}{\underset{}{E}} Nu + HX + HNu \\[2ex]
\xrightarrow{k_{gb2}(HNu)(GB)} & R \overset{\overset{Y}{\|}}{\underset{}{E}} Nu + HX + GB
\end{cases}$$

SCHEME 2.27

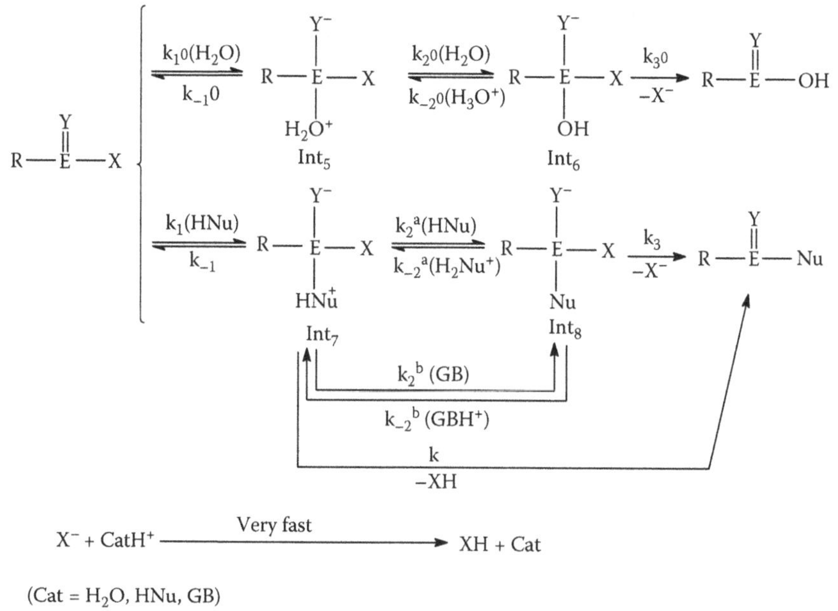

$$
\begin{array}{c}
R\!-\!\overset{\overset{Y}{\|}}{E}\!-\!X
\end{array}
\left\{
\begin{array}{l}
\xrightarrow[k_{-1}0]{k_{1}0(H_2O)} \;\; R\!-\!\overset{\overset{Y^-}{|}}{\underset{\underset{H_2O^+}{|}}{E}}\!-\!X \;\; \xrightarrow[k_{-20}(H_3O^+)]{k_{20}(H_2O)} \;\; R\!-\!\overset{\overset{Y^-}{|}}{\underset{\underset{OH}{|}}{E}}\!-\!X \;\; \xrightarrow[-X^-]{k_{30}} \;\; R\!-\!\overset{\overset{Y}{\|}}{E}\!-\!OH \\[2mm]
\qquad\qquad\qquad\qquad \mathrm{Int_5} \qquad\qquad\qquad\qquad \mathrm{Int_6} \\[4mm]
\xrightarrow[k_{-1}]{k_{1}(HNu)} \;\; R\!-\!\overset{\overset{Y^-}{|}}{\underset{\underset{H\overset{+}{N}u}{|}}{E}}\!-\!X \;\; \xrightarrow[k_{-2}{}^a(H_2Nu^+)]{k_2{}^a(HNu)} \;\; R\!-\!\overset{\overset{Y^-}{|}}{\underset{\underset{Nu}{|}}{E}}\!-\!X \;\; \xrightarrow[-X^-]{k_3} \;\; R\!-\!\overset{\overset{Y}{\|}}{E}\!-\!Nu \\[2mm]
\qquad\qquad\qquad\qquad \mathrm{Int_7} \qquad\qquad\qquad\qquad \mathrm{Int_8}
\end{array}
\right.
$$

$$
k_2{}^b\,(GB) \qquad k_{-2}{}^b\,(GBH^+)
$$

$$
\xrightarrow{k} \;\; -XH
$$

$$
X^- + CatH^+ \xrightarrow{\;\;\text{Very fast}\;\;} XH + Cat
$$

(Cat = H_2O, HNu, GB)

SCHEME 2.38

Equation 2.47 with $k_{gb2}[GB] = k_1 k_2{}^b k_3 f_b / k_{-2}{}^b f_a$ where $f_b = K_a/([H^+] + K_a)$ with $K_a = ([GB][H^+])/[GBH^+]$ and $f_a = 1 - f_b$. The two kinetically equivalent possibilities, (1) and (2), can be easily resolved by obtaining intercept c_1 of the plot of k_{obs} vs. $[Buf]_T$ in the presence and absence of GB at a constant value of $[HNu]_T$, pH, and temperature. At $[Buf]_T = 0$, $c_1 = k_0 + k_n[HNu] + k_{gb1}[HNu]^2$ and at $[Buf]_T \neq 0$, $c_1 = k_0 + k_n[HNu] + k_{gb1}[HNu]^2 + (k_1 k_2{}^b k_3 f_b[HNu])/k_{-2}{}^b f_a$.

The values of c_1, obtained at different $[HNu]_T$, at constant pH, and temperature, should vary nonlinearly with $[HNu]_T$ (Figure 2.6, O) provided $k_3 \gg k_{-2}{}^a$ $[H_2Nu^+]$. If the plot of c_1 vs. $[HNu]_T$ is linear (Figure 2.6, Δ), it implies that $k_{gb1} = 0$, that is, HNu is not acting as a GB catalyst. One may argue that the plot of c_1 vs. $[HNu]_T$ may be linear even when HNu is also acting as a GB catalyst provided $k_3 \ll k_{-2}{}^a[H_2Nu^+]$. But under such conditions, the plot of c_1 vt. $[HNu]_T$ should be sigmoid as shown in Figure 2.6. The slope (m_2) of the plot of $(c_1 - k_0)/[HNu]_T$ vs. $[HNu]_T$ gives the value of k_{gb1} from the relationship: $m_2 = k_{gb1}f_b{}^2$.

Ribonuclease A-catalyzed hydrolysis of ribonucleic acid (RNA) involves the participation of the imidazole rings of histidine-12 and histidine-119 as well as lysine-41.[167] This discovery led a series of studies carried out on the hydrolysis of RNA models (generally ribonucleotides) in the presence of imidazole buffers[168–171] and other secondary amine buffers.[172–174] These studies have produced remarkable information on the mechanistic aspects of imidazole-catalyzed hydrolysis of RNA and RNA models and led to a plausible mechanism as shown in Scheme 2.39.[175] However, despite a significant amount of data on these catalyzed reactions, the fine details of the mechanism are far from being fully under-

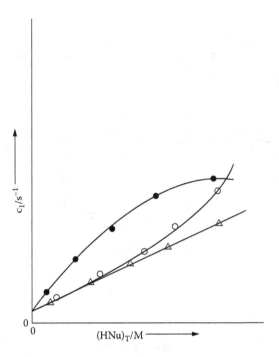

FIGURE 2.6 Plots showing the dependence of c_1 (intercept at different $[HNu]_T$, obtained from Figure 2.5) on $[HNu]_T$ under three typical limiting conditions: (i) for (O), $k_3 \gg k_{-2}^a [H_2Nu^+]$, (ii) for ($\Delta$), $k_{gb1} = 0$, and (iii) for ($\bullet$), $k_3 \ll k_{-2}^a [H_2Nu^+]$.

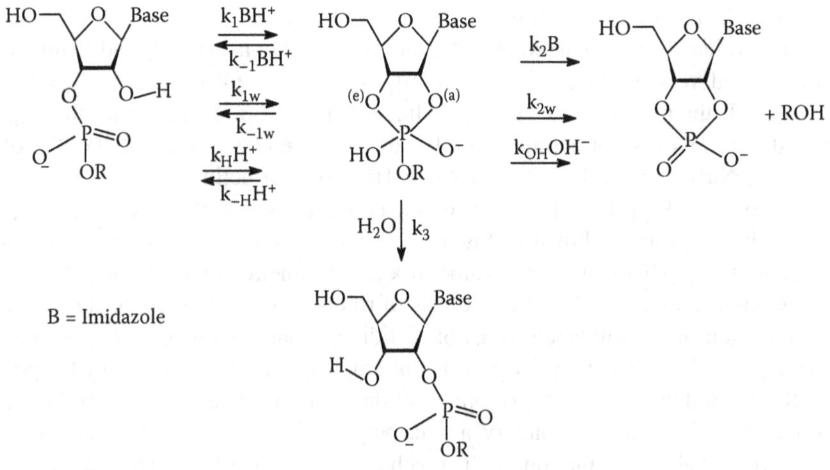

SCHEME 2.39

stood.[176] The mechanism of the hydrolysis of RNA models (usually ribonucle-otides) is primarily based upon the kinetic data, which apparently fit to kinetic Equation 2.48 and Equation 2.49 for the rate of cleavage and isomerization of RNA models, respectively.[175] However, the kinetic term $k'[\text{Im}] + k''[\text{ImH}^+]$ in Equation 2.48 (where Im and ImH$^+$ represent free and protonated imidazole, respectively) cannot be explained in terms of reaction Scheme 2.39. Kirby et al.[174]

$$k_{\text{cleavage}} = \frac{k_1 k_2 [\text{Im H}^+][\text{Im}] + k'_w}{k_{-1}[\text{Im H}^+] + k_2[\text{Im}] + k_3 + k_w} + k'[\text{Im}] + k''[\text{Im H}^+] \qquad (2.48)$$

$$k_{\text{isomerization}} = \frac{k_1 k_3 [\text{Im H}^+][\text{Im}] + k''_w}{k_{-1}[\text{Im H}^+] + k_2[\text{Im}] + k_3 + k_w} \qquad (2.49)$$

reported catalysis by imidazole in the hydrolysis of an RNA model in which catalytic second-order rate constant shows a rate maximum near 70% free base, similar to the bell-shaped curve observed by Anslyn and Breslow for imidazole catalysis of the hydrolysis of 3′,5′-uridyluridine (UpU).[169] However, Kirby et al. were unable to fit the kinetic data to the kinetic mechanism shown by Scheme 2.39: rather than describing a bell-shaped curve, the kinetic data fall on a good straight line up to 60 to 70% free base imidazole before falling off at higher pH. These authors explained a bell-shaped dependence of imidazole buffer catalytic rate constants on the buffer ratio in terms of a solvent effect on back-ground hydroxide-ion-catalyzed reaction and suggested a mechanism as shown by Scheme 2.40.

53

Pseudo-first-order rate constants (k_{obs}) for the reaction of *N*-ethoxycarbon-ylphthalimide (**53**) with *N*-methylhydroxylamine (CH$_3$NHOH) in the buffer solu-tions of CH$_3$NHOH of constant pH follow the kinetic relationship:[177]

$$k_{\text{obs}} - k_w = k_n^{\text{app}}[\text{Buf}]_T + k_b[\text{Buf}]_T^2 \qquad (2.50)$$

where $[\text{Buf}]_T = [\text{CH}_3\text{NHOH}] + [\text{CH}_3\text{NH}_2\text{OH}^+]$, k_w, k_n^{app}, and k_b represent buffer-independent first-order, apparent nucleophilic second-order, and buffer-catalyzed

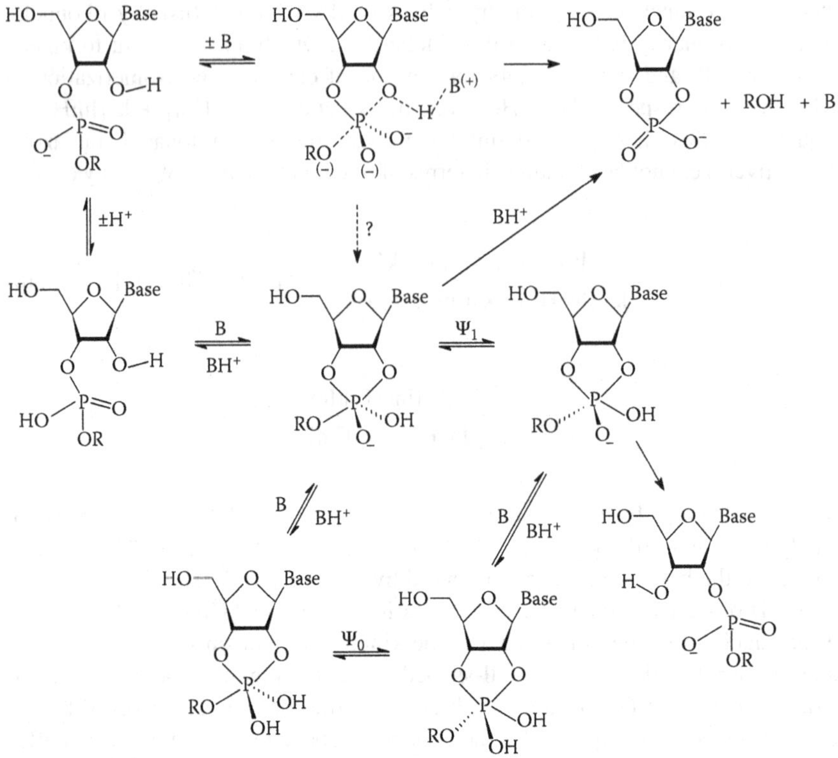

HO—O Base ⇌ (±B) HO—O Base → HO—O Base + ROH + B

±H⁺   ?   BH⁺

HO—O Base ⇌ (B / BH⁺) HO—O Base ⇌ (Ψ₁) HO—O Base

B / BH⁺   B / BH⁺

HO—O Base ⇌ (Ψ₀) HO—O Base

HO—O Base

B = Imidazole

SCHEME 2.40

third-order rate constant, respectively. The kinetic equation (of Equation 2.50) is consistent with the general rate law given as

$$\text{rate} = k_w + (k_n + k_{gb}[CH_3NHOH] + k_{ga}[CH_3NH_2OH^+])[CH_3NHOH]\}[53] \qquad (2.51)$$

where k_n, k_{gb}, and k_{ga} represent nucleophilic second-order, GB-catalyzed, and GA-catalyzed third-order rate constants, respectively. Equation 2.50 and Equation 2.51 lead to the relationship: $k_b = k_{gb}f_a^2 + k_{ga}f_af_{aH}$ where $f_{aH} = 1 - f_a$, $f_a = K_a/([H^+] + K_a)$, and K_a is the ionization constant of $CH_3NH_2OH^+$. The values of k_b at different pH fit to the relationship: $k_b = k_{gb}f_a^2 + k_{ga}f_af_{aH}$ with $k_{gb} = 54 \pm 13$ M^{-2} sec^{-1}, $k_{ga} = 4.2 \pm 2.0$ M^{-2} sec^{-1}, and $pK_a = 6.24$.

The rate of hydrolysis of N-(4-nitrobenzoyl)pyrrole has been found to be sensitive to GB catalysis.[178] In addition, carbonyl-^{18}O exchange kinetics have also been carried out at a single pH (9.48) as a function of DABCO buffer concentration. At zero buffer concentration, the measured ratio of ^{18}O exchange to hydrolysis (k_{ex}/k_{hyd}) is ~ 0.04, and this value increases and finally levels off at ~ 0.23 as the DABCO concentration is increased. These observations are consistent with the buffer acting as a GB to catalyze both the attack of H_2O to generate an anionic tetrahedral intermediate (TO⁻) and the breakdown of TO⁻ to give a hydrolysis product.

The occurrence of efficient intermolecular GB catalysis has been reported in the intramolecular nucleophilic reactions of the neutral trifluoroethyl 2-aminomethylbenzoate. [179] The rate-determining step in the GB catalyzed reaction involves proton transfer in concert with leaving-group departure.

2.5.8.1.1 Kinetic Ambiguity of the Assignment of Intermolecular General Base Catalysis

A comprehensive account of this problem is provided by Jencks[25] and, hence, a brief description of this complex problem is mentioned here. The occurrence of intermolecular GB catalysis is normally advanced owing to appearanace of a kinetic term ($k_{gb}[GB][HNu][Sub]$) in the rate law for the reaction between HNu and substrate (Sub) carried out at different total buffer concentrations ($[Buf]_T$) of constant pH. As GB is also a nucleophile and therefore if HNu is a considerably weaker nucleophile than GB, then GB may act as a nucleophile rather than as a catalyst provided the difference between [HNu] and [GB] is not very large. Under such circumstances, the [HNu] term should not appear in the rate law (i.e., the kinetic term $k_{gb}[GB][HNu][Sub]$ is replaced by $k[GB][Sub]$ in the rate law), and the reaction is simply an uncatalyzed nucleophilic reaction as well as the product must be because of nucleophilic reaction between GB and Sub. However, if the reaction product is produced because of a nucleophilic reaction between HNu and Sub and the nucleophilic site of GB is not covalently bonded to any hydrogen (such as a tertiary amine or carboxylate ion), then the kinetic term $k[GB][Sub]$ may represent nucleophilic catalysis rather than nucleophilic reaction. Under such conditions, the reaction scheme is expressed as

$$Sub + GB \xrightarrow{\text{Slow}} P_1 \xrightarrow[\text{HNu}]{\text{fast}} P_2 + GB \qquad (2.52)$$

where P_1 is more reactive than Sub toward HNu. The kinetic ambiguity between GB catalysis and nucleophilic catalysis can be qualitatively resolved by studying the rate of reaction of HNu with Sub in the buffer solution of another GB of similar pK_a but contains hydrogen covalently bonded to the basic site of GB (such as primary or secondary amine). The product under such a buffer would be P_2 if the reaction involves GB catalysis, whereas the formation of P_2 is unlikely if GB acts as a nucleophile.

The kinetic term $k_{gb}[GB][HNu][Sub]$ should also be reduced to $k[GB][Sub]$ in the rate law even if the nucleophilic reaction between HNu and Sub occurs in or before the rate-determining step provided [HNu]/[GB] > 100 and [GB]/[Sub] > 100 under the reaction conditions of the rate study. Thus, under such reaction conditions, $k = k_{gb}[HNu]$, and this relationship cannot be verified quantitatively by determining k at different [HNu] if HNu represents H_2O and water is solvent because $[H_2O] = \sim 55.5\ M$. However, the kinetic ambiguity as to whether GB acts as a GB catalyst or a nucleophilic catalyst may be resolved by studying the kinetic deuterium isotope effect in which deuterium isotopic exchange is done with reactant HNu. The presence and absence of substantial kinetic deuterium isotope effect is the indicator of GB acting as a GB catalyst and nucleophilic catalyst, respectively.

The GB-catalyzed term $k_{gb}[GB][HNu][Sub]$ is also kinetically indistinguishable from SB–GA-catalyzed term $k_{sb}{}^{ga}[HO^-][HNu][Sub][GBH^+]$. The occurrence of such a term in the rate law predicts the possible occurrence of GB–GA catalyzed term $k_{gb}{}^{ga}[GB][HNu][Sub][GBH^+]$, if the effect of the basicity difference of HO^- and GB is nearly counterbalanced by the effect of their concentration difference. Thus, the absence of GB–GA catalysis may qualitatively rule out the presence of SB–GA catalysis.

2.5.8.2 Intramolecular General-Base-Assisted Intermolecular Nucleophilic Reaction

Consider the following reactions in which the reaction of Equation 2.53 represents intramolecular GB-assisted nucleophilic cleavage of substrate (Sub$_1$), whereas the reactions of Equation 2.54 and Equation 2.55 represent nucleophilic cleavage of substrates (Sub$_2$ and Sub$_3$) in the absence of intramolecular GB assistance.

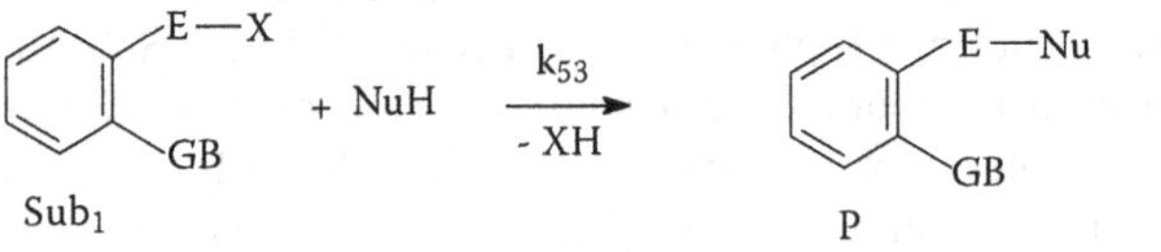

$$ (2.53) $$

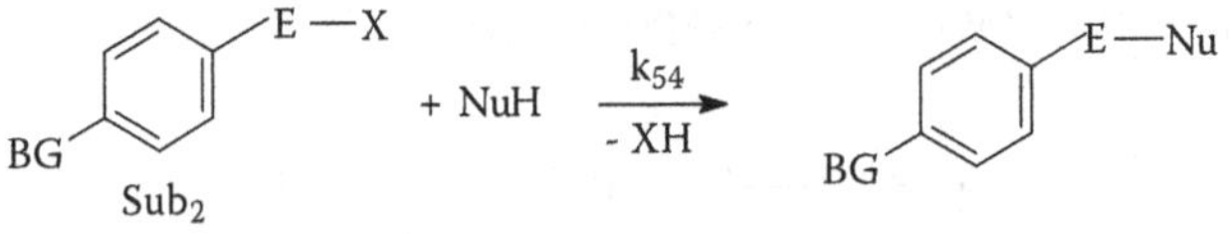

$$ (2.54) $$

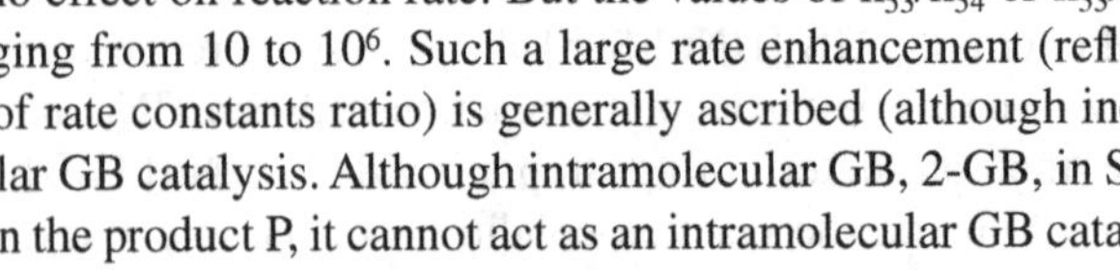

$$ (2.55) $$

Generally, the values of k_{54} and k_{55} are nearly the same if 4-GB substituent has almost no effect on reaction rate. But the values of k_{53}/k_{54} or k_{53}/k_{55} differ by a factor ranging from 10 to 10^6. Such a large rate enhancement (reflected by the large value of rate constants ratio) is generally ascribed (although incorrectly) to intramolecular GB catalysis. Although intramolecular GB, 2-GB, in Sub$_1$ remains unchanged in the product P, it cannot act as an intramolecular GB catalyst because the reaction site E–X is no longer in the same molecule (P).

The rate of hydrolysis of phenyl salicylate remains independent of [HO$^-$] within its range 0.005 to 0.060 M at 30°C.[180] The [HO$^-$]-independent hydrolysis of phenyl salicylate is expected to involve either H$_2$O and PS$^-$ (ionized phenyl salicylate) or kinetically indistinguishable HO$^-$ and PSH (nonionized phenyl salicylate) as the reactants. The occurrence of these kinetically indistinguishable reaction steps has been resolved satisfactorily and it has been concluded that pH-independent hydrolysis involves H$_2$O and PS$^-$ as the reactants.[135,181] The rates of pH-independent hydrolysis[182] and alkanolysis[183] of phenyl salicylate have been shown to be increased by ~ 10^6-fold owing to the intramolecular GB (2-O$^-$ group) assistance through an intramolecular intimate ion-pair (IIP).

IIP

The study on the rate of hydrolysis of methyl 3,5-dinitrosalicylate carried out within the pH range < 1 to ~ 12 reveals rather anomalous results. The pseudo-first-order rate constants (k$_w$) for the reactions of H$_2$O with nonionized (HMDNS) and ionized (MDNS$^-$) methyl 3,5-dinitrosalicylate are 6.5×10^{-6} sec^{-1} and 6.6×10^{-6} sec^{-1}, respectively, and these results lead to the conclusion that there is no evidence for intramolecular GB assistance for water reaction with MDNS$^-$ by the weakly basic 2 – O$^-$ (pK$_a$ of HMDNS is 2.45 at 25°C).[184] Although the conclusion appears to be correct, almost similar values of k$_w$ for hydrolysis of HMDNS and MDNS$^-$ are difficult to explain (in the absence of intramolecular GB assistance in the hydrolysis of MDNS$^-$) for the following reasons: (1) the values of $\sigma_{4\text{-OH}}$ (= –0.37) and $\sigma_{4\text{-O}^-}$ (= –0.68)[185] are very different from each other, (2) the value of k$_{OH}$ (hydroxide ion-catalyzed second-order rate constant) for hydrolysis of C$_6$H$_5$COOCH$_3$,[186] 4-$^-$OC$_6$H$_4$COOCH$_3$,[187] and 2-CH$_3$OC$_6$H$_4$COOCH$_3$[188] are 0.125, 1.7×10^{-3}, and 3.1×10^{-2} M^{-1} sec^{-1}, respectively, (3) the respective values of k$_w$ and k$_{OH}$ for hydrolysis of 4-nitrophenyl acetate are 5.5×10^{-7} sec^{-1} [189] and 10 M^{-1} sec^{-1} [190] and of MDNS$^-$ are 6.6×10^{-6} sec^{-1} and 5.3×10^{-2} M^{-1} sec^{-1}, (4) the values of k$_{OH}$ for hydrolysis of phenyl salicylate[191] and phenyl benzoate[192] are 1.24×10^{-3} M^{-1} sec^{-1} and 0.7 M^{-1} sec^{-1}, respectively. These observations show that the intramolecular GB assistance in the reaction of water with MDNS$^-$ does occur, but its effect on rate is largely offset by the rate-retarding substituent (2-O$^-$) effect (which may be owing to the presence of strong electron-withdrawing 3-NO$_2$ and 5-NO$_2$ substituents).

Recently, induced intramolecularity in the reference reaction has been concluded to be responsible for the low rate enhancement (usually ≤ 10-fold)[193] of intramolecular GA–GB-assisted intermolecular nucleophilic reactions.[194] But an approximate 10^6-fold rate enhancement of intramolecular GB-assisted hydrolysis of phenyl salicylate is certainly large enough to cast doubt on invoking the concept of induced intramolecularity of the reference reaction as a sole responsible factor for the low rate enhancement (usually one- to tenfold)[193] of intramolecular GA–GB-assisted intermolecular nucleophilic reactions.

Fife et al.[195] have reported intramolecular GB-assisted ester hydrolysis. The pseudo-first-order rate constants (k_{obs}) are pH-independent from pH 8 to pH 4. The trifluoroethyl, phenyl, and p-nitrophenyl esters of 2-aminobenzoic acid hydrolyze with similar rate constants in the pH-independent reactions, and the rates of these water reactions are ~ twofold slower in D_2O than in H_2O. The most likely mechanism involves the intramolecular GB assistance by the neighboring amino group. The rate enhancements in the pH-independent reaction in comparison with the pH-independent hydrolysis of the corresponding para-substituted esters are 50- to 100-fold.

2.5.9 GENERAL ACID CATALYSIS

All the acids except proton (SA) are called *general acid*. These GAs are also classified as Lewis acids and Bronsted acids. A *Lewis acid* is defined as a chemical species that can accept a pair of electrons from an electron donor (i.e., base), whereas a *Bronsted acid* is defined as a chemical species that can donate a proton to a base. If a reaction site in a particular reaction undergoes a very large change in basicity during the transformation of reactant state to transition state in a concerted reaction or to a reactive intermediate in a stepwise reaction, then the rate of such a reaction is enhanced by GA catalysis. GA catalyst (BH^+) can increase the rate of a nucleophilic addition–elimination reaction by increasing the electrophilicity of electrophilic reaction site by protonating the basic site bonded to electrophilic site of the reactant through an equilibrium process prior to the rate-determining step, as shown in Scheme 2.41. An alternative mechanism by which BH^+ can increase the rate of a nucleophilic addition–elimination reaction is shown in Scheme 2.42 where BH^+ enhances the rate of reaction by increasing the leaving ability of the leaving group (X) by partially protonating it during the cleavage of C–X in the rate-determining step. Scheme 2.41 and Scheme 2.42 are energetically feasible when NuH is a weak and a strong nucleophile, respectively.

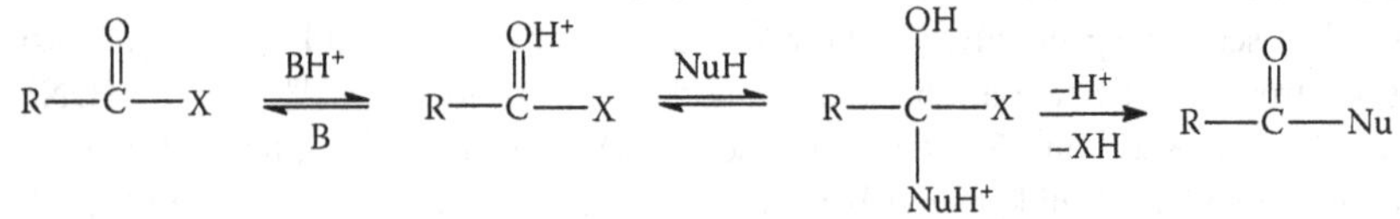

SCHEME 2.41

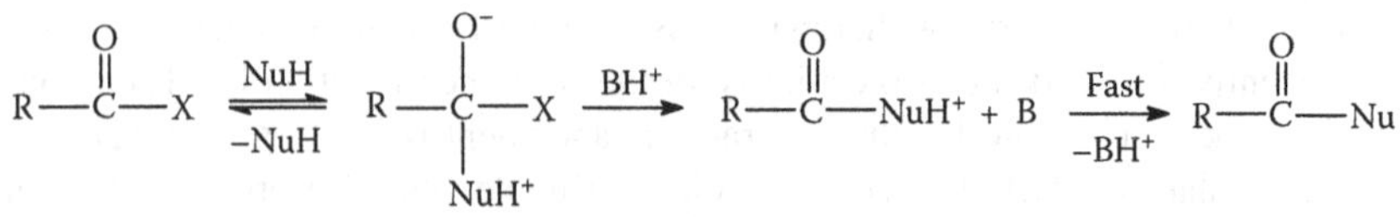

SCHEME 2.42

2.5.9.1 Intermolecular General Acid Catalysis

The discovery that several small ribozymes carry out catalytic self-cleavage at a specific phosphodiester bond generated a paradigm shift in biology in which all the catalytic biological reactions were believed to be catalyzed by only specific enzymes. Although mechanistic aspects of catalytic RNA hydrolysis are far from being fully understood, it is now almost certain that it involves limited catalytic resources, such as GA–GB catalysis, electrostatic effects, and proximity effects.[196–199] The diagnosis of the presence of GA catalysis in a simple reaction process may be carried out as follows. Consider the following general acid (BH^+)-catalyzed reaction

$$RE(=Y)X + NuH \xrightarrow{\ BH^+\ } RE(=Y)Nu + HX \tag{2.56}$$

where NuH represents a neutral nucleophile with a hydrogen atom bonded to the nucleophilic site of nucleophile, $RE(=Y)X$ is a substrate with tetravalent electrophilic site E (an atom), Y is an atom bonded by double bond to E, R is a substituent bonded to E, and X is a leaving group. In order to find out BH^+ catalytic effect on the rate of reaction, the rate study should preferably be carried out in the buffers of BH^+ at different pH. Pseudo-first-order rate constants (k_{obs}), obtained at different total buffer concentrations ($[Buf]_T$ where $[Buf]_T = [BH^+] + [B]$ with B representing conjugate base of BH^+) in aqueous solvent containing a constant value of $[NuH]_T$ (where $[NuH]_T = [NuH] + [NuH_2^+]$) and $[RE(=Y)X]$, are expected to follow Equation 2.57

$$\begin{aligned} k_{obs} - k_0 = \ &k_{n1}[NuH] + k_{n2}[B] + k_{ga1}[NuH_2^+][NuH] + \\ &k_{ga2}[BH^+][NuH] + k_{ga3}[NuH_2^+][B] + k_{ga4}[BH^+][B] \end{aligned} \tag{2.57}$$

where k_0 is the pseudo-first-order rate constant for hydrolysis of substrate at a constant pH. The rearrangement of Equation 2.57 gives Equation 2.58

$$\begin{aligned} k_{obs} - k_0 = \ &k_{n1}f_a^{NuH}[NuH]_T + k_{n2}f_a^B[Buf]_T + k_{ga1}f_a^{NuH}f_{aH}^{NuH}[NuH]_T^2 + \\ &k_{ga2}f_a^{NuH}f_{aH}^B[NuH]_T[Buf]_T + k_{ga3}f_{aH}^{NuH}f_a^B[NuH]_T[Buf]_T + \\ &k_{ga4}[f_a^Bf_{aH}^B[Buf]_T^2 \end{aligned} \tag{2.58}$$

where $f_a^{NuH} = [NuH]/[NuH]_T = K_a^{NuH}/([H^+] + K_a^{NuH})$, $f_{aH}^{NuH} = 1 - f_a^{NuH}$, $f_a^B = [B]/[Buf]_T = K_a^B/([H^+] + K_a^B)$, $f_{aH}^B = 1 - f_a^B$, $K_a^{NuH} = [NuH][H^+]/[NuH_2^+]$, and $K_a^B = [B][H^+]/[BH^+]$. At constant values of pH and $[NuH]_T$, Equation 2.58 is reduced to Equation 2.59

$$k_{obs} - k_0 = A + Q[Buf]_T + S[Buf]_T^2 \tag{2.59}$$

where $A = k_{n1}f_a^{NuH}[NuH]_T + k_{ga1}f_a^{NuH}f_{aH}^{NuH}[NuH]_T^2$, $Q = k_{n2}f_a^B + k_{ga2}f_a^{NuH}f_{aH}^B[NuH]_T + k_{ga3}f_{aH}^{NuH}f_a^B[NuH]_T$, and $S = k_{ga4}f_a^Bf_{aH}^B$.

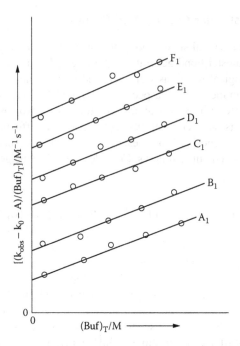

FIGURE 2.7 Plots of $\{(k_{obs}\ k_0\ A)/[Buf]_T\}$ vs. total buffer concentration ($[Buf]_T$) of conjugate base (B) of GA, BH^+, at a constant pH and different total concentrations of nucleophile ($[NuH]_T = A_1, B_1, C_1, D_1, E_1,$ and F_1 where $A_1 < B_1 < C_1 < D_1 < E_1 < F_1$, which represent the experimental validity of Equation 2.57.

The values of k_0 and A can be obtained by carrying out kinetic runs in the presence of NuH buffer at $[Buf]_T = 0$. The plot of $(k_{obs} - k_0 - A)/[Buf]_T$ vs. $[Buf]_T$ should be linear as shown in Figure 2.7 with intercept and slope as Q and S, respectively. If the values of S at different pH turn out to be nearly zero and the values of Q at a constant pH and different values of $[NuH]_T$ give a linear plot with zero intercept (Figure 2.8), then it is certain that the contribution due to $k_{n2}[B] + k_{ga4}[f_a^B f_{aH}^B[Buf]_T^2$ is insignificant compared to other terms of Equation 2.57 under the imposed reaction conditions. These observations are considered to be the consequence of insignificant nucleophilic reactivity of B compared to NuH toward the substrate and, under such circumstances, $k_{ga3}[NuH_2^+][B]$ must be negligible compared to $k_{ga2}[BH^+][NuH]$ in Equation 2.57. These conditional simplifications can reduce Equation 2.59 to Equation 2.60, which may be used to calculate the value of k_{ga2}.

$$k_{obs} - k_0 = k_{n1}[NuH] + k_{ga1}[NuH_2^+][NuH] + k_{ga2}[BH^+][NuH] \quad (2.60)$$

If the contribution of $k_{n2}[B] + k_{ga4}[f_a^B f_{aH}^B[Buf]_T^2$ is significant compared to other terms of Equation 2.57, then the values of k_{n2} and k_{ga4} can be determined

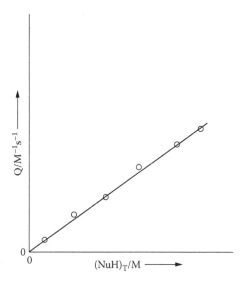

FIGURE 2.8 Plot showing the dependence of Q (intercept at different $[HNu]_T$, obtained from Figure 2.7) upon $[HNu]_T$ under the typical limiting condition where $k_{n2} = 0$.

more satisfactorily by carrying out the kinetic runs in the presence of BH^+ buffer at $[NuH]_T = 0$.

2.5.9.2 Simultaneous Occurrence of Both Intermolecular General Acid and General Base Catalysis

The magnitude of rate constants for GA- and GB-catalyzed reactions depends on (1) the sensitivity of the reaction rate to GA and GB catalysis and (2) the acidity of GA and basicity of GB, respectively. If the rate of a nucleophilic reaction is significantly and equally sensitive to both GA and GB catalysis, where GB also acts as a nucleophile, then the incorporation of GB-catalyzed term, $k_{gb}[NuH]^2$, in Equation 2.57 with $[Buf]_T = 0$, gives Equation 2.61

$$k_{obs} - k_0 = k_n[NuH] + k_{ga}[NuH_2^+][NuH] + k_{gb}[NuH]^2 \qquad (2.61)$$

where $k_n = k_{n1}$, $k_{ga} = k_{ga1}$, and k_{gb} is third-order rate constant for GB-catalyzed cleavage of the substrate. The increase in the basicity constant (K_b) of GB (= NuH) increases both k_n and k_{gb} and decreases k_{ga}. Thus, it is quite possible, depending upon the nature of the reaction, that $k_{ga}[NuH_2^+][NuH]$ is negligible compared to $k_{gb}[NuH]^2$ in Equation 2.61 even at $[NuH_2^+]/[NuH] \approx 10$. Under such circumstances, GA catalysis may not be kinetically detected. If the rate of reaction is less sensitive to GA catalysis compared to GB catalysis, then GA catalysis is unlikely to be kinetically detectable. However, GA catalysis could be detected if the reaction rate is more sensitive to GA catalysis compared to GB catalysis. The sensitivity of a reaction rate to GB catalysis appears to be proportional to its sensitivity to the nucleophilicity of the nucleophile which, in turn, is proportional to the basicity of the base, which also acts as a nucleophile. Generally GA–GB-catalyzed reactions involve either GA or GB catalysis.[200–206] Simultaneous occurrence of both GA and GB catalyses are also reported.[207–211]

In view of Equation 2.61, apparent buffer-catalyzed third-order rate constant (k_b) may be expressed as

$$k_b = k_{ga}f_a^{NuH}f_{aH}^{NuH} + k_{gb}f_a^{NuH}f_a^{NuH} \tag{2.62}$$

where $k_b = (k_{obs} - k_0 - k_n[NuH])/[Buf]_T^2$ and $[Buf]_T = [NuH]_T = [NuH] + [NuH_2^+]$. It is evident from Equation 2.62 that the plot of $k_b/(f_a^{NuH}f_a^{NuH})$ vs. a_H should yield a straight line with definite intercept (= k_{gb}) and definite slope (= k_{ga}/K_a^{NuH}).

Pseudo-first-order rate constants (k_{obs}) for the nucleophilic reaction of CH_3NHOH with N-ethoxycarbonylphthalimide (**53**), obtained at different total concentration of acetate buffer ($[Buf]_T$) of a constant pH in the presence of 0.004-M CH_3NHOH, follow the relationship $k_{obs} = k_b[Buf]_T$, which indicates that buffer-independent cleavage of **53** is insignificant compared to $k_b[Buf]_T$ under the imposed reaction conditions.[211] The general rate law for the cleavage of **53** under such experimental conditions is expressed by Equation 2.63

$$\text{rate} = (k_{gb}[CH_3NHOH][B] + k_{ga}[CH_3NHOH][BH^+])[\mathbf{53}] \tag{2.63}$$

where $B = CH_3COO^-$ and $BH^+ = CH_3COOH$. The observed rate law: rate = $k_{obs}[\mathbf{53}]$ and Equation 2.63 can lead to Equation 2.64. Comparing Equation 2.64 and the relationship: $k_{obs} = k_b[Buf]_T$ gives Equation 2.65.

$$k_{obs} = (k_{gb}f_a^B + k_{ga}f_{aH}^B)f_a^{CH3NHOH}[CH_3NHOH]_T[Buf]_T \tag{2.64}$$

$$k_b = (k_{ga}f_{aH}^B + k_{gb}f_a^B)f_a^{CH3NHOH}[CH_3NHOH]_T \tag{2.65}$$

The values of k_b obtained within pH range 5.20 to 5.97 reveal a linear plot of $k_b/(f_a^B\,f_a^{CH3NHOH}[CH_3NHOH]_T)$ against a_H. The linear least-squares calculated values of intercept (= k_{gb}) and slope (= k_{ga}/K_a^B, with K_a^B representing ionization constant of acetic acid) are $6.2 \pm 1.2\ M^{-2}\ sec^{-1}$ and $(5.3 \pm 0.3) \times 10^6\ M^{-3}\ sec^{-1}$, respectively. The value of k_{ga}/K_a^B gives k_{ga} as $111 \pm 7\ M^{-2}\ sec^{-1}$ (where $pK_a^B = 4.68$).

A kinetic study on the reaction of taurine ($TauNH_2$ = 2-aminoethanesulfonic acid) with N,N-dichlorotaurine ($TauNCl_2$), Equation 2.66, carried out in the presence of buffers

$$TauNH_2 + TauNCl_2 + BH^+ \longrightarrow TauNHCl + TauNH_2Cl^+ + B \tag{2.66}$$

$$TauNH_2 + TauNCl_2 \underset{}{\overset{K}{\rightleftharpoons}} TauNCl^- + TauNH_2Cl^+$$

$$TauNCl^- + BH^+ \xrightarrow{Slow} TauNHCl + B$$

SCHEME 2.43

reveals the presence of GA catalysis.[212] Pseudo-first-order rate constants (k_{obs}) obtained at constant pH and different total buffer concentrations ($[Buf]_T$) follow Equation 2.67

$$k_{obs} = (k_0 + k_{ga}[BH^+])[TauNH_2] \qquad (2.67)$$

where k_0 is the second-order rate constant for the solvent-catalyzed reaction and k_{ga} is the third-order rate constant for a GA-catalyzed reaction in which GA is the acid form (= BH^+) of the buffer. The values of the GA catalytic rate constants lead to the proposed termolecular transition state (TS_{24}) in the rate-determining step. The probability of the occurrence of termolecular effective collision is very low even in water solvent, especially when all three colliding molecules are polyatomic. Thus, the occurrence of TS_{24} is energetically highly improbable. An alternative mechanism for the GA-catalyzed cleavage of $TauNCl_2$ as shown in Scheme 2.43 may not be completely ruled out.

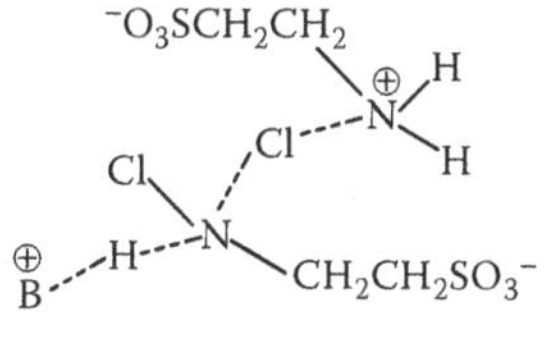

TS_{24}

Rates of hydrolysis of ($\pm$)-7β,8α-dihydroxy-9α,10α-epoxy-7,8,9,10-tetrahydrobenzo[a]pyrene (E), studied in 1:9 dioxane-H_2O solvent-containing buffer solutions of primary amines, reveal the presence of GA catalysis only for amines with pK_a values of < ~ 8 and either both GA–GB catalysis or GA catalysis-nucleophilic reaction for amines with pK_a values of > ~ 8.[213] The GA-catalyzed hydrolysis of E involves epoxide-ring opening to yield a discrete α-hydroxy carbocation as the rate-determining step. The intermediate α-hydroxycarbocation is sufficiently stable so that its reactions with external nucleophiles and bases compete with its reaction with the solvent. Curvatures in the plots of the kinetic term due to buffer (k_b) as a function of the mole fraction of buffer acid ammonium ions (BH^+) with $pK_a > 8$ are interpreted in terms of a change in rate-determining step of the GA-catalyzed pathway from epoxide-ring opening at low amine buffer base (B) concentrations to reaction of B acting as either a GB or nucleophile with an α-hydroxy carbocation at higher [B].

2.5.9.3 Kinetic Ambiguity of the Assignment of Intermolecular General Acid Catalysis

The realization of the occurrence of intermolecular GA catalysis comes from the experimental fact of the presence of a kinetic term $k_{ga}[BH^+][NuH][Sub]$ (where BH^+ is the GA, NuH is the nucleophile, and Sub is the substrate) in the rate law for the nucleophilic reaction between NuH and Sub in the presence of BH^+. But this kinetic term is kinetically indistinguishable from kinetic term $k[H^+][B][NuH][Sub]$, which represents SA–GB catalysis or SA-nucleophilic catalysis. However, these kinetically indistinguishable catalytic mechanisms can be resolved, at least qualitatively, as described in the following text.

Kinetically indistinguishable kinetic terms $k_{ga}[BH^+][NuH][Sub]$ and $k[H^+][B][NuH][Sub]$ can be resolved by selecting a nucleophile that does not contain any hydrogen covalently bonded to the nucleophilic site of the nucleophile (Nu, such as a tertiary amine). Under such experimental conditions, SA–GB catalysis cannot exist in the rate law, and if the rate of reaction is still sensitive to GA catalysis, it is because of either GA catalysis (i.e., owing to $k_{ga}[BH^+][Nu][Sub]$) or SA-nucleophilic catalysis. The ambiguity between GA catalysis and SA-nucleophilic catalysis can be resolved by characterizing the nucleophilic reaction products. The base component (B) of buffer should be appropriately selected so that its nucleophilic reaction with Sub gives stable products. If under such manipulated reaction conditions, because of nucleophilic reaction between NuH and Sub and the rate of reaction, the product is still sensitive to GA catalysis, then these observations assert the presence of GA catalysis and negate the presence of SA-nucleophilic reaction.

An interesting example of the resolution between SA-nucleophilic catalysis and SA–GB catalysis is found in the kinetic study on the rate of hydrolysis of *N*-benzyl-β-sultam in the presence of carboxylate buffers.[149] Pseudo-first-order rate constants (k_{obs}), obtained at a constant pH, ionic strength, temperature, and different total carboxylic acid buffer concentrations ([Buf]$_T$), follow Equation 2.68

$$k_{obs} = k_H a_H + k_{BH} f_{aH}^B [Buf]_T \qquad (2.68)$$

where $f_{aH}^B = [BH^+]/[Buf]_T$ and $[Buf]_T = [B] + [BH^+]$.

The observation on GA-catalyzed hydrolysis is in contrast to the GB catalysis seen in the buffer-catalyzed hydrolysis of β-lactams of penicillins.[214] Thus, the probable mechanism of buffer catalysis in this case involves SA-nucleophilic catalysis[215] (Scheme 2.44). The evidence for this catalytic proposal comes from the following observations: (1) the mixed acid anhydride intermediate (Int$_{11}$) has been trapped with aniline to give acetanilide as identified by HPLC; (2) the kinetic runs for hydrolysis of *N*-benzyl-β-sultam in acetate buffers revealed biphasic plots of observed absorbance vs. reaction time; an initial exponential burst of UV

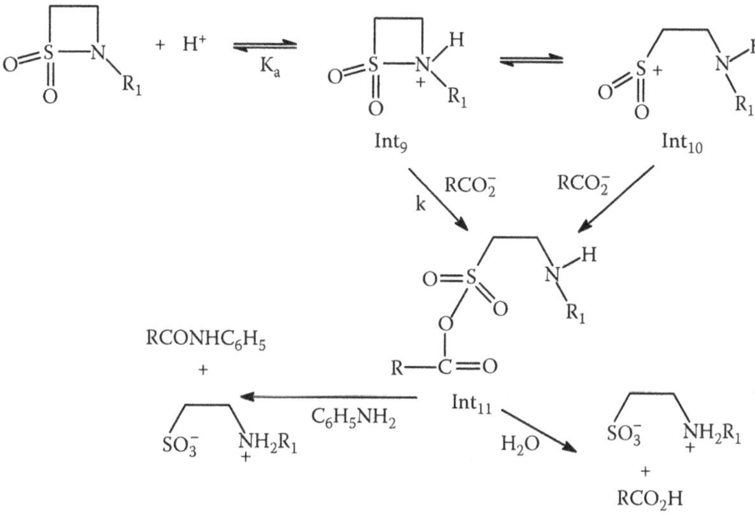

SCHEME 2.44

absorbance was followed by a much slower first-order reaction; these observations only reveal the formation of kinetically detectable intermediate which is most likely Int_{11}; (3) a good Bronsted plot of log k_{BH} against pK_a for 2-chloroacetic acid, 2-methoxyacetic acid, and acetic acid with slope (α) of 0.67 and significant positive deviation from linear Bronsted plot for formic acid are attributed to a nucleophilic pathway for catalysis; the Bronsted slope (α) of 0.67 corresponds to a β_{nuc} value of 0.33 (derived from the relationship: $\beta_{nuc}^- \alpha = 1.0$) for *the SA-nucleophilic mechanism, indicative of an early transition state in which there has been a small amount of neutralization of the negative charge on the carboxylate anion; and (4) the value of $k_{BH}^{H2O}/k_{BH}^{D2O}$ of 1.57 for chloroacetate buffer-catalyzed hydrolysis of N-benzyl-β-sultam is compatible with the SA-nucleophilic catalysis as is the observed entropy of activation of -148 J K^{-1} mol^{-1} for this reaction.

2.5.9.4 Intramolecular General-Acid-Assisted Intra- and Intermolecular Nucleophilic Reactions

Consider the following reactions

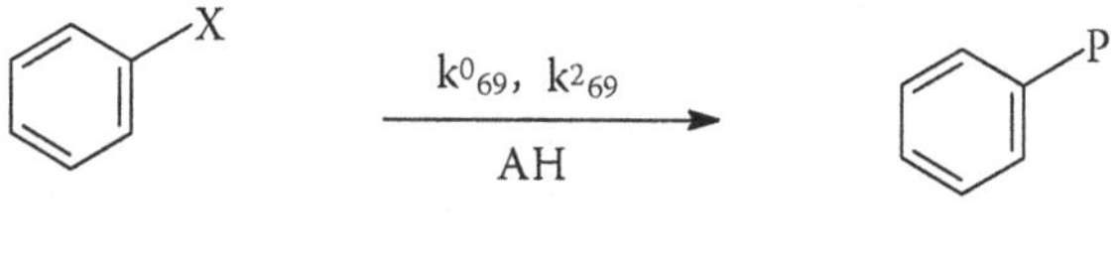

$$\qquad \xrightarrow[\text{AH}]{k^0_{69},\ k^2_{69}} \qquad \qquad (2.69)$$

---

* Scheme 2.44 and Equation 2.68 show that $k_{BH} = kK_a^{RCOOH}/K_a$ (where K_a^{RCOOH} represents ionization constant of carboxylic acid, RCOOH) and, hence, $\delta log k_{BH}/\delta(-log K_a^{RCOOH}) = \delta log k/\delta(-log K_a^{RCOOH}) + \delta(log K_a^{RCOOH})/\delta(-log K_a^{RCOOH}) - \delta(log K_a)/\delta(-log K_a^{RCOOH})$ or $\alpha = \beta_{nuc}^- 1 + 0$.

$$(2.70)$$

$$(2.71)$$

where X is the reaction site in the reactant, and AH represents GA catalyst. The rates of pH-independent and [AH buffer]-independent reactions (of Equation 2.69 and Equation 2.70) are not significantly different from each other provided polar effect of 4-AH group on the rate of reaction (of Equation 2.70) is low. The value of Hammett substituent constant ($\sigma_{4\text{-COOH}}$) for 4-COOH (= AH) is ~ 0.45 and, therefore, even if rate of reaction of Equation 2.70 is significantly sensitive to polar effect of the 4-COOH group, the values of k^0_{69} and k^0_{70} cannot differ by more than an order of one or two. However, the SA-catalyzed second-order rate constant for hydrolysis of phthalamic acid (**5**) is larger than that of 4-carboxy-benzamide (**6**) by a factor of ~ 500.[16b] A huge amount of kinetic data on intramolecular carboxylic-group-assisted hydrolysis of amide bond (both alkyl and aryl amide bonds) reveal the upper limit of rate enhancement as > 10^{14}.[66b] Similarly, the rate of intramolecular hydroxyl group-assisted lactonization of **54** is faster by a factor of 10^{15} than the rate of bimolecular reaction between phenol and aliphatic carboxylic under similar experimental conditions.[216] A rate enhancement of the order of 10^{14} is indeed higher than the catalytic limits of many enzyme-catalyzed hydrolysis of peptide bonds. Such enormous rate enhancement is some-times incorrectly described as intramolecular GA catalysis.

54

It is evident from the reaction of Equation 2.71 that the 2-AH group in 2-HAC$_6$H$_4$P cannot act as an intramolecular GA catalyst, because the reaction site (–X) is already transformed into product (2-HAC$_6$H$_4$P) and, hence, no longer available for catalytic reaction. Therefore, such rate enhancement is more appropriately described as being due to intramolecular GA-assisted reaction.

To illustrate a few examples on intramolecular GA-assisted reaction, let us consider the most extensively studied intramolecular carboxylic-group-assisted hydrolysis of amide bond. Pseudo-first-order rate constants (k_{obs}) for hydrolysis

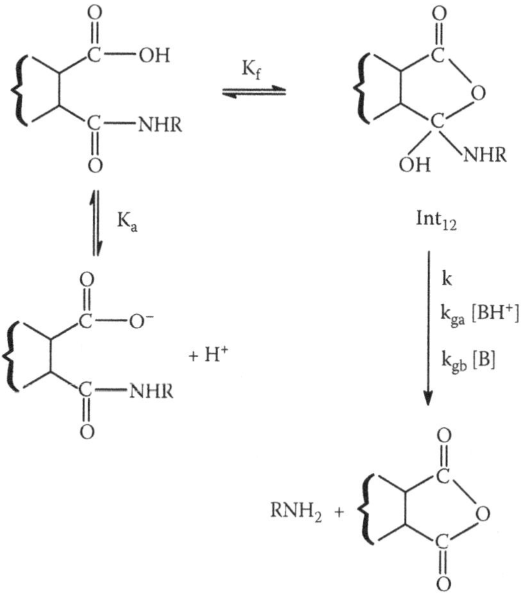

K_f

K_a

Int_{12}

k

$k_{ga} [BH^+]$

$k_{gb} [B]$

$+ H^+$

$RNH_2 +$

SCHEME 2.45

of phthalamic acid,[17a,217] N-methoxyphthalamic acid[18], and N-hydroxyphthalamic acid[19] are 5.1×10^{-5} sec^{-1} (pH 1.3–1.8, 35°C)[217], 2.0×10^{-5} sec^{-1} (0.02-M HCl, 35°C, 70.2% v/v CH$_3$CN in mixed aqueous solvent)[17a], 2.5×10^{-4} sec^{-1} (0.03-M HCl, 30°C, 30% v/v CH$_3$CN in mixed aqueous solvent)[18], and 2.3×10^{-3} sec^{-1} (0.027-M HCl, 35°C, 30% v/v CH$_3$CN in mixed aqueous solvent)[19], respectively. The accumulated kinetic data on such reactions during the last four to five decades suggest a general mechanism as shown in Scheme 2.45. But the presence of Int$_{12}$ on the reaction path has not been unequivocally proven. However, Perry and Parveen[218] did find clear evidence for a long-lived intermediate as a precursor to imide formation in the cyclization of substituted N-arylphthalamic acids in glacial acetic acid solvent. Although an attempt to isolate and subject to structural characterization of the intermediate was unsuccessful, indirect experimental evidence suggests the structure of the intermediate as shown by Int$_{13}$, which is formed owing to N-cyclization of N-arylphthalamic acid in glacial acetic acid solvent.

In order to estimate the rate enhancement due to intramolecular carboxylic group participation (Scheme 2.45) in the hydrolysis of phthalamic acid, the value of k_{obs} should be compared with the value of k_{obs} for hydrolysis of benzamide or 4-COOHC$_6$H$_4$CONH$_2$ in the presence of C$_6$H$_5$COOH or RCOOH under identical experimental conditions. However, C$_6$H$_5$COOH or RCOOH cannot react with benzamide or 4-COOHC$_6$H$_4$CONH$_2$ in the presence of ≥ 0.01 M HCl through a mechanism similar to one shown by Scheme 2.45 for the following reasons: (1) both SA catalysis and GA catalysis are bimolecular processes and [HCl] > [C$_6$H$_5$COOH] or [RCOOH] if limiting concentration of phthalamic acid is ~ 0.001

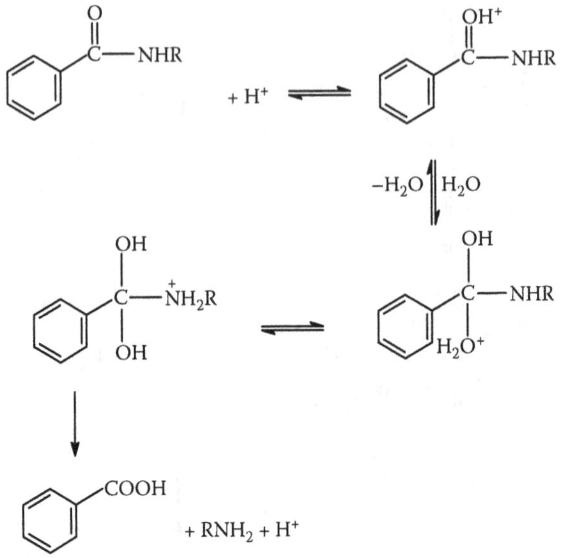

SCHEME 2.46

M, (2) the acidity of C_6H_5COOH or $RCOOH$ is significantly lower than that of HCl, (3) water is a much stronger nucleophile than C_6H_5COOH or $RCOOH$, and (4) the value of $[H_2O]/[C_6H_5COOH]$ or $[H_2O]/[RCOOH]$ is larger than 10^4, if $[C_6H_5COOH]$ or $[RCOOH] \leq 0.001$ M. Thus, the rate of hydrolysis of benzamide in the presence of 0.01 M HCl, 0.001 M C_6H_5COOH, or $RCOOH$ must be SA-catalyzed and the expected reaction mechanism is shown in Scheme 2.46.

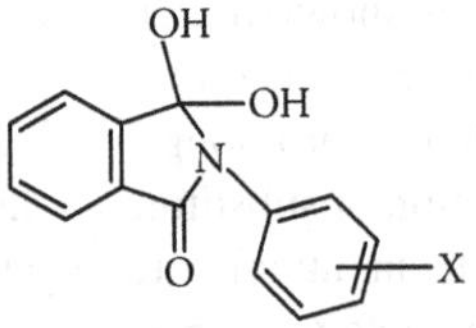

Int_{13}

It may be noted that the values of Hammett reaction constants (ρ) for SA-catalyzed hydrolysis of N-arylphthalamic acids[16b] and N-arylbenzamides[219] are − 1.2 and + 0.56, respectively, which indicate the occurrence of different reaction mechanisms in these reactions.

It is now evident that the actual rate enhancement due to intramolecular carboxylic-acid-assisted hydrolysis of phthalamic acid is ~ ∞, because k_{obs} for the hydrolysis of benzamide under identical experimental conditions is nearly zero. However, if one compares the k_{obs}^{intra} for the hydrolysis of phthalamic acid with k_{obs}^{inter} for SA-catalyzed hydrolysis of benzamide, then $k_{obs}^{intra}/k_{obs}^{inter} \approx 10^5$ where $k_{obs}^{inter} = 3.7 \times 10^{-10}$ sec^{-1} at 35°C and 0.001 M HCl.[220]

The acidic aqueous cleavages of phthalamic acid and almost all N-substituted phthalamic acids involve 100% O-cyclization as shown in Scheme 2.45. However, N-methylphthalamic acid undergoes ~ 20% N-cyclization to produce N-methylphthalimide and ~ 80% O-cyclization to produce phthalic anhydride in mild acidic aqueous solution.[32,221] Similarly, aqueous cleavage of 3- and 4-nitrophthalanilic acids yields appreciable amounts of the corresponding N-phenylphthalimide.[16a] N-cyclization, compared to O-cyclization, involves an apparently stronger nucleophile and weaker acidic hydrogen attached to the nucleophilic site. Although the exact nature of the driving force that causes predominant O-cyclization of phthalamic acid and N-substituted phthalamic acids in mild acidic aqueous solution is not fully understood, it appears that the delicate balance between nucleophilicity of intramolecular nucleophile and acidity of the hydrogen attached to nucleophilic site probably provide the driving force for the relative extent of O- and N-cyclization in these reactions.

The cleavage of N-(2-aminophenyl)phthalamic acid (**55**) in dilute aqueous acids (pH range 0 to 6) reveals the formation of N-(2-aminophenyl)phthalimide (i.e., N-cyclization) between ~ 80 and ~ 100% yields.[222] Pseudo-first-order rate constants (k_{obs}) remain unchanged with the change in pH from ~ 1 to ≤ 4 and the value of k_{obs} within this pH range is 7×10^{-4} sec^{-1} at 60°C, which represents very large rate enhancement owing to intramolecular GA [= 2-CONH(2'-NH$_2$C$_6$H$_4$)]-assisted cleavage of **55**. The suggested mechanism of this reaction is shown in Scheme 2.47. Although Scheme 2.47 is consistent with the formation of observed cyclized product (N-(2-aminophenyl)phthalimide), it does not provide an answer to the following question: why do equally probable energetic paths as shown in Scheme 2.48 for the formation of O-cyclized product not occur? An efficient intramolecular GA-assisted O-cyclization in the cleavage of N-hydroxyphthalamic acid has been recently reported.[19] Probable occurrence of intramolecular GB-assisted N-cyclization of **55** through transition state TS$_{25}$ has not been explored.

TS$_{25}$

The cleavage of N-(2-carboxybenzoyl)-L-leucine (**56**), studied in acidic aqueous solvent, reveals both N-cyclization to form imide N-phthaloylleucine and hydrolysis product phthalic acid.[223] Imide formation predominates under highly acidic conditions ($H_0 < -1$, imide > 20%) and hydrolysis product in the $H_0 > -1$ to pH 5 range (hydrolysis product > 80%). The O-cyclization of nonionized **56**

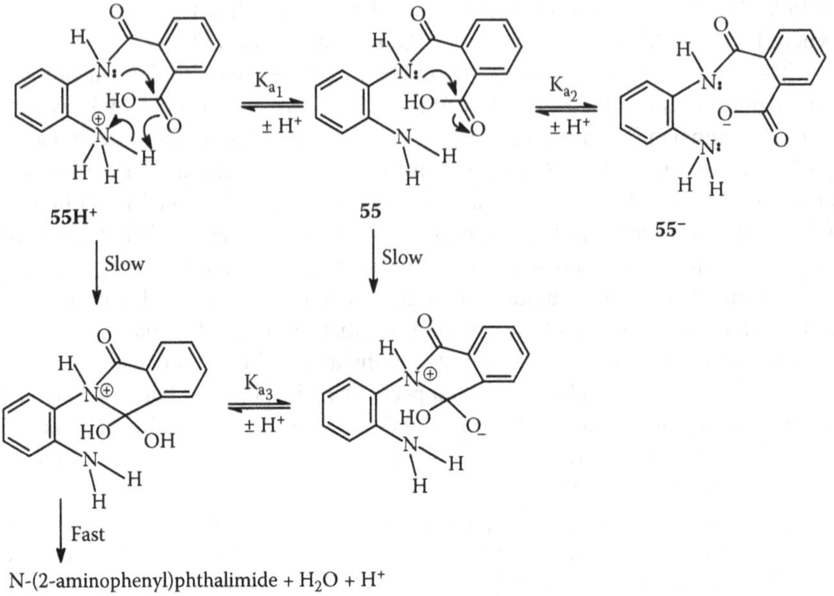

SCHEME 2.47

involves a mechanism similar to the one shown in Scheme 2.45, whereas O-cyclization of monoanion **56** involves 2-COO⁻ group acting as nucleophile and nonionized COOH group of leucine moiety provides intramolecular GA assistance for leaving-group expulsion. This proposal is however not supported by experimental evidence. Imide formation also requires participation of two carboxy groups in the pH 2 to 5 range. The proposed mechanism for hydrolysis reaction of **56** at [HCl] > 1.0 M is similar to the one shown in Scheme 2.46 despite the fact that k_{obs} (= 1 × 10⁻⁴ sec⁻¹) for hydrolysis at 50°C and 2-M HCl is ~ 10⁴-fold larger than k_{obs} (= 1.5 × 10⁻⁸ sec⁻¹) for hydrolysis of N-benzoyl-L-leucine at 50°C and 2-M HCl.[224]. It is interesting to note that the second-order rate constant (k_2) for SA-catalyzed hydrolysis of **5** (phthalamic acid), obtained within [HCl] range 1 to 5 M at 47.3°C, is about 500 times larger than that for the hydrolysis of **6** (4-COOHC₆H₄CONH₂) under similar conditions.[16b] The nearly 500-fold larger value of k_2 for **5** than that for **6** is attributed to the intramolecular carboxy group participation in the aqueous cleavage of **5** even at 5-M HCl. N-cyclization was not kinetically detected in the hydrolysis of **5** even at 6-M HCl and 47.3°C.[16b]

The aqueous cleavage of the monoanionic form of the symmetrical formaldehyde acetal of 4-hydroxybenzofuran-3-carboxylic acid (**57**) is assisted by the neighboring COO⁻ and COOH groups through the transition state TS₂₆.[225] The rate enhancement due to intramolecular GA assistance in this reaction system is 9 × 10⁴ over the rate expected if only SA catalysis occurs.

To illustrate some examples of intramolecular GA-assisted intermolecular nucleophilic reactions, consider the rate of hydrolysis of protonated diethyl 8-

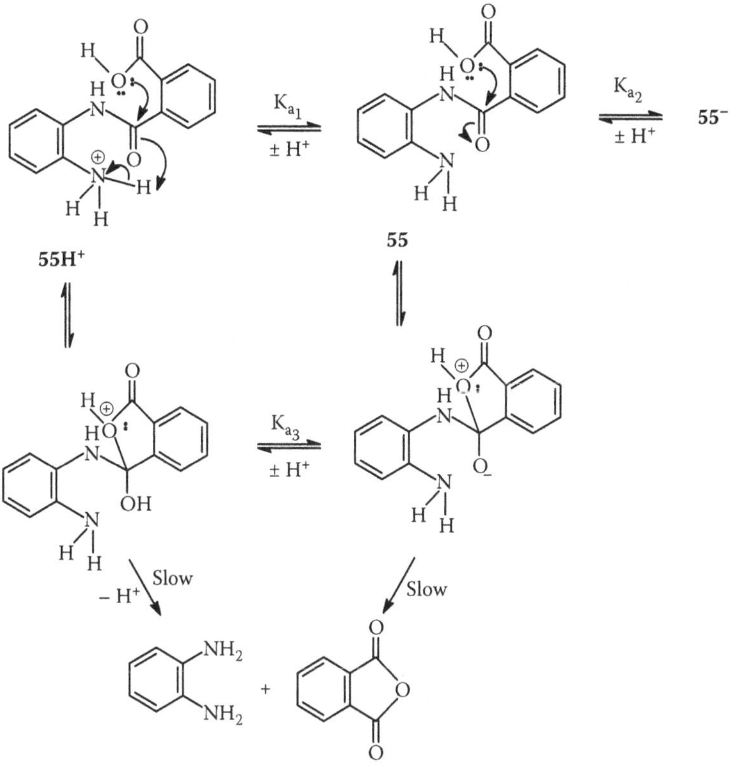

SCHEME 2.48

dimethylaminonaphthyl-1-phosphate (**58H$^+$**), which is assisted by the neighboring dimethylammonium group, with a rate enhancement compared to diethyl naphthyl-1-phosphate of almost 10^6.[226] The pK$_a$ of the naphtholate leaving group is reduced from 9.4 to 3.4 by partial protonation in the transition state. Nearly 2×10^5-fold rate acceleration is observed in the intramolecular GA, 2-COOH-assisted hydrolysis of *tert*-Bu and 1-arylethyl ethers (**59**) of salicylic acid.[227] Nearly 10^3-fold rate enhancement due to neighboring caboxy group is observed in the reaction of Br$^-$ with **60** than that with **61** as expressed by Equation 2.72 and Equation 2.73, respectively.[228]

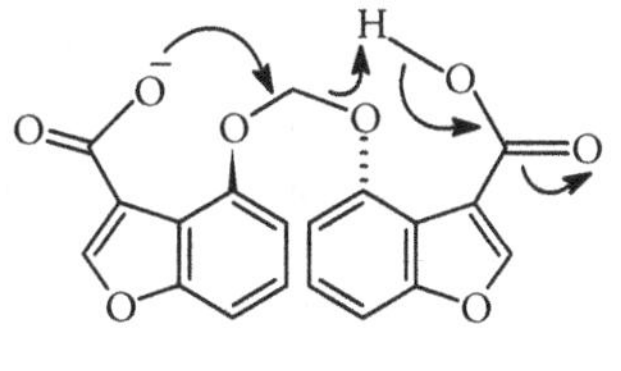

TS$_{26}$ 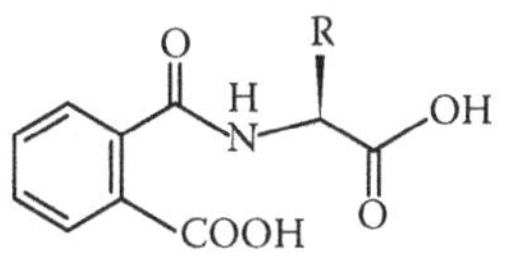   **56**

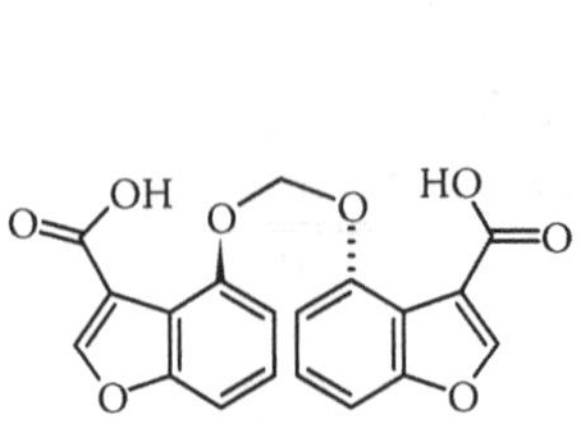

57 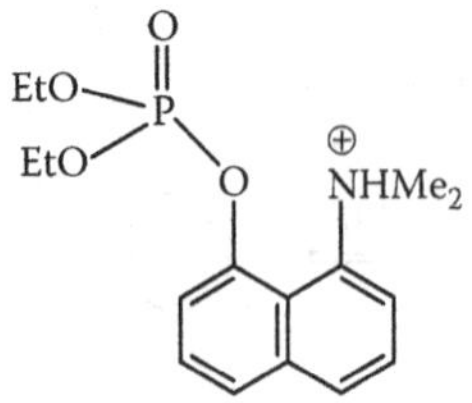   58H⁺ 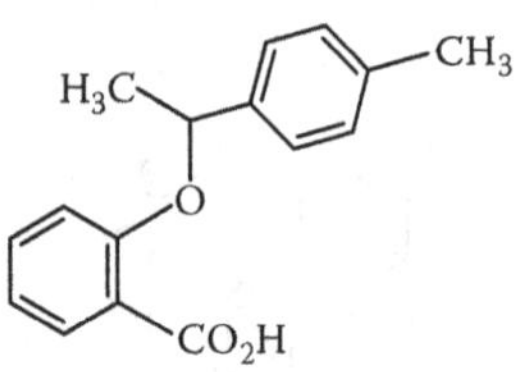   59

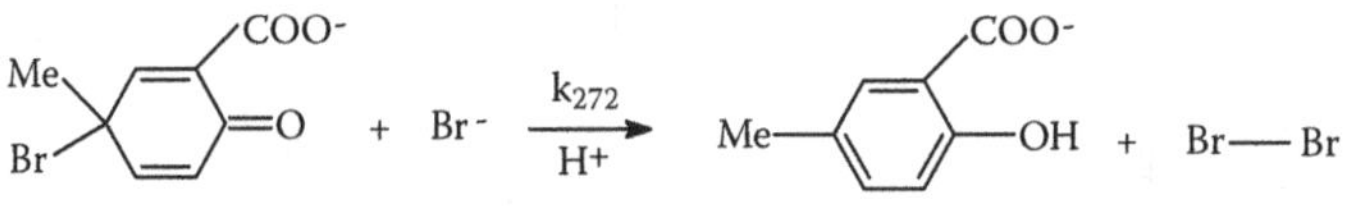

60

$$(2.72)$$

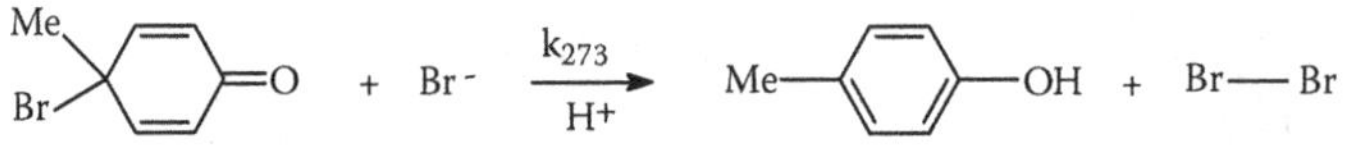

61

$$(2.73)$$

2.5.9.5 Simultaneous Occurrence of Intramolecular General Acid and General Base Catalysis

When a chemical reaction undergoes sufficiently large acidity/basicity change at the reaction site owing to bond formation and bond cleavage that makes the TS (if the reaction is concerted one) or reactive intermediate (if the reaction is stepwise) unstable enough to stop the reaction, then the rate of such a reaction becomes sensitive to SA–SB and GA–GB catalysis. The extent of catalytic efficiency of catalyst is expected to be proportional to the extent of instability of TS or reactive intermediate during the course of reaction. As the rates of intramolecular reactions are larger than those of their intermolecular counterparts by a factor ranging from 10^2 to 10^{15}, it is vital for an efficient GA, GB, or GA-GB catalyst to approximate the reaction from "intermolecular" to "very close to intramolecular" by complexing with the reaction substrate (the complex formation should involve weak bonding forces). Such characteristic of a catalyst is expected only if the catalyst has macromolecular structure with both GA and GB functionalities at geometrically appropriate positions in the catalyst–substrate complex. Perhaps this is the reason or one of the reasons why all enzyme catalysts contain large molecular structures such as protein. A catalyst with both GA and GB functionalities and small molecular structures cannot act efficiently because of extremely low probability of the formation of catalyst–substrate complex with appropriately positioned catalytic and reaction sites. Simultaneous occurrence of both GA and GB catalysis in the same reaction step (even if such a catalyst contains both GA and GB groups) is almost nonexistent. Even if the structure of a catalyst with both GA and GB functional groups is sufficiently large and has the ability to form a complex with substrate through weak bonding forces, it may not show simultaneous occurrence of both GA and GB catalysis if the difference of pK_a of GA and conjugate acid of GB is significantly large, because under such circumstances both GA and GB undergo the usual fast acid–base reaction.

Natural catalysts such as ribonuclease A and lysozyme catalyze reactions by using both GA and GB catalysis through transition states TS_{27} and TS_{28}, respectively, during the course of catalytic reactions. Whether the occurrence of GA and GB catalyses in ribonuclease A-catalyzed reactions is concurrent (TS_{27}) or stepwise is still a topic of debate.[229] But the essence of catalysis, i.e., involvement of both GA and GB catalyses is almost certain.

$$TS_{27}$$

$$TS_{28}$$

2.5.9.6 Induced Intramolecular GA–GB-Assisted Cleavage of Substrate

Certain molecules containing GA–GB units at appropriate positions in a three-dimensional molecular structural network can increase the rate of an intermolecular GA–GB-sensitive reaction by approximating its intermolecularity to intramolecularity through the formation of a stable inclusion complex with the reactant, which is sensitive to GA–GB-assisted rate enhancement. Cyclodextrins (CDs) are cyclic oligosaccharides that are composed of glucose units joined together in a fashion to have cavities of shape and size depending upon the number of glucose units in them. As a result, each CD has a cavity with which it may form inclusion complexes, and CDs can act as hosts toward a variety of organic and inorganic guests.[228] The three main CDs, designated as α-CD, β-CD, and γ-CD contain 6, 7, and 8 glucose units, respectively, so that the widths of their cavities increase in a regular manner.

Cyclodextrins have been extensively used as host molecules that reduce a large number of intermolecular nucleophilic bimolecular reactions into induced intramolecular nucleophilic reactions with rate enhancement ranging from modest (3- to 30-fold)[230] to very large ($\sim 10^5$-fold).[228] The Cu(II) complex of a β-cyclodextrin dimer with a linking bipyridyl group accelerates the rate of hydrolysis of several nitrophenyl esters by a factor of 10^4 to 10^5 with at least 50 turnovers and no sign of product inhibition.[231] A rate acceleration of 1.4×10^7 over the background reaction rate is observed with an added nucleophile, which also binds to the metal ion. The kinetic and other evidence point to a mechanism in which the metal ion plays a bifunctional GA–GB role, enforced by the binding geometry that holds the substrate functionality right on top of the catalytic metal ion. A pair of β-cyclodextrin-peptide hybrids with three functional groups, β-cyclodextrin, imidazole, and carboxylate, in this order revealed significant rate enhancement for the hydrolysis of a few activated esters.[232]

Although a search of literature does not reveal any report in which cyclodextrins cause rate enhancement due to induced intramolecular GA–GB catalysis, it seems logical that cyclodextrins with stereochemically appropriate GA and GB functionality could provide rate acceleration due to induced intramolecular GA–GB catalysis.

2.5.10 PHASE TRANSFER CATALYSIS

The role of a catalyst molecule in phase transfer catalysis (PTC) is very different from that in other types of catalysis as described earlier in Subsection 2.5.1, Subsection 2.5.2, and Subsection 2.5.3. Phase transfer catalysis is required for those bimolecular reactions in which both reactant molecules differ so much in terms of molecular characteristics such as polarity, hydrophobicity, etc., that both of them cannot solubilize in the same phase of the reaction medium. Such bimolecular reactions require two immiscible phases in which each phase solubilizes only one of two types of reactant molecule. Thus, these bimolecular reactions do not take place simply because the reactant molecules cannot come in close proximity — an essential requirement for any reaction to occur. However, such reactions might occur at the interface of the two immiscible phases provided the interface is capable (energetically) of bringing the two reactant molecules in close proximity of each other.

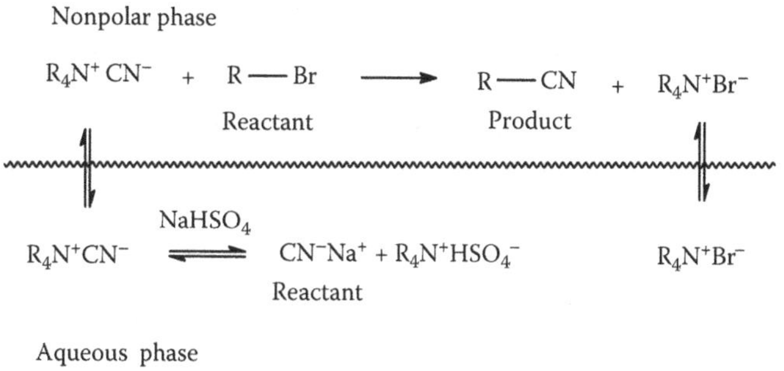

Nonpolar phase

$$R_4N^+ CN^- \ + \ R \text{---} Br \longrightarrow R \text{---} CN \ + \ R_4N^+Br^-$$

Reactant   Product

NaHSO$_4$

$$R_4N^+CN^- \rightleftharpoons CN^-Na^+ + R_4N^+HSO_4^- \qquad R_4N^+Br^-$$

Reactant

Aqueous phase

SCHEME 2.49

The close proximity of immiscible phase-separated reactants is achieved by the use of amphipathic molecules or ions (an amphipathic molecule contains both hydrophobic and hydrophilic moieties) such as tetraalkylammonium ions, which form inclusion complexes or ion-pair complexes with hydrophilic or ionic reactants. These complexes or ion-pairs can move from hydrophilic phase to hydrophobic phase with an almost insignificant energy barrier if the hydrophibic area of an amphipathic molecule or ion is significantly large. Such amphipatic molecules or ions thus enable hydrophilic or ionic reactant molecules to react with hydrophobic reactant molecules within the hydrophobic phase. A simplified and brief mechanism of this process is shown in Scheme 2.49. As this process does not involve the consumption of amphipathic molecules or ions during the progress of the reaction, they are called *phase transfer catalysts*. Although fine details of the mechanism of phase-transfer-catalyzed reactions are not fully understood, Starks[233] presented a plausible mechanism for such catalyzed reactions and proofs that the rate-determining step occurs within the hydrophobic phase. A rather more detailed description of the principles of PTC by quaternary ammonium salts is found in a review article by Brandstrom.[234] A review on progress in phase-transfer catalysts for all types of phase-transfer reactions and mechanisms was recently published.[235] The use of chiral phase-transfer catalysts in the synthesis of enantioselective amino acids was recently reviewed.[236]

2.5.11 ASYMMETRIC CATALYSIS

The catalytic synthesis of an enantioselective product from a prochiral reactant is said to involve asymmetric catalysis and the catalyst is called an *asymmetric catalyst*. Asymmetric catalysis is found in homogeneous, heterogeneous[237], and phase-transfer reactions.[238–240] This area of research gained a significant momentum after the award of a 2001 Nobel Prize to the researchers working on asymmetric catalytic chemical processes.[241–243] In the case of phase transfer catalysis, the quantitative assessment of rate enhancement is not carried out in any such catalytic reaction simply because the reaction system is just too complex for such effort. Almost all reported asymmetric catalytic reactions are involved with synthesis in which the main aim has been focused on the yield of enantioselective[244–250] or diastereoselective[251,252] products. Although mechanistic details based on selective asymmetric products, characterizations are available,[253–258] fine detailed mechanisms of these interesting and potentially important catalyzed reactions are severely lacking. However, it is almost certain that an asymmetric catalyst uses one or more than one factor (as described in Section 2.2 to Section 2.4) for its catalytic effect on the rate of a reaction.

REFERENCES

1. Moore, W.J. *Physical Chemistry*, 4th ed. Orient Longmans Ltd., New Delhi, New Impression reprinted in India **1969**, pp. 300–301.
2. Atkins, P.W. *Physical Chemistry*, Oxford University Press, Oxford, UK, **1978**, p. 260.
3. Khan, M.N. Effects of cetytrimethylammonium bromide (CTABr) micelles on the rate of the cleavage of phthalimide in the presence of piperidine. *Colloids Surf. A* **2001**, *181*, 99–114.
4. (a) Khan, M.N. The kinetics and mechanism of a highly efficient intramolecular nucleophilic reaction: the cyclization of ethyl N-[o-(N-hydroxycarbamoyl)benzoyl]carbamate to N-hydroxyphthalimide. *J. Chem. Soc. Perkin Trans. 2.* **1988**, 213–219. (b) Khan, M.N. Effect of hydroxylamine buffers on apparent equilibrium constant for reversible conversion of N-hydroxyphthalimide to o-(N-hydroxycarbamoyl)benzohydroxamic acid: evidence for occurrence of general acid-base catalysis. *Indian J. Chem.* **1991**, *30A*, 777–783.
5. Satterthwait A.C., Jencks, W.P. The mechanism of the aminolysis of acetate esters. *J. Am. Chem. Soc.* **1974**, *96*(22), 7018–7031.
6. Kanamori, K., Roberts, J.D. ^{15}N NMR studies of biological systems. *Acc. Chem. Res.* **1983**, *16*, 37–41.
7. Khan, M.N., Ohayagha, J.E. Kinetics and mechanism of the cleavage of phthalimide in buffers of tertiary and secondary amines: evidence of intramolecular general acid-base catalysis in the reactions of phthalimide with secondary amines. *J. Phys. Org. Chem.* **1991**, *4*, 547–561.
8. Fox, J.P., Jencks, W.P. General acid and general base catalysis of the methoxyaminolysis of 1-acetyl-1,2,4-triazole. *J. Am. Chem. Soc.* **1974**, *96*(5), 1436–1449 and references cited therein.
9. Bruice, P.Y., Bruice, T.C. Intramolecular general base catalyzed hydrolysis and tertiary amine nucleophilic attack vs. general base catalyzed hydrolysis of substituted phenyl quinoline-8- and -6-carboxylates. *J. Am. Chem. Soc.* **1974**, *96*(17), 5523–5532.
10. Brown, R.S., Neverov, A.A. Acyl and phoshoryl transfer to methanol promoted by metal ions. *J. Chem. Soc. Perkin Trans 2.* **2002**, 1039–1049.
11. Khan, M.N. Salt effects in the aqueous alkaline hydrolysis of N-hydroxyphthalimide: kinetic evidence for the ion-pair formation. *Int. J. Chem. Kinet.* **1991**, *23*, 561–566.
12. Khan, M.N. Salt and solvent effects on alkaline hydrolysis of N-hydroxyphthalimide: kinetic evidence for ion-pair formation. *J. Phys. Org. Chem.* **1994**, *7*, 412–419.
13. Hine, J. *Structural Effects on Equilibria in Organic Chemistry*, John Wiley & Sons, New York, **1975**, p. 29.

14. Khan, M.N., Ohayagha, J.E. Kinetic studies on the cleavage of phthalimide in methylamine buffers. *React. Kinet. Catal. Lett.* **1996**, *58*(1), 97–103.

15. Khan, M.N. Kinetics and mechanism of tertiary amine-catalyzed aqueous cleavage of maleimide. *J. Chem. Soc. Perkin Trans. 2.* **1985**, 891–897.

16. (a) Hawkins, M.D. Intramolecular catalysis. Part III. Hydrolysis of 3- and 4-substituted phthalanilic acids [o-(N-phenylcarbamoyl)benzoic acids]. *J. Chem. Soc. Perkin Trans. 2.* **1976**, 642–647. (b) Blackburn, R.A.M., Capon, B., McRitchie, A.C. The mechanism of hydrolysis of phthalamic and N-phenylphthalamic acid: the spectrophotometric detection of phthalic anhydride as an intermediate. *Bioorg. Chem.* **1977**, *6*, 71–78.

17. (a) Khan, M.N. Suggested improvement in the Ing-Manske procedure and Gabriel synthesis of primary amines: kinetic study on alkaline hydrolysis of N-phthaloylglycine and acid hydrolysis of N-(o-carboxybenzoyl)glycine in aqueous organic solvents. *J. Org. Chem.* **1996**, *61*(23), 8063–8068. (b) Khan, M.N. Kinetics and mechanism of the aqueous cleavage of N,N-dimethylphthalamic acid (NDPA); evidence of intramolecular catalysis in the cleavage. *Indian J. Chem.* Section A **1993**, *32*, 395–401.

18. Khan, M.N., Arifin, A. Kinetics and mechanism of intramolecular carboxylic acid participation in the hydrolysis of N-methoxyphthalamic acid. *Org. Biomol. Chem.* **2003**, *1*, 1404–1408.

19. Khan, M.N. Unexpected rate enhancement in the intramolecular carboxylic acid-catalyzed cleavage of o-carboxybenzohydroxamic acid. *J. Phys. Org. Chem.* **1998**, *11*, 216–222.

20. Alkorta, I., Rozas, I., Elguero, J. Non-conventional hydrogen bonds. *Chem. Soc. Rev.* **1998**, *27*, 163–170.

21. (a) Schmid, R. Recent advances in the description of the structure of water, the hydrophobic effect, and the like-dissolves-like rule. *Monatshefte fuer Chemie* **2001**, *132*(11), 1295–1326. (b) Finney, J.L., Bowron, D.T., Soper, A.K., Dixit, S.S. What really drives the hydrophobic interaction? 224th ACS National Meeting, Boston, MA, August 18–22, **2002**, PHYS-149.

22. (a) Kyte, J. The basis of the hydrophobic effect. Biophysical Chem. **2003**, *100*, 193–203. (b) Southall, N.T., Dill, K.A., Haymet, A.D.J. A view of the hydrophobic effect. *J. Phys. Chem. B* **2002**, *106*(3), 521–533. (c) Ratnaparkhi, G.S., Varadarajan, R. Thermodynamics and structural studies of cavity formation in proteins suggest that loss of packing interactions rather than the hydrophobic effect dominates the observed energetics. *Biochemistry* **2000**, *39*(40) 12365–12374. (d) Widom, B., Bhimalapuram, P., Koga, K. The hydrophobic effect. *Phys. Chem. Chem. Phys.* **2003**, *5*(15) 3085–3093.

23. Otto, S., Engberts, J.B.F.N. Hydrophobic interactions and chemical reactivity. *Org. Biomol. Chem.* **2003**, *1*, 2809–2820.

24. (a) Jencks, W.P. Binding energy, specificity and enzymic catalysis — the Circe effect. *Adv. Enzymol.* **1975**, *43*, 219–410. (b) Pratt, L.R., Pohorille, A. Hydrophobic effects and modeling of biophysical aqueous solution interface. *Chem. Rev.* **2002**, *102*(8), 2671–2691.

25. Jencks, W.P. *Catalysis in Chemistry and Enzymology*, McGraw-Hill, New York, 1969.

26. Higuchi, T., Lachman, L. Inhibition of hydrolysis of esters in solution by formation of complexes. 1. Stabilization of benzocaine with caffeine. *J. Am. Pharm. Assoc.* **1955**, *44*, 521–526.

27. Cox, M.M., Jencks, W.P. Catalysis of the methoxyaminolysis of phenyl acetate by a preassociation mechanism with a solvent isotope effect maximum. *J. Am. Chem. Soc.* **1981**, *103*(3), 572–580.

28. Pirinccioglu, N., Williams, A. Studies of reactions within molecular complexes: alkaline hydrolysis of substituted phenyl benzoates in the presence of xanthines. *J. Chem. Soc. Perkin Trans. 2.* **1998**, 37–40.

29. Menninga, L., Engberts, J.B.F.N. Hydrolysis of arylsulfonylmethyl perchlorates: a simple model to explain electrolyte effects on a water-catalyzed deprotonation reaction. *J. Am. Chem. Soc.* **1976**, *98*(24), 7652–7657.

30. Williams, D.H., Westwell, M.S. Aspects of water interactions. *Chem. Soc. Rev.* **1998**, *27*, 57–63.

31. Khan, M.N. Kinetics and mechanism of tertiary amine-catalyzed aqueous cleavage of maleimide. *J. Chem. Soc. Perkin Trans. 2.* **1985**, 891–897.

32. Brown, J., Su, S.C.K., Shafer, J.A. The hydrolysis and cyclization of some phthalamic acid derivatives. *J. Am. Chem. Soc.* **1966**, *88*(19), 4468–4474.

33. Tamura, Y., Uchida T., Katsuki, T. Highly enantioselective (OC)Ru(salen)-catalyzed sulfimidation using N-alkoxycarbonyl azide as nitrene precursor. *Tetrahedron Lett.* **2003**, *44*(16), 3301–3303.

34. (a) Otto, S., Engberts, J.B.F.N. A systematic study of ligand effects on a Lewis-acid-catalyzed Diels–Alder reaction in water: water-enhanced enantioselectivity. *J. Am. Chem. Soc.* **1999**, *121*(29), 6798–6806. (b) Otto, S., Boccaletti, G., Engberts, J.B.F.N. A chiral Lewis-acid-catalyzed Diels–Alder reaction: water-enhanced enantioselectivity. *J. Am. Chem. Soc.* **1998**, *120*(17), 4238–4239.

35. Yu. Q., Huang, L., Fong, Z., Ma, W. Study of catalytic enantioselective additions of dimethylzinc to carbonyl compounds. Abstracts, 35th Great Lakes Regional Meeting of ACS, Chicago, IL, May 31–June 2, **2003**, p. 212.

36. Ma, S., Xie, H. Steric hindrance-controlled Pd(0)-catalyzed coupling-cyclization of 2,3-allenamides and organic iodides: an efficient synthesis of iminolactones and γ-hydroxy-γ-lactams. *J. Org. Chem.* **2002**, *67*(18), 6575–6578.

37. Trehan, I., Morandi, F., Blaszczak, L.C., Shoichet, B.K. Using steric hindrance to design new inhibitors of class C β-lactamases. *Chem. Biol.* **2002**, *9*(9), 971–980.

38. Li, J.J., Tan, W. Macromolecular crowding accelerates DNA cleavage reaction catalyzed by DNA nucleases. *Polym. Preprints* **2002**, *43*(1), 712–713.

39. O'Connor, C.J., McLennan, D.J., Denny, W.A., Sutton, B.M. Substituent effects on the hydrolysis of analogs of nitracrinr [9-[[3-(dimethylamino)propyl]amino]-1-nitroacridine]. *J. Chem. Soc. Perkin Trans. 2.* **1990**, 1637–1641.

40. Khan, M.N., Malspeis, L. Kinetics and mechanism of thiolytic cleavage of the antitumor compound 4-[(9-acridinyl)amino]methanesulfon-m-anisidide. *J. Org. Chem.* **1982**, *47*(14), 2731–2740.

41. Wild, F., Young, J.M. The reaction of mepacrine with thiols. *J. Chem. Soc.* **1965**, 7261–7274.

42. Khan, M.N., Kuliya-Umar, A.F. Kinetics and mechanism of general acid-catalyzed thiolytic cleavage of 9-anilinoacridine. *Bioorg. Med. Chem.* **1995**, *3*(7), 881–890.

43. Hupe, D.J., Jencks, W.P. Nonlinear structure-reactivity correlations: acyl transfer between sulfur and oxygen nucleophiles. *J. Am. Chem. Soc.* **1977**, *99*(2) 451–464.

44. Denny, W.A., Cain, B.F., Atwell, G.J., Hansch, C., Panthananickal, A., Leo, A. Potential antitumor agents. 36. Quantitative relationships between experimental antitumor activity, toxicity, and structure for the general class of 9-anilinoacridine antitumor agents. *J. Med. Chem.* **1982**, *25*(3), 276–315.

45. Sekiguchi, S., Ohtsuka, I., Okada, K. Aromatic nucleophilic substitution. 11. Effects of ortho substituents on the rates of the Smiles rearrangements of (β-acetylamino)ethyl-2-X-4-Y-6-Z-1-phenyl ethers with potassium hydroxide in aqueous dimethyl sulfoxide. *J. Org. Chem.* **1979**, *44*(14), 2556–2560.

46. (a) Fersht, A.R., Kirby, A.J. The hydrolysis of aspirin: intramolecular general base catalysis of ester hydrolysis. *J. Am. Chem. Soc.* **1967**, *89*(19), 4857–4863. (b) Kirby, A.J., Lancaster, P.W. Structure and efficiency in intramolecular and enzymic catalysis: catalysis of amide hydrolysis by the carboxy group of substituted maleamic acids. *J. Chem. Soc. Perkin Trans.* 2 **1972**, 1206–1214.

47. Broxton, T.J., Deady, L.W., Rowe, J.E. Basic methanolysis of N-aryl-N-phenyl-benzamides. *J. Org. Chem.* **1980**, *45*(12), 2404–2408.

48. (a) Smid, J., Varma, A.J., Shah, S.C. Decarboxylation of 6-nitrobenzisoxazole-3-carboxylate in benzene catalyzed by crown ethers and their polymers. *J. Am. Chem. Soc.* **1979**, *101*(19), 5764–5769. (b) Cook, F.L., Bowers, C.N., Liotta, C.L. Chemistry of naked anions. III. Reactions of the 18-crown-6 complex of potassium cynide with organic substrates in aprotic organic solvents. *J. Org. Chem.*, **1974**, *39*(23), 3416–3418.

49. (a) Dunn, E. J., Buncel, E. Metal ion catalysis in nucleophilic displacement reactions at carbon, phosphorus, and sulfur centers.[1] I. Catalysis by metal ions in the reaction of p-nitrophenyl diphenylphosphinate with ethoxide. *Can. J. Chem.* **1989**, *67*, 1440-1448. (b) Suh, J., Mun, B.S. Crown ethers as a mechanistic probe. 1. Inhibitory effects of crown ethers on the reactivity of anionic nucleophiles toward diphenyl p-nitrophenyl phosphate. *J. Org. Chem.* **1989**, *54*(8), 2009–2010.

50. Wong, K.-H. Kinetics of the esterification of potassium p-nitrobenzoate by benzyl bromide using dicyclohexyl-18-crown-8 as phase-transfer agent. *J. Chem. Soc. Chem. Commun.* **1978**, 282–283.

51. Doddi, G., Ercolani, G., Pegna; P.L., Mencarelli, P. Remarkable catalysis by strontium ion in S_N2 and E2 reactions occurring in proximity to a crown ether structure. *J. Chem. Soc. Chem. Commun.* **1994**, 1239–1240.

52. Hine J., Khan, M.N. Internal amine-assisted attack of alcoholic hydroxyl group on esters: serine esterse models. *Indian J. Chem.* **1992**, *31B*, 427–435.

53. Bruice, T.C., Donzel, A., Huffman, R.W., Butler, A. Aminolysis of phenyl acetates in aqueous solutions. VII. Observations on the influence of salts, amine structure, and base strength. *J. Am. Chem. Soc.* **1967**, *89*(9), 2106–2121.

54. Khan M.N., Gambo, S.K. Intramolecular catalysis and the rate-determining step in the alkaline hydrolysis of ethyl salicylate. *Int. J. Chem. Kinet.* **1985** *17*, 419–428.

55. Jencks, W.P., Gilchrist, M. Nonlinear structure-reactivity correlations: the reactivity of reagents toward esters. *J. Am. Chem. Soc.* **1968**, *90*(10), 2622–2637.

56. Gresser, M.J., Jencks, W.P. Ester aminolysis: structure-reactivity relationships and the rate-determining step in the aminolysis of substituted diphenyl carbonates. *J. Am. Chem. Soc.* 1977, *99*(21), 6963–6970.

57. Werber, M.M., Shalitin, Y. Reaction of tertiary amino alcohols with active esters: acylation and deacylation steps. *Bioorg. Chem.* **1973**, *2*(3), 202–220.

58. Oakenfull, D.G., Jencks, W.P. Reactions of acetylimidazole and acetylimidazolium ion with nucleophilic reagents: structure-reactivity relationships. *J. Am. Chem. Soc.* **1971**, *93*(1), 178–188.

59. Page, M.I., Jencks, W.P. Intramolecular general base catalysis in the aminolysis of acetylimidazole and methyl formate by diamines. *J. Am. Chem. Soc.* **1972**, *94*(25), 8818–8827.

60. (a) Wolfenden, R., Jencks, W.P. Acetyl transfer reactions of 1-acetyl-3-methylimidazolium chloride. *J. Am. Chem. Soc.* **1961**, *83*, 4390–4393. (b) Bruice, T.C., Schmir, G.L. Imidazole catalysis. I. The catalysis of the hydrolysis of phenyl acetates by imidazole. *J. Am. Chem. Soc.* **1957**, *79*, 1663–1667; (c) Bruice, T.C., Schmir, G.L. Imidazole catalysis. II. The reaction of substituted imidazoles with phenyl acetates in aqueous solution. *J. Am. Chem. Soc.* **1958**, *80*, 148–156. (d) Bender, M.L., Turnquest, B.W. The imidazole-catalyzed hydrolysis of *p*-nitrophenyl acetate. *J. Am. Chem. Soc.* **1957**, *79*, 1652–1655.

61. Fersht, A.R., Jencks, W.P. The acetylpyridinium ion intermediate in pyridine-catalyzed hydrolysis and acyl transfer reactions of acetic anhydride: observation, kinetics, structure-reactivity correlations and effects of concentrated salt solutions. *J. Am. Chem. Soc.* **1970**, *92*(18), 5432–5442.

62. Hubbard, P., Brittain, W.J. Mechanism of amine-catalyzed ester formation from an acid chloride and alcohol. *J. Org. Chem.* **1998**, *63*(3), 677–683.

63. Shieh, W.-C., Dell, S., Repic, O. Nucleophilic catalysis with 1,8-diazabicyclo[5.4.0]undec-7-ene (DBU) for the esterfication of carboxylic acids with dimethyl carbonate. *J. Org. Chem.* **2002**, *67*(7), 2188–2191.

64. Shieh, W.-C., Dell, S., Bach, A., Repic, O., Blacklock, T.J. Dual nucleophilic catalysis with DABCO for the N-methylation of indoles. *J. Org. Chem.* **2003**, *68*(5), 1954–1957.

65. Page, M.I., Jencks, W.P. In defence of entropy and strain as explanations for the rate enhancement shown in intramolecular reactions. *Gazzetta Chimica Italiana* **1987**, *117*, 455–460.

66. (a) Milstien, S., Cohen, L.A. Stereopopulation control. 1. Rate enhancement in the lactonizations of *o*-hydroxyhydrocinnamic acids. *J. Am. Chem. Soc.* **1972**, *94*(26), 9158–9165. (b) Menger, F.M., Ladika, M. Fast hydrolysis of an aliphatic amide at neutral pH and ambient temperature: a peptidase model. *J. Am. Chem. Soc.* **1988**, *110*(20), 6794–6796.

67. Page, M.I. The energetics of intramolecular reactions and enzyme catalysis. *Philos. Trans. R. Soc. Lond. B* **1991**, *332*, 149–156.

68. (a) Menger, F.M. On the source of intramolecular and enzymatic reactivity. *Acc. Chem. Res.* **1985**, *18*, 128–134. (b) Menger, F.M., Venkataram, U.V. Proximity as a component of organic reactivity. *J. Am. Chem. Soc.* **1985**, *107*(16), 4706–4709.

69. Winstein S., Lindegren, C.R., Marshal, H., Ingraham, L.L. Neighboring carbon and hydrogen. XIV. Participation in solvolysis of some primary benzenesulfonates. *J. Am. Chem. Soc.* **1953**, *75*, 147–155.

70. Bruice, T.C., Benkovic, S.J. *Bioorganic Mechanisms*, Vol. 1. W.A. Benjamin, New York, **1965**, p. 199.

71. Page, M.I., Jencks, W.P. Entropic contributions to rate accelerations in enzymic and intramolecular reactions and the chelate effect. *Proc. Natl. Acad. Sci. U.S.A.* **1971**, *68*(8), 1678–1683.

72. Jencks, W.P., Page, M.I. Orbital steering, entropy and rate accelerations. *Biochem. Biophys. Res. Commun.* **1974**, *57*(3), 887–892.

73. Bruice, T.C. Proximity effects and enzyme catalysis. In *The Enzyme*, 3rd ed., Vol. 2, Ed., Boyer, P.D., Academic Press, New York, **1970**, pp. 217–279.

74. De Lisi, C., Crothers, D.M. Contribution of proximity and orientation to catalytic reaction rates. *Biopolymers* **1973**, *12*(7), 1689–1704.

75. (a) Storm, D.R., Koshland, D.E., Jr. Source for the special catalytic power of enzymes: orbital steering. *Proc. Natl. Acad. Sci. U.S.A.* **1970**, *66*(2), 445–452. (b) Dafforn, A., Koshland, D.E., Jr. Theoretical aspects of orbital steering. *Proc. Natl. Acad. Sci. U.S.A.* **1971**, *68*(10), 2463–2467. (c) Storm, D.R., Koshland, D.E., Jr. Effect of small changes in orientation on reaction rate. *J. Am. Chem. Soc.* **1972**, *94*(16), 5815–5825. (d) Storm, D.R., Koshland, D.E., Jr. Indication of the magnitude of orientation factors in esterification. *J. Am. Chem. Soc.* **1972**, *94*(16), 5805–5814.

76. Milstein, S., Cohen, L.A. Rate acceleration by stereopopulation control: models for enzyme action. *Proc. Natl. Acad. Sci. U.S.A.* **1970**, *67*(3), 1143–1147.

77. Reuben, J. Substrate anchoring and the catalytic power of enzymes. *Proc. Natl. Acad. Sci. U.S.A.* **1971**, *68*(3), 563–565.

78. Firestone, R.A., Christensen, B.G. Vibrational activation. I. Source for the catalytic power of enzymes. *Tetrahedron Lett.* **1973**, (5), 389–395.

79. Cook, D.B., McKenna, J. Contribution to the theory of enzyme catalysis: potential importance of vibrational activation entropy. *J. Chem. Soc. Perkin Trans. 2.* **1974**, 1223–1225.

80. (a) Ferreira, R., Gomes, M.A.F. Electronic aspects of enzymic catalysis. *Proc. 6th Braz. Symp. Theor. Phys.* **1980**, Vol. 2, pp. 281–293. (b) Ferreira, R., Gomes, M.A.F. Electronic aspects of enzymic catalysis. *Int. J. Quantum Chem.* **1982**, *22*(3), 537–545.

81. Low, P.S., Somero, G.N. Protein hydration changes during catalysis: a new mechanism of enzyme rate-enhancement and ion activation/inhibition of catalysis. *Proc. Natl. Acad. Sci. U.S.A.* **1975**, *72*(9), 3305–3309.

82. Warshel, A. Energetics of enzyme catalysis. *Proc. Natl. Acad. Sci. U.S.A.* **1978**, *75*(11), 5250–5254.

83. (a) Hol, W.G.J., van Duijnen, P.T., Berendsen, H.J.C. The α-helix dipole and the properties of proteins. *Nature* (London). **1978**, *273*(5662), 443–446. (b) van Duijnen, P.T., Thole, B.T., Hol, W.G.J. On the role of the active site helix in papain, an ab initio molecular orbital study. *Biophys. Chem.* **1979**, *9*(3), 273–280.

84. Henderson, R., Wang, J.H. Catalytic configurations. *Annu. Rev. Biophys. Bioeng.* **1972**, *1*, 1–26.

85. Wang, J.H. Directional character of proton transfer in enzyme catalysis. *Proc. Nat. Acad. Sci. U.S.A.* **1970**, *66*(3), 874–881.

86. Olavarria, J.M. Does the coupling between conformational fluctuation and enzyme catalysis involve a true phase transfer catalysis? *J. Theor. Biol.* **1982**, *99*(1), 21–30.

87. Dewar, M.J., Storch, D.M. Alternative view of enzyme reactions. *Proc. Natl. Acad. Sci. U.S.A.* **1985**, *82*(8), 2225–2229.

88. Mock, W.L. Torsional strain considerations in enzymology: some applications to proteases and ensuing mechanistic consequences. *Bioorg. Chem.* **1976**, *5*(4), 403–414.

89. Jencks, W.P. Binding energy, specificity and enzymic catalysis — the Circe effect. *Adv. Enzymol.* **1975**, *43*, 219–410.

90. Nowak, T., Mildvan, A.S. Nuclear magnetic resonance studies of selectively hindered internal motion of substrate analogs at the active site of pyruvate kinase. *Biochemistry* **1972**, *11*(15), 2813–2818.

91. Cullis, P.M., Harger, J.P. Intramolecular nucleophilic catalysis and the exceptional reactivity of N-benzyloxycarbonyl α-aminophosphonochloridates. *J. Chem. Soc. Perkin Trans. 2.* **2002**, 1538–1543.

92. Shashidhar, M.S., Rajeev, K.G., Bhatt, M.V. Methanolysis of ortho- and para-formylbenzenesulfonates in basic media: evidence for the intramolecular nucleophilic catalysis by the carbonyl group. *J. Chem. Soc. Perkin Trans. 2.* **1997**, 559–561.

93. Blow, D.M., Birktoft, J.J., Hartley, B.S. Role of a buried acid group in the mechanism of action of chymotrypsin. *Nature* (London) **1969**, *221*(178) 337–340.

94. Shafer, J.A., Morawetz, H. Participation of a neighboring amide group in the decomposition of esters and amides of substituted phthalamic acids. *J. Org. Chem.* **1963**, *28*(7), 1899–1901.

95. Bunton, C.A., Nayak, B., O'Connor, C. Alkaline hydrolysis of benzamide and N-methyl- and N,N-dimethylbenzamide. *J. Org. Chem.* **1968**, *33*(2), 572–575.

96. Zhou, De-Min., Taira, K. The hydrolysis of RNA: from theoretical calculations to the hammerhead ribozyme-mediated cleavage of RNA. *Chem. Rev.* **1998**, *98*(3), 991–1026.

97. (a) Oivanen, M., Kuusela, S., Lonnberg, H. Kinetics and mechanism for the cleavage and isomerization of the phosphate diester bonds of RNA by Bronsted acids and bases. *Chem. Rev.* **1998**, *98*(3), 961–990. (b) Perreault, D.M., Anslyn, E.V. Unifying the current data on the mechanism of cleavage-transesterification of RNA. *Angew. Chem., Int. Ed. Engl.* **1997**, *36*(5), 433–450.

98. Khan, M.N. Evidence for the formation and decay of o-(N,N-dimethylaminomethyl)benzyl acetate in the reaction of p-nitrophenyl acetate with o-(N,N-dimethylaminomethyl)benzyl alcohol. *J. Chem. Res. (S)* **1986**, 290–291. (M) **1986**, 2384–2394.

99. Khan, M.N. 2-endo-[(Dimethylamino)methyl]-3-endo-(hydroxymethyl)bicyclo[2.2.1]hept-5-ene: a model for serine esterases in the aqueous cleavage of p-nitrophenyl acetate. *J. Org. Chem.* **1985**, *50*(24), 4851–4855.

100. Khan, M.N. Nucleophilic reactivity of internally hydrogen bonded hydroxyl group of o-(N,N-dimethylaminomethyl)benzyl alcohol towards nonactivated esters. *Indian J. Chem.* **1986**, *25A*, 855–857.

101. Janda, K.D., Schloeder, D., Benkovic, S.J., Lerner, R.A. Introduction of an antibody that catalyzes the hydrolysis of an amide bond. *Science* **1988**, *241*(4870), 1188–1191.

102. Pollack, S.J., Hsiun, P., Schultz, P.G. Stereospecific hydrolysis of alkyl esters by antibodies. *J. Am. Chem. Soc.* **1989**, *111*(15), 5961–5962.

103. Iverson, B.L., Lerner, R.A. Sequence-specific peptide cleavage catalyzed by an antibody. *Science* **1989**, *243*(4895), 1184–1188.

104. Gibbs, R.A., Taylor, S., Benkovic, S.J. Antibody-catalyzed rearrangement of the peptide bond. *Science* **1992**, *258*(5083), 803–805.

105. Liotta, L.J., Benkovic, P.A., Miller, G.P., Benkovic, S.J. Catalytic antibody for imide hydrolysis featuring a bifunctional transition-state mimic. *J. Am. Chem. Soc.* **1993**, *115*(1), 350–351.

106. (a) Suga, H., Ersoy, O., Tsumuraya, T., Lee, J., Sinskey, A.J., Masamune, S. Esterolytic antibodies induced to haptens with a 1,2-amino alcohol functionality. *J. Am. Chem. Soc.* **1994**, *116*(2), 487–494. (b) Suga, H., Ersoy, O., Williams S.F., Tsumuraya, T., Margolies, M.N., Sinskey, A.J., Masamune, S. Catalytic antibodies generated via heterologous immunization. *J. Am. Chem. Soc.* **1994**, *116*(13), 6025–6026. (c) Tsumuraya, T., Suga, H., Meguro, S., Tsukanawa, A., Masamune, S. Catalytic antibodies generated via homologous and heterologous immunization. *J. Am. Chem. Soc.* **1995**, *117*(46), 11390–11396.

107. Ersoy, O., Fleck, R., Sinskey, A., Masamune, S. Antibody catalyzed cleavage of an amide bond using an external nucleophilic cofactor. *J. Am. Chem. Soc.* **1998**, *120*(4), 817–818.

108. March, J. *Advanced Organic Chemistry: Reactions, Mechanisms, and Structure*, 2nd ed. McGraw-Hill, Kogakusha, Tokyo, 1977, p. 227.

109. (a) Fersht, A.R. (Ed.) *Enzyme, Structure and Mechanism*, 2nd ed. Freeman, W.H. and company, New York, 1985. (b) Breslow, R. Artificial enzymes and enzyme models. *Adv. Enzymol.* **1986**, *58*, 1–60. (c) Chin, J. Developing artificial hydrolytic metalloenzymes by a unified mechanistic approach. *Acc. Chem. Res.* **1991**, *24*(5), 145–152.

110. Nagelkerke, R., Thatcher, G.R.J., Buncel, E. Alkali-metal ion catalysis in nucleophilic displacement reactions at carbon, phosphorus, and sulfur centers. IX. p-Nitrophenyl diphenyl phosphate. *Org. Biomol. Chem.* **2003**, *1*, 163–167.

111. Buncel, E., Dunn, E.J., Bannard, R.A.B., Purdon, J.P. Metal-ion catalysis in nucleophilic displacement by alkoxide ion in p-nitrophenyl diphenylphosphinate: rate retardation by crown ether and cryptand complexing agents. *J. Chem. Soc. Chem. Commun.* **1984**, 162–163.

112. (a) Buncel, E., Pregel, M.J. Reaction of ethoxide with p-nitrophenyl benzene-sulfonate. Catalysis and inhibition by alkali metal ions; contrast with p-nitrophenyl diphenylphosphinate. *J. Chem. Soc. Chem. Commun.* **1989**, 1566–1567. (b) Pregel, M.J., Dunn, E.J., Buncel, E. Metal-ion catalysis in nucleophilic displacement reactions at carbon, phosphorus, and sulfur centers. III. Catalysis vs. inhibition by metal ions in the reaction of p-nitrophenyl benzenesulfonate with ethoxide. *Can. J. Chem.* **1990**, *68*(10), 1846–1858. (c) Pregel, M.J., Buncel, E. Metal-ion catalysis in nucleophilic displacement reactions at carbon, phosphorus, and sulfur centers. 5. Alkali-metal ion catalysis and inhibition in the reaction of p-(trifluoromethyl)phenyl methanesulfonate with ethoxide ion. *J. Org. Chem.* **1991**, *56*(19), 5583–5588. (d) Pregel, M.J., Buncel, E. Bond scission in sulfur compounds. Nucleophilic displacement reactions at carbon, phosphorus, and sulfur centers: reaction of aryl methanesulfonates with ethoxide; change in mechanism with change in leaving group. *J. Chem. Soc. Perkin Trans. 2.* **1991**, 307–311. (e) Pregel, M.J., Dunn, E.J., Buncel, E. Metal-ion catalysis in nucleophilic displacement reactions at carbon, phosphorus, and sulfur centers. 4. Mechanism of the reaction of aryl benzenesulfonates with alkali-metal ethoxides: catalysis and inhibition by alkali-metal ions. *J. Am. Chem. Soc.* **1991**, *113*(9), 3545–3550.

113. Mentz, M., Modro, A.M., Modro, T.A. Solvation and metal ion effects on the structure and reactivity of phosphoryl compounds. Part 3. Alkali metal ion catalysis in the alkylation by phosphate esters. *J. Chem. Res. (S)* **1994**, 46–47.

114. Hendry, P., Sargeson, A.M. Base hydrolysis of the pentaamine (trimethylphosphate)iridium(III) ion. *J. Chem. Soc., Chem. Commun.* **1984**, 164–165.

115. Wadsworth, W.S., Jr. Lewis acid catalyzed methanolysis of a phosphate triester. *J. Org. Chem.* **1981**, *46*(20), 4080–4082.

116. Herschlag, D., Jencks, W.P. The effect of divalent metal ions on the rate and transition-state structure of phosphoryl-transfer reactions. *J. Am. Chem. Soc.* **1987**, *109*(15), 4665–4674.

117. Herschlag, D., Jencks, W.P. Catalysis of the hydrolysis of phosphorylated pyridines by $Mg(OH)^+$: a possible model for enzymic phosphoryl transfer. *Biochemistry* **1990**, *29*(21), 5172–5179.

118. Matsumura, K., Komiyama, M. Enormously fast RNA hydrolysis by lanthanide(III) ions under physiological conditions: eminent candidates for novel tools of biotechnology. *J. Biochem. (Tokyo)* **1997**, *122*(2), 387–394.

119. De Maria, P., Fontana, A., Spinelli, D. Macaluso, G. Acid-base and metal ion catalysis in the enolization of 2-acetylimidazole. *Gazzetta Chimica Italiana*, **1996**, *126*(1), 45–51.

120. Takagi, Y., Warashina, M., Stec, W.J., Yoshinari, K., Taira, K. Recent advances in the elucidation of the mechanisms of action of ribozymes. *Nucl. Acid. Res.* **2001**, *29*(9), 1815–1834.

121. Sreedhara, A., Cowan, J.A. Catalytic hydrolysis of DNA by metal ions and complexes. *J. Biol. Inorg. Chem.* **2001**, *6*(4), 337–347.

122. Ciesiolka, J. Metal ion-induced cleavages in probing of RNA structure. *RNA Biochemistry and Biotechnology*, NATO Science Series, 3-70: High Technology, **1999**, pp.111–121.

123. Liu, S., Hamilton, A.D. Rapid and highly base selective RNA cleavage by a dinuclear Cu(II) complex. *Chem. Commun.* **1999**, 587–588.

124. Ait-Haddou, H., Sumaoka, J., Wiskur, S.L., Folmer-Anderson, J.F., Anslyn, E.V. Remarkable cooperativity between a Zn^{II} ion and guanidinium/ammonium groups in the hydrolysis of RNA. *Angew. Chem. Int. Ed. Engl.* **2002**, *41*(21), 4014–4016.

125. Komiyama, M., Kina, S., Matsumura, K., Sumaoka, J., Tobey, S., Lynch, V.M., Anslyn, E. Trinuclear copper(II) complex showing high selectivity for the hydrolysis of 2-5 over 3-5 for UpU and 3-5 over 2-5 for ApA ribonucleotides. *J. Am. Chem. Soc.* **2002**, *124*(46), 13731–13736.

126. Becker, M., Lerum, V., Dickson, S., Nelson, N.C., Matsuda, E. The double helix is dehydrated: evidence from the hydrolysis of acridinium ester-labeled probes. *Biochemistry* **1999**, *38*(17), 5603–5611.

127. Roussev, C.D., Ivanova, G.D., Bratovanova, E.K., Petkov, D.D. Medium-controlled intramolecular catalysis in the direct synthesis of 5'-O-protected ribonucleoside 2- and 3-dialkylphosphates. *Angew. Chem. Int. Ed. Engl.* **2000**, *39*(4), 779–781.

128. Khan, M.N., Arifin, Z., George, A., Wahab, I.A. Basicity of some aliphatic amines in mixed H_2O-CH_3CN solvents. *Int. J. Chem. Kinet.* **2000**, *32*, 146–152.

129. Gellman, S.H., Petter, R., Breslow, R. Catalytic hydrolysis of a phosphate trimester by tetracoordinated zinc complexes. *J. Am. Chem. Soc.* **1986**, *108*(9), 2388–2394.

130. (a) Sumaoka, J., Azuma, Y., Komiyama, M. Enzymic manipulation of the fragments obtained by cerium(IV)-induced DNA scission: characterization of hydrolytic termini. *Chem.-A Eur. J.* **1998**, *4*(2), 205–209. (b) Moss, R.A. Remarkable acceleration of dimethyl phosphate hydrolysis by ceric cations. *Chem. Commun.* **1998**, 1871–1872. (c) Moss, R.A., Morales-Rojas, H. Loci of ceric cation mediated hydrolyses of dimethyl phosphate and methyl methylphosphonate. *Org. Lett.* **1999**, *1*(11), 1791–1793. (d) Ott, R., Kramer, R. DNA hydrolysis by inorganic catalysts. *Appl. Microbiol. Biotechnol.* **1999**, *52*(6), 761–767.

131. Wang, C., Choudhary, S., Vink, C.B., Secord, E.A., Morrow, J.R. Harnessing thorium(IV) as a catalyst: RNA and phosphate diester cleavage by a thorium(IV) macrocyclic complex. *Chem. Commun.* **2000**, 2509–2510.

132. Valakoski, S., Heiskanen, S., Anderson, S., Lahde, M., Mikkola, S. Metal ion-promoted cleavage of mRNA 5-cap models: hydrolysis of the triphosphate bridge and reactions of the N^7-methylguanine base. *J. Chem. Soc. Perkin Trans. 2.* **2002**, 604–610.

133. Salehi, P., Khodaei, M.M., Zolfigol, M.A., Zeinoldini, S. Catalytic Friedel-Crafts acylation of alkoxybenzenes by ferric hydrogen sulfate. *Synth. Commun.* **2003**, *33*(8), 1367–1373.

134. Khan, M.N. The mechanistic diagnosis of induced catalysis in the aqueous cleavage of phenyl salicylate in the presence of borate buffers. *J. Mol. Catal.* **1987**, *40*, 195–210.

135. Capon, B., Ghosh, B.C. The mechanism of the hydrolysis of phenyl salicylate and catechol monobenzoate in the presence and absence of borate ions. *J. Chem. Soc., (B)* **1966**, 472–478.

136. Tanner, D.W., Bruice, T.C. Boric acid esters. I. A general survey of aromatic ligands and the kinetics and mechanism of the formation and hydrolysis of boric acid esters of salicylamide, N-phenylsalicylamide, and disalicylimide. *J. Am. Chem. Soc.* **1967**, *89*(26), 6954–6972.

137. Yatsimirsky, A.K., Bezsoudnova, K.Y., Sakodinskaya, I.K. Boric acid effect on the hydrolysis of 4-nitrophenyl 2,3-dihydroxybenzoate: mimic of borate inhibition of serine proteases. *Bioorg. Med. Chem. Lett.* **1993**, *3*(4), 635–638.

138. Bruice, T.C., Tsubouchi, A., Dempcy, R.O., Olson, L.P. One- and two-metal ion catalysis of the hydrolysis of adenosine 3-alkyl phosphate esters: models for one- and two-metal ion catalysis of RNA hydrolysis. *J. Am. Chem. Soc.* **1996**, *118*(41), 9867–9875.

139. Komiyama, M., Matsumoto, Y., Takahashi, H., Shiiba, T., Tsuzuki, H., Yajima, H., Yashiro, M., Sumaoka, J. RNA hydrolysis by cobalt(III) complexes. *J. Chem. Soc. Perkin Trans. 2.* **1998**, 691–695.

140. Zhou, D.-M., Taira, K. The hydrolysis of RNA: from theoretical calculations to the hammerhead ribozyme-mediated cleavage of RNA. *Chem. Rev.* **1998**, *98*(3), 991–1026.

141. Kamitani, J., Sumaoka, J., Asanuma, H., Komiyama, M. Efficient RNA hydrolysis by lanthanide(III)-hydrogen peroxide combinations. Novel aggregates as the catalytic species. *J. Chem. Soc. Perkin Trans. 2.* **1998**, 523–527.

142. Landini, D., Maia, A., Pinna, C. Demethylation reactions of phosphate esters catalyzed by complexes of polyether ligands with metal iodides. *J. Chem. Soc. Perkin Trans. 2.* **2001**, 2314–2317.

143. Kosonen, M., Lonnberg, H. General and specific acid/base catalysis of the hydrolysis and interconversion of ribonucleoside 2- and 3-phosphotriesters: kinetics and mechanisms of the reactions of 5-O-pivaloyluridine 2- and 3-dimethylphosphates. *J. Chem. Soc. Perkin Trans. 2.* **1995**, 1203–1209.

144. Westheimer, F.H. Pseudo-rotation in the hydrolysis of phosphate esters. *Acc. Chem. Res.* **1968**, *1*(3), 70–78.

145. Kluger, R., Covitz, F., Dennis, E., Williams, L.D., Westheimer, F.H. pH-Product and pH-rate profiles for the hydrolysis of methyl ethylene phosphate: rate-limiting pseudorotation. *J. Am. Chem. Soc.* **1969**, *91*(22), 6066–6072.

146. Thatcher, G.R.J., Kluger, R. Mechanism and catalysis of nucleophilic substitution in phosphate esters. *Adv. Phys. Org. Chem.* **1989**, *25*, 99–265.

147. Kluger, R., Taylor, S.D. On the origins of enhanced reactivity of five-membered cyclic phosphate esters: the relative contributions of enthalpic and entropic factors. *J. Am. Chem. Soc.* **1990**, *112*(18), 6669–6671.

148. Taylor, S.D., Kluger, R. Heats of reaction of cyclic and acyclic phosphate and phosphonate esters: strain discrepancy and steric retardation. *J. Am. Chem. Soc.* **1992**, *114*(8), 3067–3071.

149. Baxter, N.J., Rigoreau, L.J.M., Laws, A.P., Page, M.I. Reactivity and mechanism in the hydrolysis of β-sultams. *J. Am. Chem. Soc.* **2000**, *122*(14), 3375–3385.

150. Biechler, S.S., Taft, R.W., Jr. Effect of structure on kinetics and mechanism of the alkaline hydrolysis of anilides. *J. Am. Chem. Soc.* **1957**, *79*, 4927–4935.

151. Page, M.I., Proctor, P. Mechanism of β-lactum ring opening in cephalosporins. *J. Am. Chem. Soc.* **1984**, *106*(13), 3820–3825.

152. Khan, M.N., Ismail, E. Kinetic studies on the cleavage of *N*-phthaloylglycine in the buffers of hydrazine and morpholine. *J. Chem. Res. (S)* **2001**, 554–556.

153. Khan, M.N., Ismail, N.H. Kinetics and mechanism of the cleavage of N-phtha-loylglycine in buffers of some primary amines. *J. Chem. Res. (S)* **2002**, 593–595; *J. Chem. Res. (M)* **2002**, 1243–1258.

154. Khan, M.N. Kinetic evidence for the occurrence of a stepwise mechanism in the aminolysis of maleimide: general acid and base catalysis by secondary amines. *J. Chem. Soc. Perkin Trans. 2.* **1987**, 819–828.

155. Gresser, M.J., Jencks, W.P. Ester aminolysis: partitioning of the tetrahedral addition intermediate, T$^{\pm}$, and the relative leaving ability of nitrogen and oxygen. *J. Am. Chem. Soc.* **1977**, *99*(21), 6970–6980.

156. Baxter, N.J., Laws, A.P., Rigoreau, L., Page, M.I. The hydrolytic reactivity of β-sultams. *J. Chem. Soc. Perkin Trans. 2.* **1996**, 2245–2246.

157. Wood, J.M., Hinchliffe, P.S., Laws, A.P., Page, M.I. Reactivity and the mechanisms of reactions of β-sultams with nucleophiles. *J. Chem. Soc. Perkin Trans. 2.* **2002**, 938–946.

158. Senapeschi, A.N., De Groote, R.A.M.C., Neumann, M.G. Acid catalyzed hydrolysis of substituted anilinomethanesulfonates. *Tetrahedron Lett.* **1984**, *25*(22), 2313–2316.

159. Cosimelli, B., Frenna, V., Guernelli, S., Lanza, C.Z., Macaluso, G., Petrillo, G., Spinelli, D. The first kinetic evidence for acid catalysis in a monocyclic rearrangement of heterocycles: conversion of the Z-phenylhydrazone of 5-amino-3-benzoyl-1,2,4-oxadiazole into N,5-diphenyl-2H-1,2,3-triazol-4-ylurea. *J. Org. Chem.* **2002**, *67*(23), 8010–8018.

160. (a) Jencks, W.P. General acid-base catalysis of complex reactions in water. *Chem. Rev.* **1972**, *72*(6), 705–718. (b) Jencks, W.P., Gilbert, H.F. General acid-base catalysis of carbonyl and acyl group reactions. *Pure Appl. Chem.* **1977**, *49*, 1021–1027.

161. Jencks, W.P. Enforced general acid-base catalysis of complex reactions and its limitations. *Acc. Chem. Res.* **1976**, *9*, 425–432.

162. Khan, M.N. Kinetic evidence for the occurrence of a stepwise mechanism in hydrazinolysis of phthalimide. *J. Org. Chem.* **1995**, *60*(4), 4536–4541.

163. Khan, M.N. Intramolecular general acid-base catalysis and the rate-determining step in the nucleophilic cleavage of maleimide with primary amines. *J. Chem. Soc. Perkin Trans. 2.* **1985**, 1977–1984.

164. Khan, M.N., Ohayagha, J.E. Kinetics and mechanism of the cleavage of phthalimide under the buffers of tertiary and secondary amines: evidence of intramolecular general acid-base catalysis in the reactions of phthalimide with secondary amines. *J. Phys. Org. Chem.* **1991**, *4*, 547–561.

165. Khan, M.N. Kinetic evidence for the occurrence of a stepwise mechanism in the aminolysis of maleimide: general acid and base catalysis by secondary amines. *J. Chem. Soc. Perkin Trans. 2.* **1987**, 819–828.

166. Khan, M.N. Kinetic evidence for the occurrence of a stepwise mechanism in the aminolysis of N-ethoxycarbonylphthalimide. *J. Chem. Soc. Perkin Trans. 2.* **1988**, 1129–1134.

167. Breslow, R. How do imidazole groups catalyze the cleavage of RNA in enzyme models and in enzymes? Evidence from "negative catalysis." *Acc. Chem. Res.* **1991**, *24*(11), 317–324 and references cited therein.

168. Breslow, R., Labelle, M. Sequential general base-acid catalysis in the hydrolysis of RNA by imidazole. *J. Am. Chem. Soc.* **1986**, *108*(10), 2655–2659.

169. Anslyn, E., Breslow, R. On the mechanism of catalysis by ribonuclease: cleavage and isomerization of the dinucleotide UpU catalyzed by imidazole buffer. *J. Am. Chem. Soc.* **1989**, *111*(12), 4473–4482.

170. Breslow, R., Huang, D.-L. A negative catalytic term requires a common intermediate in the imidazole buffer catalyzed cleavage and rearrangement of ribonucleotides. *J. Am. Chem. Soc.* **1990**, *112*(26), 9621–9623.

171. Beckmann, C., Kirby, A.J., Kuusela, S., Tickle, D.C. Mechanism of catalysis by imidazole buffers of the hydrolysis and isomerization of RNA models. *J. Chem. Soc. Perkin Trans. 2.* **1998**, 573–581.

172. Breslow, R., Xu, R. Recognition and catalysis in nucleic acid chemistry. *Proc. Natl. Acad. Sci. U.S.A* **1993**, *90*, 1201–1207.

173. Anslyn, E., Breslow, R. Geometric evidence on the ribonuclease model mechanism. *J. Am. Chem. Soc.* **1989**, *111*(15), 5972–5973.

174. Kirby, A.J., Marriott, R.E. General base catalysis vs. medium effects in the hydrolysis of an RNA model. *J. Chem. Soc. Perkin Trans. 2.* **2002**, 422–427.

175. Breslow, R., Xu, R. Quantitative evidence for the mechanism of RNA cleavage by enzyme mimics: cleavage and isomerization of UpU by morpholine buffers. *J. Am. Chem. Soc.* **1993**, *115*(23), 10705–10713.

176. Breslow, R. Kinetics and mechanism in RNA cleavage. *Proc. Natl. Acad. Sci. U.S.A* **1993**, *90*, 1208–1211.

177. Khan, M.N. Kinetic evidence for the occurrence of kinetically detectable intermediates in the cleavage of N-ethoxycarbonylphthalimide under N-methylhydroxylamine buffers. *Int. J. Chem. Kinet.* **2002**, *34*, 95–103.

178. Beach, L.J., Batchelor, R.J., Einstein, F.W.B., Bennet, A.J. General base-catalyzed hydrolysis and carbonyl-^{18}O exchange of N-(4-nitrobenzoyl)pyrrole. *Can. J. Chem.* **1998**, *76*(10), 1410–1417.

179. Fife, T.H., Chauffe, L. General base and general acid catalyzed intramolecular aminolysis of esters: cyclization of esters of 2-aminomethylbenzoic acid to phthalimidine. *J. Org. Chem.* **2000**, *65*(12), 3579–3586.

180. Khan, M.N. Effects of anionic micelles on the intramolecular general-base-catalyzed hydrazinolysis and hydrolysis of phenyl salicylate: evidence for a porous cluster micelle. *J. Chem. Soc. Perkin Trans. 2.* **1990**, 445–457.

181. Bender, M.L., Kezdy, F.J., Zerner, B. Intramolecular catalysis in the hydrolysis of *p*-nitrophenyl salicylates. *J. Am. Chem. Soc.* **1963**, *85*(19), 3017–3024.

182. (a) Khan, M.N., Gambo, S.K. Intramolecular catalysis and the rate-determining step in the alkaline hydrolysis of ethyl salicylate. *Int. J. Chem. Kinet.* **1985**, *17*, 419–428. (b) Lajis, N.H., Khan, M.N. A partial model for serine proteases: kinetic demonstration of rate acceleration by intramolecular general base catalysis — an experiment for a biochemistry/enzymology lab. *Pertanika* **1991**, *14*(2), 193–199.

183. Khan, M.N. Kinetics and mechanism of transesterification of phenyl salicylate. *Int. J. Chem. Kinet.* **1987**, *19*, 757–776.

184. Moozyckine, A.U., Davies, D.M. Intramolecular base catalyzed hydrolysis of *ortho*-hydroxyaryl esters: the anomalous position of methyl 3,5-dinitro salicylate on the linear free energy relationship plot. *J. Chem. Soc. Perkin Trans. 2.* **2002**, 1158–1161.

185. Hoefnagel, A.J., Hoefnagel, M.A., Wepster, B.M. Substituent effects. 6. Charged groups: a simple extension of the Hammett equation. *J. Org. Chem.* **1978**, *43*(25), 4720–4745.

186. Hegarty, A.F., Bruice, T.C. Acyl transfer reactions from and to the ureido functional group. III. The mechanism of intramolecular nucleophilic attack of the ureido functional group upon acyl groups. *J. Am. Chem. Soc.* **1970**, *92*(22), 6575–6588.

187. Khan, M.N., Olagbemiro, T.O. Kinetic evidence for the participation of the ionized form of methyl p-hydroxybenzoate in its alkaline hydrolysis. *J. Chem. Res. (S).* **1985**, 166–167.

188. Khan, M.N., Olagbemiro, T.O. The kinetics and mechanism of hydrolysis of methyl salicylate under highly alkaline medium. *J. Org. Chem.* **1982**, *47*(19), 3695–3699.

189. Jencks, W.P., Carriuolo, J. Reactivity of nucleophilic reagents toward esters. *J. Am. Chem. Soc.* **1960**, *82*, 1778–1786.

190. Kirsch, J.F., Jencks, W.P. Base catalysis of imidazole catalysis of ester hydrolysis. *J. Am. Chem. Soc.* **1964**, *86*(5), 833–837.

191. Khan, M.N. Intramolecular general base catalysis and the rate determining step in the nucleophilic cleavage of ionized phenyl salicylate with primary and secondary amines. *J. Chem. Soc. Perkin Trans. 2.* **1989**, 199–208.

192. Khan, M.N., Ismail, E., Yusoff, M.R. Effects of pure non-ionic and mixed non-ionic-cationic surfactants on the rates of hydrolysis of phenyl salicylate and phenyl benzoate in alkaline medium. *J. Phys. Org. Chem.* **2001**, *14*, 669–676.

193. Kirby, A.J. Effective molarities for intramolecular reactions. *Adv. Phys., Org. Chem.* **1980**, *17*, 183–278.

194. Pascal, R. Induced intramolecularity in the reference reaction can be responsible for the low effective molarity of intramolecular general acid-base catalysis. *J. Phys. Org. Chem.* **2002**, *15*, 566–569.

195. Fife, T.H., Singh, R., Bembi, R. Intramolecular general base catalyzed ester hydrolysis: the hydrolysis of 2-aminomenzoate esters. *J. Org. Chem.* **2002**, *67*(10), 3179–3183.

196. Lilley, D.M.J. The origins of RNA catalysis in ribozymes. *Trend. Biochem. Sci.* **2003**, *28*(9), 495–501.

197. Bevilacqua, P.C. Mechanistic considerations for general acid-base catalysis by RNA: revisiting the mechanism of the hairpin ribozyme. *Biochemistry* **2003**, *42*(8), 2259–2265.

198. Herschlag, D. Ribonuclease revisited: catalysis via the classical general acid-base mechanism or a trimester-like mechanism? *J. Am. Chem. Soc.* **1994**, *116*(26), 11631–11635.

199. Nakano, S.-I., Chadalavada, D.M., Bevilacqua, P.C. General acid-base catalysis in the mechanism of a hepatitis delta virus ribozyme. *Science* (Washington D.C.). **2000**, *287*(5457), 1493–1497.

200. Bruice, P.Y., Bruice, T.C. Aminolysis of substituted phenyl quinoline-8- and -6-carboxylates with primary and secondary amines: involvement of proton-slide catalysis. *J. Am. Chem. Soc.* **1974**, *96*(17), 5533–5542.

201. Jencks, W.P., Gilchrist, M. General base catalysis of the aminolysis of phenyl acetate by primary alkylamines. *J. Am. Chem. Soc.* **1966**, *88*(1), 104–108.

202. Morris, J.J., Page, M.I. Intra- and inter-molecular catalysis in the aminolysis of benzylpenicillin. *J. Chem. Soc. Perkin Trans. 2.* **1980**, 212–219.

203. Page, M.I., Webster, P.S., Ghosez, L. The hydrolysis of azetidinyl amidinium salts. Part 2. Substituent effects, buffer catalysis, and the reaction mechanism. *J. Chem. Soc. Perkin Trans. 2.* **1990**, 813–823.

204. Fife, T.H., DeMark, B.R. Intramolecular nucleophilic aminolysis of aliphatic esters: cyclization of methyl 2-aminomethylbenzoate to phthalimidine. *J. Am. Chem. Soc.* **1976**, *98*(22), 6978–6982.

205. Deacon, T., Steltner, A., Williams, A. Reaction of 2-hydroxy-5-nitrotoluene-α-sulfonic acid sultone with nucleophiles. *J. Chem. Soc. Perkin Trans. 2.* **1975**, 1778–1783.

206. Khan, M.N., Malspeis, L. Kinetics and mechanism of thiolytic cleavage of the antitumor compound 4-[(9-acridinyl)amino]methanesulfon-m-anisidide. *J. Org. Chem.* **1982**, *47*(14), 2731–2740.

207. Cox, M.M., Jencks, W.P. Concerted bifunctional proton transfer and general base catalysis in the methoxyaminolysis of phenyl acetate. *J. Am. Chem. Soc.* **1981**, *103*(3), 580–587.

208. Fox, J.P., Jencks, W.P. General acid and general base catalysis of the methoxyaminolysis of 1-acetyl-1,2,4-triazole. *J. Am. Chem. Soc.* **1974**, *96*(5), 1436–1449.

209. Bruice, T.C., Donzel, A., Huffman, R.W., Butler, A.R. Aminolysis of phenyl acetates in aqueous solutions. VII.[1] Observations on the influence of salts, amine structure, and base strength. *J. Am. Chem. Soc.* **1967**, *89*(9), 2106–2121.

210. Williams, A., Jencks, W.P. Acid and base catalysis of urea synthesis: nonlinear Bronsted plots consistent with a diffusion-controlled proton-transfer mechanism and the reactions of imidazole and N-methylimidazole with cyanic acid. *J. Chem. Soc. Perkin Trans. 2.* **1974**, 1760–1768.

211. Khan, M.N. Kinetic evidence for the occurrence of general acid-base catalysis in N-methylhydroxylaminolysis of N-ethoxycarbonylphthalimide (NCPH). *Int. J. Chem. Kinet.* **1997**, *29*, 647–654.

212. Antelo, J.M., Arce, F., Calvo, P., Crugeiras, J., Rios, A. General acid-base catalysis in the reversible disproportionation reaction of N-chlorotaurine. *J. Chem. Soc. Perkin Trans. 2.* **2000**, 2109–2114.

213. Lin, B., Islam, N., Friedman, S., Yagi, H., Jerina, D.M., Whalen, D.L. Change of rate-limiting step in general acid-catalyzed benzo[a]pyrenediol epoxide hydrolysis. *J. Am. Chem. Soc.* **1998**, *120*(18), 4327–4333.

214. Davis, A.M., Proctor, P., Page, M.I. Alcohol-catalyzed hydrolysis of benzylpenicillin. *J. Chem. Soc. Perkin Trans. 2.* **1991**, 1213–1217.

215. Baxter, N.J., Laws, A.P., Rigoreau, L.J.H., Page, M.I. General acid catalyzed hydrolysis of β-sultams involves nucleophilic catalysis. *Chem. Commun.* **1999**, 2401–2402.

216. Fersht, A.R., Kirby, A.J. Intramolecular catalysis and the mechanism of enzyme action. *Chem. Br.* **1980**, *16*(3), 136–142, 156 and reference cited therein.

217. Bender, M.L. General acid-base catalysis in the intramolecular hydrolysis of phthalamic acid. *J. Am. Chem. Soc.* **1957**, *79*, 1258–1259.

218. Perry, C.J., Parveen, Z. The cyclization of substituted phthalanilic acids in acetic acid solution: a kinetic study of substituted N-phenylphthalimide formation. *J. Chem. Soc. Perkin Trans. 2.* **2001**, 512–521.

219. Mandyuk, V.F., Lushina, N.P. Kinetics of the hydrolysis of carboxylic acid anilides. I. Effect of substituents on the rate of hydrolysis of benzanilide. *Ukr. Khim. Zh.* **1966**, *32*(6), 607–609; *Chem. Abs.* **1966**, *65*, 15175.

220. Bender, M.L., Chow, Y.-L., Chloupek, F. Intramolecular catalysis of hydrolytic reactions. II. The hydrolysis of phthalamic acid. *J. Am. Chem. Soc.* **1958**, *80*, 5380–5384.

221. Arifin, A., Khan, M.N. Unexpected rate retardation in the formation of phthalic anhydride from N-methylphthalamic acid in acidic H_2O-CH_3CN medium. *Bull. Korean Chem. Soc.* **2005**, *26*(7), 1077–1043.

222. Perry, C.J. A new kinetic model for the acid-catalyzed reactions of N-(2-aminophenyl)phthalamic acid in aqueous media. *J. Chem. Soc. Perkin Trans. 2.* **1997**, 977–982.

223. Onofrio, A.B., Gesser, J.C., Joussef, A.C., Nome, F. Reactions of N-(o-carboxybenzoyl)-L-leucine: intramolecular catalysis of amide hydrolysis and imide formation by two carboxy groups. *J. Chem. Soc. Perkin Trans. 2.* **2001**, 1863–1868.

224. Capindale, J.B., Fan, H.S. The hydrolysis of some N-benzoylamino acids in dilute mineral acid. *Can. J. Chem.* **1967**, *45*(17), 1921–1924.

225. Dean, K.E.S., Kirby, A.J. Conceted general acid and nucleophilic catalysis of acetal hydrolysis: a simple model for the lysozyme mechanism. *J. Chem. Soc. Perkin Trans. 2.* **2002**, 428–432.

226. Asaad, N., Kirby, A.J. Concurrent nucleophilic and general acid catalysis of the hydrolysis of a phosphate triester. *J. Chem. Soc. Perkin Trans. 2.* **2002**, 1708–1712.

227. Barber, S.E., Dean, K.E.S., Kirby, A.J. A mechanism for efficient proton-transfer catalysis. Intramolecular general acid catalysis of the hydrolysis of 1-arylethyl ethers of salicylic acid. *Can. J. Chem.* **1999**, *77*(5/6), 792–801.

228. Tee, O.S. The stabilization of transition states by cyclodextrins and other catalysts. *Adv. Phys. Org. Chem.* **1994**, *29*, 1–85.

229. Herschlag, D. Ribonuclease revisited: catalysis via the classical general acid-base mechanism or a trimester-like mechanism? *J. Am. Chem. Soc.* **1994**, *116*(26), 11631–11635.

230. Tee, O.S., Boyd, M.J. The cleavage of 1- and 2-naphthyl acetates by cyclodextrins in basic aqueous solutions. *J. Chem. Soc. Perkin Trans. 2.* **1995**, 1237–1243.

231. Zhang, B., Breslow, R. Ester hydrolysis by a catalytic cylodextrin dimer enzyme mimic with a metallobipyridyl linking group. *J. Am. Chem. Soc.* **1997**, *119*(7), 1676–1681.

232. Tsutsumi, H., Hamasaki, K., Mihara, H., Ueno, A. Rate enhancement and enantioselectivity in ester hydrolysis catalyzed by cyclodextrin-peptide hybrids. *J. Chem. Soc. Perkin Trans. 2.* **2000**, 1813–1818.

233. Starks, C.M. Phase transfer catalysis. I. Heterogeneous reactions involving anion transfer by quaternary ammonium and phosphonium salts. *J. Am. Chem. Soc.* **1971**, *93*(1), 195–199.

234. Brandstrom, A. Principles of phase-transfer catalysis by quaternary ammonium salts. *Adv. Phys. Org. Chem.* **1977**, *15*, 267–330.

235. Sasson, Y., Rothenberg, G. Recent advances in phase-transfer catalysis. Eds., Clark, J., Macquarrie, D., *Handbook of Green Chemistry and Technology.* Blackwell Science, Oxford, **2002**, pp. 206–257.

236. Maruoka, K., Ooi, T. Enantioselective amino acid synthesis by chiral phase-transfer catalysis. *Chem Rev.* **2003**, *103*(8), 3013–3028.

237. Fache, F., Dunjic, B., Games, P., Lemaire, M. Recent advances in homogeneous and heterogeneous asymmetric catalysis with nitrogen-containing ligands. *Top. Catal.* **1998**, *4*(3,4), 201–209.

238. Castle, S.L., Srikanth, G.S.C. Catalytic asymmetric synthesis of the central tryptophan residue of celogentin C. *Org. Lett.* **2003**, *5*(20), 3611–3614.

239. Ohshima, T., Gnanadesikan, V., Shibuguchi, T., Fukuta, Y., Nemoto, T., Shibasaki, M. Enantioselective syntheses of aeruginosin 298-A and its analogues using a catalytic asymmetric phase-transfer reaction and epoxidation. *J. Am. Chem. Soc.* **2003**, *125*(37), 11206–11207.

240. Kim, S., Lee, J., Lee, T., Park, H.-G., Kim, D. First asymmetric total synthesis of (-)-antofine by using an enantioselective catalytic phase transfer alkylation. *Org. Lett.* **2003**, *5*(15), 2703–2706.

241. Noyori, R. Asymmetric catalysis: science and opportunities (Prix Nobel). **2001**, Vol. Date 2002, 186–215.

242. Lohray, B.B., Asymmetric catalysis — a novel chemistry to win the Nobel prize — 2001. *Curr. Sci.* **2001**, *81*(12), 1519–1525.

243. Goossen, L.J., Baumann, K. Nobel prize for chemistry, asymmetric catalysis. *Chem. Unserer Zeit.* **2001**, *35*(6), 402–403.

244. Brunner, H., Kagan, H.B., Krewutzer, G. Asymmetric catalysis. Part 153: metal-catalyzed enantioselective α-ketol rearrangement. *Tetrahedron: Asymmetry* **2003**, *14*(15), 2177–2187.

245. Gao, J., Martell, A.E. Self-assembly of chiral and chiral macrocyclic ligands: synthesis, protonation constants, conformation and asymmetric catalysis. *Org. Biomol. Chem.* **2003**, *1*(15), 2795–2800.

246. Dieguez, M., Ruiz, A., Claver, C. Tunable furanoside diphosphite ligands: a powerful approach in asymmetric catalysis. Dalton Trans. **2003**, (15), 2957–2963.

247. Bianchini, C., Volacchi, M. Enantioselective addition of alcohols to ketenes catalyzed by a planar-chiral azaferrocene: catalytic asymmetric synthesis of arylpropionic acids — Enantioselective Staudinger synthesis of β-lactams catalyzed by a planar-chiral nucleophile. *Chemtracts* **2003**, *16*(8), 491–495.

248. Rouzaud, J., Jones, M.D., Raja, R., Johnson, B.F.G., Thomas, J.M., Duer, M.J. Potent new heterogeneous asymmetric catalysts. *Helv. Chim. Acta.* **2003**, *86*(5), 1753–1759.

249. Bartel, B., Garcia-Yebra, C., Helmchen, G. Asymmetric Irl-catalyzed allylic alkylation of monosubstituted allylic acetates with phosphorus amidites as ligands. *Eur. J. Org. Chem.* **2003**, (6), 1097–1103.

250. Bolm, C., Xiao, L., Kesselgruber, M. Synthesis of novel chiral phosphinocyrhetrenyloxazoline ligands and their application in asymmetric catalysis. *Org. Biomol. Chem.* **2003**, *1*(1), 145–152.

251. Aggarwal, V.K., Richardson, J. The complexity of catalysis: origins of enantio- and diastereocontrol in sulfur ylide mediated epoxidation reactions. *Chem. Commun.* **2003**, 2644–2651.

252. Davies, H.M.L., Venkataramani, C., Hansen, T., Hopper, D.W. New strategic reactions for organic synthesis: catalytic asymmetric C–H activation α to nitrogen as a surrogate for the Mannich reaction. *J. Am. Chem. Soc.* **2003,** *125*(21), 6462–6468.

253. Walsh, P.J. Titanium-catalyzed enantioselective additions of alkyl groups to aldehydes: mechanistic studies and new concepts in asymmetric catalysis. *Acc. Chem. Res.* **2003,** *36*(10), 739–746.

254. Mikami, K., Yamanaka, M. Symmetry breaking in asymmetric catalysis: racemic catalysis to autocatalysis. *Chem. Rev.* **2003,** *103*(8), 3369–3400.

255. Gibson, S.E., Knight, J.D. [2.2]Paracyclophane derivatives in asymmetric catalysis. *Org. Biomol. Chem.* **2003,** *1*(8), 1256–1269.

256. Mikami, K., Terada, M., Matsuzawa, H. "Asymmetric" catalysis by lanthanide complexes. *Angew. Chem., Int. Ed. Engl.* **2002,** *41*(19), 3554–3571.

257. Shibasaki, M., Kanai, M. Multifunctional asymmetric catalysis. *Chem. Pharm. Bull. (Japan).* **2001,** *49*(5), 511–524.

258. McMarthy, M., Guiry, P.J. Axially chiral bidentate ligands in asymmetric catalysis. *Tetrahedron.* **2001,** *57*(18), 3809–3844.

3 General Mechanisms of Micellar Catalysis: Kinetic Models for Micelle-Catalyzed Reactions

3.1 INTRODUCTION

Although the notion of normal micelle (hereafter mentioned as micelle) formation in the aqueous solutions of surfactants had been established beyond any doubt at the turn of the 20th century, the first systematic study on the effects of micelles on the rate and equilibrium constants of chemical reactions appeared only in 1959.[1] Until the late 1960s, the pace of progress in the reaction rate study of micellar-mediated reactions was slow. The kinetic data on the rates of micellar-mediated reactions, obtained until the mid-1960s, have been explained only qualitatively because of the lack of an acceptable kinetic micellar model based on logical and convincing mechanisms of micellar-mediated reactions. As has always been the case in the progress and expansion of scientific frontiers, theoretically/mechanistically unexplained and exciting experimental observations have been the basis for theoretical/mechanistic developments in the scientific world.

The mechanism of any catalyzed or noncatalyzed reaction is perhaps the aspect of the reaction most difficult to understand. The *mechanism* of a reaction is defined as the detailed description of the electronic, or both electronic and nuclear, reorganization during the course of a reaction. There are several methods or techniques used to diagnose the reaction mechanism, but perhaps the kinetic method is the most important one, which establishes the most refined mechanism at the molecular level for any reaction. The kinetic study provides experimentally determined reaction parameters that could be used to test a proposed reaction mechanism.

By the late 1960s, the accumulated kinetic data on micellar-mediated reactions were enough to warrant a logical kinetic micellar model to explain these kinetic data quantitatively. The kinetic studies of reaction rates provide perhaps the most extensive fine details of changes at the molecular level of chemical reactions. The effects of pure and mixed organic-aqueous solvents on the reaction rates have always been explained in terms of empirical approaches because of the lack of a completely flawless theory to explain such effects. The highly dynamic structural features of these solvent systems, which are generally sensitive to the presence of solutes of diverse structural features, pose an extremely difficult problem for anyone trying to formulate a general and perfect theory to explain quantitatively the effects of solvents on reaction rates. The micellar effects are almost similar to the mixed aqueous-organic solvent effects on reaction rates in terms of the complexity at the molecular level. However, whenever the scientific community is challenged by such molecular complexity, it is customary to understand it by developing a chain of so-called theoretical models. Although none of these theoretical models may appear perfect, they provide the basis that could lead to a so-called perfect or relatively more accurate model. As Menger[2] put it, "Models are to be used, not believed." I would prefer to say that models are to be used, but not believed in the sense of providing the ultimate truth.

Kinetic study has been one of the best mechanistic (reaction) tools to establish the most refined reaction mechanisms. In an attempt to establish such a reaction mechanism, kinetic experimental data on the reaction rate are obtained under a set of reaction conditions that could be explained by a kinetic equation derived on the basis of a proposed reaction mechanism. The study is repeated to obtain kinetic data under slightly or totally different reaction conditions, and if these kinetic data fail to fit the kinetic equation derived on the basis of the earlier reaction mechanism, a further refinement in the mechanism is suggested so that the present and earlier kinetic data could be explained mechanistically. A similar approach has been used to provide quantitative or semiquantitative explanations for the micellar effects on reaction rates. Let us now examine the micellar kinetic models developed so far for apparent quantitative explanations of the effects of micelles on reaction rates.

3.2 PREEQUILIBRIUM KINETIC MODEL OF THE MICELLE

Experimentally determined rate constants for various micellar-mediated reactions show either a monotonic decrease (i.e., micellar rate inhibition) or increase (i.e., micellar rate acceleration) with increase in $[Surf]_T$-CMC, where $[Surf]_T$ represents total micelle-forming surfactant concentration (Figure 3.1). Menger and Portnoy[3] obtained rate constants — $[Surf]_T$ plots — for hydrolysis of a few esters in the presence of anionic and cationic surfactants, which are almost similar to those plots shown in Figure 3.1. These authors explained their observations in terms of a proposed reaction mechanism as shown in Scheme 3.1 which is now called *Menger's phase-separation model, enzyme-kinetic-type model,* or *preequilibrium kinetic (PEK) model* for micellar-mediated reactions. In Scheme 3.1, K_S is the equilibrium

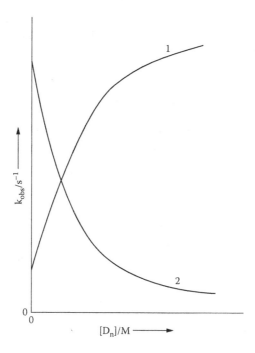

FIGURE 3.1 Plots 1 and 2 showing the effects of $[D_n]$ on experimentally determined rate constants (k_{obs}) for hypothetical reactions.

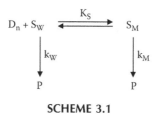

SCHEME 3.1

constant for micellization of ester molecules, P is the product, and D_n, S_W, and S_M represent micelle, free ester, and adsorbed/micellized ester molecules, respectively. The rate constants k_W and k_M stand for hydrolysis of free ester and micellized ester, respectively. In order to calculate the micelle concentration, the *phase-separation* concept was used, which assumes that the unassociated surfactant concentration remains constant above the CMC of surfactant. Thus, the *concentration of micelle* ($[D_n]$) is defined as $[D_n] = \{[Surf]_T-CMC\}/n$, where n is the aggregation number that represents the average number of surfactant monomer in a micelle. The additional assumptions involved in this model are as follows:

1. Substrate does not complex with surfactant monomer.
2. Substrate does not perturb micellization.

3. Substrate associates with the micelles in a 1:1 stoichiometry.
4. Micellization occurs exactly at the CMC rather than over a small concentration range.
5. The relationship $[D_n] = \{[Surf]_T\text{-}CMC\}/n$ is valid.

It is interesting to note that the stepwise association model or stepwise aggregation model for micelle formation, as shown in Scheme 3.2, leads to the expression $<n> = (C_t - [S])/(C_m - [S])$, where $<n>$ represents the number-average micellar aggregation number, C_t is the total analytical concentration of surfactant (i.e., $C_t \equiv [Surf]_T$), C_m is the total concentration of osmotically active particles (i.e., monomers + micelles or CMC + $[D_n]$), and $[S]$ ($\equiv$ CMC) is the monomer concentration.[4] Although the stepwise association model is not exactly similar to the phase-separation model, the basis of preequilibrium kinetic model of micelle (Scheme 3.1), the expression $<n> = (C_t - [S])/(C_m - [S])$ is exactly similar to assumption (5) in the PEK model. The mass action model or multiequilibrium model, Equation 3.1, which is equivalent to Equation 1.20 in Chapter 1

$$S_1 + S_{n-1} \underset{}{\overset{K_n}{\rightleftharpoons}} S_n \quad \text{where } n = 2, 3, 4, \ldots \tag{3.1}$$

but seems to be thermodynamically more reasonable, is different from the stepwise association model as expressed by Scheme 3.2[4-7]

$$2S \underset{}{\overset{\beta_2}{\rightleftharpoons}} S_2$$

$$3S \underset{}{\overset{\beta_3}{\rightleftharpoons}} S_3$$

$$\text{-----------------}$$

$$nS \underset{}{\overset{\beta_n}{\rightleftharpoons}} S_n$$

SCHEME 3.2

where β_2, β_3, ..., β_n represent stepwise equilibrium constants for the formation of surfactant molecular aggregates containing 2, 3, ..., and n monomers, respectively. However, both equilibrium constants, β_n and K_n, related by the relationship $\beta_n = \prod K_n - n = 2, 3, 4, \ldots$, maximum aggregation number (n_{max}).[7,8]

The PEK model, i.e., Scheme 3.1, was proposed based on a similar reaction scheme for a nonmicellar system.[9] The observed rate law, rate = $k_{obs}[Sub]_T$, (where k_{obs} and $[Sub]_T$ represent observed first- or pseudo-first-order rate constant and total substrate concentration, respectively), and Scheme 3.1 can lead to Equation 3.2. The observed data (k_{obs} vs. $[D_n]$) have

$$k_{obs} = \frac{k_W + k_M K_S [D_n]}{1 + K_S [D_n]} \tag{3.2}$$

been found to fit to Equation 3.2 reasonably well for many micellar-mediated reactions.[3,10–13] The validity of the PEK model of micelles rests merely on the reasonably good fit of experimentally observed data to Equation 3.2. It lacks physical/chemical interpretation of kinetic disposable parameters k_M and K_S because of the poorly understood structural features of micelles when the PEK model was proposed in 1967.

3.3 PSEUDOPHASE MODEL OF THE MICELLE

An aqueous solution of surfactant at $[Surf]_T$ less and greater than CMC remains transparent to UV-visible radiation and, consequently, it is defined as a single homogeneous phase. Thus, by a simple definition of a real phase, micelles cannot be considered to constitute a real phase and, for this technical reason, micelles are said to represent a pseudophase (PP). The word *pseudophase* of micelle is probably the most appropriate term, because various kinds of experimental data show that micelles are surfactant molecular aggregates with aggregation numbers varying from <100 to >100 depending on the nature and concentration of micelle-forming surfactant and additives as well as temperature[4] of aqueous surfactant solutions. Menger and Portnoy[3] proposed the concept of *micellar phase*, which does not seem to fit well within the domain of a formal definition of the real phase. A number of influential researchers in this field suggested the concept of the PP rather than real phase of micelles.[10–18] The PP model of micelles, in addition to retaining almost all the assumptions introduced in PEK model[3], considers the following assumptions:

1. Micelles and bulk aqueous solvents are regarded as distinct reaction regions.
2. Micellar effects on reaction rates and equilibria are insensitive to changes in the size and shape of micelle.
3. $k_S \gg k_W$ and $k_{-S} \gg k_M$, where k_S and k_{-S} represent rate constants for micellar incorporation and micellar exit, respectively, of solubilizate/substrate S, and hence $k_S/k_{-S} = K_S$ (Scheme 3.1).
4. Equilibrium processes or equilibrium constants for micellar incorporation/solubilization of different solubilizates are independent of each other, i.e., there is no cross-interaction between equilibrium constants of micellar incorporation of different solubilizates.
5. The equilibrium constant K_M for the formation of micelles as expressed by Equation 3.3 is independent of equilibrium constants K_S for micellar solubilization of different solubilizates, and rate constants k_M for micellar-mediated reactions. In other words, the rates of formation and disintegration of micelle are independent of the corresponding rates of micellar intake and exit of a solubilizate; $k_f^M \gg k_W$ and $k_d^M \gg k_M$ where k_f^M and k_d^M represent rate constants for micelle formation and micelle disintegration, respectively, and therefore $k_f^M/k_d^M = K_M$ (Scheme 3.1 and Scheme 3.2). In Equation 3.3,

$$\{(n-N)/N_A\} \text{ monomers} \xrightleftharpoons{K_M} (N/rN_A) \text{ micelles} \qquad (3.3)$$

where n represents total number of surfactant molecules, N is the total number of surfactant molecules used up in the formation of number of micelles (N/r), r the mean aggregation number of a micelle, and N_A is Avogadro's number.

6. For a bimolecular reaction, the reaction between a reactant (R_M) in the micellar pseudophase and the other reactant (S_W) in the aqueous pseudophase does not occur; i.e., the cross-interface reaction such as that between R_M and S_W or R_W and S_M does not take place.

In view of assumption 1, each micelle acts as a "microreactor" of one domain that provides a new reaction medium and alters the distributions of reactants in solution. Thus, the classical PP model may also be called the *two-domain pseudophase* model of micelles, in which the bulk aqueous region and the entire micellar pseudophase are considered to be two different reaction domains. A refinement of Equation 3.2 consists of including the possibility of different reaction domains within the micelle. A *three-domain pseudophase* model is one in which, for example, the bulk aqueous region, the Stern region, and the hydrophobic micellar core are treated as separate reaction regions.[19]

Despite the accumulation of a huge amount of kinetic data for micellar-mediated unimolecular reactions during the last nearly four decades, there seem to be no reported experimental observations that could unambiguously invalidate any one or all of the assumptions from 2 to 5. However, some recently observed kinetic data for an apparently unimolecular reaction could not fit into Equation 3.2, and this failure of the PP model to explain these kinetic data quantitatively is attributed to the structural modifications of micelles under these conditions.[20-22] The magnitudes of experimentally determined rate constants for micellar-mediated reactions depend largely on the values of micellar binding constants of reactants and the rates of reactions occurring in the micellar pseudophase. It is quite possible that the change in size and shape of micelles due to the increase in total surfactant concentration will not affect either the micellar binding constants of reactants or the rates of reactions occurring in the micellar pseudophase. The value of the micellar binding constant of a solubilizate is controlled by polar/ionic and van der Waals attractive as well as repulsive interactions between solubilizate and micelle. These interactions are governed by the structural features of the solubilizate and micelle-forming surfactant molecules and, consequently, they are not expected to be influenced greatly by the shape and size of micelles; these micellar characteristics depend on the structural features and concentrations of solubilizate and micelle-forming surfactant molecules. The value of the rate constant (k_M) for the reaction of micellized reactant is largely affected by the medium characteristics of the micro micellar reaction environment in which the reaction occurs. The medium characteristics of the micro micellar reaction environment are not expected to vary with the change in size and shape of the micelles. However, at an optimum micelle concentration in which the reactant molecules of a bimolecular reaction are almost completely micellized, the increase in micelle concentration will not affect k_M but will decrease the rate of reaction merely because of the dilution effect on the concentrations of micellized reactants.

The dynamics of micelle formation and disintegration as well as micellar solubilization and exit of an especially neutral solubilizate of moderate polarity/hydrophobicity indicate that assumptions 3 to 5 may easily break down for fast reactions and, under such conditions, the rate of a micellar-mediated reaction cannot be expected to follow a simple first- or second-order rate law. The rate law of such a reaction will constitute a transcendental kinetic equation because of the consecutive nature of the reaction. However, such an expected kinetic complexity has not been reported in any study on the effects of micelles on reaction rates.

As has been described in Chapter 1, the kinetics of micellization of surfactant molecules in aqueous solution clearly demonstrate the presence of two relaxation times of very different ranges. The fast relaxation time is usually in the microsecond range, whereas the other slow relaxation time is in the millisecond to second range. However, the mechanisms of these relaxation processes of micellization are not understood at the moment even at a very rudimentary level. The slow relaxation time, which falls within the millisecond to second range, is expected to affect the reaction kinetics of moderately fast reactions (with rate constant values of say >0.03 sec^{-1}). In the absence of any report in which the rate of micellization is shown to have complicated the reaction kinetics of moderately fast micellar-mediated reactions, it seems that the rate of micellization corresponding to the slow relaxation time in seconds involves micelle fusion — fission rather than micelle formation — breakup reaction as shown in Figure 3.2. Thus, the most likely mechanism of micelle formation in aqueous surfactant solution is the fusion–fission reaction mechanism (Chapter 1) in which the fragmented (fission) aggregates (A_j and A_k) are large enough to provide a micellar environment for micellized reactants similar to that afforded by fused aggregates (A_{j+k}) and, consequently, moderately fast micelle fusion–fission does not affect the micellar-mediated reaction rate. Generally, micellar structural changes do not significantly affect the rates of micellar-mediated reactions.

If a bimolecular reaction between S and R involves cross-interface reaction, i.e., the reaction occurs in violation of assumption 6, then the PP model can lead to one of the five probable reaction mechanisms described in the following text.

3.3.1 Cross-Interface Reaction between S_M and R_W

The brief reaction mechanism for such a bimolecular reaction may be shown by Scheme 3.3. Subscripts W and M represent aqueous pseudophase and micellar pseudophase, respectively. The second-order rate constant $k_M{}^W$ stands for cross-interface reaction. The observed rate law rate = $k_{obs}[Sub]_T$ and Scheme 3.3 can lead to Equation 3.4:

$$S_W + D_n \underset{\longleftarrow}{\overset{K_S}{\longrightarrow}} S_M$$

$$R_W + D_n \underset{\longleftarrow}{\overset{K_R}{\longrightarrow}} R_M$$

$$S_W + R_W \xrightarrow{k_W^2} Product(s)$$

$$S_M + R_W \xrightarrow{k_M^W} Product(s)$$

SCHEME 3.3

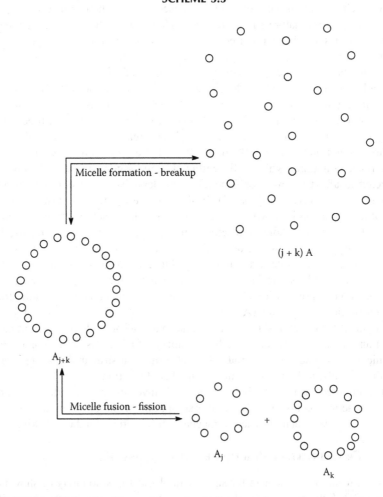

FIGURE 3.2 Schematic representation of micelle formation — breakup and micelle fusion–fission reaction mechanism.

$$k_{obs} = \frac{(k_W^2 + k_M^W K_S[D_n])[R]_T}{(1 + K_R[D_n])(1 + K_S[D_n])} \qquad (3.4)$$

where $[R]_T = [R_W] + [R_M]$. If $K_R \approx 0$, i.e., reactant R does not bind with micelles, then at a constant initial value of $[R]_T$, Equation 3.4 predicts the plots of k_{obs} vs. $[D_n]$ similar to those shown in Figure 3.1 with monotonic increase and monotonic decrease in k_{obs} with $[D_n]$ when $k_W^2 < k_M^W$ and $k_W^2 > k_M^W$, respectively. Under limiting conditions when observed rate constants (k_{obs}) become independent of $[Surf]_T$, i.e., $[D_n]$, because $k_W^2 \ll k_M^W K_S[D_n]$ and $1 \ll K_S[D_n]$, the $[Surf]_T$-independent values of k_{obs} ($= k_M^W[R]_T$) at different values of $[R]_T$ should vary linearly with $[R]_T$ with essentially zero intercept. However, if the plot of k_{obs} vs. $[Surf]_T$ does not show a $[Surf]_T$-independent region even when S molecules are fully micellized, then for the reaction system in which $k_W^2 > k_M^W$, it may be attributed to a mere possibility that $K_R \neq 0$ and the values of K_R and K_S are not very different from each other. On the other hand, if the plot of k_{obs} vs. $[Surf]_T$ does show a strong or weak $[Surf]_T$-independent region and a further increase in $[Surf]_T$ decreases k_{obs}, then such a k_{obs} vs. $[Surf]_T$ plot may be ascribed to the possibility that $K_S \gg K_R$ or K_S is moderately larger than K_R.

3.3.2 CROSS-INTERFACE REACTION BETWEEN S_W AND R_M

The reaction mechanism for such a micellar-mediated bimolecular reaction may be expressed by Scheme 3.4, which can lead to Equation 3.5.

$$S_W + D_n \xrightleftharpoons{K_S} S_M$$

$$R_W + D_n \xrightleftharpoons{K_R} R_M$$

$$S_W + R_W \xrightarrow{k_W^2} Product(s)$$

$$S_W + R_M \xrightarrow{k_W^M} Product(s)$$

SCHEME 3.4

$$k_{obs} = \frac{(k_W^2 + k_{WM}^{mr} K_R)[R]_T}{(1 + K_R[D_n])(1 + K_S[D_n])} \qquad (3.5)$$

In Equation 3.5, $k_{WM}^{mr} = k_W^M/V_M$ where V_M represents molar volume of the micellar reaction region.[12,13,16,23,24]

3.3.3 Cross-Interface Reaction: $S_M + R_W$ Parallel to $S_M + R_M$

The reaction mechanism for such micellar-mediated reactions is shown by Scheme 3.5.

$$S_W + D_n \xrightleftharpoons{K_S} S_M$$

$$R_W + D_n \xrightleftharpoons{K_R} R_M$$

$$S_W + R_W \xrightarrow{k_W^2} \text{Product(s)}$$

$$S_M + R_W \xrightarrow{k_M^W} \text{Product(s)}$$

$$S_M + R_M \xrightarrow{k_M^M} \text{Product(s)}$$

SCHEME 3.5

Scheme 3.5 and the experimentally observed rate law can lead to Equation 3.6:

$$k_{obs} = \frac{(k_W^2 + k_M^W K_S[D_n] + k_{MM}^{mr} K_S K_R[D_n])[R]_T}{(1 + K_R[D_n])(1 + K_S[D_n])} \tag{3.6}$$

where $k_{MM}^{mr} = k_M^M/V_M$; V_M represents molar volume of the micellar reaction region.[12,13,16,23,24]

3.3.4 Cross-Interface Reaction: $S_W + R_M$ Parallel to $S_M + R_M$

The reaction mechanism for such micellar-mediated reactions is shown by Scheme 3.6.

Scheme 3.6 and the experimentally observed rate law can lead to Equation 3.7:

$$k_{obs} = \frac{(k_W^2 + k_{WM}^{mr} K_R + k_{MM}^{mr} K_S K_R[D_n])[R]_T}{(1 + K_R[D_n])(1 + K_S[D_n])} \tag{3.7}$$

where $k_{WM}^{mr} = k_W^M/V_M$ and $k_{MM}^{mr} = k_M^M/V_M$; V_M represents molar volume of the micellar reaction region.[12,13,16,23,24]

$$S_W + D_n \underset{\longleftarrow}{\overset{K_S}{\rightleftharpoons}} S_M$$

$$R_W + D_n \underset{\longleftarrow}{\overset{K_R}{\rightleftharpoons}} R_M$$

$$S_W + R_W \xrightarrow{k_W^2} \text{Product(s)}$$

$$S_W + R_M \xrightarrow{k_W^M} \text{Product(s)}$$

$$S_M + R_M \xrightarrow{k_M^M} \text{Product(s)}$$

SCHEME 3.6

3.3.5 CROSS-INTERFACE REACTIONS: $S_W + R_M$ AND $S_M + R_W$ PARALLEL TO $S_M + R_M$

The reaction mechanism for such micellar-mediated reactions is shown by Scheme 3.7.

$$S_W + D_n \underset{\longleftarrow}{\overset{K_S}{\rightleftharpoons}} S_M$$

$$R_W + D_n \underset{\longleftarrow}{\overset{K_R}{\rightleftharpoons}} R_M$$

$$S_W + R_W \xrightarrow{k_W^2} \text{Product(s)}$$

$$S_W + R_M \xrightarrow{k_W^M} \text{Product(s)}$$

$$S_M + R_W \xrightarrow{k_M^W} \text{Product(s)}$$

$$S_M + R_M \xrightarrow{k_M^2} \text{Product(s)}$$

SCHEME 3.7

Scheme 3.7 and experimentally observed rate law can lead to Equation 3.8:

$$k_{obs} = \frac{(k_W^2 + k_{WM}^{mr}K_R + k_{MW}^{mr}K_S[D_n] + K_M^{mr}K_SK_R[D_n])[R]_T}{(1 + K_R[D_n])(1 + K_S[D_n])} \tag{3.8}$$

where $k_{WM}^{mr} = k_W^M/V_M$, $k_{MW}^{mr} = k_M^W/V_M$, and $k_M^{mr} = k_M^2/V_M$; V_M represents molar volume of micelle or micelle-forming surfactant.[12,13,16,23,24] Some limiting conditions of Equation 3.5 through Equation 3.8 may be discussed just as was done for Equation 3.4.

It is evident from Equation 3.4 through Equation 3.8 that, under typical experimental conditions in which one reactant or both reactants are completely micellized by micelles of nonionic or ionic surfactants with counterions as inert ions, rate constants k_{obs} should decrease with the increase in $[Surf]_T$ (total concentration of micelle-forming surfactant) at a constant $[R]_T$. There seems to be no report in the literature on the occurrence of cross-interface reactions involving nonionic or ionic micelles with inert counterions. However, cross-interface reactions with reaction schemes similar to the one represented by Scheme 3.5 were reported for the first time in the aromatic hydroxide ion nucleophilic substitution reactions in the presence of p-$C_8H_{17}OC_6H_4CH_2NMe_3^+$ HO^- where counterion HO^- was also a reactant.[25] The proposition that cross-interface reactions had occurred in this study was based on the linear increase observed in k_{obs} with increase in $[D_n]$ even when there was strong evidence that the aromatic substrate was fully micellar bound. These observations were explained by the following kinetic equation:

$$k_{obs} = \frac{\{k_W^2 + k_M^W K_S[D_n]\}\{CMC + \alpha[D_n]\} + k_M^{mr} K_S(1-\alpha)[D_n]}{1 + K_S[D_n]} \quad (3.9)$$

where $\alpha = [HO_W^-]/[D_n] = [HO^-]_T/\{[D_n](1 + K_{OH}[D_n])\}$. Equation 3.9 is similar to Equation 3.6 with $R = HO^-$. At high total surfactant concentration ($[Surf]_T$), $[Surf]_T \approx [D_n]$ because $[D_n] >> CMC$; $1 << K_S[D_n]$ and $k_W^2 << k_M^W K_S[D_n]$; under such conditions, Equation 3.9 reduces to Equation 3.10, which explains the linear increase in k_{obs} with increase in $[Surf]_T$ even when the aromatic substrate (S) is fully micellar bound.[25]

$$k_{obs} = \alpha k_M^W [Surf]_T + k_M^{mr}(1-\alpha) \quad (3.10)$$

In view of Equation 3.10, the strict linear relationship between k_{obs} and $[Surf]_T$ is expected only if α is independent of $[Surf]_T$.

Cross-interface reactions that follow reaction schemes similar to Scheme 3.5 have been reported in the dehydrochlorination of 1,1,2-trichloro-2,2-bis(p-chlorophenyl)ethane (DDT) and some of its derivatives in the presence of hydroxide ion and hexadecyltrimethylammoinum bromide (CTABr)[26] as well as hexadecyltrimethylammoinum hydroxide (CTAOH)[27] micelles. However, some of these and related experimentally observed kinetic data have later been rationalized with other approaches, which include the following:

1. The assumption that counterion binding obeys a micelle surface potential independent, mass action law[28,29]

2. Spherical cell models, in which local counterion concentrations are calculated by solving nonlinear Poisson–Boltzman equations and treating the size of CTAOH micelles as an adjustable parameter[30,31]
3. Allowance for variation in the degree of micellar ionization (α) with a change in the concentration of micelle-forming ionic surfactant[32]
4. The proposition that the interfacial local concentration of counterion in the micellar pseudophase is equal to an initial effective counter ion concentration resulting from micelle formation plus the counter ion concentration in the aqueous pseudophase[33]

3.3.6 BIMOLECULAR REACTIONS THAT DO NOT INVOLVE CROSS-INTERFACE REACTIONS

Almost all the micellar-mediated bimolecular reactions so far reported involve parallel reaction steps that constitute reactions occurring simultaneously in both aqueous phase and micellar pseudophase as expressed in Scheme 3.8. As described earlier, there are only a few reports that show an apparent failure of bimolecular reactions to obey the reaction in Scheme 3.8. However, this failure is attributed to various factors, including the possibility of cross-interface reaction.[25-33]

$$S_W + D_n \; \underset{\longleftarrow}{\overset{K_S}{\longrightarrow}} \; S_M$$

$$R_W + D_n \; \underset{\longleftarrow}{\overset{K_R}{\longrightarrow}} \; R_M$$

$$S_W + R_W \; \xrightarrow{k_W{}^2} \; \text{Product(s)}$$

$$S_M + R_M \; \xrightarrow{k_M{}^2} \; \text{Product(s)}$$

SCHEME 3.8

Scheme 3.8 and the experimentally observed rate law rate = k_{obs} [Sub]$_T$ can lead to Equation 3.11:

$$k_{obs} = \frac{(k_W^2 + k_M^{mr} K_S K_R [D_n])[R]_T}{(1 + K_R [D_n])(1 + K_S [D_n])} \tag{3.11}$$

where $k_M^{mr} = k_M^2/V_M$; V_M represents molar volume of the micellar reaction region.[12,13,16,23,24] Equation 3.11 can also be expressed as follows:

$$k_{obs} = \frac{k_W^2 [R]_T + (k_M^{mr} K_S - k_W^2) m_R^s [D_n]}{1 + K_S [D_n]} \tag{3.12}$$

where $m_R^s = [R_M]/[D_n]$.

In case $1 \gg K_R[D_n]$ under specific experimental conditions, Equation 3.11 reduces to Equation 3.13:

$$k_{obs} = \frac{k_W^2[R]_T + k_M^{mr}K_RK_S[D_n][R]_T}{1+K_S[D_n]} \qquad (3.13)$$

Equation 3.12 reduces to Equation 3.14 if either $k_M^{mr} K_S \approx k_W^2$ or $m_R^s \approx 0$; i.e., micellar affinity of R is very low (i.e., $K_R \approx 0$).

$$k_{obs} = \frac{k_W^2[R]_T}{1+K_S[D_n]} \qquad (3.14)$$

However, in case $k_M^{mr} \approx 0$, i.e., the rate of micellar-mediated reaction is negligible compared to the rate of reaction in the aqueous pseudophase, Equation 3.11 reduces to Equation 3.15.

$$k_{obs} = \frac{k_W^2[R]_T}{(1+K_S[D_n])(1+K_R[D_n])} \qquad (3.15)$$

Equation 3.11 or Equation 3.12 predicts the occurrence of maxima in the plots of k_{obs} vs. $[D_n]$ if $k_M^{mr} K_S \gg k_W^2$, whereas Equation 3.14 and Equation 3.15 predict a monotonic decrease in k_{obs} with the increase in $[D_n]$.

If the concentrations of both reactants in a bimolecular reaction are expressed, as usual, in units of moles per liter (i.e., M), then Equation 3.11 will take the form of Equation 3.16.

$$k_{obs} = \frac{(k_W^2 + k_M^2 K_S K_R[D_n]^2)[R]_T}{(1+K_R[D_n])(1+K_S[D_n])} \qquad (3.16)$$

Under limiting conditions of $K_R[D_n] \gg 1$, $K_S[D_n] \gg 1$ (i.e., when both reactants are completely micellized), and $k_W^2 \ll k_M^2 K_S K_R[D_n]^2$), Equation 3.16 predicts that k_{obs} or k_{2obs} ($= k_{obs}/[R]_T$) should be independent of $[D_n]$ or $[Surf]_T$. However, the value of a second-order rate constant depends on the concentration of the reactant, and the volume of the micellar reaction environment increases with an increase in $[D_n]$. Therefore k_{obs} or k_{2obs} can never be expected to be independent of $[D_n]$ under the limiting conditions already mentioned. Logic suggests that k_{obs} or k_{2obs} should decrease with an increase in $[D_n]$ beyond a limiting value of $[D_n]$ where both reactants are completely micellized. As early as in 1968, Bunton and Robinson[34] reported the presence of maxima in the plots of k_{2obs} vs. $[D_n]$ for the reactions of HO$^-$ with 2,4-dinitrochlorobenzene in the presence of CTABr micelles. These authors explained their observations using the kinetic equation (Equation 3.17)

$$k_{2obs} = \frac{k_W^2 + k_M^2 K_S[D_n]}{1 + K_S[D_n] + \mu[D_n]^2}$$ (3.17)

which contains an empirical term, $\mu[D_n]^2$, in the denominator with μ representing an empirical constant. Equation 3.17 is similar to Equation 3.4 provided $k_M^2 = k_M^W$, $\mu = K_R K_S$, and $K_S \approx K_S + K_R$. A qualitative rationalization of the empirical term $\mu[D_n]^2$ in Equation 3.17 is that when the number of micelles is large and virtually all the substrate is in the micellar phase, additional micelles will take up hydroxide ions, which will thereby be deactivated because of the low probability of a substrate molecule in one micelle reacting with hydroxide ion in another.[34]

The occurrence of maxima in the plots of pseudo-first-order rate constants (k_{obs}) vs. $[D_n]$ for bimolecular reactions involving a hydrophobic organic substrate and an ionic reactant is a common feature of such reaction systems.[11–13,16,23,24,35,36] Equation 3.11 and Equation 3.12 predict the occurrence of maxima in such kinetic plots provided $k_M^{mr} K_S \gg k_{2,W}$ and $K_R[D_n]$ as well as $K_S[D_n]$ are not negligible compared to 1 at considerably high values of $[D_n]$. But very often such maxima occur in the bimolecular reactions with one of the reactants (say R) being highly hydrophilic (such as R = HO⁻, HOO⁻, F⁻) and within $[D_n]$ range where it is certain that $1 \gg K_R[D_n]$.[37] Thus, these observations cannot be explained in terms of PP micellar model, i.e., simple reaction Scheme 3.8.

3.3.7 THE PSEUDOPHASE ION-EXCHANGE (PIE) MODEL

The competition between counterions (X) and other ions (Y) of similar charge for ionic micellar surface was recognized very early in the study on micellar rate effects.[38] But the theoretical interpretation of kinetic data on micellar-mediated reactions involving such ion competition or ion exchange was lacking until Romsted and others put forward a theoretical model known as the PIE model, which provides quantitative or semiquantitative interpretation of such ion exchange.[39,40] The occurrence of ion exchange at the ionic micellar surface has been unequivocally established by direct determination of the ion-exchange constant using various nonkinetic techniques.[41–46]

The PIE model is essentially an extension of the PP model and therefore contains all the assumptions involved in the PP model and a few more, as clearly expressed in several excellent reviews.[12,13,16,23,24,36,39] These additional assumptions may be summarized as follows:

1. The degree of counterion ionization remains constant (i.e., there is a strictly 1:1 ion exchange) irrespective of ion type or concentration or of surfactant concentration.

2. The micellar surface region can be thought of as an ion-exchange resin in which ion exchange processes occur in the same way as for a resin.

In view of the PIE model, the concentrations of a reactive anion, Y (anionic reactant), and an inert anion, X (counterion of cationic micelle), in the micellar and aqueous pseudophases are governed by an ion-exchange equilibrium process, Equation 3.18:

$$[Y_M^-]+[X_W^-] \underset{}{\overset{K_X^Y}{\rightleftharpoons}} [Y_W^-]+[X_M^-] \tag{3.18}$$

and

$$K_X^Y = ([Y_W^-][X_M^-])/([Y_M^-][X_W^-]) \tag{3.19}$$

The fraction of counterions bound to micelle, β, and total concentrations of X^-, $[X]_T$ and Y^-, $[Y]_T$, may be given as

$$m_X + m_Y = \beta \tag{3.20}$$

$$[X]_T = [X_W^-] + m_X[D_n] \tag{3.21}$$

$$[Y]_T = [Y_W^-] + m_Y[D_n] \tag{3.22}$$

where $m_X = [X_M^-]/[D_n]$ and $m_Y = [Y_M^-]/[D_n]$. Equation 3.19 to Equation 3.22 can yield Equation 3.23:

$$m_Y^2 + m_Y \left\{ \frac{[Y]_T + K_X^Y[X]_T}{(K_X^Y-1)[D_n]} - \beta \right\} - \frac{\beta[Y]_T}{(K_X^Y-1)[D_n]} = 0 \tag{3.23}$$

The solution of the quadratic equation (Equation 3.23) may be given as

$$m_Y = [-b + (b^2 - 4c)^{1/2}]/2 \tag{3.24}$$

where $b = \{([Y]_T + K_X^Y [X]_T)/((K_X^Y - 1) [D_n])\} - \beta$, and $c = -(\beta[Y]_T)/\{(K_X^Y - 1) [D_n]\}$.

The values of m_Y at different $[D_n]$ values can be calculated from Equation 3.24 for the given values of K_X^Y and β, and these m_Y values can be subsequently used to calculate k_M^{mr} and K_S from Equation 3.12 (with m_R^s representing m_Y), using the nonlinear least-squares technique. This is the general practice used in applying the PIE model for kinetic analysis of the rates of appropriate bimolecular reactions.[47-53]

Although the PIE model has been extensively used in kinetic data analysis for semiionic bimolecular reactions (i.e., with one of the reactants being ionic) in the presence of normal ionic micelles,[54-59] its use has been extended to such reactions in the presence of reversed micelles,[60-62] microemulsions,[63-65] cosurfactant-modified micelles,[66-67] and vesicles.[68-71]

The success of the PIE model is generally emphasized within the domain of low residual errors between observed and calculated rate constants at varying concentrations of micelles ($[D_n]$) and its ability to predict the maxima in the plots of k_{obs} vs. $[D_n]$ under experimental conditions in which the PP model cannot predict maxima in such plots. However, the PIE model breaks down with high concentrations of ions and when competing ions have very different affinities for ionic micelles.[26,39a,72] These data have been explained in terms of different kinetic models such as the mass-action or site-binding (MA-SB) model[67,73-75] and the Coulombic–Poisson–Boltzmann spherical cell (C-PBSC) model[76-79] as well as allowing different β values at different concentrations of CTAOH and CTAF (where CTA represents the cetyltrimethylammonium group).[32] It has even been reported that under some conditions the ion-exchange constant (K_X^Y) and rate constant k_M^2 are no longer independent parameters.[80]

It has been suggested that if α (= $1 - \beta$) is relatively large and/or the concentration of added counterion is high (> 0.2 M), then the assumption of one-to-one exchange in the PIE model is perhaps no longer valid.[81] Under such conditions, the interfacial local concentration of a counterion (Y_M) is shown to follow the empirical equation (Equation 3.25) provided the ionic reagent (Y) is the counterion of surfactant and no other counterions are present.[33,82,83] At $[Y_W]$ ≤ 0.2 M, $Y_M \approx (1 - \alpha)/V_M$ and at very high values of $[Y_W]$, $Y_M \approx [Y_W]$. Another problem with the PIE model is that although kinetic data

$$Y_M = [(1 - \alpha)/V_M] + [Y_W] \qquad (3.25)$$

are fitted with values of K_{Br}^{OH} generally in the range of 12 to 20, fluorescence spectroscopy shows that highly hydrophilic anions such as HO^- and F^- are singularly ineffective in displacing Br^- from cationic micellar surface.[23,84] The poor agreement between values of selectivity coefficients for exchange of highly hydrophilic counterions determined by different methods is also thought to be an indication of the breakdown of the PIE model.[84-87]

Germani et al.[88] have pointed out, without giving the details of the analytical data, that a major problem in using ion-exchange models (PIE and MA-SB) is that rate data can be fitted by using a variety of values for parameters such as β, K_X^Y (used in the PIE model), or K_X (used in the MA-SB model). It has been shown in the study on cationic micellar-mediated alkaline hydrolysis of **1** that the quality of observed data fitted to Equation 3.12 remained almost unchanged with a very large change in K_{Br}^{OH} at a constant β.[89] The quality of fitting of the same observed data in terms of the PIE model turned out to be nearly unchanged with a large change in β at a constant K_{Br}^{OH}, as evident from data summarized

TABLE 3.1
Effect of [CTABr]$_T$ on the Alkaline Hydrolysis of Securinine (1)[a]

[CTABr]$_T$ M	10^3 k_{obs} sec^{-1}	10^3 k_{calcd}[b] sec^{-1}	10^3 k_{calcd}[c] sec^{-1}	10^3 k_{calcd}[d] sec^{-1}
0.0	1.83			
0.002	1.97	1.95	1.92	1.92
0.010	2.05	2.00	1.99	2.01
0.015	1.93	1.95	1.95	1.97
0.030	1.72	1.76	1.78	1.78
0.060	1.45	1.44	1.45	1.43
0.090	1.18	1.21	1.21	1.20
0.120	1.04	1.04	1.03	1.03
0.180	0.800	0.812	0.801	0.802
0.270	0.670	0.611	0.598	0.606
$10^9 \Sigma d_i^2$ [e]		9.062	15.89	13.87

[a] $[1]_0 = 1.38 \times 10^{-4}$ M, [NaOH] = 0.05 M, λ = 290 nm, 35°C, CMC = 1×10^{-4} M, and the reaction mixture for each kinetic run contained 4% v/v CH$_3$CN.
[b] Calculated from Equation 3.12 with k_M^{ms} = $(118 \pm 5) \times 10^{-3}$ sec^{-1}, K_S = 10.9 ± 0.6 M^{-1}, K_{Br}^{OH} = 12, and β = 0.1.
[c] Calculated from Equation 3.12 with k_M^{ms} = $(18.2 \pm 0.9) \times 10^{-3}$ sec^{-1}, K_S = 13.0 ± 1.1 M^{-1}, K_{Br}^{OH} = 12, and β = 0.5.
[d] Calculated from Equation 3.12 with k_M^{ms} = $(7.32 \pm 0.20) \times 10^{-3}$ sec^{-1}, K_S = 24.1 ± 3.0 M^{-1}, K_{Br}^{OH} = 12, and β = 0.9.
[e] $d_i = k_{obs\ i} - k_{calcd\ i}$.

in Table 3.1. The general applicability of the PIE model lies in its ability to explain the maxima generally obtained in the plots of k_{obs} vs. [D$_n$] for ionic micellar-mediated semiionic reactions. But the plots of k_{obs} vs. [D$_n$] for alkaline hydrolysis of securinine (1) in the presence of TTABr micelles did not show appreciable maxima. However, in a recent study on the alkaline hydrolysis of phenyl benzoate (2) in the presence of CTACl micelles in which strong maxima were observed in the plots of k_{obs} vs. [D$_n$], it has been shown that the quality of data fitted to the kinetic equation derived from the PIE model approach remained essentially the same in terms of residual errors and least-squares values with a large change in K_{Cl}^{OH} at a constant β, and *vice versa*.[90]

Because of such inherent weakness in the PIE model, it has now been suggested that out of the usual five disposable parameters (CMC, K_X^Y, β, k_M^{mr}, and K_S) encountered in the use of PIE model, at the most only two disposable parameters, such as k_M^{mr} and K_S or k_M^{mr} and K_X^Y or any other two parameters, should be considered to be unknown, and the remaining three should be determined independently under similar conditions, using nonkinetic techniques. However, in practice, it is not so easy — and almost impossible — to determine the disposable parameters (such as CMC, β, and $K_X{}^Y$) by using nonkinetic techniques under experimental conditions exactly similar to those used for kinetic runs. It is now more clear that $K_X{}^Y$ and K_S values are technique dependent.[91–93] However, the use of CMC, $K_X{}^Y$, and β values, obtained by nonkinetic techniques under experimental conditions very close to those used for kinetic runs, in the calculation of $k_M{}^{mr}$ and K_S from Equation 3.12 using the PIE model, may produce approximate values, which may not be very far from the exact ones. A qualitative interpretation of these calculated values of $k_M{}^{mr}$ and K_S may be useful.

Initially, the PIE model was developed to give a quantitative explanation of kinetic data on ionic micellar-mediated bimolecular reactions involving a single possible ion exchange, and it has been under extensive use for such reactions. To the best of my knowledge, there are only a few reports on studies in which the PIE model was used in ionic bimolecular reactions[94–96] involving two ion-exchange processes.[96] The lack of reports on the use of the PIE model in ionic bimolecular reactions involving more than one ion-exchange process is probably due to the fact that such reaction systems increase the complexity of the kinetic equation derived in terms of the PIE model and the fact that a practically workable kinetic equation could be obtained only after introducing more assumptions and restrictive reaction conditions. For example, Oliveira et al.[96] have analyzed the kinetic data on alkaline hydrolysis of cephaclor (**3**) in terms of the PIE model by considering (1) two ion-exchange processes, Br$^-$/HO$^-$ and S$^-$/Br$^-$ (ignoring others such as S$^-$/HO$^-$, B$^-$/HO$^-$, B$^-$/Br$^-$, and B$^-$/S$^-$, where B$^-$ represents anionic buffer component) and (2) only the rate constant for micellar-mediated reaction as an unknown parameter. The other kinetic parameters such as ion-exchange constants and β were either determined using nonkinetic methods or obtained from the literature.

Vera and Rodenas[29] studied the effects of [KBr] on the rate of hydroxide-ion-catalyzed hydrolysis of a few anionic moderately hydrophobic esters (S) in the presence of CTABr micelles. The k_{obs} vs. [KBr] profiles at constant [CTABr]$_T$ were explained in terms of the PIE model considering the usual ion-exchange Br$^-$/HO$^-$ coupled with an empirical equation (Equation 3.26), which gives the measure of the ion-exchange Br$^-$/S$^-$:

$$K_S = K_S{}^0 - L\,[KBr] \qquad (3.26)$$

In Equation 3.26, K_S represents cationic micellar binding constant of S$^-$, $K_S{}^0$ = K_S at [KBr] = 0, and L is an empirical constant whose magnitude is the measure of the ability of an inert anion (Br$^-$) to expel an anionic organic solubilizate (such as S$^-$) from micellar pseudophase to the aqueous pseudophase. The possible ion-exchange process HO$^-$/S$^-$ was ignored in this study, which is plausible because (1) considerably low constant values of [HO$^-$] were maintained in the entire kinetic runs and (2) the hydrophilicity of HO$^-$ is much larger than that of S$^-$.

Pseudo-first-order rate constants (k_{obs}) for the reaction of HO$^-$ with ionized *N*-hydroxyphthalimide (**4**) in the presence of varying concentrations of CTABr micelles at a constant [NaOH] and [NaBr] were explained in terms of the PIE model, considering Br$^-$/HO$^-$ ion exchange coupled with Equation 3.26.[97] The fit of observed data to Equation 3.12 was satisfactory in view of residual errors. However, the calculated values of K_S at [NaBr] ranging from 0.001 to 0.010 M yielded a value of L that turned out to be over 4-fold larger than the one obtained from k_{obs} values derived from kinetic runs at different [KBr], as well as at a constant $[CTABr]_T$, and [NaOH]. Furthermore, the value of L seems to be dependent on $[CTABr]_T$, which is inconsistent in view of Equation 3.26. Although the value of L for Br$^-$ is nearly 2.5-fold larger than that for Cl$^-$, which is plausible, the best values for K_{Br}^{OH} and K_{Cl}^{OH} are nearly the same, which is inexplicable because Br$^-$ is certainly more hydrophobic than Cl$^-$. Such inconsistency in the calculated values of L and K_X^{OH} under different reaction conditions is probably due to the apparent weakness of the PIE model[90] and some undetected weakness of Equation 3.26.

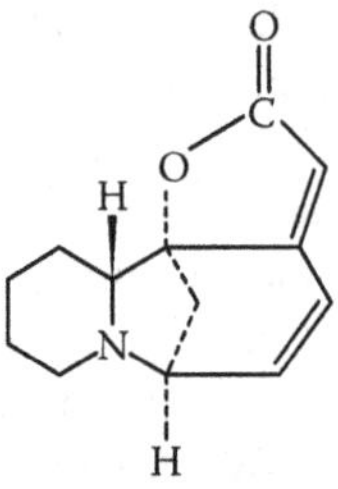

1

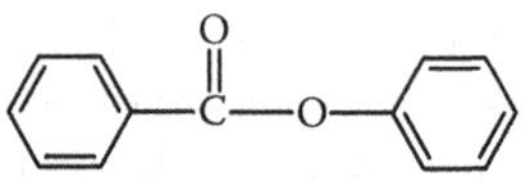

2

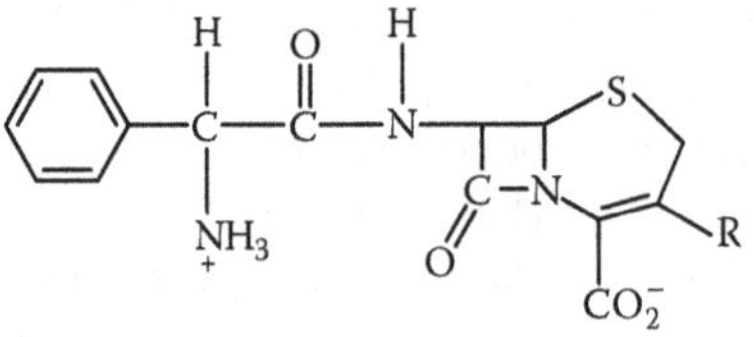

3

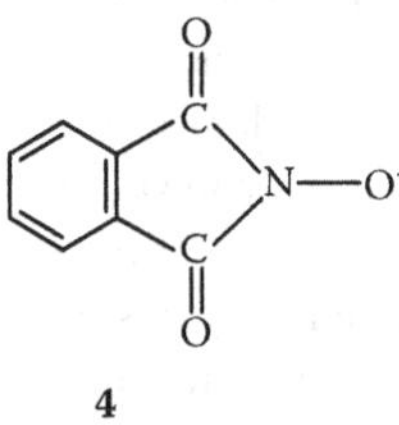

4

3.3.8 WHY THE PIE MODEL DOES NOT GIVE A BETTER DATA FIT THAN THE PP MODEL IN A REACTION SYSTEM WITH PLAUSIBLE OCCURRENCE OF AN ION-EXCHANGE

The basic difference between the PIE and PP models is the inclusion of an ion exchange in the PIE model, which consequently involves more assumptions compared to the PP model. The number of disposable parameters is also larger in the practically usable kinetic equation derived from the PIE model than from the PP model. It is a common perception that confidence in the general utility of a theoretical model decreases with an increase in the number of assumptions and disposable parameters. Thus, for some chemical or physical reasons, if the ion exchange cannot be detected kinetically, then the PP model is more reliable for use than the PIE model. The occurrence of ion exchange in the micellar-mediated bimolecular reactions may be kinetically detected if the particular ion-exchange process causes significant change in the rate of reaction, which can be easily monitored experimentally. This condition requires the following:

1. The rate of uncatalyzed reaction or solvolytic cleavage of a reactive substrate should be much slower than the rate of ionic-reactant-catalyzed reaction of the same reactive substrate (ionic reactant carries a charge similar to that of counterions of ionic micelles).
2. The value of the second-order rate constant for the ionic-reactant-catalyzed reaction of a reactive substrate should be large enough to exhibit a change in the reaction rate caused by the change in concentration of the reactive ionic reactant due to ion exchange.
3. The rate of reaction in the micellar pseudophase should not be insignificant compared to that in the aqueous pseudophase.

The rate of alkaline hydrolysis of phthalimide (**5**) in the absence of micelles has been explained in terms of Equation 3.27.

$$k_{obs} = k_0 + k_{OH}[HO^-] \qquad (3.27)$$

The value of k_{OH}/k_0 is nearly 2 M^{-1}.[98] The effects of CTABr micelles on alkaline hydrolysis of **5** revealed an insignificant rate of reaction in the micellar pseudophase compared to that in the aqueous pseudophase.[98] It has been shown in this study that the observed data (k_{obs} vs. $[D_n]$) at a constant [NaOH] and [**5**] could be explained more satisfactorily in terms of the PP model than of the PIE model; the probable reasons for such behavior are also described.

Pseudo-first-order rate constants (k_{obs}) for alkaline hydrolysis of 4-nitrophthalimide (**6**), in the absence of micelles, obeyed Equation 3.27 with $k_{OH} = 46.3 \times 10^{-3}$ M^{-1} sec^{-1} and $k_{OH}/k_0 \approx 25$ M^{-1}.[99] The values of k_{obs} showed a monotonic decrease with the increase in $[CTABr]_T$ at a constant value of [NaOH] and [**6**].[99] These results could be explained in terms of both PP and PIE models with almost equal precision (i.e., with similar residual errors and least-squares values). Although the value of k_W^{OH}/k_W^0 is nearly 25 M^{-1}, the value of k_M^{OH} is apparently so low that the increase in $[HO_M^-]$ due to ion-exchange Br$^-$/HO$^-$ has apparently no effect on the rate of alkaline hydrolysis of **6** in the micellar pseudophase. Thus, the use of PIE model in this and related reaction systems seems to be meaningless.

5$^-$ **6**

3.3.9 MASS-ACTION OR SITE-BINDING MODEL

The use of this model does not require the assumption that α (fractional ionic micellar ionization) is independent of the nature and concentrations of counterions. The basic assumption in this model is that the ionic competition between say X^- and Y^- for cationic micellar surface can be treated by using mass-action-like equations, Equation 3.28 and Equation 3.29, which have the form of Langmuir isotherms.[67,73-75,100-104]

$$K_X' = [X_M^-]/([X_W^-]([D_n] - [X_M^-] - [Y_M^-])) \qquad (3.28)$$

$$K_Y' = [Y_M^-]/([Y_W^-]([D_n] - [X_M^-] - [Y_M^-])) \qquad (3.29)$$

The relationships: $[X]_T = [X_W^-] + [X_M^-]$, $[Y]_T = [Y_W^-] + [Y_M^-]$ and Equation 3.28 and Equation 3.29 can lead to Equation 3.30 and Equation 3.31. The values of $[X_M^-]$ and $[Y_M^-]$ can be calculated simultaneously as a function of K_X and K_Y with an iterative calculation method using Equation 3.30 and Equation 3.31.

$$K_X'[X_M^-]^2 - [X_M^-](1 + K_X'[D_n] + K_X'[X]_T) + K_X'[Y_M^-][X_M^-] -$$
$$K_X'[X]_T[Y_M^-] + K_X'[[D_n][X]_T = 0 \qquad (3.30)$$

$$K_Y'[Y_M^-]^2 - [Y_M^-](1 + K_Y'[D_n] + K_Y'[Y]_T) + K_Y'[Y_M^-][X_M^-] -$$
$$K_Y'[Y]_T[X_M^-] + K_Y'[[D_n][Y]_T = 0 \qquad (3.31)$$

The calculated values of $[Y_M^-]$ at different values of $[D_n]$ for the given values of K_X' and K_Y' can be subsequently used to calculate k_M^{mr} and K_S from Equation 3.12 (with $[R_M]$ representing $[Y_M^-]$) by the use of the nonlinear least-squares technique. Vera and Rodenas[29] have used the MA-SB model for a cationic micellar-mediated reaction system containing three different counterions (negatively charged aromatic ester (S^-), HO^- and Br^-). Although apparent fits of observed data (rate–$[Surf]_T$ profiles) in terms of the MA-SB model are good in view of residual errors for those reactions that cannot be explained satisfactorily in terms of the PIE model, a major problem is that the rate data can be fitted by using a long range of values for parameters such as K_X' and K_Y'.[88]

3.3.10 COULOMBIC–POISSON–BOLTZMANN SPHERICAL CELL MODEL

In order to avoid the assumption of constancy of α (= $1 - \beta$) in the PIE model, attempts have been made to calculate the concentrations of ions at the ionic micellar surface by the use of cell model,[105,106] which involves the numerical solution of the Poisson–Boltzmann equation (PBE) under specific spherical cell boundary conditions. In this model, the distribution of counterions around an ionic micellar surface is carried out in terms of coulombic interactions between the spherical ionic micellar surface and the counterions, which are treated as point charges. This treatment neglects specific interaction between an ionic micellar surface and the counterions, and therefore does not explain the apparent differences in affinities of various counterions for ionic micelles. The allowance of specific interaction between counterions and ionic micellar surface has been incorporated in the C-PBSC model by the use of either Volmer or Langmuir isotherms.[78] The specific interaction between counterions and an ionic micellar surface has also been ascribed to partial ionic dehydration, which occurs more readily with an ion such as Br$^-$ than with strongly hydrated ions such as HO$^-$.[107] This model involves several explicit assumptions[23] whose validity is difficult to ascertain. However, many reports have appeared in the literature during the last two decades on the apparent successful use of the C-PBSC model in cases in which the PIE model apparently failed to satisfactorily fit the observed data.[76–79,81,108,109]

The validity of most of the assumptions involved in this model has not been tested rigorously. Furthermore, as with other micellar kinetic models, the success of this model is basically claimed on the basis of fairly low residual errors between experimentally determined rate constants (k_{obs}) and calculated rate constants (k_{calcd}) in terms of this model. Such a satisfactory statistical fit of observed data to such a micellar kinetic model is necessary for the apparent success of the model but is not sufficient to guarantee the reliability of the values of the calculated parameters from this model. Such a problem has been encountered in both PIE and MA-SB models.[90,88]

3.3.11 An Empirical Kinetic Approach to Studying Ion Exchange in Ionic Micellar-Mediated Reactions

The effects of [KBr] and [NaBr] on the rates of CTABr micellar-mediated methanolysis[110] and aminolysis[111] of ionized phenyl salicylate (PS$^-$) showed that the CTABr micellar binding constants (K_S) of PS$^-$ followed the following empirical relationship:

$$K_S = K_S^0/(1 + K_{X/S}[MX]) \tag{3.32}$$

where MX = KBr and NaBr, $K_S^0 = K_S$ at [MX] = 0, and $K_{X/S}$ is an empirical constant whose magnitude is the measure of the ability of X$^-$ to expel S$^-$ from cationic micellar pseudophase to the aqueous pseudophase. The values of K_S (CTABr micellar bonding constant of ionized phthalimide) obtained kinetically at different [NaBr] and [NaOH] in the study on alkaline hydrolysis of phthalimide were also found to obey Equation 3.32.[98] Recently, a spectrophotometric technique was used to determine the CTABr micellar binding constant (K_S) of PS$^-$ at different [NaBr] and [C$_6$H$_5$COONa], and these values of K_S fit reasonably well to Equation 3.32.[112] The least-squares calculated values of K_S^0 and $K_{X/S}$ are summarized in Table 3.2. Under typical experimental conditions, when $K_{X/S}$ [MX] < 1, Equation 3.32 may be approximated to Equation 3.26 with L = $K_{X/S}$ K_S^0.

TABLE 3.2
Values of K_S^0 and $K_{X/S}$ Calculated from Equation 3.32 for Different MX in the Presence of CTABr Micelles

S	MX	K_S^0/M^{-1}	$K_{X/S}{}^a/M^{-1}$
PS$^-$	KBr[b]	5140	22.8 ± 1.4[c]
	NaBr[d]	8324 ± 506	25.4 ± 4.7
	NaBr[e]	8772 ± 398	49.5 ± 5.7
	NaBr[f]	6995 ± 227	11.4 ± 1.3
	$C_6H_5COONa^f$	6841 ± 432	145 ± 24
5$^-$	NaOH[g]	3482 ± 129	3.4 ± 0.5
	NaBr[g]	2699 ± 248	101 ± 20

[a] $K_{X/S} = K_{X/S}{}^n$.
[b] Reference 110, K_S values at different [MX] were determined kinetically using methanolysis of PS$^-$.
[c] Error limits are standard deviations.
[d] Reference 98, K_S values at different [MX] were determined kinetically using piperidinolysis of PS$^-$.
[e] Reference 98, K_S values at different [MX] were determined kinetically using n-butylaminolysis of PS$^-$.
[f] Reference 112, K_S values at different [MX] were determined spectrophotometrically.
[g] Reference 98, K_S values at different [MX] were determined kinetically using hydrolysis of 5$^-$.

If a bimolecular reaction between S and R, carried out under reaction conditions of the pseudo-first-order rate law in the presence of micelles where arbitrarily $[R]_T \gg [S]_T$, follows Scheme 3.8, then pseudo-first-order rate constants (k_{obs}) are expected to obey Equation 3.11. A semiionic or fully ionic bimolecular reaction in the presence of ionic micelles is expected to involve one of the following possibilities:

1. If reactant R is neutral and reactant S is ionic with charge similar to the charge on counterions of ionic micelles, then Equation 3.11 and Equation 3.32 can lead to Equation 3.33

$$k_{obs} = \frac{k_0 + \theta K[MX]}{1 + K[MX]} \qquad (3.33)$$

where

$$k_0 = \frac{(k_W^2 + k_M^{mr} K_R K_S^0 [D_n])[R]_T}{(1 + K_S^0 [D_n])(1 + K_R [D_n])}$$ (3.34)

with $k_W^2 = k_{obs}/[R]_T$ at $[D_n] = [MX] = 0$,

$$\theta = k_W^{2,MX}[R]_T/(1 + K_R[D_n])$$ (3.35)

with $k_W^{2,MX} = k_{obs}/[R]_T$ at a typical value of $[MX]$ and $[D_n] = 0$, and

$$K = K_{X/S}/(1 + K_S^0[D_n])$$ (3.36)

2. If reactant S is neutral and reactant R is ionic with charge similar to the charge on counterions of ionic micelles, then Equation 3.11 and Equation 3.32 (with replacement of K_S and K_S^0 by K_R and K_R^0, respectively) can lead to Equation 3.37

$$k_{obs} = \frac{k_0' + \mu K[MX]}{1 + K[MX]}$$ (3.37)

where

$$k_0' = \frac{(k_W^2 + k_M^{mr} K_R^0 K_S [D_n])[R]_T}{(1 + K_S[D_n])(1 + K_R^0[D_n])}$$ (3.38)

with $k_W^2 = k_{obs}/[R]_T$ at $[D_n] = [MX] = 0$,

$$\mu = k_W^{2,MX}[R]_T/(1 + K_S[D_n])$$ (3.39)

with $k_W^{2,MX} = k_{obs}/[R]_T$ at a typical value of $[MX]$ and $[D_n] = 0$, and

$$K = K_{X/R}/(1 + K_R^0[D_n])$$ (3.40)

3. If both reactants (S and R) are ionic with charges similar to the charge on counterions of ionic micelles, then the relationship $K_R = K_R^0/(1 + K_{X/R}[MX])$, Equation 3.11, and Equation 3.32 can lead to Equation 3.41

$$k_{obs} = \frac{k_0'' + (\theta_1 K_{1S} + \mu_1 K_{1R})[MX] + \theta_1 \mu_1 K_{1S} K_{1R}[MX]^2}{(1 + K_{1S}[MX])(1 + K_{1R}[MX])}$$ (3.41)

where

$$k_0'' = \frac{(k_W^2 + k_M^{mr} K_R^0 K_S^0 [D_n])[R]_T}{(1 + K_S^0 [D_n])(1 + K_R^0 [D_n])} \tag{3.42}$$

with $k_W^2 = k_{obs}/[R]_T$ at $[D_n] = [MX] = 0$,

$$\theta_1 = k_W^{2,MX}[R]_T/(1 + K_R^0[D_n]) \tag{3.43}$$

$$\mu_1 = k_W^{2,MX}[R]_T/(1 + K_S^0[D_n]) \tag{3.44}$$

with $k_W^{2,MX} = k_{obs}/[R]_T$ at a typical value of $[MX]$ and $[D_n] = 0$,

$$K_{1S} = K_{X/S}/(1 + K_S^0[D_n]) \tag{3.45}$$

and

$$K_{1R} = K_{X/R}/(1 + K_R^0[D_n]) \tag{3.46}$$

It is well known that an increase in the number of empirical or disposable parameters in an equation decreases the reliability of the magnitude of calculated empirical parameters and, consequently, decreases the diversity of the equation. In view of this fact, the four calculated disposable parameters, θ_1, μ_1, K_{1S}, and K_{1R} from Equation 3.41 are less reliable compared to the two disposable parameters, θ and K, from Equation 3.33 or μ and K from Equation 3.37. It may be emphasized that if a reaction system involves two or more ion-exchange processes, then it is highly unlikely that all possible ion-exchange processes would kinetically be equally effective under a certain reaction condition. Relatively less effective ion-exchange processes can be safely ignored, compared to the most effective one under a specific reaction condition. The effectiveness of an ion-exchange process decreases as the difference in the hydrophobicity or hydrophilicity of exchanging ions increases. The concentrations of exchanging ions in two or more ion-exchange processes also determine the relative effectiveness of the ion-exchange processes.

If a particular reaction condition is such that $1 >> K_R [D_n]$, $k_W^{2,MX} = k_W^2$, and $\theta/k_W^{2,MX} [R]_T = F = 1$, then Equation 3.33 can be rearranged to give Equation 3.47

$$k_{obs} = \frac{(k_W^2 + k_W^2 K_{X/S}[MX] + K_M^{mr} K_R K_S^0 [D_n])[R]_T}{1 + K_S^0[D_n] + K_{X/S}[MX])} \tag{3.47}$$

The values of k_{obs} at different $[MX]$ ($MX=$ sodium benzoate,[113] sodium cinnamate, sodium 2-chlorobenzoate, sodium 4-methoxybenzoate, disodium phthalate, disodium isophthalate, disodium fumarate, and sodium sulfate[114]) for piperidinolysis of PS^- were found to fit to Equation 3.47, where $K_{X/S}$ and k_M^{mr}

TABLE 3.3
Values of $K_{X/S}$ Calculated from Equation 3.47, for Different MX in the Presence of CTABr Micelles

S	MX	$K_{X/S}{}^a/M^{-1}$
5⁻	Sodium bromide[b]	93
PS⁻	Sodium cinnamate[c]	367[d] (441)[e]
	Sodium 2-chlorobenzoate	107 (129)
	Sodium 4-methoxybenzoate	131 (157)
	Disodium phthalate	15 (19)
	Disodium isophthalate	28 (34)
	Sodium sulfate	2 (3)
	Sodium benzoate	124 (157)

[a] $K_{X/S} = K_{X/S}{}^n$ because F $(= \theta/k_{2,W} [R]_T) = 1$.
[b] Reference 113.
[c] Reference 114.
[d] Calculation of these values was carried out with $K_S{}^0 = 7000\ M^{-1}$.
[e] Calculation of parenthesized values was carried out with $K_S{}^0 = 8400\ M^{-1}$.

K_R were considered to be unknown parameters whereas $k_W{}^2$ and $K_S{}^0$ were known parameters (obtained experimentally under appropriate reaction conditions). The calculated values of $K_{X/S}$ for different MX are summarized in Table 3.3.

The limitation in the use of Equation 3.47 is that it requires that the limiting concentration of ion X be capable of expelling entire S ions (with charge similar to that of ion X) from micellar pseudophase to the aqueous pseudophase (the limiting concentration of ion X means the total concentration of ion X, $[X]_T$, at which the expulsion of ion S from micellar pseudophase to the aqueous pseudophase due to ion-exchange X/S ceased almost completely and hence an increase in $[X]_T$ beyond its limiting value has no effect on such ion exchange). It is thus apparent that Equation 3.47 can be used only in those semiionic reactions in which ions X and S are either almost equally hydrophobic or ion X (the ion that expels the other ion S of similar charge) is more hydrophobic compared to ion S. The calculated values of $K_{X/S}$ from Equation 3.47 cannot be reliable if F $(= \theta/k_W{}^{2,MX} [R]_T) < 1$ for a particular ion-exchange (X/S) process.

The validity of Equation 3.32 through Equation 3.33 has been tested in various studies on CTABr micellar-mediated hydrolysis of ionized phthalimide[98,115,116] and piperidinolysis of PS⁻ [117,118] in the presence of different nonreactive ions. The calculated values of θ and K for different nonreactive anions are summarized in Table 3.4. Equation 3.35 predicts that the value of θ should be independent of $[D_n]$ provided $1 \gg K_R[D_n]$. This prediction turned out to be true within the modest

TABLE 3.4

Values of θ and K Calculated from Equation 3.33 for Different MX in the Presence of Cationic Micelles (CTAZ with Z = Br and Cl)

S	MX	$[CTAZ]_T{}^a$ M	$10^3\,\theta$ sec^{-1}	K M^{-1}	$K_{X/S}{}^b$ M^{-1}	F^c	$K_{X/S}{}^{n\ d}$ M^{-1}
PS⁻	Sodium cinnamate[e]	0.010	27.2 ± 3.0[f]	8.1 ± 2.1[f]	575[g]	0.84	483
		0.006[h]	27.2 ± 1.4	15.8 ± 2.3	679	0.84	570
		0.008[h]	23.9 ± 1.9	13.7 ± 3.0	781	0.74	578
		0.010[h]	24.2 ± 2.5	10.5 ± 2.8	746	0.75	560
		0.020[h]	27.0 ± 6.7	3.3 ± 1.4	465	0.83	386
	Sodium acetate[e]	0.006[h]		0.045			1.9
		0.010[h]		0.032			2.3
	Sodium butanoate[e]	0.010		0.090			6.4
	Sodium 2-toluate[i]	0.005	19.1 ± 1.4	6.00 ± 1.48	216[g]	0.59	127
		0.007	15.2 ± 0.9	4.48 ± 0.72	224	0.47	105
		0.010	15.2 ± 2.5	2.95 ± 1.22	209	0.47	98
		0.015	15.8 ± 2.3	1.67 ± 0.47	177	0.49	87
		0.020	11.3 ± 1.4	1.82 ± 0.49	257	0.35	90
	Sodium 3-toluate	0.005	21.0 ± 0.9	14.6 ± 1.9	526	0.65	342
		0.007	20.5 ± 2.0	10.9 ± 3.1	545	0.63	343
		0.010	17.0 ± 1.3	7.80 ± 1.60	554	0.53	294
		0.015	16.6 ± 1.3	5.90 ± 1.20	625	0.51	319
		0.020	17.1 ± 1.9	3.60 ± 0.80	508	0.53	269
	Sodium 4-toluate	0.005	19.4 ± 0.8	17.3 ± 2.2	623	0.60	374
		0.007	18.3 ± 0.9	11.5 ± 1.5	575	0.56	322
		0.010	17.4 ± 1.3	7.12 ± 1.31	506	0.54	273
		0.015	15.8 ± 1.3	5.30 ± 1.00	562	0.49	275
		0.020	16.8 ± 2.1	3.30 ± 0.80	465	0.52	242
S⁻	Sodium bromide[j]	0.010	2.40 ± 0.06	3.01 ± 0.18	101[k]	1.0	101
		0.010	3.54 ± 0.18	2.39 ± 0.25	65[l]	1.0	65
		0.040	2.81	0.512	67[k]	1.0	67
	Potassium chloride	0.010	1.38 ± 0.07	2.24 ± 0.28	75	0.58	44
	Sodium carbonate	0.010	0.355 ± 0.009	37.4 ± 12.6	1035	0.15	155
	Sodium sulfate[m]	0.010	0.474 ± 0.012	25.6 ± 4.1	666[l]	0.19	127
		0.020	0.300 ± 0.006	19.9 ± 2.6	1035	0.12	124
	Sodium phenylacetate	0.005	2.20 ± 0.18	27.3 ± 5.3	355	0.89	316
		0.006	2.35 ± 0.07	21.2 ± 1.7	331	0.91	301
		0.010	2.45 ± 0.09	10.1 ± 0.9	263	0.99	260
		0.020	2.16 ± 0.20	5.9 ± 1.1	307	0.87	267
	Sodium benzoate[n]	0.010	2.66 ± 0.08	17.4 ± 1.3	452	1.0	452
		0.020	2.57 ± 0.10	8.88 ± 1.21	462	1.0	462
	Disodium phthalate	0.010	1.26 ± 0.04	44.2 ± 7.0	1149	0.50	575
		0.020	0.921 ± 0.050	17.7 ± 3.4	920	0.40	368
		0.010[h]	1.22 ± 0.04	43.0 ± 6.8	1118	0.50	559

TABLE 3.4 *(Continued)*

Values of θ and K Calculated from Equation 3.33 for Different MX in the Presence of Cationic Micelles (CTAZ with Z = Br and Cl)

S	MX	$[CTAZ]_T{}^a$ M	$10^3 θ$ sec^{-1}	K M^{-1}	$K_{X/S}{}^b$ M^{-1}	F[c]	$K_{X/S}{}^{n\,d}$ M^{-1}
	Disodium isophthalate	0.010	1.66 ± 0.006	55.9 ± 8.5	1453	0.70	1017
		0.020	1.20 ± 0.04	37.6 ± 6.3	1955	0.50	978
	$CH_3CO_2Na^o$	0.01[p]	8.94 ± 0.27	1.4 ± 0.1	36	0.36	13
		0.01[q]	7.84 ± 0.35	1.8 ± 0.2	47	0.31	15
		0.02	5.96 ± 0.39	0.97±0.13	49	0.24	12
	$CH_3CH_2CH_2CO_2Na$	0.01	16.2 ± 0.9	3.2 ± 0.5	83	0.65	54
		0.02	13.9 ± 0.3	1.7 ± 0.1	88	0.56	49
	$NaO_2C(CH_2)_4CO_2Na$	0.01[r]	6.49 ± 0.15	19.9 ± 2.0	520	0.26	135
		0.02	4.50 ± 0.07	14.0 ± 0.9	728	0.18	131
	$C_6H_5OCH_2CO_2Na$	0.01[q]	20.6 ± 0.7	20.4 ± 1.9	520	0.82	426
		0.02	19.9 ± 0.9	10.5 ± 1.3	546	0.80	437
	$4\text{-}CH_3OC_6H_4CO_2Na$	0.01	25.5 ± 2.1	16.4 ± 3.0	416	1.02	424
		0.02	29.1 ± 1.2	6.4 ± 0.4	338	1.16	392
	$3\text{-}CH_3C_6H_4CO_2Na$	0.005	25.2 ± 1.8	72.2 ± 13.5	936	1.01	945
		0.006	24.4 ± 0.9	68.2 ± 8.7	1066	0.98	1045
		0.010	26.4 ± 2.4	28.5 ± 6.8	754	1.06	799
		0.015	24.0 ± 2.0	19.5 ± 4.1	754	0.96	724
		0.02	26.2 ± 3.3	14.5 ± 4.2	754	1.05	792
	$4\text{-}CH_3C_6H_4CO_2Na$	0.01	26.7 ± 2.1	27.3 ± 5.2	702	1.07	751
		0.02	26.9 ± 4.4	14.4 ± 4.9	754	1.08	814
	$2\text{-}CH_3C_6H_4CO_2Na$	0.01	20.1 ± 1.1	18.6 ± 3.1	494	0.80	395
		0.02	19.1 ± 1.0	8.3±1.1	442	0.76	336

[a] CTAZ = CTABr, unless otherwise stated.
[b] $K_{X/S} = K (1 + K_S{}^0 [CTAZ]_T)$.
[c] $F = θ/k_{2,W} [R]_T$.
[d] $K_{X/S}{}^n = FK_{X/S}$.
[e] Reference 117.
[f] Error limits are standard deviations.
[g] $K_S{}^0 = 7000\ M^{-1}$ for both CTABr and CTACl.
[h] CTAZ = CTACl.
[i] Reference 118.
[j] Reference 98.
[k] $K_S{}^0 = 3262\ M^{-1}$.
[l] $K_S{}^0 = 2600\ M^{-1}$.
[m] Reference 115.
[n] Reference 116.
[o] Unpublished observations.
[p] [NaOH] = > 0.01 to ≤ 0.06 M.
[q] [NaOH] = > 0.02 to ≤ 0.06 M.
[r] [NaOH] = > 0.05 to ≤ 0.09 M.

range of $[D_n]$.[115–118] Similarly, as predicted from Equation 3.32 and Equation 3.36, the values of $K_{X/S}$ were found to be independent of $[D_n]$ within its modest range. It may be noted that a drastic micellar structural change due to large change in [MX] might violate the assumption involved in Equation 3.32 that $K_{X/S}$ is an empirical constant and is independent of [MX] at a constant $[D_n]$. Such micellar structural changes might also violate some of the assumptions in the PP model and, consequently, might affect the constancy of θ with a change in [MX] and $[D_n]$. Thus, the deviation of observed data points from the theoretical plot of k_{obs} vs. [MX] derived from Equation 3.33 at high values of [MX] may be considered the consequence of major micellar structural change under such conditions.

Pseudo-first-order rate constants (k_{obs}), obtained for alkaline hydrolysis of **2** at different [MX] in the presence of constant values of [NaOH], [**2**] and $[CTAZ]_T$ (Z = Br and Cl), were found to fit to Equation 3.37, where k_0 was considered a known parameter and μ and K were considered unknown parameters.[90,119] The nonlinear least-squares calculated values of μ and K at different [MX] are summarized in Table 3.5.

In terms of Equation 3.35 or Equation 3.39, θ/G (where $G = k_W^{2,MX} [R]_T$) or μ/G should be equal to 1 if the limiting concentration of ion X can cause 100% transfer of micellized co-ion S or R from micellar pseudophase to the aqueous pseudophase and $1 \gg K_R[D_n]$ or $1 \gg K_S[D_n]$ under the experimental conditions imposed. The limiting concentration of a salt such as MX is defined as the specific concentration of the salt at which the rate of reaction becomes independent of the salt concentration (i.e., at the limiting concentration of MX, K[MX] $\gg$ 1, and θ K[MX] $\gg k_0$ in Equation 3.33 or K [MX] $\gg$ 1 and μK[MX] $\gg k_0$ in Equation 3.37). Thus, under the kinetic conditions where $1 \gg K_R[D_n]$ or $1 \gg K_S[D_n]$, the ratio θ/G or μ/G may be considered to be the measure of the fraction of the micellized S or R ions transferred from micellar pseudophase to the aqueous pseudophase by the limiting concentration of ion X.

The values of θ/G for less hydrophobic ions X compared to ion S are considerably smaller than 1 (Table 3.4). There is no perfect technique that can measure precisely the exact locations of solubilizates of different hydrophobicity or hydrophilicity and steric requirements in the micellar pseudophase. However, among the various techniques, the ^{1}H NMR spectroscopic technique is normally used for this purpose,[46,91,120–123] although the conclusion based on this technique is not free from ambiguity.[122] The limited number of ^{1}H NMR spectroscopic studies on finding the micellar locations of solubilizates shows that the micellar locations of solubilizates are governed by the ionic (if both micelle and solubilizate are ionic), hydrophobic, steric, and hydrogen bonding interactions. Thus, a multistate model of micelle appears to be more realistic, although micellar structure is still a controversial topic.[2,13,124] It may not be unreasonable to propose that the value of θ/G is the measure of the penetration of ion X into the micellar pseudophase relative to the micellar penetration of ion S in an ion-exchange X/S. Thus, a value of θ/G of 1 indicates that both ion X and ion S can penetrate the micellar pseudophase to the same depth, and a value of θ/G of 0.5 shows that the depth of the micellar penetration of ion X is only 50% of that of ion S. A

TABLE 3.5
Values of μ and K Calculated from Equation 3.37 for Different MX in the Presence of Cationic Micelles (CTAZ with Z = Br and Cl)

S	MX	$[CTAZ]_T^a$ M	$10^3\,\mu$ sec^{-1}	K M^{-1}	$K_X^{Y\,b}$	$K_X^{Y\,c}$
					X = SO$_4$, Y = F	
HO$^-$	Sodium fluoride^d	0.003	3.16 ± 0.37^e	51 ± 7^e	6.2	
		0.006	2.82 ± 0.46	48 ± 9	5.9	
		0.010	1.81 ± 0.38	26 ± 5	9.5	14^f
			$(1.73 \pm 0.24)^g$	$(23 \pm 2)^g$	$(14)^g$	
		0.015	1.32 ± 0.44	22 ± 5	12	
			(1.41 ± 0.13)	(21 ± 2)	(11)	
		0.020	1.10 ± 0.25	20 ± 3		
			(1.25 ± 0.14)	(22 ± 2)	(9.5)	
					X = SO$_4$, Y = CH$_3$COO	
	Sodium sulfate	0.003	2.40 ± 0.23	318 ± 23	4.5	
		0.006	1.72 ± 0.16	285 ± 16	5.2	
		0.010	1.50 ± 0.14	247 ± 18	6.3	6.1^f
			(1.78 ± 0.08)	(332 ± 16)		
		0.015	0.89 ± 0.10	274 ± 17	9.8	
			(1.42 ± 0.08)	(230 ± 12)		
		0.020	(1.37 ± 0.15)	(208 ± 23)		
					X = CH$_3$COO, Y = F	
	Sodium acetate	0.003	2.64 ± 0.53	70 ± 10	1.4	
		0.006	2.62 ± 0.45	55 ± 7	1.1	
		0.010	1.53 ± 0.26	39 ± 4	1.5	2.3^f
		0.015	0.88 ± 0.16	28 ± 2	1.3	
					X = Br Y = Cl	
	Sodium chloride^h	0.007^i	0.51 ± 0.22	104 ± 4	1.8	
			0.0	88 ± 17	2.0	
		0.010^i	0.38 ± 0.15	86 ± 3	1.9	
			0.0	79 ± 14	2.1	
	Sodium bromide	0.007^i	0.01 ± 0.29	192 ± 11		
			0.0	177 ± 33		
		0.010^i	0.13 ± 0.21	167 ± 7		
			0.0	169 ± 21		

^{a} CTAZ = CTABr and aqueous solvent for each kinetic run contained 2% v/v CH$_3$CN, unless otherwise stated.

^{b} K_X^Y = K (for X)/K (for Y).

^{c} The value of K_X^Y was obtained by using nonkinetic techniques.

^{d} Reference 119.

^{e} Error limits are standard deviations.

^{f} Reference 41.

^{g} Parenthesized values were obtained in mixed aqueous solvents containing 5% v/v CH$_3$CN.

^{h} Reference 90.

^{i} CTAZ = CTACl.

skeptic might argue that a quantitative treatment that assumes reaction in a uniform medium (the PP model of micelle) should not be used as evidence against that assumption, unless the treatment does not fit the data. This argument may be resolved as follows. Davies et al.[125] have used a multiple micellar pseudophase (MMPP) model to derive a kinetic equation of the same form as that obtained from the classical PP micellar model with a modified definition of k_M and K_S in Equation 3.2 as weighted averages of constants pertaining to each micellar state or region. Thus, a change in the assumption from a two-state to a multistate micellar model is not expected to cause observed data not to fit Equation 3.2 or equations derived from Equation 3.2.

Furthermore, the mechanisms of charge–charge interaction and hydrophobic–hydrophobic interaction are different from each other, and the dielectric constant of the Stern or palisade layer is certainly not uniform. Therefore, it is unreasonable to assume that the two ionic solubilizates of different hydrophobicity, such as HO^- and PS^-, should have uniform concentration distributions throughout the Stern or palisade layer.

The values of θ/G for ion-exchange processes TA^-/PS^- (TA^- represents toluate anion) and cinnamate anion/PS^- are 0.5 to 0.6 and 0.8, respectively (Table 3.4). These results are plausible for the reasons that the negative charge on PS^- is less localized compared to those on TA^- and cinnamate anions and, probably, both cinnamate and 2-toluate anions are sterically more hindered than PS^-, and these factors seem to cause toluate and cinnamate anions to penetrate the micellar pseudophase to a lesser depth compared to PS^-. The larger value of θ/G for ion-exchange cinnamate anion/PS^- than those for ion-exchange TA^-/PS^- is due to larger hydrophobicity of cinnamate anion than that of TA^-. Similarly, the values of θ/G for ion-exchange $Br^-/5^-$ (ionized phthalimide), $Cl^-/5^-$, $CO_3^{2-}/5^-$, and $SO_4^{2-}/5^-$ are 1.0, 0.6, 0.15, and 0.15, respectively (Table 3.4). These results show that the depths of the micellar penetration of Br and 5^- are nearly the same, whereas those of Cl^-, CO_3^{2-}, and SO_4^{2-} are only ~ 60%, 15%, and 15% of the depth of 5^-. These conclusions are plausible in terms of relative hydrophobicity of Br^-, Cl^-, CO_3^{2-}, and SO_4^{2-}. The values of θ/G in Table 3.4 also show that benzoate and phenylacetate anions as well as 5^- could penetrate the micellar pseudophase to the same depth, but phthalate and isophthalate dianions show the micellar penetration to a depth that is nearly 50% of the depth of 5^-. The two anionic carboxylate groups force the phthalate and isophthalate ions to stay at the micellar surface of larger polarity and water concentration than the micellar surface of micellized benzoate ions. Such arguments have been used to describe the different locations of solubilizates in the micellar pseudophase determined by ^1H NMR spectroscopic technique.[46,91]

It seems reasonable to suggest that if the values of θ/G for respective ion-exchange X^-/Z^- and Y^-/Z^- are 0.5 and 0.8, then 50% and 80% of micellized Z_M^- ions, expelled by respective X^- and Y^- ions from micellar pseudophase to the aqueous pseudophase, reside in the micellar environment of medium characteristic (such as polarity, hydrophobicity, and water concentration) similar to the micellar environment of entire micellized X_M^- ions and Y_M^- ions at their respective limiting concentrations. This statement is graphically presented by Figure 3.3.

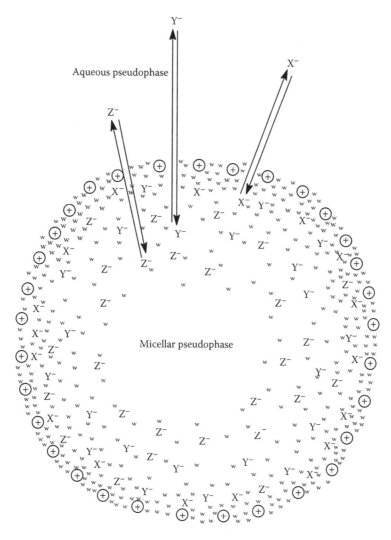

FIGURE 3.3 Schematic representation of the probable locations of X^-, Y^-, and Z^- ions in the micellar pseudophase, where these anions are assumed to have nearly the same steric requirements with hydrophobicity varying in the order $Z^- > Y^- > X^-$.

The empirical definition of $K_{X/S}$ shows that the magnitude of $K_{X/S}$ should be directly proportional to the ionic micellar binding constant (K_X) of ion X (the counterion X) and inversely proportional to the ionic micellar binding constant (K_S) of ion S (another counterion). Thus, $K_{X/S} = \delta_S K_X/K_S$, where δ_S represents the proportionality constant. The magnitude of δ_S is assumed to depend only on the molecular characteristics of ion S (the ion that is expelled by another co-ion X from micellar pseudophase to the aqueous pseudophase), and it is independent of the molecular characteristics of ion X (the ion that expels the co-ion S from micellar pseudophase to the aqueous pseudophase). If these assumptions are true, then for another ion-exchange Y/S, $K_{Y/S} = \delta_S\, K_Y/K_S$. The relationships $K_{X/S} = \delta_S\, K_X/K_S$ and $K_{Y/S} = \delta_S\, K_Y/K_S$ lead to Equation 3.48

$$K_{X/S}/K_{Y/S} = K_X/K_Y \tag{3.48}$$

where $K_X/K_Y (\equiv K_X^{Y)}$ represents the usual ion-exchange constant in PIE formalism, i.e., $K_X^Y = ([X_M[Y_W])/([X_W\,[Y_M])$. It should be noted that the relationship shown by Equation 3.48 is correct only if $K_{X/S}$ and $K_{Y/S}$ have been determined experimentally using Equation 3.32. If the values of $K_{X/S}$ and $K_{Y/S}$ have been determined experimentally using Equation 3.33 or Equation 3.37, then these values should be normalized; i.e., $K_{X/S}^n = F_X\, K_{X/S}$ and $K_{Y/S}^n = F_Y\, K_{Y/S}$, where $F = \theta/G$ with $1 \gg K_R[D_n]$ or $F = \mu/G$ with $1 \gg K_S[D_n]$ and, under such conditions, $K_{X/S}^n/K_{Y/S}^n = K_X^Y$ (with F value must be ≤ 1 and > 0). However, such a normalization process of $K_{X/S}$ presumes a uniform concentration distribution of exchanging ions S and X within their respective penetration depths of micellar pseudophase, which may not be true if ions S and X differ greatly in terms of hydrophobicity/hydrophilicity. The relationship $K_{X/S}^n/K_{Y/S}^n = K_X^Y$ has been used to calculate K_X^{Br} and K_X^{Cl} using listed values $K_{X/S}^n$ and $K_{Y/S}^n$ (Y = Br and Cl) in Table 3.2 to Table 3.4. These results are summarized in Table 3.6.

　　The calculated values of ion-exchange constants, K_X^{Br} and K_X^{Cl}, for X = benzoate, substituted benzoate, phenylacetate, and cinnamate ions are not significantly different from the corresponding K_X^{Br} and K_X^{Cl} values obtained from ^{1}H NMR spectroscopic[46,91] and UV spectrophotometric[41] techniques (Table 3.6). The values of K_{Br}^{Cl} and K_{OH}^{Br} are also comparable to those obtained from trapping of free counterions measurement[45] and UV spectrophotometric technique.[41] It is perhaps essential to note that in view of the impressive work of Magid et al.[91], the value of the ion-exchange constant is highly technique dependent. The values of K_{PhT}^{Br} and K_{IPhT}^{Br} (subscripts PhT and IPhT represent phthalate and isophthalate dianions, respectively) obtained from Equation 3.47 and Equation 3.48 are comparable to the corresponding values, obtained by ion-selective electrode measurements where the expected errors in the calculated ion-exchange constant values were ±50%.[43] However, the values of K_{PhT}^{Br} and K_{IPhT}^{Br} obtained through Equation 3.33 and Equation 3.48 are nearly 10-fold larger than those obtained through Equation 3.47 and Equation 3.48 (Table 3.6). It is noteworthy that although the respective values of K_{PhT}^{Br} and K_{IPhT}^{Br} obtained through Equation 3.47 and Equation 3.48, as well as Equation 3.33 and Equation 3.48, differ 10-fold, the values of the ratio $K_{IPhT}^{Br}/K_{PhT}^{Br}$ (≈ 2) remain independent of the choice of the combination of equations (Table 3.6). Similarly, the use of Equation 3.33 and Equation 3.48 yielded nearly 16-fold larger K_{SO4}^{Br} values than the use of Equation 3.47 and Equation 3.48 (Table 3.6). The discrepancies between the values of K_X^{Br} calculated through either Equation 3.47 and Equation 3.48 or Equation 3.33 and Equation 3.48 and determined by the use of nonkinetic physical methods are very apparent when X is a dianion (such as SO_4^{2-}, CO_3^{2-}, and phthalate dianions). The

TABLE 3.6
Values of K_X^{Br} and K_X^{Cl} Calculated from the Relationship: $K_X^Y = K_{X/S}^n / K_{Y/S}^n$ (Y = Br and Cl) for Different MX

S	MX	K_X^{Br}	$K_X^{Br\,a}$	K_X^{Cl}	$K_X^{Cl\,a}$
PS[-]	Sodium cinnamate	20.6[b]			
	Sodium cinnamate	14.7 (17.6)[c]			
	Sodium 2-toluate	4.0[b]			
	Sodium 3-toluate	12.5			
	Sodium 4-toluate	11.9			
	Sodium acetate	0.08	0.098[d]		0.50[d]
	Sodium butanoate	0.26			
	Sodium benzoate	5.8[e]			
	Sodium benzoate	5.0 (6.3)[c]			
	Sodium 2-chlorobenzoate	4.0 (5.2)			
	Sodium 4-methoxybenzoate	5.2 (6.3)			
	Disodium phthalate[f]	0.60 (0.76)	2.0 ± 1.0[g]		
	Disodium isophthalate[f]	1.1 (1.4)	3.4 ± 1.7[g]		
	Sodium sulfate[f]	0.08 (0.12)			
5[-]	Sodium hydroxide	0.044[h]	0.048[d]	0.077[i]	0.24[d]
	Sodium bromide			2.3	5.0[d]
	Sodium bromide			2.1[j]	3[k]
	Sodium bromide			1.8[l]	2.7[m]
	Potassium chloride	0.56	0.20[d]		
	Sodium carbonate	2.0	0.092[d]	3.5	0.47[d]
	Sodium sulfate	1.6	0.62[d]	2.8	3.1[d]
	Sodium phenylacetate	3.7		6.5	
	Sodium benzoate	5.9		10.4	
	Disodium phthalate	6.4		11.4	
	Disodium isophthalate	12.8		22.7	
	CH_3CO_2Na	0.2			
	$CH_3CH_2CH_2CO_2Na$	0.7			
	$NaO_2C(CH_2)_4CO_2Na$	1.7			
	$C_6H_5OCH_2CO_2Na$	5.6			
	$4-CH_3OC_6H_4CO_2Na$	4.3			
	$3-CH_3C_6H_4CO_2Na$	11			
	$4-CH_3C_6H_4CO_2Na$	9.4			
	$2-CH_3OC_6H_4CO_2Na$	4.7			
-	Sodium benzenesulfonate		11[d]		
-	Sodium p-toluenesulfonate		19[d]		
-	Sodium salicylate		20[n]		
-	Sodium o-nitrobenzoate		3.8[n]		

Continued.

TABLE 3.6 *(Continued)*
Values of K_X^{Br} and K_X^{Cl} Calculated from the Relationship: $K_X^Y = K_{X/S}^n/K_{Y/S}^n$ (Y = Br and Cl) for Different MX

S	MX	K_X^{Br}	$K_X^{Br\ a}$	K_X^{Cl}	$K_X^{Cl\ a}$
-	Sodium *m*-nitrobenzoate		11^n		
-	Sodium *o*-nitrobenzoate		3.3^n		
-	Sodium 2,6-dichlorobenzoate				$13–22^k$

^{a} The values of K_X^{Br} and K_X^{Cl} were obtained by using physical techniques.
^{b} $K_{Br/S} = 25\ M^{-1}$ and the values of $K_{X/S}^n$ are listed in Table 3.4.
^{c} $K_{Br/S} = 25\ M^{-1}$ and the values of $K_{X/S}^n$ are listed in Table 3.3.
^{d} Reference 41.
^{e} $K_{Br/S} = 25\ M^{-1}$ and the values of $K_{X/S}^n$ are listed in Table 3.2.
^{f} These values are not reliable.
^{g} Reference 43.
^{h} $K_{Br/S} = 78\ M^{-1}$ and the values of $K_{X/S}^n$ are listed in Table 3.2.
^{i} $K_{Cl/S} = 44\ M^{-1}$ and the values of $K_{X/S}^n$ are listed in Table 3.4.
^{j} $K_{Cl/S} = 44\ M^{-1}$ and the values of $K_{X/S}^n$ are listed in Table 3.3.
^{k} Reference 91.
^{l} $K_{Cl/S} = 44\ M^{-1}$ and the values of $K_{X/S}^n$ are listed in Table 3.4.
^{m} Reference 45.
^{n} Reference 46.

calculated values of $K_{X/S}$ from Equation 3.47 for X = PhT, IPhT, SO_4^{2-}, and CO_3^{2-} and S = anionic phthalimide and anionic phenyl salicylate may not be reliable because for these ion-exchange processes, the values of F $\{= \theta/(k_W^{2,MX}\ [R]_T)\}$ are expected to be < 1 (Table 3.4), whereas the derivation of Equation 3.47 involves an explicit assumption that F = 1. However, perhaps more results on related systems are needed to reach a meaningful rationalization of such results.

5

The decrease in K_S (cationic micellar binding constant of counterion S⁻) with increase in [X⁻] at a constant [S⁻] has been quantitatively described by Equation 3.32. These observations are the consequence of the competition between S⁻ and X⁻ for cationic micellar surface. Incorporation of the nonionic surfactants or *n*-butanol into the ionic micelles increases both the volume of the micellar pseudophase and α value.[126–129] The increase in α with the increase in nonionic surfactant in the cationic CTABr micelles shows the decrease in counterion affinity with cationic micelles. Thus, the increase in $[C_nE_m]_T$ (total concentration of nonionic surfactant, C_nE_m) at a constant $[CTABr]_T$-CMC is expected to decrease K_S. However, this decrease in K_S is not caused by ion-exchange processes but other factors, including the decrease in electrical surface potential.[130]

It has been concluded elsewhere [131] that the structural characteristics of mixed micelles, formed from two different surfactants, S1 and S2, change from S1 micelle type to S2 micelle type as the X values increase from very low to very high, where X = [S2]/[S1]. The values of binding constants of anionic phenyl salicylate (S⁻) with cationic CTABr and nonionic $C_{16}E_{20}$ (polyoxyethylene (20) cetyl ether, $C_{16}H_{33}(OCH_2CH_2)_{20}OH$, Brij 58) are ~7000 M^{-1} and ~60 M^{-1}, respectively.[132] Thus, a significant decrease in K_S is expected with an increase in $[C_{16}E_{20}]_T$ at constant [S⁻], $[CTABr]_T$, and [NaOH]. Although the decrease in counterion affinity to CTABr micellar surface (counterion affinity to ionic micellar surface is proportional to β or K_S) due to an increase in the concentration of nonionic surfactant is not because of the occurrence of a direct ion-exchange type of process, the essence of the effect of $[C_nE_m]_T$ on β or on the concentration of counterions (such as $[Br_M^-]$ in case of CTABr micelles) is similar to that of the effect of the concentration of another counterion, X⁻, on $[Br_M^-]$, i.e., X⁻/Br⁻ ion exchange. Hence, if it is assumed that the decrease in K_S due to increase in $[C_nE_m]_T$ follows Equation 3.32 with replacement of [MX] by $[C_nE_m]_T$ and $K_{X/S}$ by $K_{CnEm/S}$, then Equation 3.2 and Equation 3.32 can lead to Equation 3.49:

$$k_{obs} = \frac{k_0 + Fk_{obs}^{CnEm}K[C_nE_m]_T}{1 + K[C_nE_m]_T} \qquad (3.49)$$

where

$$k_0 = \frac{k_W + k_M K_S^0[D_n]}{1 + K_S^0[D_n]} \qquad (3.50)$$

with $k_W = k_{obs}$ at $[D_n] = [C_nE_m]_T = 0$,

$$K = K_{CnEm/S}/(1 + K_S^0[D_n]) \qquad (3.51)$$

and F ($= \theta/k_{obs}^{CnEm}$ with $k_{obs}^{CnEm} = k_{obs}$ at a typical value of $[C_nE_m]_T$ and $[D_n] = 0$) represents the fraction of pure cationic micellized S⁻ ions transferred to pure C_nE_m micelles by the limiting concentration of C_nE_m (the limiting concentration of C_nE_m is the optimum value of $[C_nE_m]_T$ at which $k_{obs} = F k_{obs}^{CnEm}$. Hence, ideally, the value of F should be ≤ 1.0. The magnitude of the empirical constant $K_{CnEm/S}$ is a measure of the ability of the C_nE_m surfactant to change the micellar affinity of S⁻ from pure cationic micelles to pure C_nE_m micelles (i.e., to change K_S from

$K_S^{\text{ionic micelle}} \equiv K_S^0$ to K_S^{CnEm}, where $K_S^{\text{ionic micelle}}$ and K_S^{CnEm} are the ionic and C_nE_m micellar binding constants of S^-, respectively).

For the micellar-mediated bimolecular reactions that follow the reaction scheme as shown by Scheme 3.8, if we assume that the changes in K_S and K_R due to increase in $[C_nE_m]_T$ at a constant [ionic surfactant] follow empirical equations similar to Equation 3.32 with replacement of [MX] by $[C_nE_m]_T$, then

$$K_S = K_S^0/(1 + K_{\text{CnEm/S}}[C_nE_m]_T) \tag{3.52}$$

$$K_R = K_R^0/(1 + K_{\text{CnEm/R}} [C_nE_m]_T) \tag{3.53}$$

where $K_{\text{CnEm/S}}$ and $K_{\text{CnEm/R}}$ are empirical constants. Equation 3.11, Equation 3.52, and Equation 3.53 can lead to

$$k_{\text{obs}} = \frac{k_0' + (\theta_1 K_{1S} + \mu_1 K_{1R})[C_nE_m]_T + \theta_1\mu_1 K_{1S}K_{1R}[C_nE_m]_T^2}{(1 + K_{1S}[C_nE_m]_T)(1 + K_{1R}[C_nE_m]_T)} \tag{3.54}$$

where

$$k_0' = \frac{(k_W^2 + k_M^{\text{mr}}K_R^0 K_S^0[D_n])[R]_T}{(1 + K_S^0[D_n])(1 + K_R^0[D_n])} \tag{3.55}$$

with $k_W^2 = k_{\text{obs}}/[R]_T$ at $[D_n] = [C_nE_m]_T = 0$,

$$\theta_1 = k_W^{2,\text{CnEm}} [R]_T/(1 + K_R^0[D_n]) \tag{3.56}$$

$$\mu_1 = k_W^{2,\text{CnEm}} [R]_T/(1 + K_S^0[D_n]) \tag{3.57}$$

with $k_W^{2,\text{CnEm}} = k_{\text{obs}}/[R]_T$ at a typical value of $[C_nE_m]_T$ and $[D_n] = 0$,

$$K_{1S} = K_{\text{CnEm/S}}/(1 + K_S^0[D_n]) \tag{3.58}$$

and

$$K_{1R} = K_{\text{CnEm/R}} /(1 + K_R^0[D_n]) \tag{3.59}$$

The experimental test of Equation 3.49 has been carried out by studying the effects of $[C_{16}E_{20}]_T$ on the rate of pH-independent hydrolysis of phthalimide (PT)[132,133] and phenyl salicylate (PS)[132] at constant [PT] or [PS], [NaOH], $[CTABr]_T$, and at 35°C. The observed data (k_{obs} vs. $[C_{16}E_{20}]_T$) revealed a good fit to Equation 3.49 (where $[C_nE_m]_T = [C_{16}E_{20}]_T$) and the nonlinear least-squares calculated values of F and K are summarized in Table 3.7. The calculated values

TABLE 3.7
Values of F and K, Calculated from Equation 3.49, for Hydrolysis of PS⁻, PT⁻, and PB in the Presence of Different Concentrations of CTABr

S	[CTABr]$_T$ M	F	K M^{-1}	K$_{X/S}$[a] M^{-1}
PS^{-b}	0.006	1.23 ± 0.10^c	10 ± 2^c	
		1.0	$(16 \pm 4)^d$	688
	0.010	0.98 ± 0.04	10 ± 1	
		1.0	(10 ± 1)	710
	0.020	1.01 ± 0.12	5.9 ± 1.4	
		1.0	(6.0 ± 0.8)	848
PT^{-e}	0.006	1.08 ± 0.04	16 ± 1	
		1.0	(19 ± 3)	390
	0.010	1.26 ± 0.05	5.7 ± 0.3	
		1.0	(8.0 ± 0.9)	268
	0.015	2.44 ± 0.79	1.7 ± 0.7	
		1.0	(5.2 ± 1.5)	259
	0.020	6.50 ± 4.10	0.4 ± 0.3	
		1.0	(3.6 ± 1.0)	238
PB^f	0.006	1.73 ± 0.48	651 ± 126	
		1.0	(505 ± 84)	2020
	0.010	3.35 ± 0.32	495 ± 74	
		1.0	(269 ± 55)	1614
	0.020	0.70 ± 0.53	121 ± 13	
		1.0	(142 ± 25)	1562

[a] $K_{X/S} = K (1 + K_S^0 [CTABr]_T)$ with $X = C_{16}E_{20}$, $K_S^0 = 7000$ M^{-1} for PS⁻, $K_S^0 = 3250$ M^{-1} for PT⁻ and $K_S^0 = 500$ M^{-1} for PB.
[b] PS⁻ = anionic phenyl salicylate.
[c] Error limits are standard deviations.
[d] Values in parentheses were calculated from Equation 3.60.
[e] PT⁻ = anionic phthalimide.
[f] PB = phenyl benzoate.

of F, higher than 1, are fortuitous and hence the values of K were also calculated from Equation 3.60, which is the rearranged form of Equation 3.49 with F = 1.

$$K = \frac{k_{obs} - k_0}{(k_{obs}^{CnEm} - k_{obs})/[C_nE_m]_T} \tag{3.60}$$

However, it should be noted that the calculation of K from Equation 3.60 has the disadvantage of producing less reliable values of K at both very low and very high values of $[C_nE_m]_T$, because under such conditions, $k_{obs} \rightarrow k_0$ and $k_{obs} \rightarrow$

k_{obs}^{CnEm}, respectively, and consequently, $(k_{obs} - k_0) \to 0$ and $(k_{obs}^{CnEm} - k_{obs})^{-1} \to \infty$. These mathematical limits show that under such conditions, the values of $(k_{obs} - k_0)$ and $(k_{obs}^{CnEm} - k_{obs})^{-1}$ would be very sensitive to errors.

The rate of alkaline hydrolysis of phenyl benzoate (PB) follows the second-order rate law — first-order with respect to each PB and HO⁻ reactant. The effects of $[C_{16}E_{20}]_T$ on pseudo-first-order rate constants (k_{obs}) for alkaline hydrolysis of PB at constant $[CTABr]_T$, 0.01-M NaOH, and at 35°C have been explained in terms of both Equation 3.49 and Equation 3.60 within the $[C_{16}E_{20}]_T/[CTABr]_T$ range of ~0.05 to ~1.5.[132] The calculated values of K at different $[CTABr]_T$ values are summarized in Table 3.7. Because the alkaline hydrolysis of PB is a bimolecular process, strictly speaking, the values of k_{obs} should fit to Equation 3.54 with S = PB, R = HO⁻, and $C_nE_m = C_{16}E_{20}$. However, it can be easily concluded in view of the empirical definition of $K_{CnEm/S}$ that $K_{C16E20/Pb} \gg K_{C16E20/HO}$ for $(K_{PB}^{C16E20}/K_{PB}^0) \gg (K_{OH}^{C16E20}/K_{OH}^0)$. Equation 3.56 to Equation 3.59 show that $(\theta_1 K_{1PB}/\mu_1 K_{1HO}) = (K_{C16E20/PB}/K_{C16E20/HO})$. Thus, the inequality $K_{C16E20/PB} \gg K_{C16E20/HO}$ reveals that $\theta_1 K_{1PB} \gg \mu_1 K_{1HO}$. The validity of inequality $\theta_1 K_{1PB} \gg \mu_1 K_{1HO}$ and probable fact that $\theta_1 \mu_1 K_{1PB} K_{1HO}[C_{16}E_{20}]_T^2$ is negligible compared to other terms in the numerator and $1 \gg K_{1HO}[C_{16}E_{20}]_T$ in Equation 3.54 at $[C_{16}E_{20}]_T \leq 0.02 - 0.05$ M and $[CTABr]_T$ range of $0.006 - 0.02$ M reduce Equation 3.54 to Equation 3.49 with $K = K_{1S}$ (S = PB) and $C_nE_m = C_{16}E_{20}$.

The empirical definition of empirical constant $K_{CnEm/S}$ suggests that the magnitude of $K_{CnEm/S}$ should be proportional to the magnitude of K_S^{CnEm} (= C_nE_m micellar binding constant of S) and inversely proportional to the magnitude of K_S^{CTA} (= ionic micellar binding constant of S). Thus, $K_{CnEm/S} = \delta_S (K_S^{CnEm}/K_S^{CTA})$, where δ_S is a proportionality constant with dimensions M^{-1}. The value of δ_S depends only on the nature and micellar affinity of S.

3.4 BEREZIN'S PSEUDOPHASE MODEL

In the early 1970s, the Russian school of Berezin et al.[14,15] attempted to rationalize the effects of micelles on the reaction rates. These authors adopted the pseudophase (PP) model of micelle with a slight apparent modification as expressed by Scheme 3.9

$$S_W \underset{\longleftarrow}{\overset{P_S}{\longrightarrow}} S_M$$

$$R_W \underset{\longleftarrow}{\overset{P_R}{\longrightarrow}} R_M$$

$$S_W + R_W \xrightarrow{k_W^2} \text{Product(s)}$$

$$S_M + R_M \xrightarrow{k_M^2} \text{Product(s)}$$

SCHEME 3.9

where P_S and P_R are the partition coefficients for the distribution of respective S and R molecules between aqueous and micellar pseudophases. Scheme 3.9 led to the following kinetic equation

$$k_{obs} = \frac{\{k_W^2(1-[D_n]V_M) + k_M^2 P_S P_R [D_n]V_M\}[R]_T}{\{1+(P_R-1)[D_n]V_M\}\{(1+(P_S-1)[D_n]V_M\}}$$ (3.61)

(The symbols in Equation 3.61 have been changed from those used in the original reference[134,135] to make them consistent with those used in Sections 3.3.1 to 3.3.11 in this book). For dilute surfactant solutions, the volume fraction of the micellar phase is small ($[D_n]V_M \ll 1$) and, if both S and R bind strongly to the micelles ($P_S \gg 1$ and $P_R \gg 1$), then Equation 3.61 reduces to Equation 3.11, where the relationships between binding constants and partition coefficients are expressed by[134–137]

$$K_R = (P_R - 1)V_M \text{ and } K_S = (P_S - 1)V_M$$ (3.62)

Observed pseudo-first-order rate constants for the reaction of HO^- with ethyl p-nitrophenyl chloromethyl phosphonate (7) at 0.001-M NaOH and different total mixed surfactants, cetyltrimethylammonium bromide (CTABr) – polyoxyethylene (10) oleyl ether, $C_{18}H_{35}(OCH_2CH_2)_{10}OH$ (Brij 97), concentrations and different mixing ratios have been found to fit reasonably well to Equation 3.11.[130] The competition between counterions for ionic micelle has been ascertained by a large number of experimental observations. However, a completely flawless theoretical approach to describe the counterion association with ionic micelle is still lacking. Two alternative approaches have been developed to explain counterion binding with micelle. The first is a widely used pseudophase ion-exchange (PIE) model as described in Subsection 3.3.7. Another approach, less commonly used, is to write counterion binding in terms of a micellar surface electrical potential (Ω). Both approaches or approximations are semiempirical and not free of limitations. However, additional evidence that strengthens the satisfactory fit of observed data on ionic micellar-mediated semiionic reactions comes from a comparison of the calculated values of K_R (R = HO^- in Reference 130) by the use of Equation 3.11 as well as Equation 3.62 and Equation 3.63, which is based on the electrostatic approach.[15,138]

$$P_R = \exp(-FZ_R\Omega/RT)$$ (3.63)

In Equation 3.63, F is the Faraday constant; Z_R, an ionic valence state of ion R; and RT, the multiple of gas constant and absolute temperature. An excellent agreement between the calculated values of K_{OH} from Equation 3.11 as well as Equation 3.62 and Equation 3.63 has been reported,[130] which validates the use of Equation 3.11 for ionic micellar-mediated semiionic reactions. Distribution of reactive counterions, discussed in terms of micellar surface potentials, has led to equations similar to those based on the ion-exchange model.[139,140]

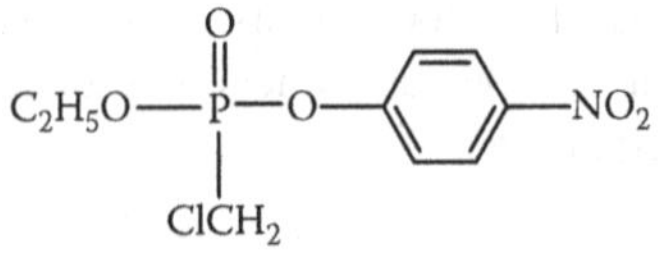

$$C_2H_5O-\overset{\overset{\displaystyle O}{\|}}{\underset{\underset{\displaystyle ClCH_2}{|}}{P}}-O-\!\!\bigcirc\!\!-NO_2$$

7

3.5 THE MULTIPLE MICELLAR PSEUDOPHASE MODEL

Various experimental observations, obtained by studies of diverse nature, indirectly suggest that micellar pseudophase is not homogeneous in terms of micropolarity, water concentration, dielectric constant, and ionic strength (for ionic micelles).[141] This fact has not been considered in the classical pseudophase kinetic model first suggested by Berezin et al.[14] and Martinek et al.[15] It is therefore logical for Davies et al.[125] to suggest that the micellar pseudophase should be divided up into an arbitrary number of pseudophases, each with a different mean partition coefficient for the reactant or reactants and each with a different mean rate constant. This generalization of the classical (Berezin's) pseudophase model is referred to as the *multiple micellar pseudophase* (MMPP) model and leads to a kinetic equation similar to Equation 3.61 or Equation 3.11 with modified definitions of kinetic parameters such as $k_M^{mr}(= (k_M^2/V_M)K_R K_S) = \Sigma(k_{M,i}^2/V_{M,i})K_{R,i}K_{S,i}$ with i = 1, 2, 3, ..., q; $K_R = \Sigma K_{R,i}$ with i = 1, 2, 3, ..., q; and $K_{S,i} = \Sigma K_{S,i}$ with i = 1, 2, 3, ..., q, where q represents an arbitrary number of micelle pseudophases.

3.6 OTHER MICELLAR KINETIC MODELS

3.6.1 PISZKIEWICZ'S KINETIC MODEL

In an attempt to establish a close analogy between micellar-mediated reactions and reactions catalyzed by many enzymes, Piszkiewicz[142] used a kinetic model for micellar-mediated reactions, as shown in Scheme 3.10. This model uses the reaction mechanism first proposed by Bruice et al.[143] for unimolecular reactions. The main assumption in this model is that a substrate S and n number of surfactant molecules (D) aggregate through positive homotropic interactions (also termed *positive cooperativity*) to form catalytic micelles, D_nS, which then react to yield the product:

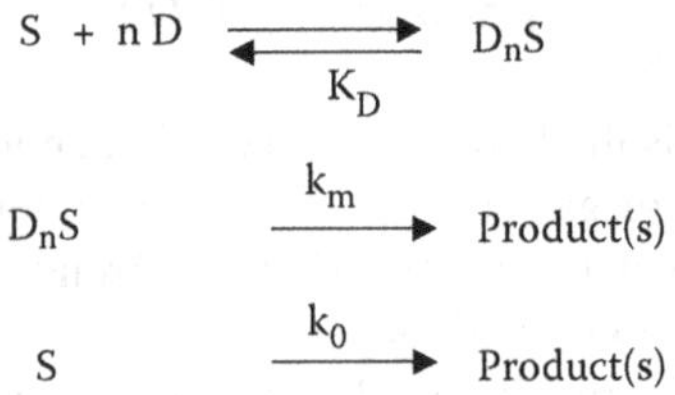

$$S + nD \underset{K_D}{\overset{}{\rightleftarrows}} D_nS$$

$$D_nS \overset{k_m}{\longrightarrow} Product(s)$$

$$S \overset{k_0}{\longrightarrow} Product(s)$$

SCHEME 3.10

where K_D is the dissociation constant of the micelle back to its free components; k_m, the rate constant for the reaction within the micelle; and k_0, the rate constant for the reaction in the absence of surfactant. The observed rate law rate = k_{obs} $[S]_T$ (where $[S]_T = [S] + [D_nS]$) and the rate law derived based on Scheme 3.10 give Equation 3.64

$$k_{obs} = \frac{k_0 K_D + k_m [D]^n}{K_D + [D]^n} \qquad (3.64)$$

For the bimolecular reactions, an additional assumption is introduced that is analogous to substrate inhibition of an enzymatic reaction with the exception that substrate and catalyst reverse roles so that the inhibition by catalyst is seen.[144] The overall reaction scheme that describes these micelle-catalyzed bimolecular reactions is as follows:

$$S + nD \;\; \underset{K_D}{\overset{}{\rightleftarrows}} \;\; D_nS$$

$$D_nS + n'D \;\; \overset{K_{n'}}{\rightleftarrows} \;\; D_nD_{n'}S$$

$$D_nS \;\; \overset{k_m}{\longrightarrow} \;\; Product(s)$$

$$S \;\; \overset{k_0}{\longrightarrow} \;\; Product(s)$$

SCHEME 3.11

In Scheme 3.11, n' is the additional number of detergent molecules that must associate with the catalytic micelle D_nS to completely inactivate it, and K_n is the association constant of this reversible process. The relationship between the second-order rate constant k_2 for a bimolecular reaction in the presence of surfactant D and surfactant concentration [D] is derived in terms of Scheme 3.11, which follows Equation 3.65.[144]

$$k_2 = \frac{k_0 K_D + k_m [D]^n}{K_D + [D]^n + K_{n'}[D_n]^n [D]^{n'}} \qquad (3.65)$$

The cooperativity, i.e., substrate (S)-induced micellization, may be expected to be significant only at surfactant concentrations very close to CMC with a substrate of large hydrophobicity yet much less compared to the hydrophobicity of surfactant. Under such conditions, the observed data may not fit to Equation 3.2. However, the observed data (k_{obs} vs. [SDS], SDS = sodium dodecyl sulfate) for the SDS micellar-mediated hydrolysis of 2-methoxymethoxy-3-methylbenzoic acid at pH 2.03 were found to fit to Equation 3.64 with n = 2 within the [SDS] range 0.004 to 0.060 M, where CMC = 0.004 M,[145] but the observed data for the hydrolysis of the same substrate at pH 5.00 in the presence of SDS and CTABr micelles obeyed Equation 3.2.[145] Although fitting of the observed data to Equation 3.64 is reasonable in terms of residual errors (= $k_{obs\,i} - k_{calcd\,i}$, where i = 1, 2, 3, 4, ..., f; f represents the total number of k_{obs}), the validity of the calculated value of K_D (= $4.0 \times 10^{-5}\ M^2$) has not been ascertained by nonkinetic means. The calculated values of n from Equation 3.64 are much smaller than the values of micellar aggregation number determined by various nonkinetic physical techniques. The validity of Equation 3.64 has not been tested by comparing the values of calculated parameters using Equation 3.64 and other nonkinetic methods.

3.6.2 THE INTERFACIAL KINETIC MODEL

The effects of oil-in-water microemulsions based on various $C_{12}E_m$ surfactants, i.e., dodecyl ethoxylate with m number of oxyethylene units, and normal micelles on the rates of bimolecular nucleophilic substitution (S_N2) reactions involving KI and 4-$XC_6H_4CH_2Br$ (X = H, CH_3, $CH(CH_3)_2$, or $C(CH_3)_3$), have been explained in terms of a so-called new interfacial kinetic model.[146] It has been argued that the common approach to describe the kinetics of such reaction systems in terms of the PP model[147,148] is not very suitable for quantitative calculations, because it is difficult to determine the partition coefficients of the reactants between the different subvolumes and it is also difficult to measure the rate constants within the subvolumes.[146] The mathematical formulation of this new interfacial model proceeds with the following explicit assumptions:

1. The reaction media containing self-assembly structures are not homogeneous at the molecular level in that they consist of microscopic domains of water and oil separated by a surfactant film.
2. The reaction may occur in either of the two domains as well as at the interface. However, depending on the structural features of the reactants such as I- and 4-$XC_6H_4CH_2Br$ with X = $CH(CH_3)_2$, or $C(CH_3)_3$, the reaction can be assumed to be a purely interfacial reaction; i.e., no reaction occurs in the bulk phases.
3. Even for reactants A (= I-) and B (= 4-$XC_6H_4CH_2Br$ with X = $C(CH_3)_3$), it is assumed that $P_{Awi} = P_{Boi} = 1$, where P_{Awi} and P_{Boi} represent respective partition coefficients of reactants A and B between bulk phase water (w) and interface (i), as well as between bulk phase oil (o) and interface (i).
4. For the reactions occurring in either of the bulk phases (such as water) as well as at the interface, it is assumed that for simplicity the microemulsion is divided into two subvolumes instead of three. The surfactant volume is split between the oil and aqueous subvolumes. (It is

assumed that the concentration of reactants in the bulk is the same as in the interface zone.)

5. There is no hydrophobic core in $C_{12}E_m$ micelles.

For only an interfacial reaction, assumptions 1 to 3 lead to Equation 3.66:

$$k = \frac{k_i \phi_s}{(\phi_s + \phi_w)(\phi_s + \phi_o)} \tag{3.66}$$

where k represents the second-order rate constant for the reaction occurring in the presence of oil-in-water microemulsions, k_i is the second-order rate constant for the reaction occurring at the interface i, $\phi_s = V_s/V$, $\phi_w = V_w/V$, and $\phi_o = V_o/V$ with total volume of the reaction mixture, $V = V_s + V_w + V_o$, where V_s, V_w, and V_o are the volumes of surfactant, water, and oil, respectively.

For substrates that have non-negligible water solubility, the reaction in the bulk water domain occurs in parallel to the interfacial reaction and the observed rate is the sum of the two processes. The rate of a reaction (v_w) occurring entirely in the aqueous phase can be written as

$$v_w = -(1/V_w)(dn_j/dt) = -\{1/(V(1 - \phi_{os}))\}(dn_j/dt) = k_w C_{Aw} C_{Bw} \tag{3.67}$$

where n_j is the amount of component j (mole); k_w, the rate constant for the reaction in the aqueous phase; C_{Aw} and C_{Bw}, the respective concentrations of reactants A (= KI) and B (= $4\text{-}XC_6H_4CH_2Br$) and, in view of assumption (4), $\phi_{os} = \{V_o + (V_s/2)\}/V$. The observed rate law $v = -(1/V)(dn_j/dt) = k_w C_A C_B$ and Equation 3.66 can lead to Equation 3.68:

$$k = \frac{k_w}{1 + \phi_{os}(P_{Bow} - 1)} \tag{3.68}$$

where $P_{Bow} = C_{Bo}/C_{Bw}$.

Most of the assumptions involved in this model are not essentially different from those in the PP model. Although this model provides an equation that allows the direct calculation of second-order rate constant (k_i) for a bimolecular reaction between reactants A_i and B_i at the interface by the use of experimentally determined parameters, it is restricted to only those reactants for which the condition $P_{Awi} = P_{Boi} = 1$ is true. Furthermore, the calculation of the volume fraction of surfactant (ϕ_s) involves uncertainty of almost the same order as that encountered in the calculation of the micellar molar volume (V_M), which is used to convert the experimentally determined first-order rate constant (k_M^{mr}) into the second-order rate constant (k_M^2) through the relationship $k_M^2 = V_M k_M^{mr}$.

3.6.3 EXTENSION OF THE PP MODEL

Minero et al.[149] proposed a micellar kinetic model to interpret quantitatively or semiquantitatively the kinetic data obtained for hexachloroiridate(IV)–iron(II) electron-transfer reactions in cationic (CTACl) micelles. The model gives a good fit to the experimental kinetic data obtained in the presence of two hydrophilic counterions over a significant range of inert salt and surfactant concentration and provides a very reasonable estimate of the degree of dissociation (α) of the cationic micelle from the fit of the kinetic data. This model is essentially an extension of the PP model and thus gives the well-known equations of Berezin and ion-exchange models under limiting conditions. Customary to any model, this particular model embraces several assumptions, including almost all those involved in the PP model. The additional assumptions are described as follows:

1. The total number of bound molecules, Σn_b, is proportional, through the proportionality constant σ, to the micellized surfactant concentration C_d; i.e., $\Sigma n_b = \sigma C_d$, where n represents number of molecules or moles, subscript b stands for "bound" (i.e., in the micellar pseudophase), and C_d is the analytical concentration of the micellized surfactant in moles L^{-1}.
2. The total charges due to bound ions and molecules on a micelle are related to its effective charge and the degree of ionization (α) through the relationship $\Sigma z_j n_b^j = (1 - \alpha) C_d$, where $j = 1, 2, ...$, and z_j is the charge of the j-th ion.
3. All the assumptions involved in the Debye–Huckel approximation (DH) are incorporated in this model because approximations have been used to establish the relationship between the electrostatic transfer constant K_j^E and $(\Upsilon_b/\Upsilon_f)_j$ (where Υ_j is the activity coefficient of the j-th species) in terms of measurable parameters.

Minero et al.[149] avoided exchange model formalism in the definition of exchange constants and used simple ratios between transfer constants because simple measurements of counterion activities showed[150] that two kinds of counterions can independently associate with or dissociate from micelles even when they have the same charge. This model contains a fairly large number of assumptions, and experimentally usable kinetic equations could be obtained after several approximations. The quality of a model is not strictly proved by the fact that it is able to fit the data, especially if the model involves many approximations and assumptions that cannot be validated with convincing evidence.

3.7 DISPERSED MEDIUM MODEL OF MICELLAR SOLUTION

Lelievre et al.[151] and Le Gall et al.[152] have argued that it is not necessary to consider micelles as defined objects, which they are not. The labile behavior of micellar aggregates implies that they develop and break down continually. Every point of the solution can thus be considered as being in turn within the micelles, then at the surface, and then in its surroundings. Obviously, such a description of the micellar solution using the static phase-separation model does not fit in with the reality. These authors have tried to show that the reactivity in a micellar medium can be treated either kinetically or thermodynamically in exactly the same way as that used for mixed solvents, such as aqueous–organic solvents, for example, by the introduction of transfer activity coefficients. The action of the micelles is then treated to a simple medium effect, which explains both the modification of the reaction rate and changes in equilibrium.[151] In the derivation of practically testable equations using DM model of micellar solution, the following explicit assumptions have been considered:

1. The micellar system can be split into two subsystems: a homogeneous bulk of constant composition b and a "phase" σ of surfactant made of σ_{n2} mol of micellized amphiphiles.
2. The number of moles of surfactant and solute i in the micellar solution are, respectively,

$$n_2 = {}^b n_2 + {}^\sigma n_2$$

$$n_i = {}^b n_i + {}^\sigma n_i$$

3. The exchange rate between the species in the two subsystems b and σ is fast with regard to the rate of the reactions studied.
4. The number of moles of i linked to the micellar aggregates is not necessarily proportional to the amount of micellized surfactant.
5. The fixation of i occurs only on the surface of the aggregate.
6. The "micellar phase" is a surface phase of area $^\sigma A$ containing $^\sigma n_2$ mol of micellized surfactant.
7. The amount of i bound to the surface $^\sigma n_i$ depends on the amount of micellized surfactant, $^\sigma n_2$, on the concentration of i in the bulk b, on the way in which the shape of micellar aggregates develops with the concentration of surfactant, $^\sigma A_2{}'$, on the nature of both phases, on their electrical potentials, and on the surface energy.
8. All the reagents and the products in a micellar solution are subject to the same thermodynamic forces wherever they are located in the medium.
9. The chemical exchange rates cannot be the same at all points of the system.

The micellar effects on the reaction rates have been modelized with the help of a new thermodynamic and kinetic analysis.[151] This modelization has been based on an approach proposed by De Donder and Defay as mentioned in References 151 and 152, which is well suited to the kinetics in a dispersed medium. This model leads to Equation 3.69, which shows the relationship between the observed bimolecular rate constant $^m k_1$ ($\equiv k_{obs}^2 = k_{obs}/[R]_T$ in Equation 3.11) and concentration of micelles.

$$^m k_1 = \frac{^b k_1 + x^\sigma k_1 \varphi_A \varphi_{OH}}{(1 + x \varphi_A)(1 + x \varphi_{OH})} \qquad (3.69)$$

In Equation 3.69, subscripts A and OH stand respectively for reactants A (= 1,3,5-trinitrobenzene) and HO$^-$, $^b k_1$ and $^\sigma k_1$ represent, respectively, second-order and first-order rate constants for reactions between A and HO in bulk phase or solution b (equivalent to aqueous pseudophase in PP model of micelle) and in micellar phase σ, x = [Surfuctant]$_T$ – CMC (equivalent to [D$_n$] in PP model of micelle) and φ_i (where i = A and HO$^-$) is the retention coefficient of i and its relationship with x is given by Equation 3.70:

$$\ln \varphi_i = \ln W + (p - 1) \ln x - (^\sigma \gamma \, \alpha_i / RT) x^{(p-1)} \qquad (3.70)$$

where W is a constant, p (which represents degree of homogeneity of micellar phase) is different from or equal to unity, $^\sigma \gamma$ is the surface tension between the micellar surface and the bulk phase b, α_i is the specific coefficient of the variable of area of the solute i and $\alpha_i x^{(p-1)} = {}^\sigma A_i'$ where $^\sigma A_i'$ is the partial area of i at the micellar surface.

The following are evident from Equation 3.70:

1. If p = 1, the φ_i's do not depend on x, and the behavior of the system can be described from the classical PP model. The retention coefficient, φ_i, is equivalent to a partition constant.
2. If p ≠ 1, the φ_i's depend on the amount of surfactant in solution, i.e., on x; the PP model does not fit.

The essence of assumptions 1 to 3 is similar to that found in some basic assumptions of the classical PP model. The calculated kinetic parameters derived from Equation 62 and Equation 63 in Reference 150, as well as Equation 21 and Equation 22 in Reference 151 have not been verified by experiments other than kinetic experiments. Some of the assumptions, such as assumptions 4 to 9, should be validated by convincing evidence. There is an apparent inconsistency in the dimension/units of φ_i. For example, φ_i is dimensionless in Equation 31, Equation 35, Equation 72, Equation 74, Table 3a, and Table 3b of Reference 151, whereas it appears to have a dimension of M^{-1} in Equation 30, Equation 33, Equation 34, Equation 36, Equation 60, Equation 62, Equation 63 of Reference 151, and Table 3 of Reference 152.

REFERENCES

1. (a) Duynstee, E.F.J., Grunwald, E. Organic reactions occurring in or on micelles. I. Reaction rate studies of the alkaline fading of triphenylmethane dyes and sulfonphthalein indicators in the presence of detergent salts. *J. Am. Chem. Soc.* **1959**, *81*, 4540–4542. (b) Duynstee, E.F.J., Grunwald, E. Organic reactions occurring in or on micelles. II. Kinetic and thermodynamic analysis of the alkaline fading of triphenylmethane dyes in the presence of detergent salts. *J. Am. Chem. Soc.* **1959**, *81*, 4542–4548.

2. Menger, F.M. The structure of micelles. *Acc. Chem. Res.* **1979**, *12*(4), 111–117.

3. Menger, F.M., Portnoy, C.E. On the chemistry of reactions proceeding inside molecular aggregates. *J. Am. Chem. Soc.* **1967**, *89*(18), 4698–4703.

4. Sugioka, H., Matsuoka, K., Moroi, Y. Temperature effect on formation of sodium cholate micelles. *J. Colloid Interface Sci.* **2003**, *259*(1), 156–162.

5. Sugioka, H., Moroi, Y. Micelle formation of sodium cholate and solubilization into the micelle. *Biochim. Biophys Acta* **1998**, *1394*(1), 99–110.

6. Ninomiya, R., Matsuoka, K., Moroi, Y. Micelle formation of sodium chenodeoxycholate and solubilization into the micelles: comparison with other unconjugated bile salts. *Biochim. Biophys Acta* **2003**, *1634*(3), 116–125.

7. Funasaki, N., Ueshiba, R., Hada, S., Neya, S. Stepwise self-association of sodium taurocholate and taurodeoxycholate as revealed by chromatography. *J. Phys. Chem.* **1994**, *98*(44), 11541–11548.

8. Mukerjee, P., Ghosh, A.K. "Isoextraction" method and the study of the self-association of methylene blue in aqueous solutions. *J. Am. Chem. Soc.* **1970**, *92*(22), 6403–6407.

9. Colter, A.K., Wang, S.S., Megerle, G.H., Ossip, P.S. Chemical behavior of charge-transfer complexes. II. Phenanthrene catalysis in acetolysis of 2,4,7-trinitro-9-fluorenyl *p*-toluenesulfonate. *J. Am. Chem. Soc.* **1964**, *86*(15), 3106–3113.

10. Cordes, E.H., Gitler, C. Reaction kinetics in the presence of micelle-forming surfactants. *Prog. Bioorg. Chem.* **1973**, *2*, 1–53.

11. Fendler, J.H., Fendler, E.J. *Catalysis in Micellar and Macromolecular Systems.* New York: Academic Press, **1975**.

12. Bunton, C.A. Reaction kinetics in aqueous surfactant solutions. *Catal. Rev. — Sci. Eng.* **1979**, *20*(1), 1–56.

13. Bunton, C.A., Savelli, G. Organic reactivity in aqueous micelles and similar assemblies. *Adv. Phys. Org. Chem.* **1986**, *22*, 213–309.

14. Berezin, I.V., Martinek, K., Yatsimirskii, A.K. Physicochemical principles of micellar catalysis. *Uspekhi Khimii* **1973**, *42*(10), 1729–1756.

15. Martinek, K., Yatsimirskii, A.K; Levashov, A.V., Berezin, I.V. The kinetic theory and the mechanisms of micellar effects on chemical reactions. In K. L. Mittal, Ed., *Micellization, Solubilization, Microemulsions*, Vol. 2, New York: Plenum Press, **1977**, pp. 489–508.

16. Bunton, C.A., Nome, F., Quina, F.H., Romsted, L.S. Ion binding and reactivity at charged aqueous interface. *Acc. Chem. Res.* **1991**, *24*, 357–364.

17. Rathman, J.F. Micellar catalysis. *Curr. Opin. Colloid Interface Sci.* **1996**, *1*(4), 514–518.

18. Tascioglu, S. Micellar solutions as reaction media. *Tetrahedron* **1996**, *52*, 11123–11152.

19. Buurma, N.J., Herranz, A.M., Engberts, J.B.F.N. The nature of the micellar Stern region as studied by reaction kinetics. *J. Chem. Soc. Perkin Trans. 2*. **1999**, 113–119.

20. Brinchi, L., Germani, R., Savelli, G., Marte, L. Decarboxylation of 6-nitrobenzisoxazole-3-carboxylate in aqueous cationic micelles: kinetic evidence of microinterface property changes. *J. Colloid Interface Sci.* **2003**, *262*(1), 290–293.

21. Brinchi, L., Germani, R., Goracci, L., Savelli, G., Bunton, C.A. Decarboxylation and dephosphorylation in new gemini surfactants: changes in aggregate structures. *Langmuir* **2002**, *18*(21), 7821–7825.

22. Cerichelli, G., Luchetti, L., Mancini, G., Savelli, G. Cyclization of 2-(ω-bromoalkyloxy)phenoxide ions in dicationic surfactants. *Langmuir* **1999**, *15*(8), 2631–2634.

23. Bunton, C.A. Micellar rate effects: what we know and what we think we know. In *Surfactants in Solution*, Vol. 11, Mittal, K.L., Shah, D.O, Eds., Plenum: New York, **1991**, pp. 17–40.

24. Bunton, C.A. Reactivity in aqueous association colloids: descriptive utility of pseudophase model. *J. Mol. Liq.* **1997**, *72*(1/3), 231–249.

25. Bunton, C.A., Romsted, L.S., Savelli, G. Test of the pseudophase model of micellar catalysis: its partial failure. *J. Am. Chem. Soc.* **1979**, *101*(5), 1253–1259.

26. Nome, F., Rubira, A.F., Franco, C., Ionescu, L.G. Limitations of the pseudophase model of micellar catalysis: the dehydrochlorination of 1,1,1-trichloro-2,2-bis(*p*-chlorophenyl)ethane and some of its derivatives. *J. Phys. Chem.* **1982**, *86*(10), 1881–1885.

27. Stadler, E., Zanette, D., Rezende, M.C., Nome, F. Kinetic behavior of cetyltrimethylammonium hydroxide. The dehydrochlorination of 1,1,1-trichloro-2,2-bis(*p*-chlorophenyl)ethane and some of its derivatives. *J. Phys. Chem.* **1984**, *88*(9), 1892–1896.

28. Rodenas, E., Vera, S. Iterative calculation method for determining the effect of counterions on acetylsalicylate ester hydrolysis in cationic micelles. *J. Phys. Chem.* **1985**, *89*(3), 513–516.

29. Vera, S., Rodenas, E. Influence of *N*-cetyl -*N,N,N*-trimethylammonium bromide counterions in the basic hydrolysis of negatively charged aromatic esters. *J. Phys. Chem.* **1986**, *90*(15), 3414–3417.

30. Bunton, C.A., Moffatt, J.R. Ionic competition in micellar reactions: a quantitative treatment. *J. Phys. Chem.* **1986**, *90*(4), 538–541.

31. Ortega, F., Rodenas, E. An electrostatic approach for explaining the kinetic results in the reactive counterion surfactants CTAOH and CTACN. *J. Phys. Chem.* **1987**, *91*(4), 837–840.

32. Neves, M. de F.S., Zanette, D., Quina, F., Moretti, M.T., Nome, F. Origin of the apparent breakdown of the pseudophase ion-exchange model for micellar catalysis with reactive counterion surfactants. *J. Phys. Chem.* **1989**, *93*(4), 1502–1505.

33. Ferreira, L.C.M., Zucco, C., Zanette, D., Nome, F. Pseudophase ion-exchange model applied to kinetics in aqueous micelles under extreme conditions: a simple modification. *J. Phys. Chem.* **1992**, *96*(22), 9058–9061.

34. Bunton, C.A., Robinson, L. Micellar effects upon nucleophilic aromatic and aliphatic substitution. *J. Am. Chem. Soc.* **1968**, *90*(22), 5972–5979.

35. Cordes, E.H., Dunlap, R.B. Kinetics of organic reactions in micellar systems. *Acc. Chem. Res.* **1969**, *2*, 329–337.

36. Bunton, C.A. Chemical kinetics in micelles. *Encyclopedia of Surface and Colloid Science,* Marcel Dekker: New York, **2002**, pp. 980–994.

37. (a) Blasko, A., Bunton, C.A., Hong, Y.S., Mhala, M.M., Moffatt, J.R., Wright, S. Micellar rate effects on reactions of hydroxide ion with phosphinate and thiophosphinate esters. *J. Phys. Org. Chem.* **1991**, *4*, 618–628. (b) Al-Lohedan, H.A., Bunton, C.A. Ion binding and micellar effects upon reactions of carboxylic anhydrides and carbonate esters. *J. Org. Chem.* **1982**, *47*(7), 1160–1166. (c) Al-Lohedan, H. A. Quantitative treatment of micellar effects upon nucleophilic substitution. *J. Chem. Soc. Perkin Trans. 2.* **1995**, 1707–1713.

38. (a) Romsted, L.R., Cordes, E.H. Secondary valence force catalysis. VII. Catalysis of hydrolysis of *p*-nitrophenyl hexanoate by micelle-forming cationic detergents. *J. Am. Chem. Soc.* **1968**, *90*(16), 4404–4409. (b) Dunlap, R.B., Cordes, E.H. Secondary valence force catalysis. VI. Catalysis of hydrolysis of methyl orthobenzoate by sodium dodecyl sulfate. *J. Am. Chem. Soc.* **1968**, *90*(16), 4395–4404. (c) Bunton, C.A., Robinson, L. Micellar effects upon the reaction of *p*-nitrophenyl diphenyl phosphate with hydroxide and fluoride ions. *J. Org. Chem.* **1969**, *34*(4), 773–780. (d) Bunton, C.A., Robinson, L. Electrolyte and micellar effects upon the reaction of 2,4-dinitrofluorobenzene with hydroxide ion. *J. Org. Chem.* **1969**, *34*(4), 780–785.

39. (a) Romsted, L.S. Micellar effects on reaction rates and equilibria. In *Surfactants in Solution*, Vol. 2, Mittal, K.L., Lindman, B., Eds., Plenum: New York, **1984**, pp. 1015–1068. (b) Romsted, L.S. A general kinetic theory of rate enhancements for reactions between organic substrates and hydrophilic ions in micellar systems. In *Micellization, Solubilization and Microemulsions*, Vol. 2, Mittal, K.L., Ed., Plenum: New York, **1977**, pp. 509–530.

40. Quina, F.H., Chaimovich, H. Ion exchange in micellar solutions. 1. Conceptual framework for ion exchange in micellar solutions. *J. Phys. Chem.* **1979**, *83*(14), 1844–1850.

41. Bartet, D., Gamboa, C., Sepulveda, L. Association of anions to cationic micelles *J. Phys. Chem.* **1980**, *84*(3), 272–275.

42. Lissi, E., Abuin, E., Cuccovia, I.M., Chaimovich, H.J. Ion-exchange between alkyl dicarboxylates and hydrophilic anions at the surface of cetyltrimethylammonium micelles. *Colloid Interface Sci.* **1986**, *112*, 513–520.

43. Menger, F.M., Williams, D.Y., Underwood, A.L., Anacker, E.W. Effect of counterion geometry on cationic micelles. *J. Colloid Interface Sci.* **1982**, *90*(2), 546–548.

44. Loughlin, J.A., Romsted, L.S. A new method for estimating counter-ion selectivity of cationic association colloid: trapping of interfacial chloride and bromide counterions by reaction with micellar bound aryldiazonium salts. *Colloids Surf.* **1990**, *48*(1–3), 123–137.

45. Cuccovia, I.M., da Silva, I.N., Chaimovich, H., Romsted, L.S. New method for estimating the degree of ionization and counterion selectivity of cetyltrimethylammonium halide micelles: chemical trapping of free counterions by a water soluble arenediazonium ion. *Langmuir* **1997**, *13*(4), 647–652.

46. Bachofer, S.J., Simonis, U. Determination of the ion exchange constants of four aromatic organic anions competing for a cationic micellar interface. *Langmuir* **1996**, *12*(7), 1744–1754.

47. Khan, M.N. Effects of salts and mixed CH_3CN-H_2O solvents on alkaline hydrolysis of phenyl benzoate in the presence of ionic micelles. *Colloids Surf. A* **1998**, *139*, 63–74.

48. Nascimento, Maria da Graca, Lezcano, M.A., Nome, F. Micellar effects on the hydrolysis of 2,2,2-trichloro-1-phenylethanone. *J. Phys. Chem.* **1992**, *96*(13), 5537–5540.

49. Berndt, D.C., Pamment, M.G., Fernando, A.C.M., Zhang, X., Horton, W.R. Micellar kinetics in aqueous acetonitrile. Part A. Sodium-hydrogen ion-exchange constant for 1-dodecanesulfonate surfactant. Part B. Substrate structural effects for micellar catalysis by perfluorooctanoic acid as reactive counterion surfactants. *Int. J. Chem. Kinet.* **1997**, *29*(10), 729–735.

50. Bravo, C., Herves, P., Leis, J. R., Pena, M. E. Basic hydrolysis of *N*-[*N*-methyl-*N*-nitroso(aminomethyl)benzamide in aqueous and micellar media. *J. Colloid Interface Sci.* **1992**, *153*(3), 529–536.

51. Al-Lohedan, H.A., Al-Hassan, M.I. Micellar effects upon spontaneous and alkaline hydrolysis of t-butyl phenyl carbonate. *Bull. Chem. Soc. Jpn.* **1990**, *63*(10), 2997–3000.

52. Malpica, A., Calzadilla, M., Linares, H. Micellar effect upon the reaction of hydroxide ion with coumarin. *Int. J. Chem. Kinet.* **1998**, *30*(4), 273–276.

53. Ruzza, A.A., Nome, F., Zanette, D., Romsted, L.S. Kinetic evidence for temperature-induced demixing of a long chain dioxolane in aqueous micellar solutions of sodium dodecyl sulfate: a new application of the pseudophase ion exchange model. *Langmuir* **1995**, *11*(7), 2393–2398.

54. Ouarti, N., Marques, A., Blagoeva, I., Rausse, M.-F. Optimization of micellar catalysis of nucleophilic substitution in buffered cetyltrimethylammonium salt solutions. 1. Buffers for the 9-10 pH range. *Langmuir* **2000**, *16*(5), 2157–2163.

55. Price, S.E., Jappar, D., Lorenzo, P., Saavedra, J.E., Hrabie, J.A., Davies, K.M. Micellar catalysis of nitric oxide dissociation from diazeniumdiolates. *Langmuir* **2003**, *19*(6), 2096–2102.

56. Rodriguez, A., Munoz, M., Graciani, M. del M., Moya, M.L. Kinetic micellar effects in tetradecyltrimethylammonium bromide-pentanol micellar solutions. *J. Colloid Interface Sci.* **2002**, *248*(2), 455–461.

57. Pazo-Llorente, R., Bravo-Diaz, C., Gonzalez-Romero, E. Ion exchange effects on the electrical conductivity of acidified (HCl) sodium dodecyl sulfate solutions. *Langmuir* **2004**, *20*(7), 2962–2965.

58. Ionescu, L.G., Trindade, V.L., De Souza, E.F. Application of the pseudophase ion exchange model to a micellar catalyzed reaction in water-glycerol solutions. *Langmuir* **2000**, *16*(3), 988–992.

59. Bobica, C., Anghel, D.F., Voicu, A. Effect of substrate hydrophobicity and electrolytes upon the micellar hydrolysis of *p*-nitrophenyl esters. *Colloids Surf. A* **1995**, *105*(2–3), 305–308.

60. El Seoud, O.A., Chinelatto, A.M. Acid-base indicator equilibriums in aerosol-OT reversed micelles in heptane: the use of buffers. *J. Colloid Interface Sci.* **1983**, *95*(1), 163–171.

61. El Seoud, O.A. Effects of organized surfactant assemblies on acid-base equilibria. *Adv. Colloid Interface Sci.* **1989**, *30*(1–2), 1–30.

62. O'Connor, C.J., Lomax, T.D., Ramage, R.E. Exploitation of reversed micelles as membrane mimetic reagents. *Adv. Colloid Interface Sci.* **1984**, *20*(1), 21–97.

63. (a) Mackay, R.A. Reactions in microemulsions: the ion-exchange model. *J. Phys. Chem.* **1982**, *86*(24), 4756–4758. (b) Pereira, R. da Rocha, Zanette, D., Nome, F. Application of the pseudophase ion-exchange model to kinetics in microemulsions of anionic detergents. *J. Phys. Chem.* **1990**, *94*(1), 356–361.

64. Athanassakis, V., Bunton, C.A., de Buzzaccarini, F. Nucleophilic aromatic substitution in microemulsions. *J. Phys. Chem.* **1982**, *86*(25), 5002–5009.

65. Perez-Benito, E., Rodenas, E. Influence of sodium dodecyl sulfate micelles on the oxidation of alcohols by chromic acid. *Langmuir* **1991**, *7*(2), 232–237.

66. Abuin, E., Lissi, E. Thallium(1+)/sodium(1+) competitive binding at the surface of dodecyl sulfate, Brij 35 mixed micelles. *J. Colloid Interface Sci.* **1992**, *151*(2), 594–597.

67. Bertoncini, C.R.A., Neves, M. de F.S., Nome, F., Bunton, C.A. Effects of 1-butanol-modified micelles on S_N2 reactions in mixed-ion systems. *Langmuir* **1993**, *9*(5), 1274–1279.

68. Herves, P., Leis, J.R., Mejuto, T.C., Perez-Juste, J. Kinetic studies on the acid and alkaline hydrolysis of *N*-methyl-*N*-nitroso-*p*-toluenesulfonamide in dioctadecyldimethylammonium chloride vesicles. *Langmuir* **1997**, *13*(25), 6633–6637.

69. Fendler, J.H., Hinze, W.L. Reactivity control in micelles and surfactant vesicles: kinetics and mechanism of base-catalyzed hydrolysis of 5,5-dithiobis(2-nitrobenzoic acid) in water, hexadecyltrimethylammonium bromide micelles, and dioctadecyldimethylammonium chloride surfactant vesicles. *J. Am. Chem. Soc.* **1981**, *103*(18), 5439–5447.

70. Cuccovia, I.M., Quina, F.H., Chaimovich, H. A remarkable enhancement of the rate of ester thiolysis by synthetic amphiphile vesicles. *Tetrahedron* **1982**, *38*(7), 917–920.

71. Kawamauro, M.K., Chaimovich, H., Abuin, E.B., Lissi, E.A., Cuccovia, I.M. Evidence that the effects of synthetic amphiphile vesicles on reaction rates depend on vesicle size. *J. Phys. Chem.* **1991**, *95*(3), 1458–1463.

72. Gonsalves, M., Probst, S., Rezende, M.C., Nome, F., Zucco, C., Zanette, D. Failure of the pseudophase model in the acid-catalyzed hydrolysis of acetals and *p*-methoxybenzldoxime esters in the presence of an anionic micelles. *J. Phys. Chem.* **1985**, *89*(7), 1127–1130.

73. Bunton, C.A., Gan, L.-H., Moffatt, J.R., Romsted, L.S., Savelli, G. Reactions in micelles of cetyl trimethylammonium hydroxide: test of the pseudophase model for kinetics. *J. Phys. Chem.* **1981**, *85* 26), 4118–4125.

74. Vera, S., Rodenas, E. Inhibiting effect of cationic micelles on the basic hydrolysis of aromatic esters. *Tetrahedron* **1986**, *42*(1), 143–149.

75. Cuenca, A. Alkaline hydrolysis of 2-phenoxyquinoxaline in reactive counterion micelles: effects of head group size. *Int. J. Chem. Kinet.* **1998**, *30*(11), 777–783.

76. Bunton, C.A., Moffatt, J.R. A quantitative treatment of micellar effects in moderately concentrated hydroxide ion. *Langmuir* **1992**, *8*(9), 2130–2134.

77. Blasko, A., Bunton, C.A., Armstrong, C., Gotham, W., He, Z.-M., Nikles, J., Romsted, L.S. Acid hydrolyses of hydrophobic dioxolanes in cationic micelles: a quantitative treatment based on the Poisson–Boltzmann equation. *J. Phys. Chem.* **1991**, *95*(18), 6747–6750.

78. Blasko, A., Bunton, C.A. Micellar charge effects on the oxidation of sulfides by periodate ion. *J. Phys. Chem.* **1993**, *97*(20), 5435–5442.

79. Ortega, F., Rodenas, E. Electrostatic approach for explaining the kinetic results in the reactive counterion surfactants CTAOH and CTACN. *J. Phys. Chem.* **1987**, *91*(4), 837–840.

80. Blasko, A., Bunton, C.A., Foroudian, H.J. Oxidation of organic sulfides in aqueous sulfobetaine micelles. *J. Colloid Interface Sci.* **1995**, *175*(1) 122–130.

81. Blasko, A., Bunton, C.A., Cerichelli, G., McKenzie, D.C. A nuclear magnetic resonance study of ion exchange in cationic micelles: successes and failures of models. *J. Phys. Chem.* **1993**, *97*(43), 11324–11331.

82. He, Z.-M., Loughlin, J.A., Romsted, L.S. Micellar effects on reactivity in concentrated salt solutions: two new approaches to interpreting counterion effects on the acid catalyzed hydrolysis of a hydrophobic ketal in aqueous solutions of cationic micelles. *Bol. Soc. Quim* **1990**, *35*(1), 43–53.

83. Rubio, D.A.R., Zanette, D., Nome, F., Bunton, C.A. Acid hydrolysis of *p*-methoxybenzaldehyde O-acyloxime in 1-butanol-modified micelles of sodium dodecyl sulfate. *Langmuir* **1994**, *10*(4), 1155–1159.

84. Abuin, E., Lissi, E., Araujo, P.S., Aleixo, R.M.V., Chaimovich, H., Bianchi, N., Miola, L., Quina, F.H. Selectivity coefficients for ion exchange in micelles of hexadecyltrimethylammonium bromide and chloride. *J. Colloid Interface Sci.* **1983**, *96*(1), 293–295.

85. Broxton, T.J. Micellar catalysis of organic reactions. VII. The effect of the micellar counter ion in nucleophilic reactions of hydroxide and nitrite ions. *Aust. J. Chem.* **1981**, *34*(11), 2313–2319.

86. Broxton, T.J., Sango, D.B. Micellar catalysis of organic reactions. X. Further evidence for the partial failure of the pseudophase kinetic model of micellar catalysis for reactions of hydroxide ions. *Aust. J. Chem.* **1983**, *36*(4), 711–717.

87. Nascimento, M.G., Miranda, S.A.F., Nome, F. Use of reactive counterion type micelles for the determination of selectivity coefficients. *J. Phys. Chem.* **1986**, *90*(15), 3366–3368.

88. Germani, R., Savelli, G., Romeo, T., Spreti, N., Cerichelli, G., Bunton, C.A. Micellar head group size and reactivity in aromatic nucleophilic substitution. *Langmuir* **1993**, *9*(1), 55–60.

89. Lajis, N.H., Khan, M.N. Effects of ionic and non-ionic micelles on rate of hydroxide ion-catalyzed hydrolysis of securinine. *J. Phys. Org. Chem.* **1998**, *11*(3), 209–215.

90. Khan, M.N., Ismail, E. An apparent weakness of the pseudophase ion-exchange (PIE) model for micellar catalysis by cationic surfactants with nonreactive counterions. *J. Chem. Soc., Perkin Trans. 2.* **2001**, 1346–1350.

91. Magid, L.J., Han, Z., Warr, G.G., Cassidy, M.A., Butler, P.W., Hamilton, W.A. Effect of counterion competition on micellar growth horizon for cetyltrimethylammonium micellar surface: electrostatics and specific binding. *J. Phys. Chem. B* **1997**, *101*(40), 7919–7927.

92. Khan, M.N. Effects of anionic micelles on the intramolecular general base-catalyzed hydrolysis of phenyl and methyl salicylates. *J. Mol. Catal. A* **1995**, *102*, 93–101.

93. Quina, F.H., Alonso, E.O., Farah, J.P.S. Incorporation of nonionic solutes into aqueous micelles: a linear salvation free energy relationship analysis. *J. Phys. Chem.* **1995**, *99*(3), 11708–11714.

94. Romsted, L.S. Quantitative treatment of benzimidazole deprotonation equilibria in aqueous micellar solutions of cetyltrimethylammonium ion (CTAX, $X^- = Cl^-$, Br^- and NO_3^-) surfactants. 1. Variable surfactant concentration. *J. Phys. Chem.* **1985**, *89*(23), 5107–5113.

95. Romsted, L.S. Quantitative treatment of benzimidazole deprotonation equilibria in aqueous micellar solutions of cetyltrimethylammonium ion (CTAX, $X^- = Cl^-$, Br^- and NO_3^-) surfactants. 2. Effects of added salt. *J. Phys. Chem.* **1985**, *89*(23), 5113–5118.

96. Oliveira, A.G., Cuccovia, I.M., Chaimovich, H. Micellar modification of drug stability: analysis of the effect of hexadecyltrimethylammonium halides on the rate of degradation of cephaclor. *J. Pharm. Sci.* **1990**, *79*(1), 37–42.

97. Khan, M.N. Effects of hydroxide ion, salts and temperature on the hydrolytic cleavage of ionized *N*-hydroxyphthalimide (NHPH) in the presence of cationic micelles. *Colloids Surf. A* **1997**, *127*, 211–219.

98. Khan, M.N., Arifin, Z. Effects of inorganic ions on rate of alkaline hydrolysis of phthalimide in the presence of cationic micelles. *J. Chem. Soc., Perkin Trans. 2.* **2000**, 2503–2510.

99. Khan, M.N., Abdullah, Z. Kinetics and mechanism of alkaline hydrolysis of 4-nitrophthalimide in the absence and presence of cationic micelles. *Int. J. Chem. Kinet.* **2001**, *33*, 407–414.

100. Gan, L.-H. Micellar effect on the formation of the 1,3,5-trinitrobenzene-cyanide complex in water. *Aust. J. Chem.* **1985**, *38*, 1141–1146.

101. Bacaloglu, R., Bunton, C.A., Ortega, F. Micellar enhancements of rates of S_N2 reactions of halide ions: the effect of headgroup size. *J. Phys. Chem.* **1989**, *93*(4), 1497–1502.

102. Bacaloglu, R., Bunton, C.A., Cerichelli, G., Ortega, F. Micellar effects upon rates of S_N2 reactions of chloride ion. *J. Phys. Chem.* **1990**, *94*, 5068–5073.

103. Munoz, M., Graciani, M. del M., Rodriguez, A., Moya, M.L. Influence of the addition of alcohol on the reaction methyl-4-nitrobenzenesulfonate + Br in tetradecyltrimethylammonium bromide aqueous micellar solutions. *J. Colloid Interface Sci.* **2003**, *266*(1), 208–214.

104. Gan, L.-H. Reaction of hydroxide ion with 1,3,5-trinitrobenzene in cationic micelles: evidence of variable counterion binding to micellar head groups. *Can. J. Chem.* **1985**, *63*(3), 598–601.

105. Mille, M., Vanderkooi, G. Electrochemical properties of spherical polyelectrolytes. 1. Impermeable sphere model. *J. Colloid Interface Sci.* **1977**, *59*(2), 211–224.

106. Gunnarsson, G., Joensson, B., Wennerstroem, H. Surfactant association into micelles: an electrostatic approach. *J. Phys. Chem.* **1980**, *84*(23), 3114–3121.

107. Morgan, J.D., Napper, D.H., Warr, G.G. Thermodynamics of ion exchange selectivity at interfaces. *J. Phys. Chem.* **1995**, *99*(23), 9458–9465.

108. Bunton, C.A., Moffatt, J.R. Ionic competition in micellar reactions: a quantitative treatment. *J. Phys. Chem.* **1986**, *90*(4), 538–541.

109. Bunton, C.A., Moffatt, J.R. Micellar effects upon substitutions by nucleophilic anions. *J. Phys. Chem.* **1988**, *92*(10), 2896–2902.

110. Khan, M.N. Effects of [NaOH] and [KBr] on intramolecular general base-catalyzed methanolysis of phenyl salicylate in the presence of cationic micelles. *J. Org. Chem.* **1997**, *62*(10), 3190–3193.

111. Khan, M.N., Arifin, Z., Ismail E., Ali, S.F.M. Effects of [NaBr] on the rates of intramolecular general base-catalyzed reactions of ionized phenyl salicylate (PS⁻) with *n*-butylamine and piperidine in the presence of cationic micelles. *J. Org. Chem.* **2000**, *65*(5), 1331–1334.

112. Khan, M.N., Ismail, E. Spectrophotometric determination of cationic micellar binding constant of ionized phenyl salicylate. *J. Chem. Res. (S)* **2001**, 143–145.

113. Khan, M.N., Arifin, Z., Ismail, E., Ali, S.F.M. Effects of [C₆H₅COONa] on rates of intramolecular general base-catalyzed piperidinolysis and *n*-butylaminolysis of ionized phenyl salicylate in the presence of cationic micelles. *Colloids Surf. A* **2000**, *161*, 381–389.

114. Khan, M.N., Yusoff, M.R. Effects of inert organic and inorganic anions on the rates of intramolecular general base-catalyzed reactions of ionized phenyl salicylate (PS⁻) with piperidine in the presence of cationic micelles. *J. Phys. Org. Chem.* **2001**, *14*, 74–80.

115. Khan, M.N., Ahmad, F.B.H. Kinetic probe to study the structure of micelle: effects of inert inorganic salts on the rate of alkaline hydrolysis of phthalimide in the presence of cationic micelles. *Colloids Surf. A* **2001**, *181*, 11–18.

116. Khan, M.N., Ahmad, F.B.H. Kinetic probe to study the structure of micelle: effects of organic salts on the rate of alkaline hydrolysis of phthalimide in the presence of cationic micelles. *React. Kinet. Catal. Lett.* **2001**, *72*, 321–329.

117. Khan, M.N., Ismail, E. An empirical approach to study the occurrence of ion-exchange in the ionic micellar-mediated semi-ionic reactions: kinetics of the rate of reactions of piperidine with ionized phenyl salicylate in the presence of cationic micelles. *Int. J. Chem. Kinet.* **2001**, *33*, 288–294.

118. Khan, M.N., Kun, S.Y. Effects of organic salts on the rate of intramolecular general base-catalyzed piperidinolysis of ionized phenyl salicylate in the presence of cationic micelles. *J. Chem. Soc., Perkin Trans. 2.* **2001**, 1325–1330.

119. Khan, M.N., Ismail, E., Misran, O. An empirical approach to study the anion selectivity at aqueous cationic micellar surface: effects of inorganic salts on kinetically determined cationic micellar binding constant of ionized phenyl salicylate. *J. Mol. Liq.* **2003**, *107*, 277–287.

120. Suratkar, V., Mahapatra, S. Solubilization site of organic perfume molecules in sodium dodecyl sulfate micelles: new insights from proton NMR studies. *J. Colloid Interface Sci.* **2000**, *225*(1), 32–38.

121. Bunton, C.A., Foroudian, H.J., Gillitt, N.D., Whiddon, C.R. Reactions of *p*-nitrophenyl diphenyl phosphinate with fluoride and hydroxide ion in nonionic micelles: kinetic salt effects. *J. Colloid Interface Sci.* **1999**, *215*(1), 64–71.

122. Otto, S., Engberts, J.B.F.N., Kwak, J.C.T. Million-fold acceleration of a Diels–Alder reaction due to combined Lewis acid and micellar catalysis in water. *J. Am. Chem. Soc.* **1998**, *120*(37), 9517–9525.

123. (a) Kreke, P.J., Magid, L.J., Gee, J.C. ¹H and ¹³C NMR studies of mixed counterion, cetyltrimethylammonium bromide/cetyltrimethylammonium dichlorobenzoate, surfactant solutions: the interaction of aromatic counterions. *Langmuir* **1996**, *12*(3), 699–705. (b) Yoshino, A., Yoshida, T., Okabayashi, H., Kamaya, H., Ueda, I. ¹⁹F and ¹H NMR and NOE study on halothane-micelle interaction: residence location of anesthetic molecules. *J. Colloid Interface Sci.* **1998**, *198*(2), 319–322.

124. Menger, F.M., Mounier, C.E. A micelle that is insensitive to its ionization state: relevance to the micelle wetness problem. *J. Am. Chem. Soc.* **1993**, *115*(25), 12222–12223.

125. Davies, D.M., Gillitt, N.D., Paradis, P.M. Catalytic and inhibition of the iodide reduction of peracids by surfactants: partitioning of reactants, product and transition state between aqueous and micellar pseudophases. *J. Chem. Soc., Perkin Trans. 2.* **1996**, 659–666.

126. Blasko, A., Bunton, C.A., Toledo, E.A., Holland, P.M., Nome, F. S_N2 Reactions of a sulfonate ester in mixed cationic/phosphine oxide micelles. *J. Chem. Soc., Perkin Trans. 2.* **1995**, 2367–2373.

127. (a) Bunton, C.A., Wright, S., Holland, P.M., Nome, F. S_N2 Reactions of a sulfonate ester in mixed cationic/nonionic micelles. *Langmuir* **1993**, *9*(1), 117–120. (b) Foroudian, H.J., Bunton, C.A., Holland, P.M., Nome, F. Nucleophilicity of bromide ion in mixed cationic/sulfoxide micelles. *J. Chem. Soc., Perkin Trans. 2.* **1996**, 557–561. (c) Froechner, S.J., Nome, F., Zanette, D., Bunton, C.A. Micellar-mediated general acid catalyzed acetal hydrolysis: reactions in micelles. *J. Chem. Soc., Perkin Trans. 2.* **1996**, 673–676.

128. Friere, L., Iglesias, E., Bravo, C., Leis, J.R., Pena, M.E. Physicochemical properties of mixed anionic-non-ionic micelles: effects on chemical reactivity. *J. Chem. Soc., Perkin Trans. 2.* **1994**, 1887–1894.

129. Calvaruso, G., Cavasino, F.P., Sbriziolo, C., Liveri, M.L.T. Interaction of the chloropentaaminecobalt(III) cation with mixed micelles of anionic and non-ionic surfactants: a kinetic study. *J. Chem. Soc., Faraday Trans.* **1995**, *91*(7), 1075–1079.

130. Zakharova, L., Valeeva, F., Zakharov, A., Ibragimova, A., Kudryavtseva, L., Harlampidi, H. Micellization and catalytic activity of the cetyltrimethylammonium bromide — Brij 97 — water mixed micellar system. *J. Colloid Interface Sci.* **2003**, *263*, 597–605.

131. Khan, M.N. Effects of mixed anionic and cationic surfactants on rate of transesterification and hydrolysis of esters. *J. Colloid Interface Sci.* **1996**, *182*(2), 602–605.

132. Khan, M.N., Ismail, E. Effects of non-ionic and mixed non-ionic-cationic micelles on the rate of aqueous cleavages of phenyl benzoate and phenyl salicylate in alkaline medium. *J. Phys. Org. Chem.* **2004**, *17*, 376–386.

133. Khan, M.N., Ismail, E. Effects of non-ionic and mixed cationic-non-ionic micelles on the rate of alkaline hydrolysis of phthalimide. *J. Phys. Org. Chem.* **2002**, *15*, 374–384.

134. Martinek, K., Osipov, A.P., Yatsimirskii, A.K., Berezin, I.V. Mechanism of micellar effects in imidazole catalysis: acylation of benzimidazole and its *N*-methyl derivative by *p*-nitrophenyl carboxylates. *Tetrahedron* **1975**, *31*(7), 709–718.

135. Yatsimirskii, A.K., Martinek, K., Berezin, I.V. Mechanism of micellar effects on acylation of aryl oximes by *p*-nitrophenyl carboxylates. *Tetrahedron* **1971**, *27*(13), 2855–2868.

136. Martinek, K., Levashov, A.V., Berezin, I.V. Mechanism of catalysis by functional micelles containing a hydroxyl group: model of action of serine proteases. *Tetrahedron Lett.* **1975** (15), 1275–1278.

137. Fendler, J.H. *Membrane Mimetic Chemistry.* Wiley-Interscience: New York, **1982**.

138. Shirahama, K. Formulation of chemical kinetics in micellar solutions. *Bull. Chem. Soc. Jpn.* **1975**, *48*(10), 2673–2676.

139. Funasaki, N. Micellar effects on the kinetics and equilibrium of chemical reactions in salt solutions. *J. Phys. Chem.* **1979**, *83*(15), 1998–2003.

140. Almgren, M., Rydholm, R. Influence of counterion binding on micellar reaction rates: reaction between *p*-nitrophenyl acetate and hydroxide ion in aqueous cetyltrimethyl-ammonium bromide. *J. Phys. Chem.* **1979**, *83*(3), 360–364.

141. Cordes, E.H. Kinetics of organic reactions in micelles. *Pure Appl. Chem.* **1978**, *50*, 617–625.

142. Piszkiewicz, D. Positive cooperativity in micelle-catalyzed reactions. *J. Am. Chem. Soc.* **1977**, *99*(5), 1550–1557.

143. Bruice, T.C., Katzhendler, J., Fedor, L.R. Nucleophilic micelles. II. Effect on the rate of solvolysis of neutral, positively, and negatively charged esters of varied chain length when incorporated into nonfunctional and functional micelles of neutral, positive, and negative charge. *J. Am. Chem. Soc.* **1968**, *90*(5), 1333–1348.

144. Piszkiewicz, D. Cooperativity in bimolecular micelle-catalyzed reactions: inhibition of catalysis by high concentrations of detergent. *J. Am. Chem. Soc.* **1977**, *99*(23), 7695–7697.

145. Dunn, B.M., Bruice, T.C. Further investigation of the neighboring carboxyl group catalysis of hydrolysis of methyl phenyl acetals of formaldehyde: electrostatic and solvent effects. *J. Am. Chem. Soc.* **1970**, *92*(4), 6589–6594.

146. Hager, M., Olsson, U., Holmberg, K. A nucleophilic substitution reaction performed in different types of self-assembly structures. *Langmuir* **2004**, *20*(15), 6107–6115.

147. Schomaecker, R., Stickdorn, K., Knoche, W. Chemical reactions in microemulsions: kinetics of the alkylation of 2-alkylindan-1,3-diones in michroemulsions and polar organic solvents. *J. Chem. Soc., Faraday Trans.* **1991**, *87*(6), 847–851.

148. Oh, S.-G., Kizling, J., Holmberg, K. Microemulsions as reaction media for the synthesis of sodium decyl sulfonate 1: role of microemulsion composition. *Colloids Surf. A* **1995**, *97*(2), 169–179.

149. Minero, C., Pramauro, E., Pelizzetti, E. Generalized two-pseudophase model for ionic reaction rates and equilibria in micellar systems: hexachloroiridate(IV)-iron(II) electron-transfer kinetics in cationic micelles. *J. Phys. Chem.* **1988**, *92*(16), 4670–4676.

150. Moroi, Y., Matuura, R. Size distribution of anionic surfactant micelles. *J. Phys. Chem.* **1985**, *89*(13), 2923–2928.

151. Lelievre, J., Le Gall, M., Loppinet-Serani, A., Millot, F., Letellier, P. Thermodynamic and kinetic approach of the reactivity in micellar media: reaction of 1,3,5-trinitrobenzene upon hydroxide ion in aqueous solutions of cationic surfactants. *J. Phys. Chem. B* **2001**, *105*(51), 12844–12856 and references cited therein.

152. Le Gall, M., Lelievre, J., Loppinet-Serani, A., Letellier, P. Thermodynamic and kinetic approach of the reactivity in micellar media — reaction of hydroxide ion with 1,3,5-trinitrobenzene in aqueous solutions of neutral nonionic surfactant: effect of the concentration of background electrolyte. *J. Phys. Chem. B* **2003**, *107*(13), 8454–8461.

4 Normal Micelles: Effects on Reaction Rates

4.1 INTRODUCTION

The first systematic kinetic study on the effects of normal micelles (hereafter referred to as *micelles*) on reaction rates appeared in 1959 in the classic papers by Duynstee and Grunwald.[1] By the early 1980s, a huge amount of kinetic data on the effects of micelles on reaction rates, obtained under a variety of reaction conditions, had appeared in the literature. Most of these kinetic data have been discussed in a quantitative or semiquantitative manner by using one of the various micellar kinetic models described in Chapter 3. These studies have been reviewed in various articles[2] and two books.[3] In view of these reported studies, it may be concluded that the rate of a micellar-mediated reaction may be influenced by one or more than one of the following factors: (1) micellar medium effect, (2) micellar effect on keeping reactant molecules apart from each other, that is, proximity effect, (3) electrostatic effect, (4) hydrophobic effect, (5) ionic strength effect of ionic micellar surface, (6) ion exchange between two counterions of ionic micelles, and (7) the effect of ion-pair formation between counterions or ionic head groups of ionic micelles and ionic reactants.

In this chapter, an attempt is made to include critical discussions on some of the kinetic studies on the effects of micelles on the reaction rates that appeared, especially during the 1980s through 2003.

4.2 UNIMOLECULAR AND SOLVOLYTIC REACTIONS

The rate constants for unimolecular and solvolytic reactions generally show a monotonic decrease (i.e., micellar inhibition)[4–7] or a monotonic increase (i.e., micellar catalysis)[8–10] or insensitivity (i.e., micellar-independent rate)[11–14] to an increase in micellar concentration. There seems to be no exception to this generalization and, if there is one, it is owing to some specific chemical or physical reasons. For example, the unimolecular decarboxylation of 6–nitrobenzisoxazole–3–carboxylate ion (**1**) in CTABr micelles is enhanced by the salts of hydrophilic anions and slowed by the salts of hydrophobic anions, whereas salts such as sodium tosylate increased reaction rate when in low concentration, and retarded it when in high concentration.[8] The first theoretical model, known as the

preequilibrium kinetic (PEK) model, which is also equivalent to the pseudophase
(PP) model of micelle in terms of the mechanistic description of micellar-mediated reaction as shown in Scheme 3.1 (Chapter 3), to provide a quantitative or
semiquantitative interpretation of observed data for such reactions in which
increase in the micelle concentration ($[D_n]$) decreases or increases k_{obs} (pseudo-
first-order rate constant for the reaction) monotonically, was proposed by Menger
and Portnoy.[15]

As described in Chapter 3, the rate law for the reaction in view of the reaction
steps shown in Scheme 3.1 and the observed first-order or pseudo-first-order rate
law (rate = $k_{obs} [Sub]_T$ where $[Sub]_T = [S_W] + [S_M]$) can lead to Equation 3.2,
which predicts (1) a monotonic increase and decrease in k_{obs} with the increase in
$[D_n]$ if $k_M > k_W$ and $k_W > k_M$, respectively, (2) a linear increase in k_{obs} with the
increase in $[D_n]$ if $1 >> K_S [D_n]$, and (3) $k_{obs} = k_M$ (i.e., rate constants k_{obs} become
independent of $[D_n]$ if $[S_W] \approx 0$ or $1 << K_S [D_n]$ and $k_W << k_M K_S [D_n]$ or $k_W \approx k_M$).

For a set of observed data, k_{obs} vs. $[D_n]$, it is always possible to determine k_W
by carrying out the experiment in the absence of micelles under the experimental
conditions used for kinetic runs in the presence of micelles. This value of k_W can
be used to calculate k_M and K_S from Equation 3.2 provided k_W value is insensitive
to $[Surf]_T$ at its value < CMC. Equation 3.2 requires appropriate modification if
k_W becomes sensitive to the concentration of monomers of the surfactant. Such
complicated micellar-mediated reactions are rare. Thus, the actual unknown
parameters, k_M and K_S, can be calculated from Equation 3.2 using the nonlinear
least-squares technique, which can be carried out with easy access to computers
these days. It should be noted that the use of nonlinear least-squares technique to
calculate k_M and K_S from Equation 3.2 may not yield reliable values of k_M and
K_S if the observed data do not include some data points in or very near to plateau
region of k_{obs} vs. $[D_n]$ plots. It also seems essential to point out that the calculated
value of K_S using Equation 3.2 may not be exactly similar to the one obtained by
any of the various nonkinetic techniques. It is now more apparent that the values
of micellar binding constants, K_S, are technique dependent. It is generally not
possible to maintain exactly the same experimental conditions in the determination
of K_S by using both kinetic and nonkinetic methods.

Pseudo-first-order rate constants, k_{obs}, for hydrolysis of ionized phenyl salicylate in the presence of different concentrations of CTABr[16] were used to calculate $k_M K_S$ and K_S (considered to be unknown parameters) from Equation 3.2
and such calculated values of $k_M K_S$ and K_S are 0.61 ± 0.24 M^{-1} sec^{-1} and $(6.3
\pm 1.1) \times 10^3$ M^{-1}, respectively. These values of $k_M K_S$ and K_S yielded k_M as 9.6
$\times 10^{-5}$ sec^{-1}, which is exactly the same as the one calculated from Equation 3.2
considering k_M and K_S as unknown parameters. The quality of the data fit of a
set of observed data to Equation 3.2 remained unchanged with the change in the
choice of unknown parameters from k_M and K_S to $k_M K_S$ and K_S. This analysis
thus rules out the inherent perception of a possible compensatory effect between
the calculated values of k_M and K_S when they, rather than $k_M K_S$ and K_S, are
considered unknown parameters in using Equation 3.2 for data analysis.

Although the values of k_M and K_S should be calculated from nonlinear Equation 3.2, very often the linearized form of Equation 3.2 (i.e., Equation 4.1) is used, probably because it does not require an approximation method, such as nonlinear least-squares method, and computer for data analysis; and, furthermore, Equation 4.1 gives the exact solution.

$$\frac{1}{k_W - k_{obs}} = \frac{1}{k_W - k_M} + \frac{1}{(k_W - k_M)K_S[D_n]} \tag{4.1}$$

But the main problem in using Equation 4.1 is the fact that the statistical reliability of $k_W - k_{obs}$ or $k_{obs} - k_W$ decreases as $[D_n] \rightarrow 0$ and $1/(k_W - k_{obs})$ or $1/(k_{obs} - k_W) \rightarrow \infty$ as $[D_n] \rightarrow 0$. Thus, the use of Equation 4.1 for calculation of k_M and K_S suffers from the disadvantages of placing a very high emphasis on the values of k_{obs} as $[D_n] \rightarrow 0$ and being very sensitive to even small errors when $k_{obs} \rightarrow k_W$.

Another practically usable linearized form of Equation 3.2 is Equation 4.2, which has been used to calculate K_S and CMC with known values of k_W and k_M (the value of k_M can be easily obtained from the plateau region of the k_{obs} vs. $[Surf]_T$ plot).[17-20] The disadvantage of using Equation 4.2 is similar to that described for using Equation 4.1.

$$\frac{k_W - k_{obs}}{k_{obs} - k_M} = K_S[Surf]_T - K_S CMC \tag{4.2}$$

There is a large uncertainty in the values of $(k_W - k_{obs})/(k_{obs} - k_M)$ when $k_{obs} \rightarrow k_M$ and $k_{obs} \rightarrow k_W$.

The use of Equation 3.2 and Equation 4.1 requires the value of CMC of a micelle-forming surfactant obtained under strictly reaction kinetic conditions. Very often, the values of CMC used are obtained by various physical methods, under experimental conditions not strictly similar to the reaction kinetic conditions. It is well known that CMC values are affected by various factors such as ionic strength, specific salt, specific solubilizate, and mixed aqueous–nonaqueous solvent.[21] Recently, attempts have been made to determine CMC under strictly reaction kinetic conditions by the use of a graphical method[22-27] and an iterative method.[28,29] However, similar to other physical methods, graphical and iterative methods also have limitations and, hence, one has to use intelligent choice to use the correct method for the determination of CMC value under reaction kinetic conditions.

4.2.1 MICELLES OF NONIONIC NONFUNCTIONAL SURFACTANTS

Extensive studies on the kinetics of the effects of micelles on the rates of organic reactions contain only a few studies that involve nonionic micelles.[2,3] The most common nonionic micelle-forming surfactants are linear alkyl ethoxylates (Brij, Igepal) with general chemical formula $C_mH_{2m+1}O(CH_2CH_2O)_{n-1}CH_2CH_2OH$ (designated as C_mE_n) and the others are branched alkyl ethoxylates, octylphenol ethoxylates, alkane diols, alkyl mono- and disaccharide ethers and esters, ethoxylated alkyl amines and amides, and fluorinated linear ethoxylates and amides.[30] The rates of unimolecular reactions involving anionic phosphate[10,31] and sulfonate[32] esters were unaffected and slightly catalyzed, respectively, by the presence of nonionic micelles. Similarly, the rate of unimolecular cleavage of 1 was only slightly catalyzed by nonionic micelles.[8a]

Pseudo-first-order rate constant, k_M, for pH-independent hydrolysis of 2 and 3 are $(5.8 \pm 0.3) \times 10^{-5}$ sec^{-1} and $(0.6 \pm 0.4) \times 10^{-5}$ sec^{-1}, respectively, in the presence of nonionic $C_{12}E_7$ micelles.[4] These values of k_M represent ~22- and 510-fold inhibition for pH-independent hydrolysis of 2 and 3, respectively, in $C_{12}E_7$ micellar pseudophase compared to that in the aqueous pseudophase. It is interesting to note that the standard deviation associated with k_M value for 3 is so high that it may be regarded as statistically not different from zero. Thus, a value of k_M more than 510-fold lower compared to k_W is considered to be amazing if the reactants for micellar-mediated reaction are solvent molecules (H_2O) and 3. The rate of hydrolysis of 3 in aqueous pseudophase is reported to be pH-independent within the pH range 1 to 5.5. Reaction rates were studied at aqueous pH 4, and it is certain that the pH of micellar environment of micellized 3 molecules cannot be the same as the pH of the aqueous pseudophase.[33] Although the authors have argued the rate inhibition of micellar-mediated hydrolysis of 2 and 3 in terms of the effects of the concentration of head groups in palisade layer in which micellar-bound 2 and 3 molecules exist, it is quite possible that k_M values are affected, at least partly, by the change in pH of micellar environment of micellized 2 and 3 molecules compared to pH of the aqueous pseudophase. The kinetically determined micellar binding constants of 2 and 3 are summarized in Table 4.1.

The values of k_{obs} for pH-independent hydrolysis of 4-nitrophenyl chloroformate were lower in the nonionic micelles of polyoxyethylene (9) nonylphenyl ether compared to that in the aqueous pseudophase.[34] These results were explained in terms of lower microscopic polarity at the reaction site in the micelle. An ^1H NMR study of the solubilization revealed that the average solubilization site of solubilizate (4-nitrophenyl chloroacetate) did not depend on the charge of the micelle or the length of the surfactant hydrophobic tail.[34]

The increase in $[C_{12}E_{23}]_T$ (total concentration of $C_{12}E_{23}$) from 0 to 0.01 and 0.03 M did not show any appreciable change in k_{obs} for hydrolysis of fully ionized phenyl salicylate (PS$^-$) at 0.01 and 0.03 M NaOH, respectively.[35] However, significant depletion of hydroxide ions from the micellar environment of micellized PS$^-$ ions was noticed at 0.02 M $C_{12}E_{23}$, and both the concentration of PS$^-$ and the value of k_{obs} became almost zero at 0.03 M $C_{12}E_{23}$ in the presence of 0.01 M NaOH. The observed results showed nearly 70% of total initial concentration of PS$^-$ became nonionized phenyl salicylate (PSH) at 0.02 M $C_{12}E_{23}$ and 0.01 M NaOH, and these entire PSH molecules were irreversibly trapped by the micellar pseudophase. Alternatively, one can say that as the effect of increasing concen-

TABLE 4.1
Values of Micellar Binding Constants (K_S) of Solubilizates

Surfactant	Solubilizate	K_S/M^{-1}	Method	Reference
$C_{12}E_7$	**2**	1100	Kinetic	4
	3	520		4
$C_{12}E_8$	**18**	21		112
$C_{12}E_{23}$	**18**	29		112
$C_{12}E_9$	Ethyl cyclohexanone-2-carboxylate	310		47
$C_{12}E_{10}$	**8**	15		99
$C_{12}E_{23}$	Nonanoic acid	1170	Potentiometric	36
	Nonanoate ion	37		36
	3-Chlorobenzoic acid	349		36
	3-Chlorobenzoate ion	5		36
	Pernonanoic acid	1010	Kinetic	36
	Pernonanoate ion	282		36
	3-Chloroperbenzoic acid	172		36
	3-Chloroperbenzoate ion	Negative		36
	n-Nonanoyloxybenzenesulfonate	3.9×10^3		110
	Hydrogen peroxide ion	0		110
	Hydrogen peroxide	0		110
	9	880		35
	10	20		101
$C_{16}E_{10}$	**18**	27		112
$C_{16}E_{20}$	**18**	38		112
	14$^-$	7		105
	Ethyl cyclohexanone-2-carboxylate	480		47
Triton X-100	Nonanoic acid	1060	Potentiometric	36
	Nonanoate ion	Negative		36
	3-Chlorobenzoic acid	480		36
	3-Chlorobenzoate ion	Negative		36
	Pernonanoic acid	518	Kinetic	36
	Pernonanoate ion	22		36
	3-Chloroperbenzoic acid	152		36
	3-Chloroperbenzoate ion	4		36
SDS	Nonanoic acid	1190	Potentiometric	36
	Nonanoate ion	1		36
	3-Chlorobenzoic acid	350		36
	3-Chlorobenzoate ion	9		36
	Pernonanoic acid	950	Kinetic	36
	Pernonanoic acid	1350		110
	Pernonanoate ion	0		36
	Pernonanoate ion	3.2		110
	3-Chloroperbenzoic acid	94		36

Continued.

TABLE 4.1 *(Continued)*
Values of Micellar Binding Constants (K_S) of Solubilizates

Surfactant	Solubilizate	K_S/M^{-1}	Method	Reference
	3-Chloroperbenzoate ion	0		36
	n-Nonanoyloxybenzenesulfonate	2.7×10^3		110
	Phenyl nonanoate	2.6×10^6		110
	Hydrogen peroxide ion	0		110
	Hydrogen peroxide	0		110
2		1719		4
3		484		4
PS⁻		6		68
8		32		99
PSH		2150		68
PSH		1890· 1130		71
MS⁻		< 0.5		71
PS⁻		4	UV	70
PSH		2350		70
MS⁻		6		70
MSH		100		70
	Methylethylamine	0.6[a] (0.6)[b]		113
	Diethylamine	3.4 (3.4)		113
	Methylisopropylamine	1.5 (1.5)		113
	Methylbutylamine	22 (22)		113
	Dipropylamine	36 (36)		113
	Methylcyclohexylamine	63 (64)		113
	Hexamethylenimine	65 (64)		113
	Dibutylamine	138 (152)		113
	Diisobutylamine	150 (150)		113
	Dipentylamine	1320 (1346)		113
	Pyrrolidine	(1.4)		113
	2-Bromoethyl nitrite	8	Kinetic	113
	1-Phenylethyl nitrite	80		113
	2-Methylbenzenediazonium ion	240		61
	3-Methylbenzenediazonium ion	325		61
	4-Methylbenzenediazonium ion	1200		61
	Vitamin C	6.1 (pH 2)	UV	61
	Vitamin C	3.1 (pH 4)		61
	Ethyl cyclohexanone-2-carboxylate	313	Kinetic	86
18		36–50		112
19a		9		112
19b		34		112
19c		66		112
17		9–23		116
19a		6–13		116
19b		30–39		116

TABLE 4.1 *(Continued)*
Values of Micellar Binding Constants (K_S) of Solubilizates

Surfactant	Solubilize	K_S/M^{-1}	Method	Reference
	19c	63–69		116
	20	23–28		116
	21	2500		116
HDS	Ethyl cyclohexanone-2-carboxylate	328		86
DTABr	Ethyl cyclohexanone-2-carboxylate	219		47
	Phenyl chloroformate	76		77
	29a	3.1		78
	29b	3.4		78
	29c	1.5		78
	29d	4.2		78
	29e	6.0		78
	29f	1.4		78
	Methyl 4-nitrobenzenesulfonate	39		119
TTABr	Methylethylamine	2.5[a] (2.5)[b]		113
	Diethylamine	4.1 (4.2)		113
	Methylisopropylamine	3.5 (8)		113
	Methylbutylamine	10 (10)		113
	Dipropylamine	19 (19)		113
	Methylcyclohexylamine	29 (30)		113
	Hexamethylenimine	25 (23)		113
	Dibutylamine	86 (79)		113
	Diisobutylamine	50 (50)		113
	Dipentylamine	220 (265)		113
	Pyrrolidine	-(0.4)		113
	2-Bromoethyl nitrite	15		113
	1-Phenylethyl nitrite	300		113
	8	9		99
	Ethyl cyclohexanone-2-carboxylate	312		47
	Phenyl chloroformate	100		77
	Methyl 4-nitrobenzenesulfonate	71		119
TTACl	Ethyl cyclohexanone-2-carboxylate	331		47
	27	47[c],76[d]		128
	28	74[c],83[d]		128
CTABr	**2**	636		4
	3	1000		4
	8	22		99
	Ethyl cyclohexanone-2-carboxylate	446		47

Continued.

TABLE 4.1 *(Continued)*
Values of Micellar Binding Constants (K$_S$) of Solubilizates

Surfactant	Solubilizate	K$_S$/M^{-1}	Method	Reference
	PS$^-$	6990	UV	28
	PS$^-$	6710	Kinetic	16
	14$^-$	2500–3300		e
	10$^-$	6710		103
	Phenyl chloroformate	104		77
	29a	9.3		78
	29b	10.3		78
	29c	3.4		78
	29d	14		78
	29e	21		78
	29f	6.4		78
	22	1245		117
	23	1738		117
	24	852		117
	25	830		117
	Methyl 4-nitrobenzenesulfonate	76		119
	18	22–27		12
	19a	7–11		112
	19b	145–290		112
	19c	66		112
	27	55[c]		128
	28	39[c]		128
CTAOAc	27	50[c],90[d]		128
	28	58[c],47[d]		128
CTBABr	1	700		84
C$_{12}$E$_9$+DTABr	Ethyl cyclohexanone-2-carboxylate	222		47

Note: DTABr = Dodecyltrimethylammonium bromide; CTBABr = Cetyltri-*n*-butylammonium bromide, HDS = Hydrogen dodecyl sulfate.

[a] These values were obtained in the presence of 2-bromoethyl nitrite.
[b] These values were obtained in the presence of 1-phenylethyl nitrite.
[c] 0.02-*M* acetate buffers 50% base.
[d] 0.2-*M* acetate buffers 50% base.
[e] Reference 98 in Chapter 3.

tration of C$_{12}$E$_{23}$ on the depletion of hydroxide ions in micellar pseudophase increases, the value of K$_S$^{app} increases where K$_S$^{app} = ([PS$_M^-$] + [PSH$_M$])/{[D$_n$]([PS$_W^-$] + [PSH$_W$])} and K$_S$^{PSH} (= [PSH$_M$/([D$_n$][PSH$_W$]) should be much larger than K$_S$$^{PS-}$ (= [PS$_M^-$]/([D$_n$][PS$_W^-$]).^{36} Pseudo-first-order rate constant for the reaction of H$_2$O with PSH is nearly 10^5-fold smaller than that with PS$^-$ owing to occurrence of intramolecular general base catalysis in the nucleophilic

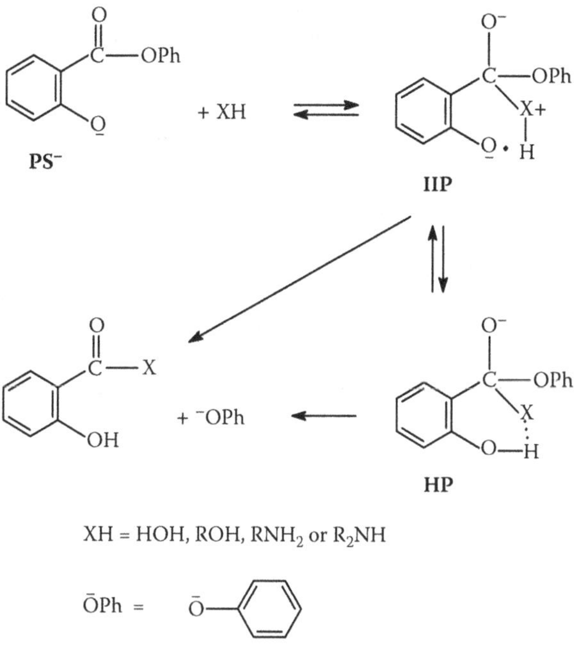

XH = HOH, ROH, RNH$_2$ or R$_2$NH

$\bar{O}$Ph = $\bar{O}$—⟨ ⟩

SCHEME 4.1

reaction of H$_2$O with PS$^-$.[37] Such unusual results were not obtained at ≤ 0.03 *M* C$_{12}$E$_{23}$ in the presence of 0.03-*M* NaOH.

The rate of hydrolysis of phenyl salicylate in aqueous solvent was found to be [HO$^-$]-independent within the [HO$^-$] range ~0.002 to 0.070 *M*,[38] and under such conditions, the hydrolysis reactions of salicylate esters involved ionized salicylate ester and H$_2$O as the reactants.[39–41] A brief mechanism for [HO$^-$]-independent hydrolysis of PS$^-$ is shown in Scheme 4.1, in which IIP and HP represent intramolecular intimate ion-pair (highly reactive intermediate) and internally hydrogen-bonded highly reactive intermediate, respectively. Although micelles accelerate or decelerate the rate of a reaction, it is very rare that they change the mechanism of the reaction with the change in the reaction environment from aqueous pseudophase to the micellar pseudophase.[42] It is, therefore, reasonable to assume that the mechanism of hydrolysis of PS$^-$ remains the same in both the aqueous and micellar pseudophases.

The absence of a kinetically detectable effect of C$_{12}$E$_{23}$ micelles on k$_{obs}$ for hydrolysis of PS$^-$ shows that the micellized PS$^-$ ions remain in the palisade layer of C$_{12}$E$_{23}$ micelle in which water concentration is not significantly different from water concentration in the aqueous pseudophase. Recent studies[43–46] reveal that the ethylene oxide moieties of C$_{12}$E$_{23}$ surfactant in the palisade layer are highly hydrated and, based on the hydration number of three of each ethylene oxide moiety, Buurma et al.[4] have shown that the estimated concentration of water in the palisade layer of $C_{12}E_7$ is 48 M. Although it is a very rough estimate in view of various aspects, it does show the high concentration of water in the palisade layer.

First-order rate constants (k_{obs}) for hydrolysis of ethyl cyclohexanone-2-carboxylate (ECHC) decrease monotonically with increase in the concentration of nonionic surfactants, $C_{12}E_9$ and $C_{16}E_{20}$, at 0.050 M HCl. These observed data fit to Equation 3.2 (Chapter 3) with $k_W = 13.95 \times 10^{-4}$ sec^{-1} and calculated respective values of k_M and K_S are $(3.05 \pm 0.16) \times 10^{-5}$ sec^{-1} and 310 ± 5 M^{-1} for $C_{12}E_9$ micelles and $(3.24 \pm 0.15) \times 10^{-5}$ sec^{-1} and 480 ± 4 M^{-1} for $C_{16}E_{20}$ micelles.[47] The experimentally determined value of k_{obs} (= 3.05×10^{-5} sec^{-1}) at 0.083 M HCl and 83% v/v propanol in mixed water–propanol solvent is similar to the k_M values. The rate of hydrolysis of ECHC is sensitive to both uncatalyzed and specific-acid-catalyzed reaction steps and, therefore, close similarity between k_M values at 0.050-M HCl and k_{obs} value at 0.083-M HCl and 83% v/v propanol may not reflect the nonionic micellar rate-retarding effect as due to only micellar reaction medium effect.

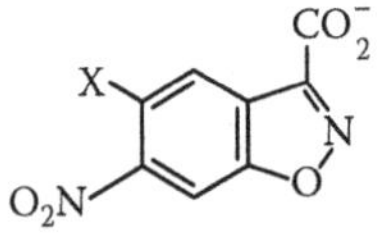

1 ; X = H
1, OMe ; X = OMe
1, OTD ; X = n-C$_{14}$H$_{29}$O

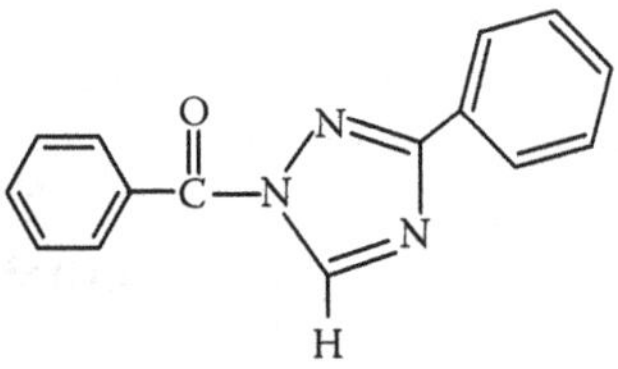

2

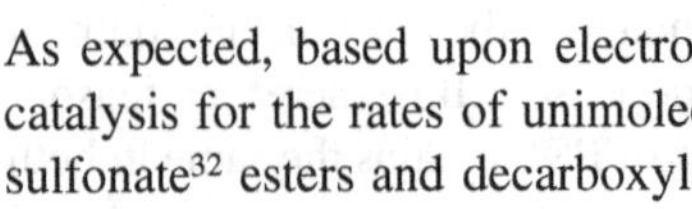

3

$Ph_2PO.OC_6H_4NO_2$ (4–)

4

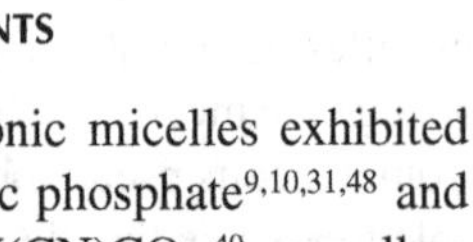

ECHC

4.2.2 MICELLES OF IONIC NONFUNCTIONAL SURFACTANTS

As expected, based upon electrostatic interactions, cationic micelles exhibited catalysis for the rates of unimolecular cleavage of anionic phosphate[9,10,31,48] and sulfonate[32] esters and decarboxylation of $\mathbf{1}$[6] and $C_6H_5CH(CN)CO_2{}^{-}$[49] as well as inhibition for the rates of carbocation formation in the cleavage of $C_6H_5CH(Br)CH_2CO_2{}^{-}$.[49] Similarly, anionic micelles did not affect the rates of unimolecular cleavage of anionic phosphate[10,31,48] and sulfonate[32] esters and decarboxylation of $\mathbf{1}$.[6] In an extensive and well-executed study on the effects of cationic and anionic micelles on the rates of reactions of water with reactive organic compounds such as $(4\text{-}YC_6H_4O)_2CO$ (Y = t-Bu, H, NO$_2$, CN), $(4\text{-}YC_6H_4O)_2CO$ and $4\text{-}YC_6H_4OCOCl$ (Y = H,NO$_2$), C_6H_5COCl, $C_6H_5CH_2Br$, $C_6H_5SO_3CH_3$, and $o\text{-}C_6H_4(CO)_2O$, it has been concluded that micellar effects on the rates of these reactions may be attributed to (1) lower water activity in the micellar pseudophase compared to that in the aqueous pseudophase and (2) electrostatic interaction between ionic head groups and transition state, which is apparently more polar than the ground state in almost every hydrolysis reaction of the study.[50] These results favor the so-called porous-cluster micellar model.[51–53] However, these authors[50] have suggested that although the hydrophobicity of the substrates or solubilizates controls the average location of the substrates or solubilizates in the micelle, the reaction rates involving such substrates cannot be considered a very useful probe for finding out the micellar location of the substrate, because the substrate may reside largely in an apolar region of the micelle, but the reaction may occur in a water-rich region. Thus, the kinetic probe can give information about the characteristic nature of the micellar environment in which micellar-mediated reaction occurs, but that may or may not be the actual micellar location of the reactant (i.e., substrate).

A recent study[34] on the effects of cationic [alkyltrimethylammonium chlorides, alkyldimethylbenzylammonium chlorides (alkyl group = cetyl and dodecyl)] and anionic [SDS and sodium dodecylbenzene sulfonate] micelles on the rate of uncatalyzed hydrolysis of 4-nitrophenyl chloroformate reaffirmed the conclusions reached in an earlier study.[50] Cationic [n-$C_{16}H_{33}NR_3X$: CTAX, CTEAX, CTBAX, R = CH_3, C_2H_5, n-C_4H_9, respectively, X = Cl, Br, and OMs] and anionic [SDS] micelles exhibited inhibitory effects on the rate of uncatalyzed hydrolysis of phenyl chloroformate, but the inhibition by anionic micelles was much greater than the inhibition by cationic micelles.[54] Furthermore, the inhibition by cationic micelles increased with the increasing head group size and cationic micellar affinity of the counterions. These cationic head group and counterion affinity effects were attributed to the combined effects of lower microscopic water activity in the micellar region of micellar-mediated reaction compared to that in the aqueous pseudophase and electrostatic interaction between ionic head group and more polar transition state compared to ground state.[54] The values of Hammett reaction constant, ρ, obtained for the reactions of water with p-substituted methyl benzenesulfonates in the cationic micelles of varying head groups and counterions (n-$C_{16}H_{33}NR_3X$, X = HO, Br, R = CH_3, C_2H_5, n-C_3H_7, n-C_4H_9), confirmed that the micellar interfacial regions are less polar than bulk water and polarities decrease with increasing size of the surfactant head group.[55]

Engberts has studied the effects of micelles of 1-alkyl-4-alkylpyridinium halide on the rate of unimolecular decarboxylation of 1 and found substantial catalytic effect of all micelle-forming surfactants.[56] Interestingly, the surfactant, which undergoes a transition from spherical to rodlike micelles at a specific concentration, shows that the respective values of k_M and K_S are nearly 1.5- and 6-fold larger in rodlike micelles compared to that in spherical micelles. Vesicles-forming surfactants exhibited the largest catalytic effect, whereas the value of K_S turned out to be almost lowest among a large number of surfactants.[56]

The rates of unimolecular decarboxylation of anionic substrates[8a,57] are faster in cationic and zwitterionic micelles than in water, and the values of rate enhancement range up to ~10^3. These rate enhancements are understandable, because rates of these reactions are accelerated by aprotic solvents and by a decrease in the water content of mixed solvents.[58] Micellar surfaces are less polar than water, there is less hydration of bromide ion,[59] and water activity is lower than that of the aqueous pseudophase.[60] Furthermore, the rate of unimolecular decarboxylation of **1** is favored by cationic surfactants that have bulky head groups.[57] Experimentally observed first-order rate constants (k_{obs}) for unimolecular dediazoniation of either 2-, 3-, or 4-MBD (MBD, methylbenzenediazonium) are mildly depressed with the increase in concentration of anionic micelles $[SDS]_T$.[61] The observed data fit to Equation 3.2 (Chapter 3) with the calculated values of unknown parameters k_M (= 5.7 ± 0.3) $\times 10^{-4}$ sec^{-1} and $K_S = 325 \pm 12$ M^{-1} with known k_W (= 8×10^{-4} sec^{-1}) for 3-MBD.

Cationic micelles $[C_{16}H_{33}N(CH_3)_3X$, X = Cl, Br, $(SO_4)_{0.5}$; $(CH_3)_3N^+(CH_2)_nN^+(CH_3)_3$ $2X^-$, n = 22, X = Br, $(SO_4)_{0.5}$, n = 16, 12, X = Br; $(C_{12}H_{25})_2N(CH_3)_2X$, X = Cl, Br, $(SO_4)_{0.5}$ and $(C_{16}H_{33})_2N(CH_3)_2X$, X = Cl, $(SO_4)_{0.5}]$ caused the nonlinear increase in k_{obs} for uncatalyzed hydrolysis of 2,4-dinitrophenyl phosphate dianion (2,4-DNPP) with the increase in concentration of micelle, $[D_n]$, at its relatively low values, and the values of k_{obs} became independent of $[D_n]$ at its value where 2,4-DNPP was fully micellar bound.[62,63] The k_{obs} vs. $[D_n]$ profiles appeared to be in accordance with Equation 3.2. The values of k_M/k_W for various cationic micelles were in the range of 20 to 110, and k_M values appeared to be insensitive to the nature of the surfactant, whereas the extent of micellar binding of 2,4-DNPP was sensitive to the nature of the surfactant. The decrease in polarity and water content in mixed aqueous–organic solvents increased k_{obs} in the absence of micelles.[64–65] Thus, the cationic micellar effect on k_{obs} is attributed to lower polarity and lower water concentration in the micellar environment of the occurrence of the micellar-mediated hydrolysis compared to the corresponding polarity and water concentration in the aqueous pseudophase. The effects of counterions upon reaction rates in CTAX, $X = SO_4^{2-}$, Cl$^-$, Br$^-$, where k_M increases in the sequence $SO_4^{2-} >$ Cl$^- >$ Br$^-$, were partially related to the relative stability or strength of hydration shell of X$^-$, which is expected to vary in the order $SO_4^{2-} >$ Cl$^- >$ Br$^-$, and the increase in the stability of the hydration shell of X$^-$ decreases the availability of water at the micellar surface.[62]

The rates of hydrolysis of phenyl and methyl salicylates became independent of [HO$^-$] within the [HO$^-$] range of ~ 0.002 to 0.070 M in the absence of micelles.[37] The values of ionization constants, K_a, for phenyl salicylate[66] and methyl salicylate[67] in water solvent are 5.67×10^{-10} M and 2.48×10^{-10} M, respectively. Thus, within the [HO$^-$] range of ~ 0.002 to 0.070 M, both phenyl and methyl salicylates exist in fully ionized forms. The pH-independent rate of hydrolysis of salicylate esters involve ionized salicylate ester and H$_2$O as the reactants.[39–41] Intramolecular general base catalysis has been shown to occur in the aminolysis of ionized phenyl salicylate in both aqueous pseudophase and micellar pseudophase of SDS.[68,69] The brief reaction scheme for hydrolysis or alkanolysis of salicylate ester or any substrate containing an easily ionizable proton in the presence of micelles, D_n, may be given in terms of PP model of micelle as depicted in Scheme 4.2

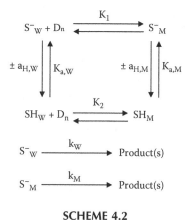

SCHEME 4.2

where K_1 and K_2 represent micellar binding constants of ionized (S^-) and non-ionized salicylate ester or substrate, respectively, k_W and k_M are respective pseudo-first-order rate constants for solvolysis of S^- in the aqueous and micellar pseudophases, and K_a is the ionization constant.

As has been pointed out earlier in the text, the rate of reaction of H_2O with ionized (S^-) salicylate ester is nearly 10^5-fold larger than that with nonionized (SH) salicylate ester. In view of these results, the reaction steps for the solvolysis of SH_W and SH_M have been omitted from Scheme 4.2.

The observed rate law, rate $= k_{obs} [Sub]_T$ (where $[Sub]_T = [S^-]_T + [SH]_T$ with $[S^-]_T = [S_W^-] + [S_M^-]$ and $[SH]_T = [SH_W] + [SH_M]$) and Scheme 4.2 give Equation 4.3:

$$k_{obs} = \frac{k_W f_W^{S^-} + k_M f_M^{S^-} K_S^{app}[D_n]}{1 + K_S^{app}[D_n]} \qquad (4.3)$$

where $f_M^{S^-} = K_{a,M}/(a_{H.M} + K_{a,M})$, $f_W^{S^-} = K_{a,W}/(a_{H.W} + K_{a,W})$, $a_{H,W}$ and $a_{H,M}$ are activity of proton in aqueous pseudophase and micellar pseudophase, respectively, and $K_S^{app} = ([S_M^-] + [SH_M])/\{[D_n]([S_W^-] + [SH_W])\}$. The values of $f_M^{S^-}$ were found to be almost unchanged with the change in $[SDS]_T$ (total concentration of SDS) at a constant [NaOH].[70]

Pseudo-first-order rate constants, k_{obs}, for hydrolysis of phenyl salicylate decreased by nearly 9- to 3-fold with the increase in $[SDS]_T$ from 0.0 to 0.4 M at different [NaOH] ranging from 0.003 to 0.060 M.[71] The values of k_{obs}, obtained within $[SDS]_T$ range of 0.0 to 0.4 M at a constant [NaOH] obeyed Equation 4.3, and the nonlinear least-squares calculated values of $k_M f_M^{S^-}$ were not statistically different from zero at [NaOH] ≥ 0.005 M. The value of $k_M f_M^{S^-}$ ($= 4.5 \times 10^{-5}$ sec^{-1}) at 0.003 M NaOH, although associated with nearly 50% standard deviation, gave $k_M = 1.5 \times 10^{-4}$ sec^{-1} (because $f_M^{S^-} = 0.29$ under such conditions[70]), which is nearly 6-fold smaller than k_W. Significantly lower rate of hydrolysis of ionized phenyl salicylate (PS$^-$) in SDS micellar pseudophase compared to that in the aqueous pseudophase is attributed to (1) the lower water activity in the micellar environment of micellized PS^- ions (PS_M) compared to that in the aqueous pseudophase and (2) the reduction in the efficiency of the intramolecular general base catalysis due to probable formation of ion pair (IP) in the micellar environment of considerably low water activity.

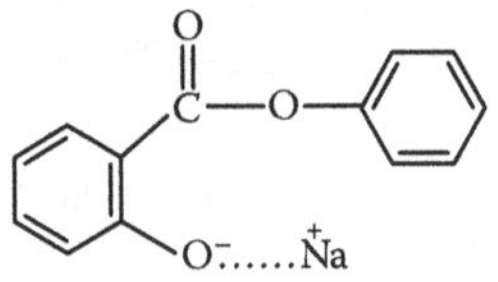

IP

The values of K_S^{app} decreased from 22.9 to 2.9 M^{-1} with the increase in [NaOH] from 0.003 to 0.060 M.[71] The relationship: $K_S^{app} = ([S_M^-] + [SH_M])/\{[D_n]([S_W] + [SH_W])\}$ and Scheme 4.2 can lead to Equation 4.4

$$K_S^{app} = \frac{K_1 K_{a,W} + K_2 a_{H,W}}{a_{H,W} + K_{a,W}} \qquad (4.4)$$

But $K_{a,W} \gg a_{H,W}$ under the experimental conditions imposed,[70] and this inequality reduced Equation 4.4 to Equation 4.5*

$$K_S^{app} = K_1 + \frac{K_2 K_{w,W}}{K_{a,W} a_{OH,W}} \qquad (4.5)$$

where $K_{w,W} = a_{H,W} a_{OH,W}$. The values of K_S^{app} at different [NaOH] were used to calculate K_1 and K_2 from Equation 4.5 and such calculated values of K_1 and K_2 are 2.9 ± 0.4 M^{-1} and 1130 ± 54 M^{-1}, respectively. Spectrophotometric coupled with potentiometric technique yielded K_1 and K_2 as 4.0 ± 2.0 M^{-1} and 2350 ± 90 M^{-1}, respectively.[70]

Unlike the effects of the concentration of SDS micelles on k_{obs} (~900 to 300% decrease) for hydrolysis of PS^-, the values of k_{obs} for hydrolysis of ionized methyl salicylate (MS^-) decreased only slightly (~25%) with the increase in $[SDS]_T$ from 0.0 to 0.4 M as evident from the observed results summarized in Table 4.2. These results could not fit to Equation 4.3 apparently owing to extremely low decrease in k_{obs} within the $[SDS]_T$ range covered in the study.[71] However, these data showed statistically good fit to Equation 4.6 (as evident from data shown in Table 4.2), which is derived from Equation 4.3 with condition $k_M f_M^S K_S^{app} [D_n] \ll k_W f_W^{S-}$.

---

* It may also be shown that under conditions where $K_{a,W} \gg a_{H,W}$, that is, when $[SH_W]$ is negligible compared to $[S_W^-]$, $K_S^{app} = K_1 + K_1 a_{H,M}/K_{a,M}$.

TABLE 4.2
Effects of $[SDS]_T$ on Pseudo-First-Order Rate Constants, k_{obs}, for Hydrolysis of Methyl Salicylate at 0.03-M NaOH and 37°C[a]

$[SDS]_T$ M	$10^4 k_{obs}$[b] sec^{-1}	E_{app}[b] $M^{-1} cm^1$	A_∞	$10^3 Y_{obs}$[c]	$10^3 Y_{calcd}$[d]
0.0	2.47 ± 0.02[e]	3187 ± 9[e]	0.007 ± 0.002[e]		
0.02	2.46 ± 0.02	3231 ± 9	0.001 ± 0.003	4.06	6.52
0.07	2.39 ± 0.02	3224 ± 9	0.001 ± 0.003	33.5	33.7
0.10	2.34 ± 0.03	3071 ± 13	0.019 ± 0.003	55.6	50.0
0.14	2.28 ± 0.02	3096 ± 9	0.023 ± 0.003	83.3	71.7
0.20	2.27 ± 0.02	3103 ± 9	0.026 ± 0.003	88.1	104
0.30	2.11 ± 0.02	2996 ± 10	0.037 ± 0.002	171	159
0.40	1.97 ± 0.02	2985 ± 10	0.060 ± 0.003	254	213

[a] Conditions: [Methyl salicylate]$_0$ = 3 × 10^{-4} M, λ = 350 nm, aqueous reaction mixture contained 1% v/v CH_3CN.

[b] Calculated from the relationship: $A_{obs} = E_{app}$ [Methyl salicylate]$_0$ $\exp(k_{obs} t) + A_\infty$, where A_{obs} represents observed absorbance at reaction time t, E_{app} is the apparent molar extinction coefficient of reaction mixture, and A_∞ is absorbance at t = ∞.

[c] $Y_{obs} = (k_W - k_{obs})/k_{obs}$ where $k_W = k_{obs}$ at $[SDS]_T = 0$.

[d] Calculated from the relationship: $Y_{obs} = K_S^{app} [D_n]$ with $K_S^{app} = 0.54$ M^{-1} and CMC = 0.008 M.

[e] Error limits are standard deviations.

$$(k_W - k_{obs})/k_{obs} = K_S^{app} [D_n] \qquad (4.6)$$

Although the calculated value of K_S^{app} (= 0.54 ± 0.11 M^{-1}) is associated with reasonably low standard deviation, it may not be very reliable because of low range of $(k_W - k_{obs})/k_{obs}$ (= 0.004 to 0.254) under the experimental conditions imposed.[71] The spectral data revealed the absence of nonionized methyl salicylate (MSH) within the $[SDS]_T$ range of 0.0 to 0.4 M at 0.03 M NaOH.[70] Thus, Equation 4.5 shows that $K_S^{app} = K_1$ at 0.03 M NaOH.

It is perhaps noteworthy that K_1 (= 5.7 ± 1.2 M^{-1}), obtained by the spectrophotometric coupled with potentiometric technique,[70] is nearly 10 times larger than K_1 (= 0.54 ± 0.11 M^{-1}) obtained by kinetics or rate measurement technique.[71] But these two different techniques gave almost the same K_1 value for PS$^-$. Such technique-dependent and technique-independent values of K_1 for MS$^-$ and PS$^-$, respectively, may be explained as follows. Although the structure of micelles and the extent of water penetration into the micellar pseudophase have been the subject of unsettled debate and controversy,[51] it is becoming increasingly clear that the porous cluster model of micelles,[51] suggests a decrease in water activity with an increase in the distance from the outer surface (i.e., the outermost region of Stern layer or palisade layer) to the center or innermost region of the micelle. Proton NMR spectroscopic study has revealed, at least qualitatively, different solubilization sites of organic perfume molecules of different hydrophobicity in the SDS micellar pseudophase.[72]

Micellar location of a solubilizate (which is difficult to determine with some degree of certainty by any experimental technique) depends largely upon the nature and structural features of solubilizate as well as head groups of micelle-forming surfactant. Purely on energetic grounds (which include all or most of the following molecular interactions: ionic or electrostatic, polar, resonance, hydrogen bonding, ion-pair formation, hydrophobic, and packing constraint or steric interactions), it is reasonable to assume that MS^- ion, being less hydrophobic than PS^- ion, remains in the micellar region in which water activity is almost similar to the water activity in the aqueous pseudophase. But the PS^- ion, being more hydrophobic than MS^- ion, is dragged deeper inside the micellar region of lower water activity compared with water activity of the aqueous pseudophase. These statements can best be represented graphically as in Figure 4.1, in which the presentation of the spherical shape of the micelle is a trivial one. As the rate of hydrolysis of MS^- and PS^- involves H_2O and MS^- or PS^- as the reactants, $k_W \approx k_M$ if the micellar-mediated hydrolysis occurs in the micellar region of water activity similar to water activity of the aqueous pseudophase. It is evident from Equation 4.3 that if $k_W \approx k_M f_M^{S-}$ (where $f_M^{S-} \approx 1$), then k_{obs} should be independent of $[D_n]$ even if $K_S^{app} \neq 0$. This explains why k_{obs} values are independent of $[SDS]_T$ within the $[SDS]_T$ range of 0.0 to 0.2 M at 0.03 M NaOH (Table 4.2).

Spectrophotometric coupled with potentiometric determination of SDS micellar binding constants of methyl and phenyl salicylates involves pH and initial absorbance measurements of SDS micellar solution. In the pH measurement of micellar solution, the pH electrode gives response to the hydronium ion activity of the aqueous pseudophase only (i.e., $a_{H,W}$), whereas initial absorbance depends on the concentration of MS^- or PS^- in both aqueous pseudophase and micellar pseudophase. At a constant [NaOH], the increase in $[SDS]_T$ increases $a_{H,M}$ or decreases $a_{OH,M}$ because of the electrostatic interaction between anionic head group and H^+ or HO^- even in the micellar region where water activity is similar to the water activity of aqueous pseudophase. Thus, the increase in $a_{H,M}$ (in the micellar region where MS_M^- ions exist) with the increase in $[SDS]_T$ at a constant [NaOH] increases the concentration of nonionized methyl salicylate ([MSH]), which, in turn, decreases the initial absorbance at 350 nm (molar extinction coefficients of MSH and MS are ~ 0 and 3200 M^{-1} cm^{-1} at 350 nm, respectively). But the kinetic technique is unable to detect the presence of MS^- ions in the micellar region of water activity similar to the water activity of the water bulk phase even under the conditions in which [MSH]/([MSH]+[MS$^-$]) is nearly 0.15, because under such conditions, rate constants, k_{obs}, become independent of $[SDS]_T$.[14] On the other hand, PS_M^- ions reside in the micellar environment where water activity is considerably lower than the water activity of water bulk phase and that is the reason why K_1 values for PS^- are almost independent of spectrophotometric coupled with potentiometric and rate measurement techniques.

The respective values of k_M for CTABr micellar-mediated hydrolysis of PS^- and MS^- are nearly 8- and 6-fold smaller compared to the corresponding k_W

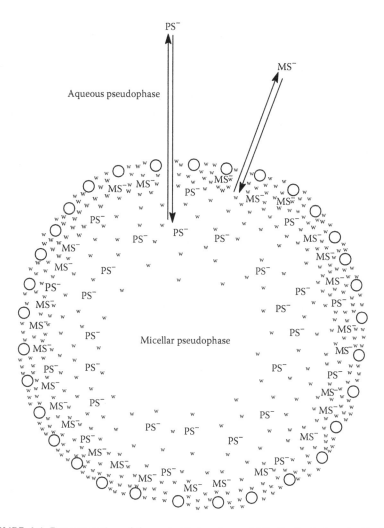

FIGURE 4.1 Representation of the probable locations of PS⁻ (ionized phenyl salicylate) and MS⁻ (ionized methyl salicylate) ions in the micellar pseudophase.

values.[16] These inhibitions are attributed to (1) the decrease in the efficiency of intramolecular general base catalysis due to probable ion-pair formation between cationic head groups of micelle and anionic site of the substrate (PS_M^- or MS_M^-), (2) lower water activity in the region of micellar-mediated hydrolysis compared to that in the aqueous pseudophase, and (3) the probable difference in the water structure for water molecules in the aqueous pseudophase and the micellar region where micellar-mediated hydrolyses occur. Significantly larger inhibition in the rate of micellar-mediated hydrolysis of PS_M^- compared to that of MS_M^- shows the higher water activity in the region of CTABr micelle for hydrolysis of MS_M^- compared to that of PS_M^-.

The rate of methanolysis of PS$^-$ was found to be independent of [HO$^-$] within its range 0.01 to 0.15 M in mixed aqueous solvents containing 80% v/v CH$_3$OH and 0.8% v/v CH$_3$CN.[73] The pH-independent rate of alkanolysis of phenyl salicylate has been shown to involve IGB catalysis and reaction mechanism[74] similar to the one shown in Scheme 4.1. Pseudo-first-order rate constants, k_{NM}, obtained for methanolysis of PS$^-$ in the presence of 0.01 M NaOH and 10% v/v CH$_3$OH in mixed H$_2$O-CH$_3$OH-CH$_3$CN solvents decreased with the increase in [CTABr]$_T$, and the amount of this decrease increased with the decrease in [CH$_3$CN].[27] The subscript NM represents nonmicellar pseudophase. The increase in the acetonitrile content from 5 to 20% v/v increased k_M (pseudo-first-order rate constant for methanolysis of PS$^-$ in the CTABr micellar pseudophase) from 1.42×10^{-3} to 2.04×10^{-3} sec^{-1}, whereas the values of k_{NM} decreased from 6.67×10^{-3} to 5.51×10^{-3} sec^{-1}. The most plausible source of micellar inhibition of rate of methanolysis of PS$^-$ is the lower concentration of reactive methanol ([CH$_3$OH$_M$]) in the micellar environment in which PS$_M^-$ ions exist (when compared with [CH$_3$OH$_{NM}$]), whereas the other possible sources such as polarity, ionic strength, and ion-pair (similar to IP) formation have been concluded to be less important. The increase in k_M from 1.42×10^{-3} to 2.04×10^{-3} sec^{-1} with the increase in CH$_3$CN content from 5 to 20% v/v at a constant content (10% v/v) of CH$_3$OH has been attributed to (1) the effect of solvation energy of monomeric methanol in the micellar pseudophase, which decreases with the increase in the CH$_3$CN content owing to decrease in the number of water molecules in the solvation shell and (2) the possibility of the gradual increase in [CH$_3$OH$_M$] (in the micellar region where PS$_M^-$ ions exist) with the increase in the content of CH$_3$CN in mixed aqueous solvents containing 10% v/v CH$_3$OH.

The values of k_{NM}/k_M for methanolysis[24] and ethanediolysis[75] of PS$^-$ obtained in mixed aqueous–alkanol solvents containing 2% v/v CH$_3$CN, 0.01 M NaOH, and CTABr micelles, vary from ~ 6 to ~ 2 with the increase in alkanol content from 10 to 50% v/v. These CTABr micellar inhibitory effects are similar to ones obtained in the related study[27] on the effects of CTABr micelles on the rate of methanolysis of PS$^-$ at varying contents of CH$_3$CN in H$_2$O-CH$_3$OH-CH$_3$CN solvents. The plot of k_{NM} vs. alkanol content exhibited positive slope, which decreased gradually with the increase in alkanol content. But the plot of k_M vs. alkanol content showed positive slope that increased gradually with the increase in content of alkanol. The gradual decrease in the positive slope of the plot of k_{NM} vs. alkanol content has been explained in terms of self association of alkanol molecules.[24,75,76] The continuous increase in the positive slope of the plot of k_M vs. alkanol content indicates that the value of [ROH$_M$] in the micellar region where PS$_M^-$ ions exist increases nonlinearly with the increase in the content of ROH in mixed ROH–H$_2$O solvents containing ≤ 2% v/v CH$_3$CN. These results show that unlike in aqueous pseudophase, the self-association of ROH molecules in micellar pseudophase is kinetically insignificant. It seems that the micellized mixed solvent molecules (H$_2$O$_M$ and ROH$_M$) do not maintain the same ratio R (= [ROH$_M$]/[H$_2$O$_M$]) in both aqueous and micellar pseudophases and, also, the value of R increases with the increase in distance from outermost of the Stern layer to the interior of micellar pseudophase.

Pseudo-first-order rate constants, k_{obs}, for methanolysis of PS$^-$ show a decrease of ~ three- to ~ fivefold with the increase in CH$_3$CN content from 2 to 60 or 70% v/v in mixed aqueous solvents containing 0.01 M LiOH and a constant content of CH$_3$OH.[7a] However, at 0.01 M KOH, the rate constants k_{obs} show a decrease of 15 to 20% and an increase of 7 to 130% with the increase in CH$_3$CN content from 2 to 30% v/v and from 30 to 60 or 70% v/v, respectively. Such characteristic difference of Li$^+$ and K$^+$ ions on k_{obs} has disappeared for the rate of methanolysis of PS$^-$ in the CTABr micellar pseudophase containing 2% v/v CH$_3$CN. This shows that neither Li$^+$ nor K$^+$ ions entered the microenvironment of low polarity in which the micellar-mediated reactions occurred. The counterions for HO$_M^-$ and PS$_M^-$ in this micellar environment are perhaps cationic micellar head groups.

The effects of [cationic micelles] (micelle-forming surfactants are hexadecyltrimethylammonium bromide, CTABr, tetradecyltrimethylammonium bromide, TTABr, and dodecyltrimethylammonium bromide, DTABr) on the rates of spontaneous hydrolysis of phenyl chloroformate in water–ethanediol (ED) solvents have been rationalized in terms of pseudophose (PP) model, i.e., Equation 3.2 with k_W replaced by k_{NM}.[77] Experimentally observed values of k_{NM} decrease from 14.1×10^{-3} to 13.8×10^{-3} sec^{-1} with the increase in percent weight ED from 0 to 50. The calculated values of k_M remain essentially unchanged with the change in percent weight ED from 0 to 50 in the presence of all three cationic micelles (CTABr, $10^3 k_M = 4.1 \pm 0.2$ sec^{-1}, TTABr, $10^3 k_M = 4.2 \pm 0.1$ sec^{-1}, and DTABr, $10^3 k_M = 4.9 \pm 0.1$ sec^{-1}). The calculated values of K_S at different percent weight ED are summarized in Table 4.3. The mechanism of hydrolysis of phenyl chloroformate is concluded to be same in both the absence and presence of micelles based upon the observation that the rate of hydrolysis is substantially slower in anionic than in cationic water–ED micellar solutions (as was found in aqueous micellar solutions). An insignificant decrease (~2%) in k_{NM} with the increase in percent weight ED from 0 to 50 in mixed aqueous solvent may be the cause for almost similar calculated values of k_M under the same mixed solvent systems for all three cationic micelles. The cationic micellar binding constant, K_S, of phenyl chloroformate decreases by increasing the amount of ED in mixed aqueous solvent, because the water–ED mixed solvent is a better solvent for organic compounds than water.

Nearly 19- and 26-fold lower values of k_M than k_W for pH-independent hydrolysis of **2** in CTABr and SDS micelles, respectively, are explained in terms of high concentration of ionic head groups in Stern layer and electrostatic effect on partially anionic transition state. However, such an electrostatic effect cannot explain nearly 190- and 65-fold lower values of k_M compared to k_W for pH-independent hydrolysis of **3**.[4] It has been suggested that the influence of hydrophobic chains is more pronounced for **3** than for **2**. But the nearly 3-fold larger value of k_M for **3** in SDS micelles than in CTABr micelles remained unexplained. The decrease in k_M for **2** from 4.8×10^{-5} to 2.4×10^{-5} sec^{-1} with the increase in [NaCl] from 0.0 to 0.5 M in SDS micelles has been attributed to increased counterion binding (i.e., β value in pseudophase ion-exchange [PIE] model for-

TABLE 4.3
Effects of Mixed Aqueous–Organic Solvents on Micellar Binding Constants (K_S) of Solubilizates

Solubilizate	Surfactant	Temperature/ °C	X/(% v/v)	K_S/M^{-1}	$K_S^{calcd\ a}/M^{-1}$
9			X = CH_3CN^b		
	SDS	35	2	566 ± 7^c	579
			6	425 ± 2	399
			10	271 ± 3	274
			14	180 ± 4	189
			30	31 ± 2	43
PS⁻			X = CH_3CN^d		
	CTABr	35	5	4500 ± 710	4624
			8	2900 ± 250	2896
			10	2600 ± 440	2120
			15	620 ± 90	972
			20	190 ± 30	446
PS⁻			X = $HOCH_2CH_2OH^e$		
	CTABr	25	20	7140 ± 480	7042
			30	4330 ± 560	4629
			40	3270 ± 100	3043
		30	15	7450 ± 600	7877
			20	6960 ± 730	5723
			25	3540 ± 810	4158
			30	2480 ± 520	3021
			40	1740 ± 230	1595
			50	1150 ± 100	842
		35	20	4180 ± 630	4141
			30	2180 ± 350	2318
			40	1420 ± 190	1297
		40	20	3510 ± 420	3514
			30	1920 ± 340	1906
			40	1020 ± 200	1033
		45	20	3060 ± 570	3175
			30	2210 ± 300	1805
			40	671 ± 75	1027
PS⁻			X = CH_3OH^f		
	CTABr	25	10	10200 ± 1000	10343
			20	6130 ± 380	5603
			30	2550 ± 133	3036
		30	10	8080 ± 560	7951
			15	4460 ± 600	5021
			20	4010 ± 240	3171
			25	1540 ± 300	2002
			30	1290 ± 106	1265
			35	835 ± 96	799
			40	202 ± 24	
			45	80 ± 10	
			50	39 ± 6	
		35	10	6390 ± 720	6394

TABLE 4.3 *(Continued)*
Effects of Mixed Aqueous–Organic Solvents on Micellar Binding Constants (K$_S$) of Solubilizates

Solubilizate	Surfactant	Temperature/ °C	X/(% v/v)	K$_S$/M^{-1}	K$_S$$^{calcd\ a}$/ M^{-1}
			20	3030 ± 270	3012
			30	1400 ± 62	1419
		40	10	4070 ± 480	4129
			20	2320 ± 320	2088
			30	828 ± 81	1057
		45	10	4930 ± 690	4938
			20	1690 ± 420	1643
			30	476 ± 65	547
Methyl 4-nitrobenzenesulfonate			X = HOCH$_2$CH$_2$OH^g		
	CTABr	25.2	12	62	63
			25	42	40
			35	27	29
	TTABr	25.2	12	42	43
			25	30	25
			35	13	17
	DTABr	25.2	12	27	27
			25	17	16
			35	10	11
Phenyl chloroformate			X = HOCH$_2$CH$_2$OH^h		
	CTABr	25.2	20	77	74
			35	41	49
			50	39	33
	TTABr	25.2	20	80	81
			35	56	52
			50	30	33
	DTABr	25.2	20	55	55
			35	34	34
			50	20	20
1			X = CH$_3$CH$_2$OH^i		
	CTBABr	25	0.020	300	298
			0.045	100	110
			0.090	40	18

[a] Calculated from Equation 4.14 using the values of K$_S$^{0} and δ listed in Table 4.4.

[b] Data are from Reference 23.

[c] Error limits are standard deviations.

[d] Data are from Reference 27 and each mixed H$_2$O – X reaction solution contained 10% v/v CH$_3$OH.

[e] Data are from Reference 75 and each mixed H$_2$O – X reaction solution contained 2% v/v CH$_3$CN.

[f] Data are from Reference 24 and each mixed H$_2$O – X reaction solution contained 2% v/v CH$_3$CN.

[g] Data are from Reference 119.

[h] Data are from Reference 77 and the concentration of organic cosolvent is expressed in percent weight (w/w).

[i] Data are from Reference 84 and the concentration of organic cosolvent is expressed in mole fractions.

malism). It is interesting to note that although the value of $k_W(2)/k_W(3) = 0.4$, the values of $k_M(2)/k_M(3) = 4$ and 1 for CTABr and SDS micelles, respectively. Similarly, the values of k_M for 2 are 6.7×10^5, 12.6×10^{-5}, and 14.5×10^{-5} sec^{-1} in CTABr, DTABr, and CTACl micelles, respectively.

The rate-retarding effects of cationic micelles (CTABr and DTABr) on the rate of uncatalyzed hydrolysis of p-substituted 1-benzoyl-1,2,4-triazoles and 2 have been shown to be caused by the high concentration of head groups as well as by hydrophobic tails in the Stern region where micellized 1-benzoyl-1,2,4-triazoles and 2 molecules are assumed to exist.[78] This conclusion is derived by the use of a fairly new model of solutions for CTABr and DTABr micelles containing both tetramethylammonium bromide, mimicking micellar head groups, and 1-propanol, mimicking micellar hydrophobic tails in Stern region. Rate constants of spontaneous hydrolysis of methyl benzenesulfonate, 2-adamantyl and pinacoyl 4-nitrobenzenesulfonate, phenyl and 4-nitrophenyl chloroformate, and bis(4-nitrophenyl) carbonate in the presence of zwitterionic, cationic, and anionic micelles are smaller in micellar pseudophase than in aqueous phase.[79] Reactions with extensive bond-breaking in the transition state (S_N1 hydrolysis) are faster in SDS than in cationic and sulfobetaine micelles, but the other cases of hydrolysis, which involve significant bond-making, are slower in SDS. Rate constants are similar in cationic and sulfobetaine micelles. These micellar charge effects are ascribed to interactions of the polar transition states with the asymmetric charged interfacial region, which complement effects of the lower polarities of micelles relative to water.

The presence of cationic micelles (CTABr, CTEABr, CTPABr and CTBABr, alkyl = Me, Et, n-Pr and n-Bu, respectively) enhances the rate of unimolecular decarboxylation of 6-nitro-5-alkoxybenzisoxazole-3-carboxylate ion (**1,OMe** and **1,OTD**, alkoxy = MeO and n-C$_{14}$H$_{29}$O, respectively).[80] First-order rate constants (k_{obs}) for cleavage of **1,OMe** increase monotonically with [surfactant] and become independent of [surfactant] when the substrate is fully micelle bound and $k_{obs} = k_M$. The values of k_{obs} for **1,OTD** increase sharply with increasing [surfactant] and reach well-defined maxima at [surfactant] at or below the CMC, before decreasing to values corresponding to $k_{obs} = k_M$. The magnitude of rate maxima and increase in k_M are in the sequence CTABr < CTEABr < CTPABr < CTBABr. The rate maxima are due to formation of premicellar complexes of substrate with one or a few surfactant monomers, but they dissolve in micelles at higher [surfactant]. Various probable reasons for the lower reactivities in micelles as compared with the premicellar species have been ruled out. The most likely reason for these observations involves the assumption that in premicelles the carboxylate ion is shielded from water, and in the transition state charge dispersion is favored by interaction with the quaternary ammonium ion. Arrangements that fit these conditions are shown in cartoon form in Figure 4.2. The alkyl groups are shown in linear, extended conformations, although this is a crude approximation. Probably, even in micelles and premicelles, the alkyl groups have globular-like conformations to decrease water–hydrocarbon contact. Figure 4.2 shows two surfactant monomers associating with the substrate, but the stoichiometry is uncertain

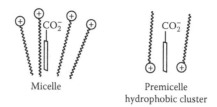

Micelle | Premicelle
hydrophobic cluster

FIGURE 4.2 A molecular cartoon showing the location of anionic solubilizate in the cationic micelle and cationic premicelle hydrophobic cluster.

and assemblies of various compositions can exist. Association between very dilute surfactants and ionic dyes is well established,[81] and 1:1 stoichiometry of assemblies is often observed. There are premicellar rate enhancements of unimolecular hydrolysis of a 2,4-dinitrophenyl phosphate dianion,[82] and the explanation of these results is essentially similar to one described for cationic micellar-mediated unimolecular decarboxylation of **1,OTD**.[80]

The rate of spontaneous hydrolysis of phenyl chloroformate decreases with increase in the concentration of micelles of various anionic, nonionic, and cationic surfactants.[83] A comparison of the kinetic data in nonionic micellar solutions to those in anionic and zwitterionic micellar solutions makes it clear that charge effects of micelles is not the only factor responsible for the variations in the reaction rates. Depletion of water in the interfacial region and its different characteristics as compared to bulk water, the presence of high ionic concentration in the Stern layer of ionic micelles, and differences in the stabilization of the reactant state and the transition state by hydrophobic interactions with surfactant tails can also influence reactivity.

The effects of ethanol on the rates of decarboxylation of **1** in aqueous cationic surfactants of different counterions and head groups could be explained in terms of PP model only up to $X_{EtOH} = 0.1$ (where X_{EtOH} represents mole fraction of ethanol in mixed aqueous solvent).[84] The calculated values of k_M and K_S from Equation 3.2 (Chapter 3) decrease from 8.5×10^{-3} to 5.2×10^{-3} sec^{-1} and 700 to 40 M^{-1}, respectively, whereas k_M/k_W values decrease from 2800 to 650 with increase in X_{EtOH} from 0 to 0.099. In water, the increase in the size of the surfactant head group leads to an increase in micellar catalysis owing to decreased water content at the interface[57c]; in ethanol-modified micelles, the head group growth leads to similar effects. Modifications of the micellar structure induced by ethanol, such as an increase in micellar charge, could also be responsible for this rate effect. The reactant state carboxylate ion is hard and does not interact strongly with soft quaternary ammonium ion, if water is available. However, transition state formation generates charge-delocalization soft anion, which interacts more readily with the cationic head groups.[80] These arguments predict that k_M should increase with the increase in X_{EtOH}, but the values of k_M show decrease (~40%) with increase in X_{EtOH} from 0 to 0.1.[84] First-order rate constants (k_{obs}) for the pH-independent hydrolysis of 4-nitrophenyl 2,2-dichloropropionate (NPDCP) in the presence of aqueous micelles of sodium dodecyl sulfate, sodium dodecylbenzenesulfonate, alkyltrimethylammonium chlorides, and alkyldimethylbenzylammonium chlorides (alkyl = cetyl and dodecyl) decrease with increase in [micelles], and this decrease is ascribed to preferential destabilization of transition state by micellar reaction medium.[85] These observations are different from those observed for pH-independent hydrolysis of 4-nitrophenyl chloroformate (NPCF) where rate of hydrolysis is faster in cationic micelles than in bulk water. The observed results of $k_M < k_W$ and $k_M > k_W$ for pH-independent hydrolysis of NPDCP and NPCF, respectively, may be rationalized in terms of different locations of carbonyl groups of NPDCP and NPCF in the cationic micelles.

First-order rate constants (k_{obs}) for hydrolysis of ethyl cyclohexanone-2-carboxylate (ECHC) in acidic medium (at 0.05-M HCl) are 20- to 40-fold and ~10-fold (in aqueous phase) larger than those in cationic[47] and anionic[86] micellar pseudophase, respectively. These rate-retarding micellar effects may partially or fully be attributed to micellar medium effect because a 34-fold decrease in k_{obs} for hydrolysis of ECHC has been observed with increase in dioxane from 0 to 10 M in mixed aqueous solvents containing 0.083-M HCl.[47] The rate of hydrolysis of ECHC in the absence of micelles involves uncatalyzed, specific-acid-catalyzed, and general-base-catalyzed reaction steps, and the reaction mechanism does not change with change in solvent from pure water to mixed water–organic solvents with organic solvent content as large as 83% v/v. The value of first-order rate constant (k_0) for uncatalyzed hydrolysis of ECHC is 9.87×10^{-4} sec^{-1} at 25°C which is ~ 2-fold larger than k_0 (at 30°C) for hydrolysis of ionized ethyl salicylate in which the occurrence of intramolecular general base catalysis has been unequivocally ascertained.[87] The value of k_0 for hydrolysis of 4-nitrophenyl acetate is 5.5×10^{-7} sec^{-1} at 25°C[88], and the value of k_0 for hydrolysis of ethyl α-cyclohexylacetate (ECA) must be much smaller than 5.5×10^{-7} sec^{-1}. Thus, more than 10^4-fold larger value of k_0 for ECHC than that for ECA is most likely owing to the occurrence of intramolecular general acid catalysis through enol form of ECHC as shown by **TS**.

TS

4.3 BIMOLECULAR REACTIONS

Within the domain of PP model, the general reaction scheme for a bimolecular reaction between reactants, S and R, in the presence of micelle, D_n, is shown in Scheme 3.8 (Chapter 3). Almost all the kinetic data, obtained for micellar-mediated bimolecular reactions, have been discussed in terms of PP or extended

PP model of micelle. This model generally gives a satisfactory fit of observed data in terms of residual errors (= $k_{obs\ i} - k_{calcd\ i}$, where $k_{obs\ i}$ and $k_{calcd\ i}$ are, at the i-th independent reaction variables such as $[D_n]$, experimentally determined and calculated [in terms of micellar kinetic model] rate constants, respectively). The model also provides plausible values of kinetic parameters such as micellar binding constants of reactant molecules and rate constants for the reactions in the micellar pseudophase. The deviations of observed data points from reasonably good fit to a kinetic equation derived in terms of PP model for a specific bimolecular reaction under a specific reaction condition are generally understandable in view of the known limitations of the model. Such deviations provide indirect information regarding the fine, detailed structural features of micelles.

For bimolecular reactions involving counterions as one of the reactants, the micellar rate enhancements, relative to reaction in water, decrease significantly as the total concentration of reactive ion is increased, because pseudo-first-order rate constant (k_{obs}) in water increases approximately linearly with the total concentration of reactive counterion, but ionic-reactant concentration and, therefore, the rate constant for the reaction at the micellar interface increase more slowly.[2f,89] Reactions of dilute hydrophilic co-ions are strongly micellar inhibited, but inhibition decreases significantly as the total co-ion concentration is increased. The kinetic results are readily fitted by theoretical treatments.[2f,90] Although most kinetic data have been obtained for organic reactions, the general treatment used for organic reactions also fits results on inorganic reactions.[91]

4.3.1 MICELLES OF NONIONIC NONFUNCTIONAL SURFACTANTS

The rate constants for the reactions of a variety of polar molecules are only weakly affected by the nonionic micelles as predicted by Hartley-formulated rules for micellar-mediated reactions.[3,92] However, although the effect of nonionic micelles (dinonylphenol condensed with 24 ethylene oxide units) on the rate of reaction of HO^- with 2,4-dinitrochlorobenzene is very small,[93] the rates of reactions of HO^- and F^- with neutral p-nitrophenyl diphenyl phosphinate (**4**) are strongly inhibited by the micelles of $C_{12}E_{10}$ ($C_{12}E_{10}$ represents polyoxyethylene 10 lauryl ether, $C_{12}H_{25}(OCH_2CH_2)_9OCH_2CH_2OH$), and $C_{12}E_{23}$.[94] The rate constants (k_{obs}) decrease sharply in very dilute surfactant solution and become approximately constant as **4** becomes fully micellar bound. These inhibitions are explained in terms of greater ground state stabilization of the reactant ester in the micellar pseudophase than in the aqueous pseudophase. Strong inhibition of the alkaline hydrolysis of bis-2,4-dinitrophenyl phosphate (**5**) by a nonionic detergent (Triton X-114) has also been observed.[95]

Rates of reactions of HO^- and F^- with neutral ester p-nitrophenyl diphenyl phosphate (**6**) are also strongly inhibited by very dilute micellar solutions of $C_{12}E_{10}$ and $C_{12}E_{23}$, but the rate constants (k_{obs}) become independent of surfactant concentrations within $[C_{12}E_{10}]_T$ or $[C_{12}E_{23}]_T$ range of 0.003 to $\leq 0.20\ M$ at [KOH] or [NaF] ranging from ≥ 0.01 to 0.1 M.[96] Inert anions such as Cl^- and Br^- have essentially no effect, whereas ClO_4^- anions inhibit the rate of hydrolysis of **6** in the presence of $C_{12}E_{10}$ micelles at 0.1 M NaOH. Based on the binding constant of **6** to the micelles of Igepal[97], the assumption is made that **6** will be fully micellar bound at 0.002 M surfactant. The constancy of k_{obs} over a range of $[C_{12}E_{10}]_T$ and $[C_{12}E_{23}]_T$ indicates that concentrations of HO$^-$ and F$^-$ in the micellar environment in which $\mathbf{6}_M$ molecules reside do not decrease with increasing surfactant concentration, which is against the prediction of PP model of micelle; i.e., Equation 3.11 to Equation 3.15 (Chapter 3) do not predict a constancy in k_{obs} with the increase in the concentration of micelle-forming surfactant after a substrate such as **6** is fully micellar bound (a situation where $K_S [D_n] \gg 1$). The nonionic micellar inhibition of the rates of reactions of HO$^-$ and F$^-$ with **6** is suggested to depletion of HO$^-$ and F$^-$ in the palisade layer relative to water.

At high pH, the terminal OH of $C_{12}E_{10}$ or $C_{12}E_{23}$ surfactant is largely deprotonated and, as such, it should compete with HO$^-$ in nucleophilic reactivity towards reactive molecules such as **4** to **6** and 2,4-dinitrochlorobenzene (**7**). But such reaction of terminal alkoxide ion of $C_{12}E_{10}$ or $C_{12}E_{23}$ surfactant with **6**[96,98] was not observed. However, considerable amounts of ether are formed in the aqueous cleavage of **7** in $C_{12}E_{10}$ and $C_{12}E_{23}$ at high pH by the reaction of alkoxide ion with **7**. The differences in the chemistries of reactions of **6** and **7** in $C_{12}E_{10}$ and $C_{12}E_{23}$ micelles at high pH are ascribed to the differences in the locations of these substrates in the nonionic micelles.[96] The **6** molecules, being highly hydrophobic, are dragged deeper inside the micelles in which the approach of terminal alkoxide ion is sterically inaccessible, whereas **7** molecules, being relatively less hydrophobic, remain in the micellar environment where the approach of terminal alkoxide ion is easily accessible.

The rate of aqueous cleavage of securinine (**8**), the major alkaloid from *Breynia coronata*, showed substantial decrease with the increase in $[C_{12}E_{10}]$ from 0.0 to 0.2 M in the presence of 0.05 M NaOH at 35°C.[99] The rate of alkaline hydrolysis of **8** was found to be first order with respect to each of the two reactants, **8** and HO$^-$, and the rate of uncatalyzed (i.e., solvent-assisted) hydrolysis was negligible compared with the HO$^-$ ion-catalyzed one even at [NaOH] = 0.02 M in the absence of micelles.[100] Thus, the hydrolysis of **8** in the presence of $C_{12}E_{10}$ micelles at 0.05 M NaOH should involve **8** and HO$^-$ as the reactants. The values of k_{obs} strictly followed Equation 3.14 (Chapter 3), which revealed that either m_R^s (R = OH) $\approx$ 0 or $k_{2,W} \approx k_M^{mr} K_S$. The least-squares fit of k_{obs} to Equation 3.14 gave K_S and CMC as 14.8 M^{-1} and 0.006 M, respectively. Although it is difficult to provide definite reasons for ruling out the possibility that $k_{2,W} \approx k_M^{mr} K_S$, it seems to be an unlikely possibility for the following reason. The effects of SD, CTABr, and TTABr micelles were also studied under similar experimental conditions, and the observed kinetic data obeyed Equation 3.14 with least-squares calculated values of K_S for SDS, CTABr, and TTABr as 32, 22, and 9 M^{-1}, respectively. If the possibility of $k_{2,W} \approx k_M^{mr} K_S$ is true, then it is difficult to explain 32-, 22-, 9-, and 15-fold smaller values of k_M^{mr} compared to $k_{2,W}$ in SDS, CTABr, TTABr, and $C_{12}E_{10}$ micelles, respectively. Thus, the most obvious possibility is concluded to be m_R^s (R = OH) $\approx$ 0. But it is perhaps essential to note that kinetically $m_R^s \approx 0$ means essentially complete absence of HO ions in the micellar region of micellized **8** molecules (8_M). It does not show whether or not [HO⁻] = 0 in the micellar region where 8_M molecules do not exist.

Absence or kinetically insignificant rate of micellar-mediated hydrolysis of **8** is attributed to (1) the different average locations of reactants, HO_M^- and 8_M, in the micellar pseudophase and (2) the location of 8_M molecules in the micellar region of low dielectric constant. The considerably low value of K_S for highly hydrophobic **8** molecules is ascribed to probable significant steric interaction between highly hindered **8** molecule and methylene units of micelle-forming straight-chain surfactant ($C_{12}E_{10}$).

$Ph_2PO.OC_6H_4NO_2$ (4–)

4

$(ArO)_2PO_2^-$ where Ar =

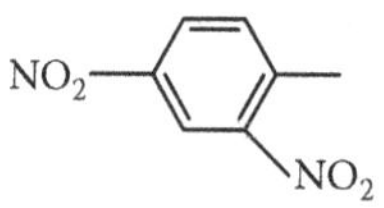

5

$(PhO)_2PO.OC_6H_4NO_2$ (4 –)

6

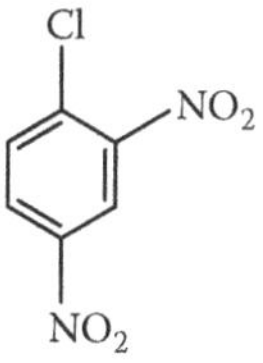

7

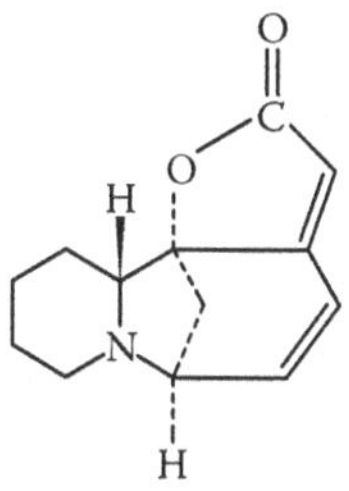

8

Pseudo-first-order rate constants (k_{obs}) for alkaline hydrolysis of phenyl benzoate (**9**), obtained at different $[C_{12}E_{23}]_T$ ranging from 0.0 to 0.01 M in the presence of 0.01-M NaOH, were found to fit to Equation 3.14 (Chapter 3) with least-squares calculated values of K_S and CMC as 880 M^{-1} and 4 × 10⁻⁵ M, respectively, at 35°C.[35] An attempt to obtain k_{obs} at ≥ 0.02 M $C_{12}E_{23}$ in the presence of 0.01-M NaOH failed because the absorbance at 290 nm (the wavelength at which the rate of formation of the product phenolate ion was monitored, and the molar extinction coefficient for nonionized phenol is nearly zero at 290 nm) increased only slightly (~10%) within the reaction period of nearly 3 h. Spectral study revealed that [HO⁻] in the micellar environment of micellized product phenol ($PhOH_M$) was enough to produce only ~50% product phenolate ion (PhO^-) at 0.015-M $C_{12}E_{23}$ and 0.01-M NaOH.[35] It is certain that [HO⁻] must be lower in the micellar environment of 9_M molecules than that of micellized products, phenol and benzoic acid, because these products are less hydrophobic compared to **9**. It has been shown elsewhere[23] that the hydrolysis reaction, under such conditions, involves **9** and HO⁻ as the reactants. Thus, the fitting of k_{obs} to Equation 3.14 shows that either $k_{2,W} \approx k_M^{mr} K_S$ or m_R^s (R = OH) ≈ 0. The possibility that $k_{2,W} \approx k_M^{mr} K_S$ gives $k_{2,W} /k_M^{mr} = K_S = 880$ M^{-1}, which shows the insignificant rate of hydrolysis in micellar pseudophase compared with the aqueous pseudophase. The possible reason for the much slower rate of hydrolysis, or its total absence, in micellar pseudophase is attributed to kinetically insignificant amount of HO^- ions in the vicinity of 9_M molecules. The 9_M molecules, being highly hydrophobic, are expected to be dragged deeper inside the micellar pseudophase, whereas HO^- ions, being highly hydrophilic, are expected to remain in considerably polar and hydrophilic regions of micellar pseudophase.

The value of calculated rate constant (k_{calcd}) using Equation 3.14 with $K_S =$ 880 M^{-1}, $k_{2,w} = 0.722$ M^{-1} sec^{-1}, $[HO^-] = 0.01$ M, and 10^5 CMC = 4 M is 3.9 $\times$ 10^{-4} sec^{-1} at 0.02 M $C_{12}E_{23}$. This value of k_{calcd} indicates that the absorbance (A_{obs}) at 290 nm should increase from ~0.10 to ~ 0.58 with the increase in the reaction time (t) from 0 to 3 h at 0.02 M $C_{12}E_{23}$, provided the product phenol is fully ionized. It may be shown qualitatively that the product phenol must be fully ionized under the conditions in which $k_{obs} = 3.9 \times 10^{-4}$ sec^{-1} and initial concentration of **9** is 2×10^{-4} M. But the observed values of A_{obs} increased from 0.10 to 0.15 with the increase in t from 0 to 3 h. It should be noted that phenol is apparently more hydrophilic than **9** and, hence, the location in micellar pseudophase for phenol should be more polar than that for **9**. This analysis shows that the K_S value no longer remains independent of $[C_{12}E_{23}]_T$ at its value ≥ 0.015 M, which is unusual in the domain of PP model of micelle.[2c–2g]

The increase in $[C_{12}E_{23}]_T$ from 0 to ≤ 0.07 M resulted in a monotonic decrease in k_{obs} for hydrolysis of 4-nitrophthalimide (**10**) at 0.01 and 0.03-M NaOH in aqueous solvents containing 1% v/v CH_3CN. But k_{obs} could not be obtained at $[C_{12}E_{23}]_T \geq 0.03$ M in the presence of 0.01-M NaOH, because the rate of hydrolysis of **10** became too slow to determine k_{obs} with an acceptable degree of precision under such conditions. However, k_{obs} could be conveniently obtained until the maximum value of 0.07 M of $[C_{12}E_{23}]_T$ at 0.03-M NaOH.[101]

The rate of aqueous cleavage of **10** at [NaOH] ≥ 0.01 M in the absence of micelles follows the reaction mechanism, which can briefly be shown as in Scheme 4.3 where $K_i' = [H_2O]$ K_a/K_w with $K_a = [10^-]$ $[H^+]/[10]$ and $K_w =$ $[HO][H^+]$. Observed pseudo-first-order rate law and Scheme 4.3 can lead to Equation 4.7

$$k_{obs} = k_0 + k_{OH} [HO^-] \qquad (4.7)$$

where $k_0 = k_{OH}'$ K_w/K_a. It should be noted that the term k_{H2O} $[H_2O][10^-]$ in the rate law, which is kinetically indistinguishable from k_{OH}' $[HO^-][10]$, may be ruled out based on evidence described elsewhere.[102] Thus, one of the reactants in the alkaline hydrolysis of 4-nitrophthalimide is HO^-. It is evident from a huge amount of kinetic data on micellar-mediated organic reactions that the reaction mechanism of a particular reaction remains generally the same in both aqueous and micellar pseudophases. The values of k_0 (= 2.0×10^{-3} sec^{-1}) and k_{OH} (= 46.3×10^{-3} M^{-1} sec^{-1}), obtained in the absence of micelles,[103] show that the contribution of k_0 toward k_{obs} at 0.01 and 0.03 M NaOH are ~80% and ~60%, respectively.

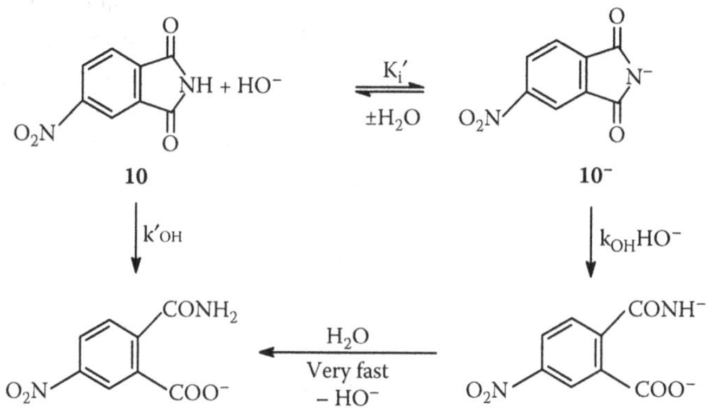

SCHEME 4.3

$$S_W^- + D_n \underset{}{\overset{K_1}{\rightleftharpoons}} S_M^-$$

$$S_W^- \underset{K_{a,W}}{\overset{\pm a_{H,W}}{\rightleftharpoons}} SH_W \qquad S_M^- \underset{K_{a,M}}{\overset{\pm a_{H,M}}{\rightleftharpoons}} SH_M$$

$$SH_W + D_n \overset{K_2}{\rightleftharpoons} SH_M$$

$$R_W + D_n \overset{K_R}{\rightleftharpoons} R_M$$

$$SH_W + R_W \xrightarrow{k'_W{}^2} \text{Product(s)}$$

$$SH_M + R_M \xrightarrow{k'_M{}^2} \text{Product(s)}$$

$$S_W^- + R_W \xrightarrow{k_W{}^2} \text{Product(s)}$$

$$S_M^- + R_M \xrightarrow{k_M{}^2} \text{Product(s)}$$

SCHEME 4.4

For reactions that follow the reaction mechanism shown by Scheme 4.3 in the presence of micelles, Scheme 3.8 (Chapter 3) should be modified as shown in Scheme 4.4.

Observed rate law (rate = k_{obs} $[S]_T$) and Scheme 4.4 can lead to Equation 4.8

$$k_{obs} = \frac{(k_W'^2 f_W^{SH} + k_W^2 f_W^{S-})[R_W] + (k_M^{mr,SH} f_M^{SH} + k_M^{mr,S-} f_M^{S-}) M_R K_S^{app}[D_n]}{1 + K_S^{app}[D_n]} \tag{4.8}$$

where $k_M^{mr,SH} = k_M'^2/V_M$, $k_M^{mr,S-} = k_M^2/V_M$, V_M is the molar volume of the micellar reaction region, $f_M^{S-} = K_{a,M}/(a_{H.M} + K_{a,M}) = [S_M^-]/[S_M]_T$ with $[S_M]_T = [SH_M] + [S_M^-]$, $f_W^{S-} = K_{a,W}/(a_{H.W} + K_{a,W}) = [S_W^-]/[S_W]_T$ with $[S_W]_T = [SH_W] + [S_W^-]$, $f_M^{SH} = 1 - f_M^{S-}$, $f_W^{SH} = 1 - f_W^{S-}$, $a_{H,W}$ and $a_{H,M}$ are activity of proton in the aqueous pseudophase and the micellar pseudophase, respectively, and $K_S^{app} = [S_M^-]_T/\{[D_n][S_W]_T\} = ([S_M] + [SH_M])/\{[D_n]([S_W^-] + [SH_W])\}$. Equation 4.8 can be rearranged to give Equation 4.9 provided $1 >> K_R [D_n]$ under specific experimental conditions.

$$k_{obs} = \frac{(k_W'^2 f_W^{SH} + k_W^2 f_W^{S-})[R]_T + (k_m^{mr,SH} f_M^{SH} + k_M^{mr,S-} f_M^{S-})K_R[R]_T K_S^{app}[D_n]}{1 + K_S^{app}[D_n]} \quad (4.9)$$

It is noteworthy that when $R = HO^-$, then kinetic terms $k_W'^2[SH_W][HO_W^-]$ and $k_M'^2[SH_M][HO_M^-]$ are kinetically indistinguishable from $k_W^2[S_W^-][H_2O_W]$ and $k_M^2[S_M^-][H_2O_M]$, respectively. For such reaction systems, and under a specific reaction condition, if $f_W^{SH} \to 0$, then $k_W'^2 f_W^{SH} [HO_W^-] \to k_W'^2 K_{w,w}/K_{a,W}$ or $\to k_W^2 [H_2O_W]$ and similarly, if $f_M^{SH} \to 0$, then $k_M'^2 f_M^{SH} [HO_M^-] \to k_M'^2 K_{w,M}/K_{a,M}$ or $\to k_M^2 [H_2O_M]$ where K_w represents ionic product of water. Mathematically, if $f_M^{SH} \to 0$, or $k_M'^2 \to 0$ then $k_M'^2 f_M^{SH} \to 0$. However, in terms of physical reality, if $f_M^{SH} \to 0$, then $k_M^2 f_M^{SH}$, depending upon the magnitude of $k_M'^2$, may or may not tend to 0, but if $k_M'^2 \to 0$ then $k_M'^2 f_M^{SH}$ must tend to 0.

The values of k_{obs}, obtained within $[C_{12}E_{23}]_T$ range of 1×10^{-4} to 0.07 M at 0.03 M NaOH, were found to fit to Equation 4.9 with CMC ≈ 0 and the nonlinear least-squares calculated values of $(k_M^{mr,SH} f_M^{SH} + k_M^{mr,S-} f_M^{S-}) K_R [R]_T$ (where $[R]_T = [NaOH] = 0.03$ and $[S]_T = [10]_T = 1 \times 10^{-4}$ M) and K_S^{app} are $(5.7 \pm 2.2) \times 10^{-4}$ sec^{-1} and 33 ± 6 M^{-1}, respectively.[89] The value of $(k_M^{mr,SH} f_M^{SH} + k_M^{mr,S-} f_M^{S-}) K_R [R]_T$ is associated with a considerably large standard deviation ($\sim$40%), which may be due to very low contribution of $(k_M^{mr,SH} f_M^{SH} + k_M^{mr,S-} f_M^{S-}) K_R [R]_T K_S^{app} [D_n]$ compared to $(k_{2,W}' f_W^{SH} + k_{2,W} f_W^{S-})[R]_T$ in Equation 4.9. The maximum contribution of $(k_M^{mr,SH} f_M^{SH} + k_M^{mr,S-} f_M^{S-}) K_R [R]_T K_S^{app} [D_n]$, which is at the maximum value of $[C_{12}E_{23}]_T$ (= 0.07 M) attained in the study, is only 30%. Therefore, the observed data were also treated with Equation 4.10,

$$\{(k_W'^2 f_W^{SH} + k_W^2 f_W^{S-})[R]_T\} / k_{obs} = A + K_S^{app}[C_{12}E_{23}]_T \quad (4.10)$$

which may be derived from Equation 4.9 with $(k_M^{mr,SH} f_M^{SH} + k_M^{mr,S-} f_M^{S-}) K_R [R]_T K_S^{app} [D_n] = 0$. The data seem to fit reasonably well to Equation 4.10 as is evident from the plots of $\{(k_W'^2 f_W^{SH} + k_W^2 f_W^{S-})[R]_T\}/k_{obs}$ vs. $[C_{12}E_{23}]_T$ in Figure 4.3. The least-squares calculated value of A and K_S^{app} are 1.07 ± 0.02 and 19 ± 1 M^{-1}, respectively, with k_{obs} values within $[C_{12}E_{23}]_T$ range of 1×10^{-4} to 0.07 M.

The respective value of A and K_S^{app} changed to 1.05 ± 0.01 and 20 ± 1 M^{-1} when the k_{obs} values were considered within $[C_{12}E_{23}]_T$ range of 1×10^{-4} to 0.04 M.

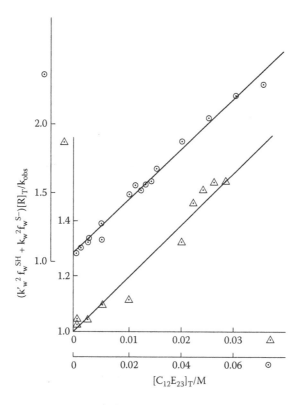

FIGURE 4.3 Plots showing the dependence of $(k_w'^2 f_w^{SH} + k_w^2 f_w^{S-})[R]_T/k_{obs}$ upon total concentration of nonionic surfactant $([C_{12}E_{23}]_T)$ for alkaline hydrolysis of **10** at 0.01-*M* NaOH (Δ) and 0.03-*M* NaOH (O). Solid lines are drawn through the least-squares calculated data points using Equation 4.10 as described in the text.

The values of k_{obs}, obtained within $[C_{12}E_{23}]_T$ range of 1×10^{-4} to 0.028 *M* at 0.01 *M* NaOH, could not fit to Equation 4.9 apparently owing to insufficient data points for the nonlinear least-squares data fit to Equation 4.9. However, these data fit well to Equation 4.10 as is evident from the plot of Figure 4.3. The least-squares calculated values of A and K_S^{app} are 1.01 ± 0.02 and 19 ± 1 M^{-1}, respectively. The calculated values of K_S^{app} from Equation 4.10 are the same at both 0.01 and 0.03 *M* NaOH. The most obvious reason for the lack of hydrolysis of **10** at $[C_{12}E_{23}]_T \geq 0.03$ *M* in the presence of 0.01-*M* NaOH is described as follows.

The calculated value of K_S^{app} is a complex function of K_1 and K_2 as described by Equation 4.4. But the micellar binding constant (K_2) of less polar **10** should be many times larger than that (K_1) of more polar **10⁻** in view of the reported values of $C_{12}E_{23}$ micellar binding constants of nonionized and ionized peracids.[36,104] Because of the larger polarity of **10⁻** compared to that of **10**, the micellar location of **10** should be more hydrophobic compared to that of **10⁻**. It is obvious that at a constant [NaOH], the increase in the concentration of $C_{12}E_{23}$ micelles

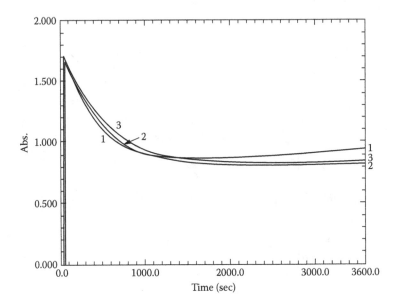

FIGURE 4.4 Plots of absorbance, Abs. (of reaction mixture at 260 nm) vs. Time (reaction time) for hydrolysis of **10** at 0.010 (**1**), 0.020 (**2**), and 0.022-M $C_{12}E_{23}$ (**3**) in mixed aqueous solvent containing 1% v/v CH_3CN, 0.03 M NaOH, and 1×10^{-4} M **10**.

should decrease m_{OH}^s ($= [HO_M^-]/[D_n]$) and probably $[HO_M^-]$ at larger values of $[C_{12}E_{23}]_T$. In terms of PP model of micelle, the value of $[HO_M^-]$ should be proportional to the total concentration of HO^- (i.e., $[HO^-]_T$). It appears that at $[C_{12}E_{23}]_T$ ≥ 0.03 M and 0.01 M NaOH, the magnitude of $[HO_M^-]$ in the vicinity of 10_M, and probably 10_M^-, became too low to cause a kinetically detectable hydrolysis.

The plots of Abs. (observed absorbance) vs. Time (reaction time) as shown in Figure 4.4 show a rapid monotonic decrease in Abs. in the initial phase of the reaction followed by a very slow monotonic increase in Abs. with Time in the latter phase of the reaction at 260 nm.

The rapid decrease in the initial phase of the reaction is due to the formation of immediate product 4-nitrophthalamic acid (**11**) as shown in Scheme 4.3. However, such a minimum is distinctly clear only at 0.01-M $C_{12}E_{23}$, whereas it is not so clear at 0.02- and 0.022-M $C_{12}E_{23}$ (Figure 4.4). The occurrence of minimum in the plot of Figure 4.4 demonstrates the formation of a stable intermediate product, which absorbs significantly at 260 nm. The most probable intermediate is 4-nitrophthalic anhydride (**12**), because in a related study on phthalimide, evidence has been given for the formation of phthalic anhydride in the hydrolytic cleavage of phthalimide in the presence of mixed $C_{12}E_{23}$/CTABr micelles at 0.02-M NaOH.[105] The brief mechanism for the formation of **12** from hydrolysis product (**11**) of **10** may be shown in Scheme 4.5.[106] The decrease in the magnitude or size of minima in the plots of Figure 4.4 with the increase in $[C_{12}E_{23}]_T$ is attributed to the consequence of the effect of $[C_{12}E_{23}]_T$ (**1**) on the

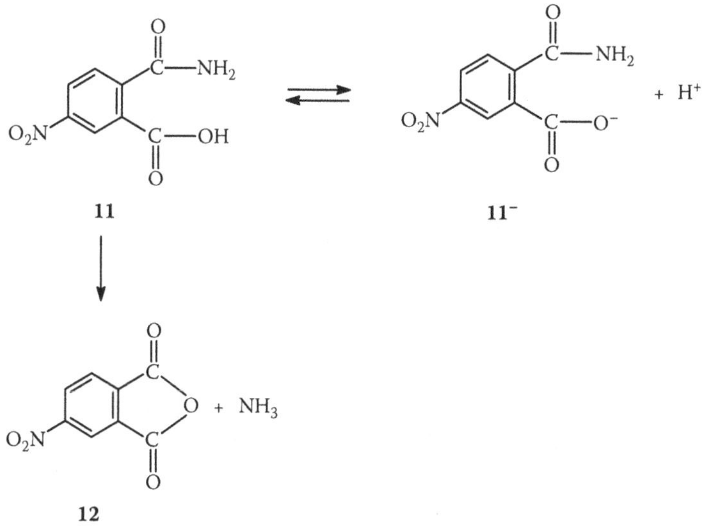

SCHEME 4.5

rate of formation of **11** from **10** and (2) on the pH of the micellar environment where 11_M molecules reside. The formation of **12** is an indirect evidence for the proposal that the increase in $[C_{12}E_{23}]_T$ causes the depletion of hydroxide ions from the micellar environment of micellized hydrophobic solubilizate or alternatively causes hydrophobic solubilizate to move from a more to a less polar micellar environment.

The presence of 0.01- to 0.03-M NaOH in $C_{12}E_{23}$ micellar solution can cause ionization of hydroxyl group of $C_{12}E_{23}$ surfactant. Although the apparent concentration of ionized $C_{12}E_{23}$ surfactant should be extremely low under such conditions, owing to the fact that the effective or local concentration of these anionic groups at the micellar surface may be considerably high and methoxide ion reacts nearly 60-fold larger than HO$^-$ with carbonyl carbon of esters[107], the probable reaction between ionized hydroxyl group of $C_{12}E_{23}$ surfactant and 10_M as shown in Scheme 4.6 may not be completely ruled out. However, it has been shown elsewhere[108] that $k_{-1}^8/k_1^8 > 10^3$ for the reaction of CH_3O^- with phthalimide owing to occurrence of intramolecular nucleophilic addition–elimination mechanism in k_{-1}^8-step.[109] Thus, it is apparent that even if the reaction in Scheme 4.6 is significant, the amount of **13** would be negligible compared with that of 10_M during the aqueous cleavage of 10_M.

The effect of $[C_{16}E_{20}]_T$ on k_{obs} for aqueous cleavage of phthalimide (**14**) at 0.02 M NaOH is explained in terms of pseudophase model of micelle where rate of reaction in the micellar pseudophase is negligible compared with that in the aqueous pseudophase.[105] On the other hand, the values of k_{obs} for $C_{12}E_{23}$ show a sharp decrease at very low values of $[C_{12}E_{23}]_T$ followed by a very slow decrease in k_{obs} with the increase in $[C_{12}E_{23}]_T$ at its relatively high values. The rate of

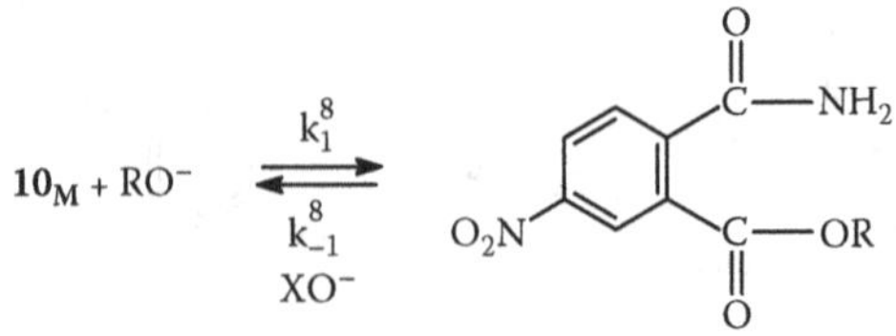

SCHEME 4.6

$XO^- = HO^-$ and RO^-

$RO^- = C_{12}H_{25}O(CH_2CH_2O)_{22}CH_2CH_2O^-$

hydrolysis of **14** becomes too slow to monitor at $[C_{12}E_{23}]_T \geq 0.04\ M$ in the presence of $0.02\ M$ NaOH, whereas such a characteristic behavior is kinetically absent with $C_{16}E_{20}$ surfactant. The insignificant rate of hydrolysis of **14** in the micellar pseudophase compared to that in the aqueous pseudophase is attributed to (1) extremely low concentration of hydroxide ions in the vicinity of 14_M, (2) different average locations of HO^- and **14** molecules in the micellar pseudophase, and (3) micellar medium polarity effect. The most unusual observation is the depletion of hydroxide ions and water molecules from $C_{12}E_{23}$ micellar environment of 14_M molecules and, consequently, irreversible trapping of **14** and its hydrolysis product, phthalamic acid (**15**) by micelles at higher concentration of $C_{12}E_{23}$.

Pseudo-first-order rate constants for bimolecular iodide reduction of peracids in the presence of nonionic surfactants ($C_{12}E_{23}$ and Triton X-100) show monotonic increase with the increase in $[C_{12}E_{23}]_T$ and $[\text{Triton X-100}]_T$ from 0.0 to 0.04 M at pH ranging from 5.69 to 8.39.[36] These data have been explained in terms of multiple micellar pseudophase (MMPP) kinetic model that is a generalization of the classic PP model. The kinetic equation derived based upon MMPP model is of the same form as that Equation 3.11 obtained from the PP model. The values of k_{obs} become independent of $[C_{12}E_{23}]_T$ and $[\text{Triton X-100}]_T$ at optimum concentrations of surfactants where substrates (peracids) are almost fully micellar bound. These experimental results are explained by assuming that the micellar association constant with iodide is small (i.e., $1 \gg K_S [D_n]$) within the $[D_n]$ range imposed in the study. A nearly 3- to 4-fold larger value of k_{obs} in micellar pseudophase compared to that in the aqueous pseudophase (i.e., $k_M{}^{mr}K_R/k_{2,W} \approx 3$ to 4) is attributed to the stabilization of transition state and destabilization of the ground state by the terminal OH groups due to the polydisperse nature of Triton X-100 and most likely $C_{12}E_{23}$ surfactants in micelles. The micellar association constants for nonionized and ionized peracids and the corresponding parent acids are shown in Table 4.1.

Pseudo-first-order rate constants (k_{obs}) for the reaction of *n*-nonanoyloxybenzenesulfonate and hydrogen peroxide are strongly inhibited by $C_{12}E_{23}$ micelles. The observed data (k_{obs} vs. $[C_{12}E_{23}]_T$) were treated with a kinetic equation similar to Equation 3.11 (Chapter 3).[110] The micellar binding constants for reactants, obtained kinetically, are mentioned in Table 4.1. The plots of pseudo-first-order rate constants (k_{obs}) for the reaction of pernonanoic acid and *n*-nonanoyloxybenzenesulfonate vs. $[SDS]_T$ at different pH exhibited maxima and obeyed a kinetic equation similar to Equation 3.12 (Chapter 3).[110] That the occurrence of maxima is qualitatively described as the initial increase is partly due to the increased local concentration of reactants in the micelle, and the subsequent fall off in k_{obs} is due to the dilution of two reactants when they are fully micellar bound. However, k_{obs} for the reaction of hydrogen peroxide and n-nonanoyloxybenzenesulfonate showed a monotonic decrease with the increase in $[SDS]_T$. The rate of Diels–Alder reaction of 3-(para-substituted phenyl)-1-(2-pyridyl)-2-propen-1-ones (**16a–16g**), containing neutral, cationic, or anionic substituents, and cyclopentadiene (**17**) in the absence of Lewis acids is retarded slightly (~2%) by micelles of $C_{12}E_7$.[111] The kinetic data have been interpreted by the use of PP model. In a situation in which **16** does not bind to the micelle, the rate of reaction is inhibited because of micellar uptake of **17**. However, rate inhibitions, in a situation in which both reactants, **16** and **17**, are completely micelle bound, are attributed to different micellar locations of **16** and **17**.

Bimolecular rate constants ($k_{app} = k_{obs}/[R]_T$ where $[R]_T$ represents the total concentration of one of the two reactants and whose concentration is larger than that of the second reactant, S, by a factor of more than 5) for 1,3-dipolar cycloaddition reactions of benzonitrile oxide (**18**) with a series of N-substituted maleimides (**19a–19c**) in micelles of nonionic surfactants, $C_{12}E_8$, $C_{12}E_{23}$, $C_{16}E_{10}$, and $C_{16}E_{20}$, fit reasonably well to a kinetic equation similar to Equation 3.61 (Chapter 3).[112] The calculated values of k_M^2/k_W^2 vary from 0.30 to 0.39 for all four nonionic micelles. Nearly 3-fold micellar deceleration effects have been shown to be similar to the effect of mixed water/1-propanol solvent with $[H_2O]$ about 15 M on k_{app}. However, a comparison of k_M^2 with k_{app} in such water–organic solvents does not provide detailed information about the exact nature of the reaction environment in the micellar pseudophase. Instead, it provides information about the question insofar as the micellar reaction environment is satisfactorily mimicked by such mixed water–organic solvents.

16a	16b	16c	16d	16e	16f	16g
X = NO$_2$	Cl	H	CH$_3$	OCH$_3$	CH$_2$SO$_3$Na	CH$_2$NMe$_3$B

17   18

19a, R = Et
19b, R = *n* - Bu
19c, R = CH$_2$Ph

4.3.2 MICELLES OF IONIC NONFUNCTIONAL SURFACTANTS

Unlike nonionic micelles, the ionic micelles exert strong electrostatic effect on rates of semiionic or fully ionic bimolecular reactions. Electrostatic interactions are much stronger than other nonelectrostatic interactions (such as hydrophobic, polar, and steric interactions) between micelle-forming surfactants and solubilizates. A large amount of kinetic data on ionic micellar-mediated bimolecular reactions in which one of the reactants is a considerably hydrophilic counterion have been analyzed in terms of PIE model,[2e–2g] that is, Equation 3.12 and Equation 3.18 to Equation 3.24 as described in Chapter 3, Section 3.3.7. In the majority of these reactions, the micellar rate constant, k_M^2, is not significantly different from those in bulk water, k_W^2; that is, k_M^2/k_W^2 values are not very different from unity. Accordingly, it has been emphasized that micellar rate acceleration or deceleration effects are mainly due to an increase or decrease in one of the two reactants' concentrations in the micellar pseudophase; that is, they do not originate from significant variation in the rate constants for bimolecular reactions.[2e–2g]

The PP micellar model (i.e., Equation 3.11 and Equation 3.12 and their various limiting forms [Equation 3.13 to Equation 3.15, Chapter 3]) is applicable for the semiionic or fully ionic bimolecular reactions in which the rate of micellar-mediated reaction is negligible compared with the rate of reaction in the aqueous pseudophase, i.e., $k_W^2 \gg (k_M^{mr} K_S - k_W^2) m_R^s [D_n]$, under the experimental conditions imposed. Pseudo-first-order rate constants for the reactions of several secondary amines with 2-bromoethyl and 1-phenylethyl nitrites in the presence of TTABr micelles obeyed Equation 3.11[113] and Equation 3.13.[114]

The effects of the concentration of SDS micelles on second-order rate constants (k_n) for the reaction of hydrazine with PS⁻ follow Equation 4.11, in which it turns out that $k_{2,W} \approx k_M^{mr} K_S^{app} f_M^{S-}$.[67] The rate constant for hydrazinolysis of PS⁻ in micellar pseudophase is either insignificant or

$$k_n = \frac{k_W^2 + k_M^{mr} K_R K_S^{app} f_M^{S-} [D_n]}{(1 + K_S^{app}[D_n])(1 + K_R[D_n])}$$ (4.11)

significantly smaller than the corresponding rate constant in aqueous pseudophase. This is attributed largely to considerably low water activity in the specific micellar environment in which PS$_M^-$ ions exist. The UV spectral observations reveal that the micellar bound reactants (PSH and PS⁻) and products (nonionized and ionized phenol and N-hydrazinylsalicylamide) exist in the micellar environment of different water activity. The values of k_{obs} remain unchanged with the increase in concentration of 1,4-diazabicyclo[2.2.2]octane from 0.06 to 0.60 M at 0.03-M SDS, 0.05-M NaOH, and 37°C. This shows that the presence of SDS micelles does not change the reaction mechanism of aqueous pseudophase hydrolysis and aminolysis of PS⁻.[67]

The values of k_n for the reactions of n-butylamine, piperidine, and pyrrolidine with PS⁻ in the presence of SDS micelles obey Eq. (4.12)[115]

$$k_n = \frac{k_W^2 f_W^R + k_M^{mr} f_M^R K_R^{app} K_1[D_n]}{(1 + K_1[D_n])(1 + K_R^{app}[D_n])} \qquad (4.12)$$

where $f_W^R = K_{a,W}^R/(a_{H,W} + K_{a,W}^R)$, $f_M^R = K_{a,M}^R/(a_{H,M} + K_{a,M}^R)$, $K_R^{app} = ([R_M] + [RH^+_M])/\{(R_W] + [RH^+_W])[D_n]\}$ where R represents free amine base and other symbols have their usual meanings. The observed results show that $k_W^2 f_W^R \approx k_M^{mr} f_M^R K_1$ for n-butylamine, whereas $k_W^2 f_W^R \gg k_M^{mr} f_M^R K_R^{app} K_1[D_n]$ for piperidine and pyrrolidine.[115]

The values of bimolecular rate constants (k_M^2) for Diels–Alder reactions of cyclopentadiene, sorbyl alcohol (20), and sorbyltrimethylammonium bromide (21) with a series of N-substituted maleimides are 20 to 40 times lower in SDS micelles than the corresponding rate constants (k_W^2) in aqueous phase.[116] The low micellar rate constants (k_M^2) have been ascribed to the relatively apolar region of micelle, in which the reactions take place. The observed apparent second-order rate constants (k_{app}) for the reactions of 18 with 19a to 19c are significantly increased, up to a factor of 17 (for 18 + 19c /CTABr).[112] These observed data have been explained in terms of PP model (i.e., Equation 3.61, Chapter 3). However, the values of k_M^2/k_W^2 remain constant at ~ 0.23 and ~ 0.45 for the reactions of 18 with 19a to 19c in SDS and CTABr micelles, respectively. Nearly 4-fold and 2-fold lower values of k_M^2 compared to k_W^2 in respective SDS and CTABr micelles are largely attributed to ionic micellar reaction environment.[112]

4,4'-(Dialkylamino)pyridine-based compounds 22 to 25 catalytically cleave 4-nitrophenyl hexanoate (PNPH) and 4-nitrophenyl diphenyl phosphate (6) in aqueous cationic micellar medium and cationic Gemini surfactant based micellar aggregates provide more than one order of magnitude better reaction medium than their conventional single-chain, single-charge, cationic cetyl trimethylammonium bromide (CTABr) micelles.[117] Dephosphorylation or deacylation reactions are generally facilitated by a decrease in the water content of the reaction medium. Thus, the kinetic benefits associated with Geminis may be due to the fact that the spacer chain at the head group level decreases the extent of water penetration at the micellar surface. These conclusions are consistent with the report that proximity of the positive charges increases anion binding at the expense of binding of water.[118]

20   **21**   **22**   **23**

24   **25**

The effects of cationic micelles on pseudo-first-order rate constants (k_{obs}) for the S_N2 substitution reaction between methyl 4-nitrobenzenesulfonate and bromide ion in pure water and mixed water–ethylene glycol (EG) solvents have been discussed in terms of PP and mass-action models and the increase in EG content from 0 to 35% v/v decreases the values of calculated k_M^{mr} from 5.87×10^{-3} to 3.80×10^{-3} sec^{-1} for CTABr, 4.13×10^{-3} to 3.60×10^{-3} sec^{-1} for TTABr, and 3.92×10^{-3} to 3.30×10^{-3} sec^{-1} for DTABr micelles.[119] But all these values of k_M^2 (= $V_M k_M^{mr}$, V_M is micellar molar volume) are larger than k_W^2 (= 4.5×10^{-4} M^{-1} sec^{-1}). For a given [surfactant], k_{obs} decreases as the volume percent of EG increases. This is due to (1) a decrease in the equilibrium binding constant, which results in a decrease of the contribution of the reaction occurring in the micellar pseudophase and (2) a decrease in the bromide ion concentration in the interfacial region. However, the values of k_W^2 increase from 4.50×10^{-4} to 6.27×10^{-4} sec^{-1} with increase in vol% EG from 0 to 35. The values of k_M^2 for alkaline hydrolysis of nine 4-substituted phenyl esters of 4-substituted benzoates in CTABr micelles, calculated by using a pseudophase ion-exchange model, constitute Hammett plots with ρ values larger than the ρ values of the corresponding Hammett plots for reactions in aqueous phase by 0.8 units.[120] The effective low dielectric constant in the micellar surface may be responsible for the larger ρ values obtained for the reactions in micellar pseudophase.

The rate of nucleophilic substitution reaction between F$^-$ and **6** have been studied at constant pH (attained by Tris and ethanolamine buffers), 25°C, and at different [CTACl], and the observed data have been analyzed in terms of PIE model. The pH- and buffer-independent micellar rate constant, k_M^2 (= 0.096 M^{-1} sec^{-1}) is similar to that measured in bulk water, that is, $k_M^2/k_W^2 \approx 1$. The value of k_M^2 is satisfactorily reproduced only in a medium containing water and Me$_4$NCl concentrations similar to those measured at the interfacial region of CTACl micelles.[121] The concentration factor is mainly responsible for the acceleration of the rates of nucleophilic substitution reaction of **6** with 2-aminomethylphenols and the basic hydrolysis of ethyl 4-nitrophenyl ethylphosphonate in the presence of micelles, whereas the micellar microenvironment negatively affects the reactivity.[122]

The plots of pseudo-first-order rate constants (k_{obs}) for the reactions between anionic 6-O-octanoyl-L-ascorbic acid (VC8) and 3-methylbenzenediazonium cation (3MBD), under reaction conditions of apparent constant pH and [3MBD] << [VC8], vs. [micelles] (micelles of SDS, CTABr, and TTABr) show distinct maxima with $k_{obs}^{max}/k_{obs}^{[Dn]=0} \geq 3$ for SDS and ≥ 11 for CTABr as well as TTABr micelles where k_{obs}^{max} is the value of k_{obs} at the point of maximum in the k_{obs} – [Surf]$_T$ plot and $k_{obs}^{[Dn]=0} = k_{obs}$ at [Surf]$_T$ = 0.[123] These observations are apparently unexpected because the two reactants, 3MBD and VC8$^-$, of this bimolecular reaction are counterion and co-ion for both anionic and cationic micelles, and the rates of such micellar-mediate bimolecular reactions are normally strongly inhibited owing to predominant unfavorable and favorable electrostatic interactions between ionic micellar head groups/co-ions and ionic micellar head groups/counterions, respectively. Unexpected significant micellar rate enhancements in these reactions may be the consequence of mixed micelles formation in which mixed micelle component VC8$^-$ causes mixed micelles to behave as functional micelles. Functional micelles are known to catalyze reactions much more strongly compared to nonfunctional micelles, because functional micelles result in induced intramolecularity in the reaction due to partial loss of translational motion of reactant VC8$^-$. The favorable effect of this induced intramolecularity dominates over the unfavorable effect of electrostatic interaction between co-ions and ionic micellar head groups. This conclusion is supported by the observed inhibitory effects of SDS micelles on the rate of reaction of 3MBD with L-ascorbic acid or vitamin C (VC) where the formation of mixed micelles is impossible.[61]

4.4 INTRAMOLECULAR REACTIONS

Intramolecularity or induced intramolecularity is an important feature of many enzyme-catalyzed reactions.[124] Micellar systems provide models for understanding the effects of changes in the microreaction environment on reaction rates of bioorganic reactions. There are only a limited number of reports on systematic studies on the effects of micelles on the rates of intramolecular reactions.

The intramolecular S_N2 reactions such as cyclization of o-3-halopropylox-yphenoxide ions (26) follow first-order kinetics and micellar effect on the rates of such reactions may be explained quantitatively in terms of Scheme 3.1 (Chapter 3). The cyclization of 26 involves substitution of halide by aryloxide ion as shown in Scheme 4.7.[125] In the presence of CTABr micelles, the values of k_M/k_W are 1.8 and 3.9 for Y = Br and I, respectively.[126] The values of k_M/k_W become 7.2 and 29 for Y = Br and I, respectively, in the presence of cetyltributylammonium bromide.[126] In the absence of micelles, the rate constant for cyclization of 26 is slightly higher in ethanol than in water, but is much higher in a dipolar aprotic solvent. Mild micellar effects on the rate of cyclization of 26 may be caused by

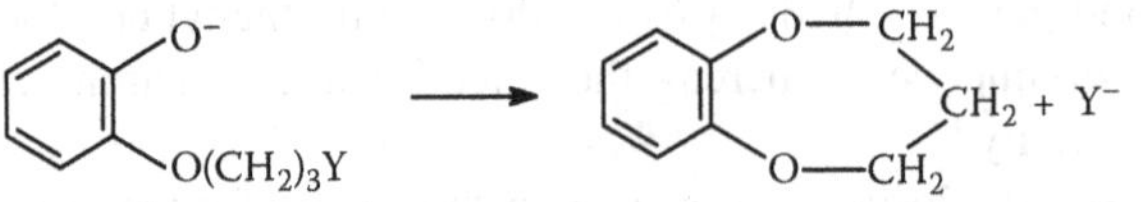

26

SCHEME 4.7

micellar reaction medium effect. An increase in the size of the surfactant head group increases k_M by a factor of about 7, which could be due to partial exclusion of water from the interface by the large head groups. It may be worth it to mention that, in the absence of micelles, first-order rate constants, k_{obs}, for cyclization of *N,N*-dimethylphthalamic acid to form phthalic anhydride, which follows addition–elimination mechanism as shown in Scheme 4.8, decreased from 8.8×10^{-4} to 5.1×10^{-4} sec^{-1} with the increase in ethanol content from 10 to 80% v/v in mixed aqueous solvents. But the values of k_{obs} increased from 4.2×10^{-4} to 6.4×10^{-4} sec^{-1} with the increase in acetonitrile content from 10 to 80% v/v in mixed aqueous solvents.[127]

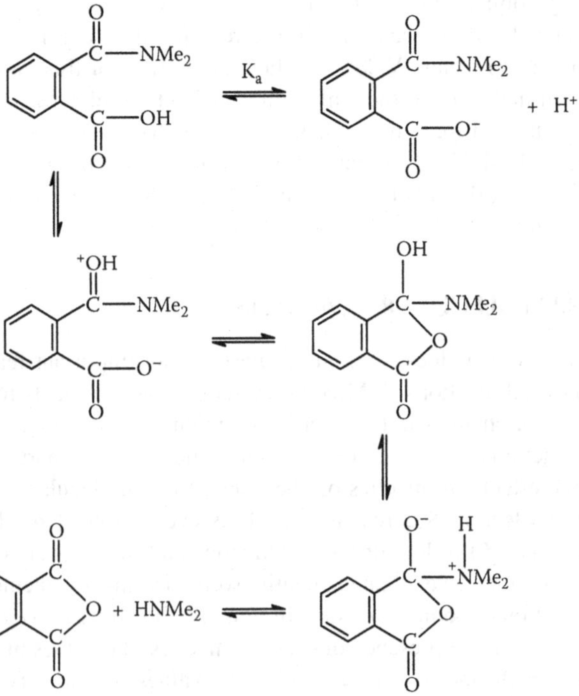

SCHEME 4.8

The rates of hydroxide-ion- and acetate-ion-catalyzed cyclization of two ethyl hydantoates, **27** and **28**, to the corresponding hydantoins have been studied at varying concentrations of cationic micelles of three cationic surfactants, CTAX, $X^- = Br^-$, Cl^-, and CH_3COO^- (AcO$^-$) and the plot of pseudo-first-order rate constants (k_{obs}) vs. [CTAX]$_T$ at constant aqueous pH as well as total acetate buffer concentration and 25°C reveals well defined maximum.[128] These observed data have been analyzed in terms of PIE model, that is Equation 4.13

$$k_{obs} = \frac{k_W^{HO}[HO_W^-] + k_W^{AcO}[AcO_W^-] + \{k_M^{HO}[HO_M^-] + k_M^{AcO}[AcO_M^-]\}K_S[D_n]}{1 + K_S[D_n]} \quad (4.13)$$

in which k's are second-order rate constants, with concentrations in square brackets referring to molarity in the total solution volume.

The data treatment involves the calculation of K_S and k_M^{HO} with $k_M^{AcO} = 0$ and K_S and k_M^{AcO} with $k_M^{HO} = 0$ from the same set of the observed data. These are two extreme possibilities, which may or may not be correct. The reacting system is too complex, because it involves three possible ion-exchange processes, X^-/HO^-, X^-/AcO^-, and AcO^-/HO^- when $X^- = Br^-$ and Cl^-. An attempt to consider all these three plausible ion-exchange processes will face enormous mathematical and data simulation complexities in terms of the large number of disposable parameters and, consequently, the values of calculated kinetic parameters, K_S, k_M^{AcO}, and k_M^{HO}, may be very unreliable. The unreliability of the calculated values of K_S and k_M^{AcO} or k_M^{HO} is evident from these values obtained at 0.02 M and 0.2 M acetate buffers with 50% free base for different cationic micelles.[128] Thus, although the calculated values of K_S and k_M^{AcO} or k_M^{HO} are not very reliable, they do show that k_M^{HO}/k_W^{HO} and k_M^{AcO}/k_W^{AcO} may not be very different from respective values of 0.2 and 10 for **27** as well as 0.45 and 3 for **28**. These results show that cationic micelles have only moderate effects on second-order rate constants for hydroxide ion- and acetate ion-catalyzed cyclization of **27** and **28**.

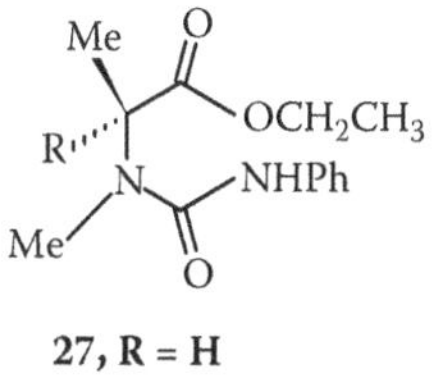

27, R = H

28, R = Me

A certain class of RNA molecules called ribozymes catalyze aqueous degradation of their own as well as other RNAs through an efficient induced intramolecular general acid (GBH$^+$) and general base (GB)-catalyzed intramolecular

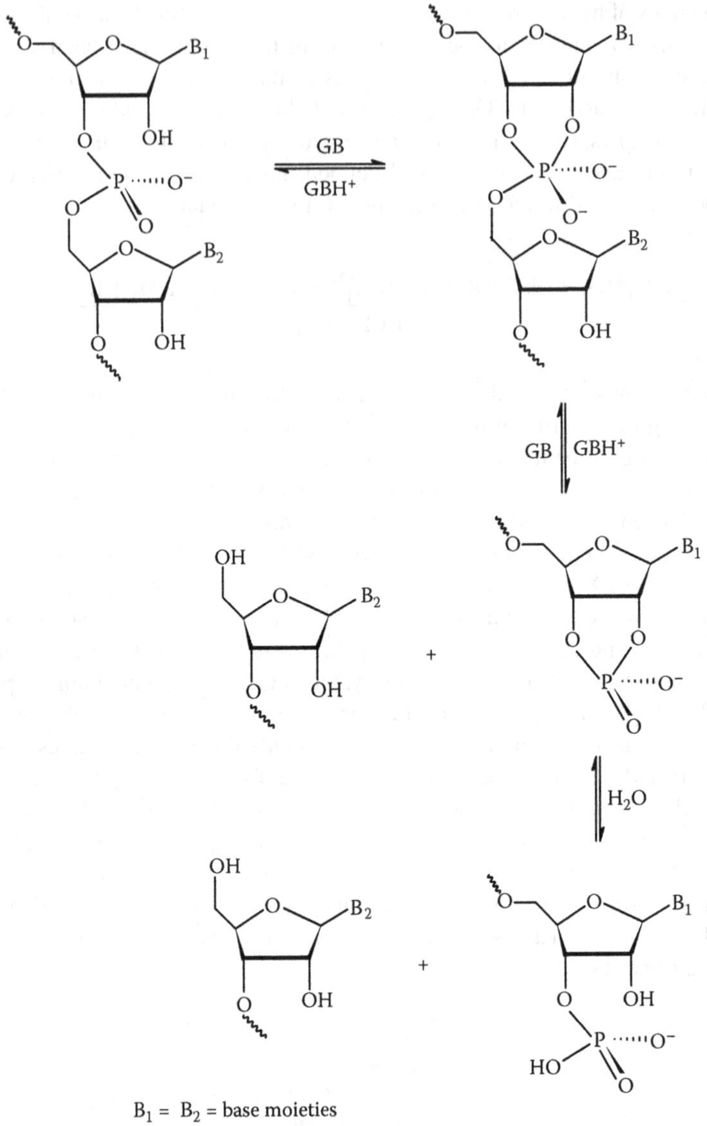
GB
GBH⁺
B₁
B₂
OH
H₂O
B₁ = B₂ = base moieties

SCHEME 4.9

nucleophilic-assisted cleavage as shown in Scheme 4.9.[129] The fine details of the mechanism are still unknown. However, GB and GBH^+ catalyst molecules are presumably in the complexed form in which they have lost their translational degrees of freedom to a great extent and become so-called induced intramolecular general base and general acid catalysts.

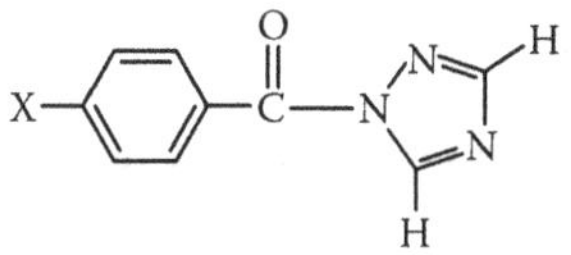

29a, X = OMe
29b, X = Me
29c, X = H
29d, X = Cl
29e, X = OCF$_3$
29f, X = NO$_2$

It has been reported recently that naturally occurring catalytic RNAs like hammerhead and hairpin ribozymes do not require metal ions for efficient catalysis.[130] This discovery led to the speculation that the folded tertiary structure of the RNA contributes more to the catalytic function that was initially recognized. This speculation is supported by the finding that a highly specific self-cleavage reaction can occur within a small bulge loop of four nucleotides in a minisubstrate derived from *Arabidopsis thaliana* intron-containing pre-tRNATyr in the absence of metal ions. Nearly 100-fold rate enhancement of RNA self-cleavage has been observed in the presence of ammonium cations and nonionic or zwitterionic detergents at or above their CMC.[130] Ammonium ions have been suggested to help the entry of the negatively charged RNA into the hydrophobic interior of a micelle. The rate enhancement (~100-fold) is most likely caused by a change of pattern of hydration or hydrogen bonds owing to the hydrophobic surrounding of micelle. These findings suggest that highly structured RNAs may shift pK$_a$ values toward neutrality via the local environment and thereby enhance their ability to perform GA–GB catalysis without the participation of metal ions.

4.5 EFFECTS OF MIXED AQUEOUS–ORGANIC SOLVENTS ON MICELLAR BINDING CONSTANTS (K$_S$) OF SOLUBILIZATES

The increase in the concentration of neutral organic cosolvent (such as acetonitrile, methanol, and 1,2-ethanediol) in mixed aqueous–organic solvents decreases the cationic and anionic micellar binding constants (K$_S$) of hydrophobic neutral and anionic solubilizates such as **9** and PS$^-$ (Table 4.3). Such an effect is explained by the following empirical equation

$$K_S = K_S^0 \exp(-\delta X) \tag{4.14}$$

where X is the concentration of added neutral organic cosolvent and K$_S{}^0$ as well as δ are unknown empirical parameters. The magnitude of δ is the measure of the ability of organic cosolvent to decrease K$_S$. Electrostatic interactions are

TABLE 4.4

Values of K_S^0 and δ, Calculated from Equation 4.14, for Various Solubilizates[-] in the Presence of SDS and CTABr Micelles

Solubilizate	Surfactant	X	Temperature/ °C	K_S^0/M^{-1}	$10^2 \delta/$ (% v/v)$^{-1}$
9	SDS	CH_3CN	35	697 ± 27[a]	9.3 ± 0.6[a]
PS[-]	CTABr	CH_3CN	35	10087 ± 1832	15.6 ± 2.5
		$HOCH_2CH_2OH$	25	16297 ± 2639	4.2 ± 0.6
			30	20535 ± 4934	6.4 ± 1.2
			35	13224 ± 2032	5.8 ± 0.6
			40	11949 ± 237	6.1 ± 0.1
			45	9819 ± 5667	5.6 ± 2.4
		CH_3OH	25	19090 ± 2843	6.1 ± 1.0
			30	19939 ± 3135	9.2 ± 1.1
			35	13574 ± 130	7.5 ± 0.1
			40	8161 ± 1444	6.8 ± 1.2
			45	14840 ± 796	11.0 ± 0.4
Methyl 4-nitro-benzenesulfonate	CTABr	$HOCH_2CH_2OH$	25.2	94 ± 7	3.4 ± 0.4
	TTABr		25.2	70 ± 19	4.1 ± 1.4
	DTABr		25.2	44 ± 4	4.0 ± 0.4
Phenyl chloro-formate	CTABr	$HOCH_2CH_2OH$	25.2	128 ± 41	2.7 ± 1.1[b]
	TTABr		25.2	148 ± 22	3.0 ± 0.5
	DTABr		25.2	107 ± 3	3.3 ± 0.1
1	CTABr	CH_3CH_2OH	25	659 ± 138	40 ± 8[c]

[a] Error limits are standard deviations.
[b] Unit of δ is (w/w)$^{-1}$.
[c] Unit of δ is (mole fraction)$^{-1}$.

certainly the major factors for strong and weak binding between ionic solubilizate and micelles of opposite charges and same charges, respectively. However, it is tempting to suggest that the increase in X in aqueous ionic micellar solution increases the nonmicellar pseudophase affinity of ionic and neutral solubilizates, which, in turn, decreases the apparent ionic micellar affinity of such solubilizates, and such a decrease is empirically expressed by Equation 4.14.

The reported values of K_S at different values of X, as summarized in Table 4.3, were used to calculate the values of K_S^0 and δ from Equation 4.14 using nonlinear least-squares technique. The calculated values of K_S^0 and δ for **9** and PS[-] in the presence of SDS and CTABr micelles, respectively, at different temperature are summarized in Table 4.4. The values of δ appear to be independent of temperature and nearly 45% larger for CH_3OH than for $HOCH_2CH_2OH$ (Table 4.4).

4.6 MICELLES AS MODIFIERS OF REACTION RATE

Pseudo-first-order rate constants (k_{obs}) for the reaction of piperidine (Pip) with **14**, obtained within total concentration of Pip ($[Pip]_T$) ranging from 0.02 to 0.08 M at 35°C, obeyed Equation 4.15

$$k_{obs} - k_0 = k_n [Pip]_T \qquad (4.15)$$

where k_0 is the pseudo-first-order rate constant for pH-independent hydrolysis and k_n is the nucleophilic second-order rate constant for the reaction of Pip with **14**.[131] The values of k_n (= 0.30 M^{-1} sec^{-1}) were independent of [NaOH] and $[CTABr]_T$ within their respective concentration range of 0.0 to 0.005 M and 0.0 to 0.0001 M. But the pseudo-first-order rate constants (k_{obs}) obtained at 0.01-M CTABr, 0.005-M NaOH, and 35°C were found to increase from 1.69×10^{-4} to 3.38×10^{-4} sec^{-1} with the increase in $[Pip]_T$ from 0.02 to 0.80 M. These results give an approximate value of k_n as 2.2×10^{-4} M^{-1} sec^{-1}. This value of k_n is nearly 1500-fold smaller than k_n obtained in the absence of CTABr micelles. The nearly 1500-fold decreasing effect of CTABr micelles on the rate of piperidinolysis of **14** is apparently inexplicable because the rate of piperidinolysis of ionized phenyl salicylate (PS$^-$) decreased only 13-fold with the increase in $[CTABr]_T$ from 0 to 0.01 M at 35°C.[132] These observations apparently show that $k_n = 0$ under such experimental conditions.

The possibility that the lack of piperidinolysis of **14** in the presence of 0.01 M CTABr might be due to preferential micellar incorporation of only one reactant (i.e., relatively more hydrophobic **14** or **14**$^-$) has been ruled out based upon experimental evidence.[132] The values of k_{obs} at 100% v/v CH$_3$CN, obtained in the absence of CTABr, indicate small but definite nucleophilic reactivity of piperidine toward **14**.[133] Thus, the absence of apparent nucleophilic reactivity of Pip toward **14** in the presence of 0.01 M CTABr cannot be attributed to merely micellar medium effect.

The other plausible explanation for the lack of piperidinolysis of **14** in the presence of 0.01 M CTABr may be attributed to the different average locations of **14** and **14**$^-$ as well as Pip in the micellar pseudophase. However, this possibility is unlikely within the domain of the following observations. There is a small but kinetically detectable nucleophilic reactivity of Pip toward PS$^-$ in the CTABr micellar pseudophase.[131,132] Both PS$^-$ and **14**$^-$ are expected to remain in essentially the same CTABr micellar environment, because electrostatic interaction is the dominant factor for the binding affinity of anionic substrates with cationic micelles. It may be argued that the lack of detectable piperidinolysis of **14** in the presence of 0.01-M CTABr is due to negligible concentration of free base Pip in aqueous phase (i.e., almost entire Pip molecules in aqueous phase exist in acid form). Although this is plausible, it has been ruled out based upon UV spectral evidence.[131] The most plausible reason for the lack of kinetically detectable piperidinolysis of **14** in the presence of 0.01-M CTABr is described as follows.

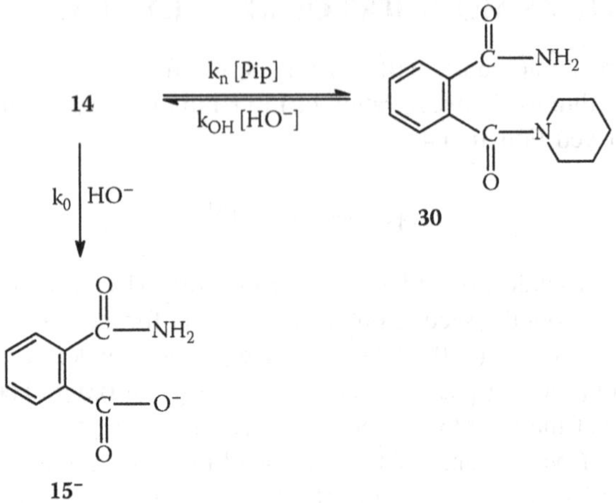

SCHEME 4.10

The brief mechanism of the cleavage of **14** in aqueous solution containing different concentrations of Pip, 0.005-M NaOH and 2% v/v CH_3CN at $[CTABr]_T$ = 0 is shown in Scheme 4.10. It is well known that the reaction mechanism of a particular reaction remains generally the same in both aqueous pseudophase and micellar pseudophase. The value of CTABr micellar binding constant (K_S = 2500 to 3300 M^{-1}) shows the presence of > 96% micellized **14** (i.e., $[14_M]/[14]_T$ > 0.96) at 0.01-M CTABr and [NaOH] $\geq$ 0.005 M. Thus, in the presence of CTABr micelles, the reaction steps shown in Scheme 4.10 occur simultaneously both in aqueous and micellar pseudophases. However, the expected very low value of K_S^{Pip}, CTABr micellar binding constant of Pip, (K_S^{Pip} = 0.3 M^{-1} with TTABr micelle[114]) coupled with significantly high value of $[HO_M^-]$ because of the occurrence of ion-exchange Br^-/HO^- result in $k_{OH,M}[HO_M^-] \gg k_{n,M} [Pip_M]$ even at the highest concentration of Pip in the presence of 0.005-M NaOH. Such conditions caused very fast attainment of equilibrium between **14** and **30** in both aqueous and micellar pseudophases because $k_{0,w} \ll k_{n,w} [Pip_w]$ and presumably $k_{0,M} \ll k_{n,M} [Pip_M]$. The observed rate law (rate = $k_{obs} [14]_T$) and Scheme 4.10 with inequalities $k_{0,w} \ll k_{n,w} [Pip_w]$ and $k_{0,M} \ll k_{n,M} [Pip_M]$ can lead to Equation 4.16

$$k_{obs} = \frac{k_{0,w}}{\{1+((k_{n,w}[Pip_w])/(k_{OH,w}[HO_w^-]))\}K_S[D_n]} +$$

$$\frac{k_{0,M}}{1+((k_{n,M}[Pip_M])/(k_{OH,M}[HO_M^-]))}$$

$$(4.16)$$

However, in the presence of 0.01-M CTABr, the rate of cleavage of **14** in the aqueous pseudophase is negligible compared to that in the micellar pseudophase, because under such conditions, $[14]_T = [14_M] + [14_w] \approx [14_M]$ owing to the fact that $K_S = 2500 - 3300\ M^{-1}$. It has also been concluded earlier in the text that 1 $\gg (k_{n,M}\ [Pip_M])/(k_{OH,M}[HO_M^-])$ at 0.01-M CTABr. Thus, under such experimental conditions, Equation 4.16 reduced to Equation 4.17. Nearly 2-fold increase in k_{obs} with the increase in $[Pip]_T$ from 0.02 to 0.80 M at 0.01-M CTABr has been ascribed to the significant contribution of

$$k_{obs} = k_{0,M} \tag{4.17}$$

$k_{2,M}[HO_M^-][\mathbf{14_M^-}]$ compared to $k_{0,M}\ [H_2O_M][\mathbf{14_M^-}]$ at higher values of $[Pip]_T$ in the presence of 0.005 M NaOH.[131]

Micelles modify the rate of reaction in aqueous solvent by merely causing unequal molar distribution of reactants and by micellar medium effects. Such behavior of micelles is described by recently coined, exotic terms such as "smart catalysts"[134] and "smart aqueous reaction medium."[135] Smart catalysts and substrates that have thermoresponsive polymers covalently attached to them have been found to cause anti-Arrhenius behavior in which their reaction rates slow down considerably with increasing temperature merely owing to phase separation.[134] Davies and Stringer[135] describe an aqueous solution of the thermoresponsive poloxamer, $H(OCH_2CH_2)_{27}(OCH(CH_3)CH_2)_{61}(OCH_2CH_2)_{27}OH$, (**P104**), with a concentration of 5 or 15 g l^{-1} as a smart aqueous reaction medium because in the absence of **P104**, the relationship between the logarithm of the rate constant of a bimolecular reaction and inverse temperature is linear, which constitutes Arrhenius behavior. When one reactant is hydrophobic and the other hydrophilic, anti-Arrhenius behavior occurs because the hydrophobic reactant partitions into the thermally induced poloxamer micelles whereas the hydrophilic reactant remains in the bulk aqueous phase, causing a decrease in the rate. Hyper-Arrhenius behavior occurs when both the reactants are hydrophobic and thus partition into thermally induced **P104** micelles, causing a much greater increase in rate with temperature than in the absence of **P104**.[133] Arrhenius and apparent anti-Arrhenius behavior have been observed in the acid-catalyzed hydrolysis of 2-(*p*-tetradecoxyphenyl)-1,3-dioxolane (*p*-TPD) in the presence of SDS micelles and this apparent anti-Arrhenius behavior, obtained at > 32°C, is attributed to temperature-induced demixing of *p*-TPD in aqueous micellar solutions of SDS.[136]

It is perhaps noteworthy that the use of the term "anti-Arrhenius behavior" in such reaction systems is probably inappropriate or misleading. When one says that the rate constants, k, follow anti-Arrhenius behavior, it apparently means that k corresponds to a reaction step that has negative energy of activation and by definition and physical reality, energy of activation can never be negative for any elementary reaction step. The apparent anti-Arrhenius plots are obtained[135,136] when the rate constants used to construct the Arrhenius plots are the functions of various rate and equilibrium constants for various irreversible and reversible reaction steps in the overall reactions. It is almost certain for most micellar-mediated reactions that micellar effects on reaction rates are not caused by the decrease or increase in the energy of activation for the rate-determining steps of these reactions. For example, reported anti-Arrhenius behavior of reaction rate[135] is not caused by the negative energy of activation of the rate-determining step of the reaction. It is therefore more appropriate to use the term "apparent anti-Arrhenius" rather than "anti-Arrhenius."

REFERENCES

1. (a) Duynstee, E.F.J., Grunwald, E. Organic reactions occurring in or on micelles. I. Reaction rate studies of the alkaline fading of triphenylmethane dyes and sulfonphthalein indicators in the presence of detergent salts. *J. Am. Chem. Soc.* **1959**, *81*, 4540–4542. (b) Duynstee, E.F.J., Grunwald, E. Organic reactions occurring in or on micelles. II. Kinetic and thermodynamic analysis of the alkaline fading of triphenylmethane dyes in the presence of detergent salts. *J. Am. Chem. Soc.* **1959**, *81*, 4542–4548.

2. (a) Cordes, E.H., Dunlap, R.B. Kinetics of organic reactions in micellar systems. *Acc. Chem. Res.* **1969**, *2*, 329–337. (b) Fendler, E.J., Fendler, J.H. Micellar catalysis in organic reactions: kinetic and mechanistic implications. *Adv. Phys. Org. Chem.* **1970**, *8*, 271–406. (c) Cordes, E.H., Gitler, C. Reaction kinetics in the presence of micelle-forming surfactants. *Prog. Bioorg. Chem.* **1973**, *2*, 1–53. (d) Bunton, C.A. Reaction kinetics in aqueous surfactant solutions. *Catal. Rev. — Sci. Eng.* **1979**, *20*(1), 1–56. (e) Bunton, C.A., Savelli, G. Organic reactivity in aqueous micelles and similar assemblies. *Adv. Phys. Org. Chem.* **1986**, *22*, 213–309. (f) Romsted, L.S. Micellar effects on reaction rates and equilibria. In *Surfactants in Solution*, Vol. 2, Mittal, K.L., Lindman, B., Eds., Plenum: New York, **1984**, 1015–1068. (g) Bunton, C.A. Reactivity in aqueous association colloids: descriptive utility of pseudophase model. *J. Mol. Liq.* **1997**, *72*(1/3), 231–249.

3. (a) Fendler, J.H., Fendler, E.J. *Catalysis in Micellar and Macromolecular Systems.* Academic Press: New York, **1975**. (b) Fendler, J.H. *Membrane Mimetic Chemistry*, Wiley-Interscience: New York, **1982**.

4. Buurma, N.J., Herranz, A.M., Engberts, J.B.F.N. The nature of the micellar Stern region as studied by reaction kinetics. *J. Chem. Soc., Perkin Trans. 2.* **1999**, 113–119.

5. Bunton, C.A., Kamego, A.A., Minch, M.J. Micellar effects upon the decarboxylation of 3-bromo and 2-cyano carboxylate ions. *J. Org. Chem.* **1972**, *37*(9), 1388–1392.

6. Bunton, C.A., Romsted, L.S. *Handbook of Microemulsion Science and Technology*, Kumar, P., Mittal, K.L., Eds., Marcel Dekker: New York, **1999**, chap. 15 and references cited therein.

7. (a) Khan, M.N., Arifin, Z. Effects of Li^+ and K^+ ions on the rate of intramolecular general base-catalyzed methanolysis of ionized phenyl salicylate in the absence and presence of cationic micelles in mixed H_2O-CH_3CN solvents. *J. Phys. Org. Chem.* **1996**, *9*, 301–307. (b) Khan, M.N., Arifin, Z. Effects of cationic micelles on rates and activation parameters for the hydrolysis of phthalimide in an alkaline medium. *Malays. J. Sci.* **1994**, *15B*, 41–48.

8. (a) Bunton, C.A., Minch, M.J., Hidalgo, J., Sepulveda, L. Electrolyte effects on the cationic micelle catalyzed decarboxylation of 6-nitrobenzisoxazole-3-carboxylate anion. *J. Am. Chem. Soc.* **1973**, *95*(10), 3262–3272. (b) Bunton, C.A., Minch, M., Sepulveda, L. Enhancement of micellar catalysis by added electrolytes. *J. Phys. Chem.* **1971**, *75*(17), 2707–2709.

9. Bunton, C.A., McAneny, M. Catalysis of reactions of *p*-nitrobenzoyl phosphate by functional and nonfunctional micelles. *J. Org. Chem.* **1977**, *42*(3), 475–482.

10. Bunton, C.A., Fendler, E.J., Sepulveda, L., Yang, K.-U. Micellar-catalyzed hydrolysis of nitrophenyl phosphates. *J. Am. Chem. Soc.* **1968**, *90*(20), 5512–5518.

11. Kurz, J.L. Effects of micellization on the kinetics of the hydrolysis of monoalkyl sulfates. *J. Phys. Chem.* **1962**, *66*, 2239–2245.

12. Bunton, C.A., Huang, S.K. Micellar effects upon the reaction of the tri-*p*-anisyl carbonium ion with nucleophiles. *J. Org. Chem.* **1972**, *37*(11), 1790–1793.

13. Bunton, C.A., De Buzzaccarini, F. Oil-in-water microemulsions as reaction media. *J. Phys. Chem.* **1981**, *85*(21), 3139–3141.

14. Khan, M.N., Naaliya, J., Dahiru, M. Effects of anionic micelles on intramolecular general base-catalyzed hydrolyses of salicylate esters: evidence for a porous cluster micelle. *J. Chem. Res. (S)* **1988**, 116–117; *J. Chem. Res. (M)*, 1168–1176.

15. Menger, F.M., Portnoy, C.E. On the chemistry of reactions proceeding inside molecular aggregates. *J. Am. Chem. Soc.* **1967**, *89*(18), 4698–4703.

16. Khan, M.N., Arifin, Z. Effects of cationic micelles on the rate and activation parameters of intramolecular general base-catalyzed hydrolysis of ionized salicylate esters. *J. Colloid Interface Sci.* **1996**, *180*(1), 9–14.

17. Rafique, M.Z.A., Shah, R.A., Kabir-ud-Din, Khan, Z. Kinetics of the interaction of Cd(II)-histidine complex with ninhydrin in absence and presence of cationic and anionic micelles. *Int. J. Chem. Kinet.* **1997**, *29*(2), 131–138.

18. Rajanna, K.C., Reddy, K.N., Kumar, U.U., Saiprakash, P.K.A. A kinetic study of electron transfer from L-ascorbic acid to sodium perborate and potassium peroxy disulfate in aqueous acid and micellar media. *Int. J. Chem. Kinet.* **1996**, *28*(3), 153–164.

19. Matha, M., Sundari, L.B.T., Rajanna, K.C., Saiprakash, P.K. Kinetics and mechanism of hydrogen peroxide oxidation of chromone-3-carboxaldehydes in aqueous acid and micellar media. *Int. J. Chem. Kinet.* **1996**, *28*(9), 637–648.

20. Bunton, C.A., Robinson, L., Sepulveda, L. Structural effects upon catalysis by cationic micelles. *J. Org. Chem.* **1970**, *35*(1) 108–114.

21. Mukerjee, P. *Solution Chemistry of Surfactants*, Vol. 1, Mittal, K.L., Ed., Plenum: New York, **1979**, pp. 153–174.

22. Broxton, T.J., Christie, J.R., Chung, R.P.-T. Micellar catalysis of organic reactions. 23. Effect of micellar orientation of the substrate on the magnitude of micellar catalysis. *J. Org. Chem.* **1990**, *53*(13), 3081–3084.

23. Khan, M.N., Effects of salts and mixed CH_3CN-H_2O solvents on alkaline hydrolysis of phenyl benzoate in the presence of ionic micelles. *Colloids Surf. A* **1998**, *139*, 63–74.

24. Khan, M.N., Arifin, Z. Kinetics and mechanism of intramolecular general base-catalyzed methanolysis of ionized phenyl salicylate in the presence of cationic micelles. *Langmuir* **1996**, *12*(2), 261–268.

25. Senz, A., Gsponer, H.E. Micellar binding of phenoxide ions to cetyltrimethylammonium chloride (CTAC). *J. Colloid Interface Sci.* **1994**, *165*(1), 60–65.

26. Rico, I., Halvorsen, K., Dubrule, C., Lattes, A. Effects of micelles on cyclization reactions: the use of *N*-hexadecyl-2-chloropyridinium iodide as an amphiphilic carboxyl-activating agent in lactonization and lactamization. *J. Org. Chem.* **1994**, *59*(2), 415–420.

27. Khan, M.N., Arifin, Z. Effects of mixed CH_3CN-H_2O solvents on intramolecular general base-catalyzed methanolysis of ionized phenyl salicylate in the presence of cationic micelles. *Colloids Surf. A* **1997**, *125*, 149–154.

28. Khan, M.N., Arifin, Z., Wahab, I.B., Ali, S.F.M., Ismail, E. Effects of cationic micelles on rate of intramolecular general base-catalyzed reaction of ionized phenyl salicylate with tris-(hydroxymethyl)aminomethane (Tris). *Colloids Surf. A* **2000**, *163*, 271–281.

29. Khan, M.N., Arifin, Z. Effects of cationic micelles on the rate of intramolecular general base-catalyzed hydrolysis of ionized phenyl salicylate at different temperatures. *J. Chem. Res. (S)* **1995**, 132–133.

30. Huibers, P.D.T., Lobanov, V.S., Katritzky, A.R., Shah, D.O., Karelson, M. Prediction of critical micelle concentration using a quantitative structure-property relationship approach. 1. Nonionic surfactants. *Langmuir* **1996**, *12*(6), 1462–1470.

31. Buist, G.J., Bunton, C.A., Robinson, L., Sepulveda, L., Stam, M. Micellar effects upon the hydrolysis of bis-2,4-dinitrophenyl phosphate. *J. Am. Chem. Soc.* **1970**, *92*(13), 4072–4078.

32. (a) Fendler, E.J., Liechti, R.R., Fendler, J.H. Micellar effects on the hydrolysis of 2,4-dinitrophenyl sulfate. *J. Org. Chem.* **1970**, *35*(5), 1658–1661. (b) Fendler, J.H., Fendler, E.J., Smith, L.W. Micellar effects on hydrolysis and aminolysis of 2,4-dinitrophenyl sulfate. *J. Chem. Soc., Perkin Trans 2.* **1972**, 2097–2104.

33. Bunton, C.A., Wolfe, B. Problem of pH in micellar catalyzed reactions. *J. Am. Chem. Soc.* **1973**, *95*(11), 3742–3749.

34. Possidonio, S., Sivierro, F., El Seoud, O.A. Kinetics of the pH-independent hydrolysis of 4-nitrophenyl chloroformate in aqueous micellar solutions: effects of the charge and structure of the surfactant. *J. Phys. Org. Chem.* **1999**, *12*(4), 325–332.

35. Khan, M.N., Ismail, E., Yussof, M.R. Effects of pure nonionic and mixed nonionic-cationic surfactants on the rates of hydrolysis of phenyl salicylate and phenyl benzoate in the alkaline medium. *J. Phys. Org. Chem.* **2001**, *14*(10), 669–676.

36. Davies, D.M., Gillitt, N.D., Paradis, P.M. Catalytic and inhibition of the iodide reduction of peracids by surfactants: partitioning of reactants, product and transition state between aqueous and micellar pseudophases. *J. Chem. Soc., Perkin Trans. 2.* **1996**, 659–666.

37. Lajis, N.H., Khan, M.N. A partial model for serine proteases: kinetic demonstration of rate acceleration by intramolecular general base catalysis: an experiment for a biochemistry/enzymology lab. *Pertanika* **1991**, *14*, 193–199.

38. Khan, M.N., Fatope, I.L., Isaak, K.I., Zubair, M.O. Solvent effect on activation parameters for intramolecular general base-catalyzed hydrolysis of salicylate esters and hydroxide ion-catalyzed hydrolysis of methyl *p*-hydroxybenzoate. *J. Chem. Soc., Perkin Trans. 2.* **1986**, 655–659.

39. Bender, M.L., Kezdy, F.J., Zerner, B. Intramolecular catalysis in the hydrolysis of *p*-nitrophenyl salicylates. *J. Am. Chem. Soc.* **1963**, *85*(19), 3017–3024.

40. Capon, B., Gosh, B.C. The mechanism of the hydrolysis of phenyl salicylate and catechol monobenzoate in the presence and absence of borate ions. *J. Chem. Soc. B* **1966**, 472–478.

41. Khan, M.N. The mechanistic diagnosis of induced catalysis in the aqueous cleavage of phenyl salicylate in the presence of borate buffer. *J. Mol. Catal.* **1987**, *40*, 195–210.

42. Broxton, T.J., Wright, S. Micellar catalysis of organic reactions. 18. Basic hydrolysis of diazepam and some *N*-alkyl derivatives of nitrazepam. *J. Org. Chem.* **1986**, *51*(15), 2965–2969.

43. Jonstroemer, M., Joensson, B., Lindman, B. Self-diffusion in nonionic surfactant-water systems. *J. Phys. Chem.* **1991**, *95*(8), 3293–3300.

44. Phillies, G.D.J., Yambert, J.E. Solvent and solute effects on hydration and aggregation numbers of Triton X-100 micelles. *Langmuir* **1996**, *12*(14), 3431–3436.

45. Romsted, L.S., Yao, J. Arenediazonium salts: new probes of the interfacial compositions of association colloids. 4. Estimation of the hydration numbers of aqueous hexaethylene glycol monododecyl ether, $C_{12}E_6$, micelles by chemical trapping. *Langmuir* **1996**, *12*(10), 2425–2432.

46. Van Os, N.M., Haak, J.R., Rupert, L.M.A. (Eds.) *Physico–Chemical Properties of Selected Anionic, Cationic and Nonionic Surfactants*, Elsevier: Amsterdam, **1993**.

47. Iglesias, E. Ethyl cyclohexanone-2-carboxylate in aqueous micellar solutions. 1. Ester hydrolysis in cationic and nonionic micelles. *J. Phys. Chem. B* **2001**, *105*(42), 10287–10294.

48. Del Rosso, F., Bartoletti, A., Di Profio, P., Germani, R., Savelli, G., Blasko, A., Bunton, C.A. Hydrolysis of 2,4-dinitrophenyl phosphate in normal and reverse micelles. *J. Chem. Soc., Perkin Trans. 2.* **1995**, 673–678.

49. Bunton, C.A., Kamego, A.A., Ng, P. Micellar effects upon the decomposition of 3-bromo-3-phenylpropionic acid: effect of changes in surfactant structure. *J. Org. Chem.* **1974**, *39*(24), 3469–3471.

50. Al-Lohedan, H., Bunton, C.A., Mhala, M.M. Micellar effects upon spontaneous hydrolysis and their relation to mechanism. *J. Am. Chem. Soc.* **1982**, *104*(24), 6654–6660.

51. Menger, F.M. The structure of micelles. *Acc. Chem. Res.* **1979**, *12*(4), 111–117.

52. Menger, F.M., Yoshinaga, H., Venkatasubban, K.S., Das, A.R. Solvolysis of a carbonate and a benzhydryl chloride inside micelles: evidence for a porous cluster micelle. *J. Org. Chem.* **1981**, *46*(2), 415–419.

53. (a) Menger, F.M. Laplace pressure inside micelle. *J. Phys. Chem.* **1979**, *83*(7), 893. (b) Menger, F.M., Chow, J.F. Testing theoretical models of micelles: the acetylenic probe. *J. Am. Chem. Soc.* **1983**, *105*(16), 5501–5502. (c) Menger, F.M. On the structure of micelles. *J. Am. Chem. Soc.* **1984**, *106*(4), 1109–1113.

54. Brinchi, L., Di Profio, P., Micheli, F., Germani, R., Savelli, G., Bunton, C.A. Structure of micellar head-groups and the hydrolysis of phenyl chloroformate — the role of perchlorate ion. *Eur. J. Org. Chem.*, **2001**, (6) 1115–1120.

55. Brinchi, L., Di Profio, P., Germani, R., Savelli, G., Spreti, N., Bunton, C.A. The Hammett equation and micellar effects on S_N2 reactions of methyl benzenesulfonates — the role of micellar polarity. *Eur. J. Org. Chem.*, **2000**, (23), 3849–3854.

56. Engberts, J.B.F.N. Catalysis by surfactant aggregates in aqueous solutions. *Pure Appl. Chem.* **1992**, *64*(11), 1653–1660.

57. (a) Di Profio, P., Germani, R., Savelli, G., Cerichelli, G., Spreti, N., Bunton, C.A. Cyclization and decarboxylation in zwitterionic micelles: effects of head group structure. *J. Chem. Soc., Perkin Trans. 2.* **1996**, 1505–1509. (b) Bunton, C.A., Kamego, A.A., Minch, M.J., Wright, J.L. Effect of changes in surfactant structure on micellarly catalyzed spontaneous decarboxylations and phosphate ester hydrolysis. *J. Org. Chem.* **1975**, *40*(9), 1321–1327. (c) Germani, R., Ponti, P.P., Savelli, G., Spreti, N., Cipiciani, A., Cerichelli, G., Bunton, C.A. *J. Chem. Soc., Perkin Trans. 2.* **1989**, 1767–1771. (d) Germani, R., Ponti, P.P., Romeo, T., Savelli, G., Spreti, N., Cerichelli, G., Luchetti, L., Mancini, G., Bunton, C.A. Decarboxylation of 6-nitrobenzisoxazole-3-carboxylate in cationic micelles: effect of head group size. *J. Phys. Org. Chem.* **1989**, *2*(7), 553–558. (e) Brinchi, L., Germani, R., Savelli, G., Spreti, N. Decarboxylation of 6-nitrobenzisoxazole-3-carboxylate as kinetic probe for piperazinium-based cationic micelles. *J. Colloid Interface Sci.* **2004**, *274*(2), 701–705.

58. (a) Kemp, D.S., Paul, K.G. The physical organic chemistry of benzisoxazoles. III. The mechanism and the effects of solvents on rates of decarboxylation of benzisoxazole-3-carboxylic acids. *J. Am. Chem. Soc.* **1975**, *97*(25), 7305–7312. (b) Thomson, A. Decarboxylations of 2-cyano-2-phenylacetate ions in water, aqueous ethanol, and aqueous dioxane. *J. Chem. Soc. B* **1970**, 1198–1201.

59. Bacaloglu, R., Bunton, C.A., Cerichelli, G., Ortega, F. NMR study of the location of bromide ion and methyl naphthalene-2-sulfonate in cationic micelles: relation to reactivity. *J. Phys. Chem.* **1989**, *93*(4), 1490–1497.

60. Angeli, A.D., Cipiciani, A., Germani, R., Savelli, G., Cerichelli, G., Bunton, C.A. Hydration of *p*-alkoxy-α,α,α-trifluoroacetophenone and water activity at a micellar surface. *J. Colloid Interface Sci.* **1988**, *121*(1), 42–48.

61. Costas-Costas, U., Bravo-Diaz, C., Gonzalez-Romero, E. Sodium dodecyl sulfate micellar effects on the reaction between arenediazonium ions and ascorbic acid derivatives. *Langmuir* **2003**, *19*(13), 5197–5203.

62. Bunton, C.A., Dorwin, E.L., Savelli, G., Si, V.C. Hydrolysis of 2,4-dinitrophenyl phosphate catalyzed by single-chain, twin-tailed and bolaform surfactants. *Recl. Trav. Chim. Pays-Bas* **1990**, *109*(2), 64–69.

63. Bunton, C.A., Diaz, S., Hellyer, J.M., Ihara, Y., Ionescu, L.G. Micellar effects upon the reactions of 2,4-dinitrophenyl phosphate and ethyl *p*-nitrophenyl phosphate with amines. *J. Org. Chem.* **1975**, *40*(16), 2313–2317.

64. (a) Kirby, A.J., Varvoglis, A.G. Reactivity of phosphate esters: monoester hydrolysis. *J. Am. Chem. Soc.* **1967**, *89*(2), 415–423. (b) Ramirez, F., Marecek, J.F. Oxyphosphorane and monomeric metaphosphate ion intermediates in phosphoryl transfer from 2,4-dinitrophenyl phosphate in aprotic and protic solvents. *J. Am. Chem. Soc.* **1979**, *101*(6), 1460–1465.

65. Bunton, C.A., Fendler, E.J., Fendler, J.H. Hydrolysis of dinitrophenyl phosphates. *J. Am. Chem. Soc.* **1967**, *89*(5), 1221–1230.

66. Khan, M.N. Structure-reactivity relationships and the rate-determining step in the nucleophilic cleavage of phenyl salicylate with primary and secondary amines. *J. Org. Chem.* **1983**, *48*(12), 2046–2052.

67. Khan, M.N. Structure-reactivity relationships and the rate-determining step in the reactions of methyl salicylate with primary amines. *Int. J. Chem. Kinet.* **1987**, *19*(5), 415–434.

68. Khan, M.N. Effects of anionic micelles on the intramolecular general base-catalyzed hydrazinolysis and hydrolysis of phenyl salicylate. *J. Chem. Soc., Perkin Trans. 2.* **1990**, 445–457.

69. Khan, M.N., Dahiru, M., Naaliya, J. Effects of anionic micelles on intramolecular general base-catalyzed aminolysis of phenyl and methyl salicylates. *J. Chem. Soc., Perkin Trans. 2.* **1989**, 623–628.

70. Khan, M.N. Spectrophotometric determination of anionic micellar binding constants of ionized and nonionized phenyl and methyl salicylates. *J. Phys. Org. Chem.* **1996**, *9*(5), 295–300.

71. Khan, M.N. Effect of anionic micelles on the intramolecular general base-catalyzed hydrolysis of phenyl and methyl salicylates. *J. Mol. Catal. A* **1995**, *102*, 93–101.

72. Suratkar, V., Mahapatra, S. Solubilization site of organic perfume molecules in sodium dodecyl sulfate micelles: new insights from proton NMR studies. *J. Colloid Interface Sci.* **2000**, *225*(1), 32–38.

73. Khan, M.N. Kinetics and mechanism of transesterification of phenyl salicylate. *Int. J. Chem. Kinet.* **1987**, *19*(8), 757–776.

74. Khan, M.N. Kinetic probe to study the structure of the mixed aqueous-organic solvents: kinetics and mechanism of nucleophilic reactions of *n*-propanol and D-(+)-glucose with ionized phenyl salicylate. *J. Phys. Chem.* **1988**, *92*(22), 6273–6278.

75. Khan, M.N., Arifin, Z. Effects of cationic micelles on rate of intramolecular general base-catalyzed ethanediolysis of ionized phenyl salicylate (PS⁻). *Langmuir* **1997**, *13*(25), 6626–6632.

76. (a) Khan, M.N. Kinetic probe to study the structural behavior of mixed aqueous-organic solvents and micelles: kinetics and mechanism of intramolecular general base-catalyzed alkanolysis of phenyl salicylate in the presence of anionic micelles. *Int. J. Chem. Kinet.* **1991**, *23*(9), 837–852. (b) Khan, M.N., Audu, A.A. Kinetic probe to study the structural behavior of the mixed aqueous-organic solvents: effects of the inorganic and organic salts on the kinetics and mechanism intramolecular general base-catalyzed methanolysis of ionized phenyl salicylate. *J. Phys. Org. Chem.* **1992**, *5*(3), 129–141. (c) Khan, M.N. Transesterification of phenyl salicylate. II. Kinetics and mechanism of intramolecular general base-catalyzed cleavage of phenyl salicylate under the presence of 1,2-ethanediol and 2-ethoxy-ethanol. *Int. J. Chem. Kinet.* **1988**, *20*(6), 443–454. (d) Khan, M.N., Audu, A.A. Kinetic probe to study the structural behavior of the mixed aqueous-organic solvents: salt effects on the kinetics and mechanism of intramolecular general base-catalyzed ethanolysis of ionized phenyl salicylate. *Int. J. Chem. Kinet.* **1990**, *22*(1), 37–57. (e) Khan, M.N. Effects of urea, Na⁺, and Li⁺ ions on the kinetics and mechanism of intramolecular general base-catalyzed glycolysm of ionized phenyl salicylate in HOCH₂CH₂OH-CH₃CN solvents at a constant water concentration. *J. Phys. Org. Chem.* **1998**, *11*(2), 109–114.

77. Rodriguez, A., Graciani, M. del M., Munoz, M., Moya, M.L. Water-ethylene glycol alkyltrimethylammonium bromides micellar solutions as reaction media: study of spontaneous hydrolysis of phenyl chloroformate. *Langmuir* **2003**, *19*(18), 7206–7213.

78. Buurma, N.J., Serena, P., Blandamer, M.J., Engberts, J.B.F.N. The nature of the micellar Stern region as studied by reaction kinetics. 2. *J. Org. Chem.* **2004**, *69*(11), 3899–3906.

79. Bunton, C.A., Gillitt, N.D., Mhala, M.M., Moffatt, J.R. Spontaneous hydrolysis in sulfobetaine micelles: dependence of micellar charge effects upon mechanism. *Croatica Chemica Acta* **2001**, *74*(3), 559–573.

80. Brinchi, L., Di Profio, P., Germani, R., Giacomini, V., Savelli, G., Bunton, C.A. Surfactant effects on decarboxylation of alkoxynitrobenzisoxazole-3-carboxylate ions: acceleration by premicelles. *Langmuir* **2000**, *16*(1), 222–226.

81. Buwalds, R.I., Janker, J.M., Engberts, J.B.F.N. Aggregation of azo dyes with cationic amphiphiles at low concentrations in aqueous solution. *Langmuir* **1999**, *15*(4), 1083–1089.

82. Brinchi, L., Di Profio, P., Germani, R., Savelli, G., Tugliani, M., Bunton, C.A. Hydrolyses of dinitroalkoxyphenyl phosphates in aqueous cationic micelles: acceleration by premicelles. *Langmuir* **2000**, *16*(26), 10101–10105.

83. Munoz, M., Rodriguez, A., Graciani, M. del M., Moya, M.L. Micellar medium effects on the hydrolysis of phenyl chloroformate in ionic, zwitterionic, nonionic, and mixed micellar solutions. *Int. J. Chem. Kinet.* **2002**, *34*(7), 445–451.

84. Brinchi, L., Di Profio, P., Germani, R., Savelli, G., Spreti, N. Effect of ethanol on micellization and on decarboxylation of 6-nitrobenzisoxazole-3-carboxylate in aqueous cationic micelles. *J. Colloid Interface Sci.* **2002**, *247*(2), 429–436.

85. El Seoud, O.A., Ruasse, M.-F., Possidonio, S. pH-independent hydrolysis of 4-nitrophenyl 2,2-dichloropropionate in aqueous micellar solutions: relative contributions of hydrophobic and electrostatic interactions. *J. Phys. Org. Chem.* **2001**, *14*(8), 526–532.

86. Iglesias, E. Ethyl cyclohexanone-2-carboxylate in aqueous micellar solutions. 2. Enol nitrosation in anionic and cationic micelles. *J. Phys. Chem. B* **2001**, *105*(42), 10295–10302.

87. Khan, M.N., Gambo, S.K. Intramolecular catalysis and the rate-determining step in the alkaline hydrolysis of ethyl salicylate. *Int. J. Chem. Kinet.* **1985**, *17*(4), 419–428.

88. Jencks, W.P., Carriuolo, J. Reactivity of nucleophilic reagents toward esters. *J. Am. Chem. Soc.* **1960**, *82*, 1778–1786.

89. Bunton, C.A., Foroudian, H.J. A quantitative treatment of micellar effects upon dephosphorylation by the hydroxide anion. *Langmuir* **1993**, *9*(11), 2832–2835.

90. Bunton, C.A., Mhala, M.M., Moffatt, J.R. Reactions of anionic nucleophiles in anionic micelles: a quantitative treatment. *J. Phys. Chem.* **1989**, *93*(23), 7851–7856.

91. (a) Koerner, T.B., Brown, R.S. The hydrolysis of an activated ester by a tris(4,5-di-n-propyl-2-imidazolyl)phosphine-Zn^{2+} complex in neutral micellar medium as a model for carbonic anhydrase. *Can. J. Chem.* **2002**, *80*(2), 183–191. (b) Mondai, S.K., Das, M., Kar, D., Das, A.K. Micellar effect on the reaction of chromium(VI) oxidation of formaldehyde in the presence and absence of picolinic acid in aqueous acid media: a kinetic study. *Indian J. Chem.* **2001**, *40A*(4), 352–360. (c) Drummond, C.J., Grieser, F. A study of competitive counterion binding to micelles using the acid-catalyzed reaction of hydrogen peroxide with iodide ions. *J. Colloids Interface Sci.* **1989**, *127*(1), 281–291.

92. Hartley, G.S. State solution of colloidal electrolytes. *Quart. Rev. Chem. Soc.* **1948**, 2, 152–183.

93. Bunton, C.A., Robinson, L. Micellar effects upon nucleophilic aromatic and aliphatic substitution. *J. Am. Chem. Soc.* **1968**, *90*(22), 5972–5979.

94. Bunton, C.A., Foroudian, H.J., Gillitt, N.D., Whiddon, C.R. Reactions of *p*-nitrophenyl diphenyl phosphinate with fluoride and hydroxide ion in nonionic micelles: kinetic salt effects. *J. Colloid Interface Sci.* **1999**, *215*(1), 64–71.

95. Bunton, C.A., Kamego, A., Sepulveda, L. Inhibition of the hydrolysis of bis-2,4-dinitrophenyl phosphate by a nonionic detergent. *J. Org. Chem.* **1971**, *36*(17), 2571–2572.

96. Bunton, C.A., Foroudian, H.J., Gillitt, N.D., Whiddon, C.R. Dephosphorylation and aromatic nucleophilic substitution in nonionic micelles: the importance of substrate location. *Can. J. Chem.*, **1998**, *76*(6), 946–954.

97. Bunton, C.A., Robinson, L. Micellar effects upon the reaction of *p*-nitrophenyl diphenyl phosphate with hydroxide and fluoride ions. *J. Org. Chem.* **1969**, *34*(4), 773–778.

98. Gellman, S.H., Petter, R., Breslow, R. Catalytic hydrolysis of a phosphate trimester by tetracoordinated zinc complexes. *J. Am. Chem. Soc.* **1986**, *108*(9), 2388–2394.

99. Lajis, N.H., Khan, M.N. Effects of ionic and nonionic micelles on alkaline hydrolysis of securinine: evidence for an unusually weak micellar incorporation of securinine molecules. *J. Phys. Org. Chem.* **1998**, *11*(3), 209–215.

100. Lajis, N.H., Noor, H.M., Khan, M.N. Kinetics and mechanism of the alkaline hydrolysis of securinine. *J. Pharm. Sci.* **1995**, *84*(1), 126–130.

101. Khan, M.N., Ismail, E. Effects of nonionic and mixed cationic-nonionic micelles on the rate of alkaline hydrolysis of 4-nitrophthalimide. *J. Colloid Interface Sci.* **2001**, *240*(2), 636–639.

102. Khan, M.N. Effect of anionic micelles on alkaline hydrolyses of N-hydroxyphthalimide: evidence for the probable occurrence of the reaction in between Gouy-Chapman and exterior boundary of Stern layers of the micelles. *Int. J. Chem. Kinet.* **1991**, *23*(7), 567–578.

103. Khan, M.N., Abdullah, Z. Kinetics and mechanism of alkaline hydrolysis of 4-nitrophthalimide in the absence and presence of cationic micelles. *Int. J. Chem. Kinet.* **2001**, *33*(7), 407–414.

104. Davies, D.M., Gillitt, N.D. Reaction of deuteroferrihaem and *m*-chloroperoxybenzoic acid in surfactant micelles. *J. Chem. Soc., Dalton Trans.* **1997**, 2819–2823.

105. Khan, M.N., Ismail, E. Effects of non-ionic and mixed cationic-non-ionic micelles on the rate of alkaline hydrolysis of phthalimide. *J. Phys. Org. Chem.* **2002**, *15*, 374–384.

106. (a) Bender, M.L. General acid-base catalysis in the intramolecular hydrolysis of phthalamic acid. *J. Am. Chem. Soc.* **1957**, *79*, 1258–1259. (b) Bender, M.L., Chow, Y.-L., Chloupek, F. Intramolecular catalysis of hydrolytic reactions. II. The hydrolysis of phthalamic acid. *J. Am. Chem. Soc.* **1958**, *80*, 5380–5384. (c) Kirby, A.J., McDonald, R.S., Smith, C.R. Intramolecular catalysis of amide hydrolysis by two carboxy groups. *J. Chem. Soc., Perkin Trans. 2.* **1974**, 1495–1504. (d) Aldersley, M.F., Kirby, A.J., Lancaster, P.W., McDonald, R.S., Smith, C.R. Intramolecular catalysis of amide hydrolysis by carboxy group: rate-determining proton transfer from external general acids in the hydrolysis of substituted maleamic acids. *J. Chem. Soc., Perkin Trans. 2.* **1974**, 1478–1495. (e) Hawkins, M.D. Intramolecular catalysis. Part III. Hydrolysis of 3- and 4-substituted phthalanilic acids. *J. Chem. Soc., Perkin Trans. 2.* **1976**, 642–647. (f) Kluger, R., Lam, C.-H. Carboxylic acid participation in amide hydrolysis: external general base catalysis and general acid catalysis in reactions of norbornenylanilic acids. *J. Am. Chem. Soc.* **1978**, *100*(7), 2191–2197. (g) Menger, F.M., Ladika, M. Fast hydrolysis of an aliphatic amide at neutral pH and ambient temperature: a peptidase model. *J. Am. Chem. Soc.* **1988**, *110*(20), 6794–6796. (h) Blackburn, R.A.M., Capon, B., McRitchie, A.C. The mechanism of hydrolysis of phthalamic and N-phenylphthalamic acid: the spectrophotometric detection of phthalic anhydride as an intermediate. *Bioorg. Chem.* **1977**, *6*(1), 71–78. (i) Khan, M.N. Suggested improvement in the Ing-Manske procedure and Gabriel synthesis of primary amines: kinetic study on alkaline hydrolysis of N-phthaloylglycine and acid hydrolysis of N-(*o*-carboxybenzoyl)glycine in aqueous organic solvents. *J. Org. Chem.* **1996**, *61*(23), 8063–8068.

107. (a) Hupe, D.J., Jencks, W.P. Nonlinear structure-reactivity correlations: acyl transfer between sulfur and oxygen nucleophiles. *J. Am. Chem. Soc.* **1977**, *99*(2), 451–464. (b) Khan, M.N., Gleen, P.C., Arifin, Z. Effects of inorganic salts and mixed aqueous-organic solvents on the rates of alkaline hydrolysis of aspirin. *Indian J. Chem.* **1996,** *35A*(9), 758–765.

108. Khan, M.N. Effects of mixed CH_3OH-H_2O and CH_3OH solvents on rate of reaction of phthalimide with piperidine. *Int. J. Chem. Kinet.* **2001**, *33*(1), 29–40.

109. Shafer, J.A., Morawetz, H. Participation of a neighboring amide group in the decomposition of esters and amides of substituted phthalamic acids. *J. Org. Chem.* **1963**, *28*(7), 1899–1901.

110. Davies, D.M., Foggo, S.J., Paradis, P.M. Micellar kinetics of acyl transfer from *n*-nonanoyloxybenzenesulfonate and phenyl nonanoate bleach activators to hydrogen peroxide and pernonanoic acid: effect of charge on the surfactant and activator. *J. Chem. Soc., Perkin Trans. 2.* **1998**, 1597–1602.

111. Otto, S., Engberts, J.B.F.N., Kwak, J.C.T. Million-fold acceleration of a Diels–Alder reaction due to combined Lewis acid and micellar catalysis in water. *J. Am. Chem. Soc.* **1998**, *120*(37), 9517–9525.

112. Rispens, T., Engberts, J.B.F.N. A kinetic study of 1,3-dipolar cycloadditions in micellar media. *J. Org. Chem.* **2003**, *68*(22), 8520–8528.

113. Fernandez, A., Iglesias, E., Garcia-Rio, L., Leis, J.R. Chemical reactivity and basicity of amines modulated by micellar solutions. *Langmuir* **1995**, *11*(6), 1917–1924.

114. Iglesias, E., Leis, J.R., Pena, M.E. Micellar effects on the nitrosation of piperidines by 2-bromoethyl nitrite and 1-phenylethyl nitrite in basic media. *Langmuir* **1994**, *10*(3), 662–669.

115. Khan, M.N. Effects of anionic micelles on intramolecular general base-catalyzed aminolysis of ionized phenyl salicylate (PS⁻). *J. Phys. Org. Chem.* **1999**, *12*, 1–9.

116. Rispens, T., Engberts, J.B.F.N. Micellar catalysis of Diels–Alder reactions: substrate positioning in the micelle. *J. Org. Chem.* **2002**, *67*(21), 7369–7377.

117. Bhattacharya, S., Kumar, V.P. Evidence of enhanced reactivity of DAAP nucleophiles toward dephosphorylation and deacylation reactions in cationic Gemini micellar media. *J. Org. Chem.* **2004**, *69*(2), 559–562.

118. Menger, F.M., Keiper, J.S., Mbadugha, B.N.A., Caran, K.L., Romsted, L.S. Interfacial composition of gemini surfactant micelles determined by chemical trapping. *Langmuir* **2000**, *16*(23), 9095–9098.

119. Graciani, M.M., Rodriguez, A., Munoz, M., Moya, M.L. Water-ethylene glycol alkyltrimethylammonium bromide micellar solutions as reaction media: study of the reaction methyl 4-nitrobenzenesulfonate + Br⁻. *Langmuir* **2003**, *19*(21), 8685–8691.

120. Correia, V.R., Cuccovia, I.M., Chaimovich, H. Effect of hexadecyltrimethylammonium bromide micelles on the hydrolysis of substituted benzoate esters. *J. Phys. Org. Chem.* **1991**, *4*(1), 13–18.

121. Tada, E.B., Ouarti, N., Silva, P.L., Blagoeva, I.B., El Seoud, O.A., Ruasse, M.-F. Nucleophilic reactivity of the CTACl-micelle-bound fluoride ion: the influence of water concentration and ionic strength at the micellar interface. *Langmuir* **2003**, *19*(26), 10666–10672.

122. Zakharova, L. Ya., Kudryavtseva, L.A., Shagidullina, R.A., Valeeva, F.G. The factor determining the micellar effects on nucleophilic substitution reactions. *J. Mol. Liq.* **2001**, *94*(1), 79–86.

123. Costas-Costas, U., Bravo-Diaz, C., Gonzalez-Romero, E. Micellar effects on the reaction between an arenediazonium salt and 6-O-octanoyl-L-ascorbic acid: kinetics and mechanism of the reaction. *Langmuir* **2004**, *20*(5), 1631–1632.

124. Jencks, W.P. *Catalysis in Chemistry and Enzymology*, McGraw-Hill: New York, **1969**.

125. (a) Cerichelli, G., Luchetti, L., Savelli, G., Bunton, C.A. Specific micellar rate effects on unimolecular decarboxylation and cyclization. *J. Phys. Org. Chem.* **1991**, *4*(2), 71–76. (b) Mandolini, L. Intramolecular reactions of chain molecules. *Adv. Phys. Org. Chem.* **1986**, *22*, 1–111.

126. Cerichelli, G., Luchetti, L., Mancini, G., Muzzioli, M.N., Germani, R., Ponti, P.P., Spreti, N., Savelli, G., Bunton, C.A. Solvent and micellar effects upon the cyclization of *o*-3-halopropyloxyphenoxide ions. *J. Chem. Soc., Perkin Trans. 2.* **1989**, 1081–1085.

127. Khan, M.N. Kinetics and mechanism of the aqueous cleavage of *N,N*-dimethylphthalamic acid (NDPA): evidence of intramolecular catalysis in the cleavage. *Indian J. Chem.* **1993**, *32A*, 395–401.

128. Blagoeva, I.B., Toteva, M.M., Ouarti, N., Ruasse, M.-F. Changes in the relative contribution of specific and general base catalysis in cationic micelles: the cyclization of substituted ethyl hydantoates. *J. Org. Chem.* **2001**, *66*(6), 2123–2130.

129. Takag, Y., Warashina, M., Stec, W.J., Yoshinari, K., Taira, K. Recent advances in the elucidation of the mechanisms of action of ribozymes. *Nucl. Acid. Res.* **2001**, *29*(9), 1815–1834.

130. Riepe, A., Beier, H., Gross, H.J. Enhancement of RNA self-cleavage by micellar catalysis. *FEBS Lett.* **1999**, *457*(2), 193–199.

131. Khan, M.N. Effects of cetyltrimethylammonium bromide (CTABr) micelles on the rate of the cleavage of phthalimide in the presence of piperidine. *Colloids Surf. A* **2001**, *181*, 99–114.

132. Khan, M.N., Arifin, Z., Lasidek, M.N., Hanifiah, M.A.M., Alex, G. Effects of cationic micelles on rate of intramolecular general base-catalyzed aminolysis of ionized phenyl salicylate. *Langmuir* **1997**, *13*(15), 3959–3964.

133. Khan, M.N. Effects of CH_3CN-H_2O and CH_3CN solvents on rates of piperidinolysis of phthalimide. *J. Phys. Org. Chem.* **1999**, *12*(3), 187–193.

134. (a) Bergbreiter, D.E., Zhang, L., Mariagnanam, V.M. Smart ligands that regulate homogeneously catalyzed reactions. *J. Am. Chem. Soc.* **1993**, *115*(20), 9295–9296. (b) Bergbreiter, D.E., Caraway, J.W. Thermoresponsive polymer-bound substrates. *J. Am. Chem. Soc.* **1996**, *118*(25), 6092–6093. (c) Bergbreiter, D.E., Case, B.L., Liu, Y.-S., Caraway, J.W. Poly(*N*-isopropylacrylamide) soluble polymer supports in catalysis and synthesis. *Macromolecules* **1998**, *31*(18), 6053–6062.

135. Davies, D.M., Stringer, E.L. Smart aqueous reaction medium. *Langmuir* **2003**, *19*(6), 1927–1928.

136. Ruzza, A.A., Nome, F., Zanette, D., Romsted, L.S. Kinetic evidence for temperature-induced demixing of a long chain dioxolane in aqueous micellar solutions of sodium dodecyl sulfate: a new application of the pseudophase ion exchange model. *Langmuir* **1995**, *11*(7), 2393–2398.

5 Mixed Normal Micelles: Effects on Reaction Rates

5.1 INTRODUCTION

Mixed normal micelle is defined as the normal micelle formed from monomers of two different (in terms of either head groups or hydrophobic tails or both) micelle-forming surfactants in aqueous solvent. Mixed micellar systems have been the subject of growing theoretical and practical interest because such systems are encountered in nearly all practical uses of surfactants.[1-3] Surface-active compounds used in commercial applications typically consist of a mixture of surfactants because they can be produced at a cost relatively lower than that of pure surfactants. In addition, mixed surfactants generally provide better performance than a single pure surfactant in practical fields,[1a,4] where the compositions and concentrations can be optimized for each practical application. Because different types of surfactants exist, various kinds of combinations are possible to produce different properties and application fields. Mixed aqueous solutions of nonionic and ionic surfactants have been intensively investigated during the last nearly two decades.[5-18] Most studies on mixed surfactant systems are concerned with physicochemical aspects such as CMC measurements,[1a,19-23] micelle composition, micellar size and aggregation numbers,[20,23,25-28] micelle demixing, modeling, etc., of these systems.[1,21,27,28] Physical properties of mixed micellar solutions could be understood in terms of various theoretical models such as Clint,[29] Rubingh,[1a,30,31] Motomura,[32] or Georgiev,[33] Nagarajan,[34,35] or Puwada's model.[36,37] Relatively recent investigations, however, have pointed out serious limitations of the use of regular solution treatment in various theoretical models.[38,39]

The behavior of the mixed system dodecylethyldimethylammonium bromide (DEDABr)/dodecyltrimethylammonium bromide (DTABr)/H_2O has been studied by conductivity and steady-state fluorescence measurements where a critical review of various theoretical models is also described.[40] A new experimental approach has been considered to determine the partial contribution of each surfactant to the mixed micellization process, through their critical micellar concentrations, CMC_1^* and CMC_2^* and their aggregation numbers, N_1^* and N_2^*. The values of CMC_1^* and CMC_2^* were obtained from the break points in the plots of their respective specific conductivities, κ, as a function of [DEDABr], at several premicellar [DTABr] and κ as a function of [DTABr], at several premicellar [DEDABr], whereas the values of CMC* (total critical mixed micellar concentration) were obtained from the break points in the plots of κ vs. total mixed surfactant concentration, $[S]_T$ (= [DEDABr] + [DTABr]), at various fixed values of the molar fraction, X_1, for mixed system DEDABr(1) + DTABr(2). These observed data fit reasonably well to Equation 5.1

$$X_1^{CMC^*} = \frac{CMC_1^*}{CMC_1^* + CMC_2^*} = \frac{CMC_1^*}{CMC^*} \qquad (5.1)$$

where $X_1^{CMC^*}$ is defined as the *molar fraction* just at the CMC*. The concentration range of both surfactants for which the mixed micellization may occur is best shown by a linear plot of CMC_2^* vs. CMC_1^*, which not only allows one to obtain CMC_1^* or CMC_2^*, and from them, the CMC* at any surfactant composition, but also divides the composition space in two regions: the mixed micelle region (above the linear plot) and the nonmicelle region (below the linear plot). The linear relationship between CMC_2^* and CMC_1^* is known as the *limit mixed micellization line*.[40]

The study usually done on mixed micelles is analysis of a physicochemical property of the mixed micellar system as a function of total surfactant concentration, keeping constant the molar fraction of the system. These kinds of studies permit one to obtain global information on both monomeric and micellar phases, through the values of CMC* and N* (total aggregation number), respectively, and are necessary to check theoretical models.[1a] However, these results do not give information about the separated contribution of surfactants constituting the mixed system, which can be obtained only by performing studies in which the molar fraction of the system is changing throughout the experiment, because in these cases, the physicochemical property is measured as a function of the concentration of one of the surfactants while that of the other one is kept constant.[40]

The aggregation numbers of mixed SDS/$C_{12}E_8$ micelles vary monotonically with the composition from the value of the aggregation number of pure $C_{12}E_8$ to that of pure SDS.[15] Addition of 0.1 M NaCl to a mixed micellar system SDS/dodecylmalono-bis-N-methylglucamide has no effect on the CMC, micelle size, and shape.[12] The formation of rodlike micelles in mixed CTABr/$C_{12}E_6$ micellar solution containing salt is observed.[41] Tablet-shaped and ribbon-like micellar structures have been discovered in SDS/dodecyltrimethylammonium bromide (DTABr).[42] Electron spin-echo modulation measurements show that DTACl head groups are located deeper inside the mixed DTACl/$C_{12}E_6$ micelle than SDS head groups in mixed SDS/$C_{12}E_6$ micelle, perhaps reflecting specific ineractions.[43] The small-angle neutron scattering (SANS) measurements on mixtures of SDS and $C_{12}E_6$ of varying composition in heavy water reveal the formation of mixed micelles.[44] Micelles become smaller, and the degree of dissociation of SDS molecules in the micelles decreases from 0.78 to 0.28 with increase in X_{SDS} {=

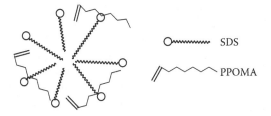

FIGURE 5.1 Structure of SDS-PPOMA mixed micellar system.

[SDS]/([SDS] + [$C_{12}E_8$])} from 0.05 to 1.0 at 25 mM ([SDS] + [$C_{12}E_8$]) and 15°C.[44] It has been concluded that the formation of mixed micelles is subjected to the delicate balance of different contributors such as the size and charge of polar groups, length of the hydrophobic chains of surfactants, and solution characteristics (temperature, pH, added salt, etc.).[44]

Poly(propylene oxide) methylacrylate (PPOMA) (of molecular weight = 434 g mol^{-1}) forms mixed micelles when mixed with SDS, and the total aggregation numbers of mixed micelle (N_{SDS} + N_{PPOMA}) and N_{SDS} decrease nonlinearly from respective 58 to 38 and from 58 to 21, whereas the degree of ionization (α) of SDS increases from 0.371 to 0.696 with the increase in the weight ratio WR (= PPOMA/SDS) from 0 to 1.0.[45] Fluorescence quenching and ^{1}H NMR measurements on mixed SDS/PPOMA systems reveal that: (1) the hydrophobic core of mixed micelle is primarily composed of the C_{12} chains of SDS, (2) the mixed micellar surface is made up of the polar head groups of SDS as well as methylacrylate functions of PPOMA, and (3) the poly(propylene oxide) chains of PPOMA monomers are adsorbed preferentially at the surface as shown in Figure 5.1.[45] The relative arrangement of surfactant molecules in the mixed nonionic/cat-

$$CH_2{=}C\begin{array}{l}CH_3\\[4pt]C{-}O{-}\left[CH{-}CH_2{-}O\right]_6{-}H\\ \parallel\quad\ \ |\\ O\quad\ \ CH_3\end{array}$$

PPOMA

SCHEME 5.1

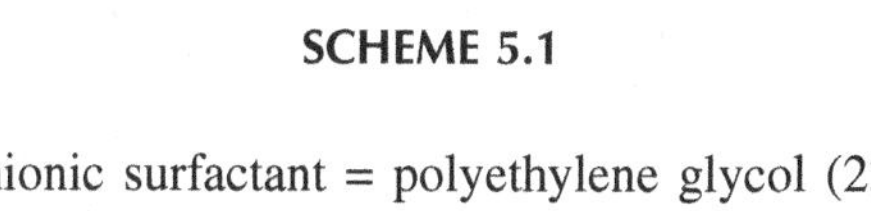

ionic micelles (nonionic surfactant = polyethylene glycol (23) lauryl ether and Brij35 and ionic surfactant = SDS and CTABr) has been studied by various techniques including ^{1}H NMR, and the observed results show that: (1) for CTABr/Brij35 mixed micelles, the hydrophobic chains of both surfactants are coaggregated in the micellar core, (2) the trimethyl groups of CTABr molecules are located between the first oxyethylene groups next to the hydrophobic chains of the Brij35 molecules, (3) the CTABr head groups cause the first oxyethylene groups of the Brij35 molecules to extend outward from the hydrophobic core with an increase in [CTABr]/[Brij35] owing to steric hindrance, (4) the conformation of the long hydrophilic polyoxyethylene chains remain unchanged, and (5) intermolecular interaction among the Brij35 molecules gradually weakens in the mixed micelles as the [CTABr]/[Brij35] increases. For the SDS/Brij35 mixed micelles, the hydrophobic chains of both SDS and Brij35 constitute the mixed micellar core, and there is almost no variation in conformation of the hydrophilic chains of Brij35 molecules with the increase in [SDS]/[Brij35].[46]

Although a fairly large number of structural studies of mixed micellar systems have been carried out during the last nearly two decades by using various kinds of physical techniques,[40,44–46] fine details of structural features of such micellar systems are far less understood compared to those of pure normal micellar systems. Quantitative interpretation of kinetic data on mixed micellar-mediated reactions requires information on the detailed mixed micellar structure. Thus, the lack of mixed micellar structural details could be the reason for the limited number of kinetic studies on the mixed micellar-mediated reactions.[47,48] However, it is well known that the kinetic studies on micelle formation and micellar-mediated reactions have provided some valuable information about the fine details of the dynamic structural features of micelles, which could not be otherwise obtained by any other physical technique. Although it has been known for a long time that surfactant mixtures show a very different behavior in comparison to their components, the mechanism of such behavior is not fully understood at the molecular level. However, significantly different and unexpected effects of mixed micelles compared to those of pure micelles on the reaction rates are yet to be discovered.

5.2 UNIMOLECULAR AND SOLVOLYTIC REACTIONS

Studies on the effects of mixed micelles on rates of unimolecular reactions appear to be nonexistent, whereas only a few papers on the effects of mixed micelles on the rates of solvolytic reactions seem to have appeared in the literature until the end of 2004.

Pseudo-first-order rate constants (k_{obs}) for spontaneous hydrolysis of phenyl chloroformate at different mixed SDS–Brij35 concentration ([MS]$_T$) with a constant molar fraction X_{SDS} {= [SDS]/([SDS] + [Brij35])} have been explained in terms of PP model (i.e., Equation 3.2, Chapter 3).[49] The respective calculated values of k_M and K_S are $(0.9 \pm 0.5) \times 10^{-3}$ sec^{-1} and 104 ± 15 M^{-1} at $X_{SDS} = 0.67$, $(2.0 \pm 0.4) \times 10^{-3}$ sec^{-1} and 216 ± 37 M^{-1} at $X_{SDS} = 0.50$, and $(2.1 \pm 0.4) \times 10^{-3}$ sec^{-1} and 223 ± 25 M^{-1} at $X_{SDS} = 0.33$, whereas the respective values of k_M and K_S, obtained in pure micelles, are $(0.2 \pm 0.2) \times 10^{-3}$ sec^{-1} and 61 ± 3 M^{-1} at $X_{SDS} = 1.0$ or $X_{Brij35} = 0$, and $(2.7 \pm 0.2) \times 10^{-3}$ sec^{-1} and 209 ± 15 M^{-1} at $X_{SDS} = 0$ or $X_{Brij35} = 1.0$. These results show a continuous change in the mixed SDS/Brij35 micellar characteristics from pure SDS micelles to pure Brij35 micelles as the value of X_{SDS} decreases from 1 to 0. However, the values of K_S in mixed micellar solutions with $X_{SDS} = 0.50$ and 0.33 are comparable with the K_S value obtained in pure Brij35 micellar solutions. These observations are attributed to synergism in these mixed micellar solutions.[49] However, *synergism* in mixed micellar solutions is defined as an existing property in a mixed surfactant system when a given property of the mixture can reach a more desirable value than that attained by either surfactant component of the mixture by itself.

The effects of mixed SDS and CTABr surfactants on the rate of methanolysis of PS$^-$ in mixed aqueous solvents containing 2×10^{-4} M PS$^-$, 0.01 M NaOH, 2% v/v CH$_3$CN, and 10% v/v CH$_3$OH reveal that the rate constants k$_{obs}$ at [SDS]$_T$/[CTABr]$_T \leq 0.3$ (where [CTABr]$_T$ was kept constant at 0.005 M) are similar to the corresponding k$_{obs}$ obtained under similar conditions with [SDS]$_T$ = 0. Similarly, the values of k$_{obs}$ at [SDS]$_T$/[CTABr]$_T \geq 4.0$ (where [CTABr]$_T$ was kept constant at 0.005 M) are not significantly different from k$_{obs}$, obtained under similar conditions with [CTABr]$_T$ = 0.[50] The CTABr micellar binding constant of PS$^-$ in aqueous solution containing 10% v/v CH$_3$OH and 2% v/v CH$_3$CN is 6400 M^{-1}.[51] Thus, PS$^-$ ions are nearly 97% micellized at 0.005 M CTABr in the absence of SDS micelles. These observations show the change in the ionic character of mixed micelles from cationic when [SDS]$_T$/[CTABr]$_T \leq 0.3$ to anionic when [SDS]$_T$/[CTABr]$_T \geq 4.0$. The values of k$_{obs}$ could not be obtained within the [SDS]$_T$/[CTABr]$_T$ range of 0.4 to 2.0 owing to development of extremely high turbidity in the reaction mixture under such conditions. It has been shown that rodlike micelles were present at the point of maximum turbidity in the aqueous mixtures of SDS and DTABr.[52] It is evident from the results of the study on the effects of mixed SDS-CTABr surfactants on transesterification of PS$^-$ [50] that (1) at a constant value of [SDS]$_T$/[CTABr]$_T$, the turbidity is significantly higher in the presence of 2×10^{-4} M PS$^-$ than in the presence of 2×10^{-4} M phenyl benzoate, (2) at a constant value of [SDS]$_T$/[CTABr]$_T$ with different values of [SDS]$_T$ and [CTABr]$_T$, the turbidity is highly sensitive to [SDS]$_T$/[CTABr]$_T$ and less sensitive to [SDS]$_T$ and [CTABr]$_T$, and (3) the turbidity of a well-stirred reaction mixture containing all reaction components (with mix SDS and CTABr) except PS$^-$ at room temperature ($\approx 26°C$) becomes almost double with the increase in the mixing period from < 10 min to 1 to 2 h.

The increase in [C$_{12}$E$_{23}$]$_T$ from 0.005 to 0.015 M at constant [CTABr]$_T$ (= 0.01 M) and 0.01 M NaOH caused a linear increase (~2-fold) in k$_{obs}$ for hydrolysis of PS$^-$. Similar observations were obtained in the presence of 0.02 and 0.03 M CTABr.[53] The values of k$_{obs}$ vary from 1.7×10^{-4} to 1.2×10^{-4} sec^{-1} with the change in [CTABr]$_T$ from 0.01 to 0.03 M at 0.005 M C$_{12}$E$_{23}$ and PS$^-$ ions are almost 100% CTABr micellar bound under such conditions (K$_S$ = 7000 M^{-1} [54]). Nearly 6-fold lower values of k$_{obs}$ at [CTABr]$_T$ ranging from 0.01 to 0.03 M compared to k$_{obs}$ at [CTABr]$_T$ = 0 is primarily attributed to lower concentration of water in the micellar environment where PS$_M^-$ ions exist than that of [H$_2$O] in the aqueous pseudophase.[55] In view of recent studies on mixed nonionic/cationic micellar systems,[56] it is evident that the increase in [C$_{12}$E$_{23}$]$_T$ at a constant [CTABr]$_T$ should increase both α (degree of fractional micellar ionization) and the volume of micellar pseudophase owing to so-called swelling of micelle. These two effects are expected to increase the water concentration in the vicinity of micellized PS$^-$ ions, which could be the cause for the increase in k_{obs} with the increase in $[C_{12}E_{23}]_T$ at constant $[CTABr]_T$.

Pseudo-first-order rate constants (k_{obs}) for pH-independent hydrolysis of phenyl salicylate and phthalimide, obtained at constant post-CMC $[CTABr]_T$ and varying values of $[C_{16}E_{20}]_T$, have been found to fit to Equation 3.49, which is derived by the use of PP model (i.e., Equation 3.2) coupled with an empirical equation (Equation 3.32) as described in Chapter 3. The linear increase in k_{obs} with increase in $[C_{12}E_{23}]_T$ from 0.0 to 0.015 M at constant $[CTABr]_T$ (= 0.01, 0.02, and 0.03 M)[53] indicates that $1 >> K[C_nE_m]_T$ (where n = 12 and m = 23) and, consequently, Equation 3.49 is reduced to Equation 5.2 under such conditions.

$$k_{obs} = k_0 + F \, k_{obs}^{CnEm} \, K[C_nE_m]_T \qquad (5.2)$$

The values of k_{obs}^{CnEm} are almost independent of $[C_nE_m]_T$ within its range 0.005 to 0.015 and empirical constant $F \leq 1$ and > 0.[53] The values of K for $C_{16}E_{20}$ are 10 and 6 M^{-1} at 0.01 and 0.02 M CTABr, respectively, and if the values K for $C_{12}E_{23}$ are similar to the corresponding values of K for $C_{16}E_{20}$, then the values of K $[C_{12}E_{23}]_T \leq 0.15$ at $[C_{12}E_{23}]_T \leq 0.015$ M and $[CTABr]_T \geq 0.01$ M.

The values of rate constants for oxidation of rosaniline hydrochloride in mixed micellar solutions (CTABr + Tween-20, CTABr + Tween-80) are less than those in pure CTABr solution but are higher than those in pure nonionic surfactant solutions, whereas the values of rate constants in mixed micellar solutions (SDS + Triton X-100, SDS + Triton X-102) are less than those in pure anionic as well as in nonionic surfactants.[57] The effects of mixed micelles of cationic–cationic, cationic–nonionic, and anionic–nonionic surfactants on the rate of alkaline hydrolysis of N-phenylbenzohydroxamic acid have been studied at 55°C where the addition of cationic surfactant to nonionic surfactant accelerates the rate of hydrolysis, and the kinetic data have been analyzed by the Menger's enzyme-kinetic-type model.[58]

The effects of mixed micelles of two surfactants, Surf$_1$ and Surf$_2$, on k_{obs} for solvolysis of S can be studied either by varying total concentration of one of the two surfactant components, say, Surf$_2$ at a constant post-CMC concentration of the other component Surf$_1$ or by varying total mixed surfactant concentration, $[Surf]_T$ (= $[Surf_1]_T + [Surf_2]_T$) at a constant molar fraction of one of the two surfactant components, say, Surf$_2$ {$X_{sm2} = [Surf_2]_T/([Surf_1]_T + [Surf_2]_T)$}. In the first experimental approach in which $[Surf_1]_T$ (with $[Surf_1]_T > CMC_1$, critical micelle concentration of pure Surf$_1$) is kept constant while $[Surf_2]_T$ is varied, the k_{obs} vs. $[Surf_2]_T$ data should be treated with Equation 3.49 (Chapter 3) because under such experimental conditions, the value of X_{sm2} is increasing with the increase in $[Surf_2]_T$ and, consequently, Surf$_1$ micellar binding constant of S (K_S^{sm1}) is no longer constant. In the second experimental approach in which the value of X_{sm2} is kept constant while $[Surf]_T$ is varied, the k_{obs} vs. $[Surf]_T$ data should be treated with an equation similar to Equation 3.2 (Chapter 3) with K_S replaced by K_S^{mix}, because under such experimental conditions, mixed micellar binding constant of S (K_S^{mix}) remains constant. If the mixing of Surf$_1$ and Surf$_2$ is ideal in the sense that the effects of $[Surf]_T$ $(=[Surf_1] + [Surf_2])$ on mixed micellar binding constant (K_S^{mix}) of substrate S allow one to express the variation of K_S^{mix} with X_{sm1} and X_{sm2} $(=1 - X_{sm1})$ as shown by Equation 5.3,

$$K_S^{mix} = X_{sm1} K_S^{sm1} + X_{sm2} K_S^{sm2} \tag{5.3}$$

where K_S^{sm1} and K_S^{sm2} represent the pure $Surf_1$ and pure $Surf_2$ micellar binding constants, respectively; Equation (5.3) predicts that the variation K_S^{mix} with X_{sm1} or X_{sm2} must be linear. But, if the variation K_S^{mix} with X_{sm1} or X_{sm2} is monotonic nonlinear, then the change in K_S^{mix} as a function of X_{sm1} or X_{sm2} may be expressed by the following empirical equation:

$$K_S^{mix} = \frac{K_S^{sm1}}{1 + K_{sm2/sm1}X_{sm2}} = \frac{K_S^{sm2}}{1 + K_{sm1/sm2}X_{sm1}} \tag{5.4}$$

where $K_{sm2/sm1}$ and $K_{sm1/sm2}$ are empirical constants and the magnitude of $K_{sm2/sm1}$ and $K_{sm1/sm2}$ is the measure of the ability of $Surf_2$ and $Surf_1$ to change K_S^{sm1} and K_S^{sm2}, respectively. The empirical relationship $K_S^{mix} = K_S^{sm1}/(1 + K_{sm2/sm1}X_{sm2})$ and Equation 3.2 can lead to Equation 5.5

$$k_{obs} = \frac{k_0 + F^{sm2}k_{obs}^{sm2}KX_{sm2}}{1 + KX_{sm2}} \tag{5.5}$$

where

$$k_0 = \frac{k_W + k_M K_S^{sm1}[D_n^{sm1}]}{1 + K_S^{sm1}[D_n^{sm1}]} \tag{5.6}$$

with $k_W = k_{obs}$ at $[D_n^{sm1}] = ([Surf_1]_T - CMC_1) = 0$ and $[Surf_2]_T = 0$,

$$K = K_{sm2/sm1}/(1 + K_S^{sm1} [D_n^{sm1}]) \tag{5.7}$$

and F^{sm2} $(= \theta/k_{obs}^{sm2}$ with $k_{obs}^{sm2} = k_{obs}$ at a typical value of $[Surf_2]_T$ and $[D_n^{sm1}]$ $= 0)$ represents the fraction of pure $Surf_1$ micellized S molecules/ions transferred to pure $Surf_2$ micelles (D_n^{sm2}) by the limiting concentration of $Surf_2$ (the limiting concentration of $Surf_2$ is the optimum value of $[Surf_2]_T$ at which $k_{obs} = F^{sm2} k_{obs}^{sm2}$).

The empirical relationship $K_S^{mix} = K_S^{sm2}/(1 + K_{sm1/sm2}X_{sm1})$ and Equation 3.2 can lead to Equation 5.8

$$k_{obs} = \frac{k_0 + F^{sm1}k_{obs}^{sm1}KX_{sm1}}{1 + KX_{sm1}} \tag{5.8}$$

where

$$k_0 = \frac{k_W + k_M K_S^{sm2}[D_n^{sm2}]}{1 + K_S^{sm2}[D_n^{sm2}]} \tag{5.9}$$

with $k_W = k_{obs}$ at $[D_n^{sm2}] = ([Surf_2]_T - CMC_2) = 0$ and $[Surf_1]_T = 0$,

$$K = K_{sm1/sm2}/(1 + K_S^{sm2}[D_n^{sm2}]) \tag{5.10}$$

and $F^{sm1} (= \theta/k_{obs}^{sm1}$ with $k_{obs}^{sm1} = k_{obs}$ at a typical value of $[Surf_1]_T$ and $[D_n^{sm2}]$ = 0) represents the fraction of pure $Surf_2$ micellized S molecules/ions transferred to pure $Surf_1$ micelles (D_n^{sm1}) by the limiting concentration of $Surf_1$ (the limiting concentration of $Surf_1$ is the optimum value of $[Surf_1]_T$ at which $k_{obs} = F^{sm1} k_{obs}^{sm1}$).

5.3 BIMOLECULAR REACTIONS

Perhaps one of the first of the very few systematic kinetic studies on the effects of mixed micelles on the bimolecular reaction rates is one on the effects of mixed CTABr/$C_{10}E_4$ {where $C_{10}E_4 = C_{10}H_{21}(OCH_2CH_2)_3OCH_2CH_2OH$} on the rate of an S_N2 reaction of methyl naphthalene-2-sulfonate (1) with counterions of cationic surfactant (Br-).[59] Pseudophase (PP) model, i.e., Equation 5.11 has been used to explain quantitatively the decrease in k_{obs} with increase in X_{CnEm} {= $[C_{10}E_4]_T/([C_{10}E_4]_T + [CTABr]_T)$}. In Equation 5.11, K_S is CTABr micellar binding constant of nonionic substrate 1,

$$k_{obs} = \frac{k'_W + k_M^{mr} m_{Br}^s K_S[D_n]}{1 + K_S[D_n]} \tag{5.11}$$

$k'_W = k_W^2 [Br_W^-]$, $k_M^{mr} = k_M^2/V_M$ with V_M representing the molar volume of micellar reaction region, $m_{Br}^s = [Br_M^-]/[D_n]$ with $[D_n] = [CTABr]_T - CMC$, $[Br_W^-]$ and $[Br_M^-]$ represent the concentrations of reactant Br- in aqueous phase and normal CTABr micellar phase, respectively, and respective k_W^2 and k_M^2 represent second-order rate constants for the bimolecular reaction in aqueous phase and normal CTABr micellar phase. In order to reduce some expected complexities such as the effects of mixed micelles on K_S and the rate of reaction in aqueous pseudophase, the kinetic study was carried out under restricted reaction conditions where $K_S[D_n] \gg 1$ and $k'_w \ll k_M^{mr} m_{Br}^s K_S[D_n]$ and, consequently, under such restricted reaction conditions, Equation 5.11 is reduced to Equation 5.12

$$k_{obs} = k_M^{mr} m_{Br}^s = k_M^{mr}\beta \tag{5.12}$$

Addition of nonionic inert surfactant or hydrophobic nonionic solutes to cationic micelles decreases counterionic concentrations at ionic micellar surfaces and consequently reduces fractional ionic micellar coverage (β),[60] and these inert additives also increase the volume of the micellar pseudophase,[61] which caused dilution effect on reactant concentration in micelles. These two effects have been attributed to the decrease in k_{obs} for the reaction of **1** with Br$^-$ with increase in the concentration of nonionic surfactant.[59,62–64] The inhibitory effect of nonionic surfactant additive on the rate of reaction of **1** with Br$^-$ in the presence of cationic micelles is treated empirically by modifying Equation 5.12 to Equation 5.13[59,62–64]

$$k_{obs} = k_M{}^{mr}\beta X_{ion} \qquad (5.13)$$

where $X_{ion} = [CTABr]_T/([NIA]_T + [CTABr]_T)$ with NIA representing as nonionic inert additive such as C_nE_m (n = 10 and m = 4).

The values of $k_M{}^{mr}$, calculated from Equation 5.13, are almost independent of $[C_{10}E_4]_T$.[59,62] However, the inhibition by dodecyl(dimethyl)phosphine oxide ($C_{12}PO$) is less than that by $C_{10}E_4$, for example, with 0.05 M CTABr and 0.05 M $C_{12}PO$, $10^4 k_{obs} = 4.83$ sec^{-1} as compared to $10^4 k_{obs} = 3.0$ sec^{-1} with added 0.05 M $C_{10}E_4$, and rate differences are similar over a range of conditions.[64] These qualitative observations indicate that the simple treatment with constant $k_M{}^{mr}$ probably does not fit the effects of $C_{12}PO$. The simplest explanation of this failure is that the rate constant $k_M{}^{mr}$ increases on addition of $C_{12}PO$, and this increase is found to fit the following empirical equation:[64]

$$k_M{}^{mr} = k_M{}^{mr}{}_0 + \phi(1 - X_{CATBr}) \qquad (5.14)$$

where ϕ is an empirical constant and $\phi = 0.0017$ sec^{-1} for NIA = $C_{12}PO$[64] and $\phi = 0.0014$ sec^{-1} for NIA = $C_{10}SO$ (n-decyl methylsulfoxide).[63]

The fit of k_{obs} to Equation 5.13 coupled with Equation 5.14 is reasonably good under only certain specific reaction conditions. The treatment underpredicts the rate constants at high $[C_{10}SO]$ and variable surfactant for several reasons: (1) uncertainty in the correction for reaction with water; (2) the method of slopes[65] is least reliable at high α where it tends to overestimate α so that β values are too low, and k_{obs} is therefore underestimated; (3) the neglect of the concentration of monomeric surfactant. Furthermore, in water, for strongly micellar-interacting counterions, (e.g., Br$^-$), the counterion concentration at the micellar–water interface, as given by β, increases only modestly on addition of counterion,[66] (e.g., the assumption of constant β and constant [Br$^-$] at the micellar surface is reasonably satisfactory for CTABr and dilute added Br$^-$). However, values of β increase with increasing [counterion] as micelle–ion affinities decrease, (e.g., owing to addition of a nonionic solute,[67] or with very hydrophilic counterion,[68] or as the bulk of the surfactant head group is increased.[69]) Thus, one may underestimate values of [Br$^-$] at the micellar surface from values of α determined conductimetrically without added NaBr for mixtures of CTABr and $C_{10}SO$.

The increase of ion concentration at the ionic micellar surface can be treated by solving the Poisson–Boltzmann equation in the appropriate symmetry,[70] or more simply in terms of Equation 5.15,

$$K'_{Br} = [Br_M^-] / \{[Br_W^-]([CTABr_M] - [Br_M^-])\} \qquad (5.15)$$

which has the form of a Langmuir isotherm,[71] and is satisfactory over a limited range of counterion concentration.[64] The association constant K_{Br}' is related to fractional micellar ionization α through Equation 5.16,

$$K'_{Br} = \beta / (\alpha^2 [CTABr_M]) \qquad (5.16)$$

which is another form of Equation 5.15.

Pseudo-first-order rate constants (k_{obs}) for the reactions of n-dodecyl [2-(hydroximino)-2-phenylethyl]dimethylammonium bromide (DHDBr) with p-nitrophenyl diphenyl phosphate (2) in comicelles with inert surfactants (ID) and CTABr, at pH such that the oximate zwitterions (DHD) are formed quantitatively, show good fit to Equation 5.17 under experimental conditions where 2 is fully micelle bound.[72] The values of k_M^{mr} for the reactions of 2 with DHD in mixed micelles of DHDBr,

$$k_{obs} = (k_M^{mr} [DHD])/([DHD] + [CTABr] + [ID]) \qquad (5.17)$$

CTABr and other third surfactant ID approximately follow Equation 5.18,

$$k_M^{mr} = k_M^{mr,CTA} X_{CTA} + k_M^{mr,ID}(1 - X_{CTA}) \qquad (5.18)$$

where $X_{CTA} = [CTABr]_T/([CTABr]_T + [ID]_T)$ as well as $k_M^{mr,CTA}$ and $k_M^{mr,ID}$ represent k_M^{mr} for reaction in CTABr and ID micelles, respectively. The linear relationship, expressed by Equation 5.18, fails with polyoxyethylene surfactants where there is curvature in the plots of k_M^{mr} vs. X_{CTA} or X_{ID} (= $1 - X_{CTA}$). This curvature in the plots could be due to nonuniform mixing.[72]

Pseudo-first-order rate constants (k_{obs}) for alkaline hydrolysis of 2,4-dinitrophenyl acetate (3) and octanoate (4) and benzoic anhydride (5) at 0.1 M total mixed surfactant solution of sodium dodecanoate (SDOD = $C_{11}H_{23}CO_2Na$) and sulfobetaines (SB3 – n = $C_nH_{2n+1}N^+Me_2(CH_2)_3SO_3^-$, n = 10, 12, 14, and 16) go through maxima at a mole fraction of SDOD (X_{SDOD}) of about 0.5 for 3 and 4 and 0.8 for 5.[73] Quantitative treatment of kinetic data is carried out in terms of PP model with the following assumptions:

1. The effective molarity of HO⁻ in the mixed micelles, [HO_M⁻], is given by $pH_{app} = log([HO_M^-]/K_w^W)$ and $K_w^W \approx K_w^{mix}$ where K_w is the auto-

protolysis constant of water and superscripts "W" and "mix" refer to aqueous phase and mixed micellar phase, respectively.

2. The rate constant for the reaction of HO_M^- with neutral substrates (**3**, or **4**, or **5**) in mixed micelles is unaffected by dodecanoate ion (DOD).

For reactions of the strongly bound substrates, **4** and **5**, the reaction in aqueous phase is neglected, and k_{obs} is given by Equation 5.19

$$k_{obs} = k_M^{OH}[HO_M^-] + k_M^{DOD}X_{SDOD} \qquad (5.19)$$

where k_M^{OH}, M^{-1} sec^{-1}, is second-order rate constant in the micelles with respect to the molarity of HO_M^- at the micellar surface as estimated by indicator measurements.[74]

For alkaline hydrolysis of **3**, which is not strongly bound by the mixed micelles, k_{obs} is given by Equation 5.20.

$$k_{obs} = \frac{k_W^2[HO_W^-] + (k_M^{OH}[HO_M^-] + k_M^{DOD})X_{SDOD}K_S^{mix}[D_n]}{1 + K_S^{mix}[D_n]} \qquad (5.20)$$

The values of k_M^{DOD} as function of X_{SDOD}, were calculated from Equation 5.20 by using K_S^{mix} value (= 55 M^{-1})[75] obtained in pure SDOD micelles. Thus, the assumption introduced here[73] is that the value of K_S^{mix} is independent of [SB3-n], which may be true only if K_S value with SDOD micelles is approximately similar to K_S value with SB3-n micelles. Almost similar observations and data analysis have been reported on the effects of mixed micelles of sulfobetaine and sodium decyl phosphate surfactants on the rate of alkaline hydrolysis of **3**, **4**, and **5**.[76]

The first-order rate constant (k_{obs}) for general acid-catalyzed hydrolysis of di-*tert*-butyl benzaldehyde acetal in mixed micelles of SDS and sodium decyl hydrogenphosphate (NaDeHP = $C_{10}H_{21}PO_3HNa$) increases linearly with increasing mole fraction of the phosphate surfactant (NaDeHP), which is a general acid catalyst in the micellar pseudophase, and thus the mixed SDS/NaDeHP systems act as functional comicelles. It has also been shown that the PP model can be applied quantitatively to acidic functional comicelles as well as nonfunctional micelles.[77]

The effects of total concentration of mixed surfactants, $[S]_T$ (= $[SDS]_T$ + $[C_{16}E_{20}]_T$) on pseudo-first-order rate constants (k_{obs}) for the acid-catalyzed denitrosation of 1-phenylethyl nitrite and hexyl nitrite at a constant mole ratio, MR (= $[C_{16}E_{20}]_T/[SDS]_T$) and [HCl] show the presence of maxima in the plots of k_{obs} vs. $[S]_T$ at MR $\leq$ 0.5.[78] These observed data have been treated by the PIE model (i.e., Equation 3.12, Chapter 3) developed by Romsted[79] for pure normal micellar systems, by using a single value of ion-exchange constant K_{Na}^H (= 0.75) for MR range 0 to 0.5. The kinetic data obtained at MR > 0.75 cannot be explained by the PIE model; these data can be fitted quantitatively by an equation similar to Equation 3.14, Chapter 3. It is interesting to note that although the values of

cal mixed micelle concentration of mixed SDS/C_nE_m micelles at MR = 0.03
0 are < CMC^{SDS} and > CMC^{CnEm} where superscripts "SDS" and "C_nE_m" refer
ure SDS micelles and C_nE_m micelles, respectively, the calculated values of
ed micellar binding constant (K_S^{mix}) of 1-phenylethyl nitrite are $\leq K_S^{SDS}/2$ and
$^{CnEm}/10$ and the values of k_M^{mr} are < $k_M^{mr, SDS}$ and > $k_M^{mr, CnEm}$.[78,80] It is difficult
xplain that the binding affinity of 1-phenylethyl nitrite becomes significantly
ker with mixed SDS/C_nE_m micelles compared to those with pure SDS and
$_m$ micelles. Although these unusual results have been attributed to the forma-
of mixed micelles, this argument is not substantiated by any other indepen-
experimental observation. The values of k_{obs} for acid-catalyzed hydrolysis
ster functional group of ethyl cyclohexanone-2-carboxylate in mixed
$Br/C_{12}E_9$ at MR (= $[C_{12}E_9]_T/[DTABr]_T$) = 1, fit to Equation 3.2 of Chapter
nd the calculated values of K_S^{mix} and $k_M^{mr, mix}$ are well within the respective
esponding values of K_S^{C12E9} and K_S^{DTABr} and $k_M^{mr,C12E9}$ and $k_M^{mr,DTABr}$.[81]
Treiner[82] used Equation 5.21 to calculate empirical constant B from experi-
tally determined mixed surfactants' ($surf_1 + surf_2$) micellar binding constant
, K_S^{mix}, as well as pure $surf_1$ and pure $surf_2$ micellar binding constants of S,
^{1} and K_S^{sm2}, respectively. In Equation 5.21, X_{sm1} and X_{sm2} represent

$$\ln K_S^{mix} = X_{sm1} \ln K_S^{sm1} + X_{sm2} \ln K_S^{sm2} + X_{sm1} X_{sm2} B^S \qquad (5.21)$$

ar fraction (X) of respective $surf_1$ and $surf_2$ (i.e., $X_{sm1} = (Surf_1]_T/([surf_1]_T +$
$f_2]_T)$ and $X_{sm2} = 1 - X_{sm1}$), and empirical constant B^S accounts for the devi-
ns from ideal behavior in the partitioning of solutes (S) between mixed
elles and the bulk aqueous phase. The value of B^S for S = m-chloroperbenzoic
obtained in mixed micelles of Brij 35 (= $C_{12}E_{23}$) and SDS is $- 0.9 \pm 0.3$.[83]
ie value of B^S for S = 1-phenylethyl nitrite and mixed micelles of $C_{16}E_{20}$ and
5 is assumed to be similar to B^S (= -0.9) for m-chloroperbenzoic acid and
ed micelles $C_{12}E_{23}$/SDS by considering the fact that although both solutes
ht differ in terms of hydrophobicity, they are nonionic, then the calculated
ies of K_S^{mix} for S = 1-phenylethyl nitrite range from 81 to 226 M^{-1} with
rease in X_{SDS} (= X_{sm1}) from 0.97 to 0.33 with K_S^{sm1} = 79 M^{-1} and K_S^{sm2} = 510
.[78] These values of K_S^{mix} are significantly different from K_S^{mix} values obtained
ising the PIE model.[78]

Zakharova et al.[47] have studied the effects of mixed CTABr/Brij 97 (where
97 = $C_{18}H_{35}(OCH_2CH_2)_{10}OH$) micelles on the rate of hydroxide ion-catalyzed
rolysis of ethyl p-nitrophenyl chloromethyl phosphate, and the plots of exper-
ntally determined pseudo-first-order rate constants (k_{obs}) against total mixed
actant concentration, $[S]_T$ (= $[CTABr]_T + [Brij 97]_T$) at a constant molar
tion, X_{CTA} {= $[CTABr]_T/([CTABr]_T + [Brij 97]_T)$}, reveal maxima at $X_{CTA} \geq$
7. These observed data have been satisfactorily explained by Equation 3.11
$K_S = K_S^{mix}$, $K_R = K_R^{mix}$, and R = HO^-, derived by using Berezin's
idophase (BPP) model, (i.e., Equation 3.61 as described in Chapter 3). The

calculated values of K_S^{mix} remain almost independent of X_{CTA}, whereas the values of K_{OH}^{mix} decrease from 69 to 12 M^{-1} with decrease in X_{CTA} from 1.0 to 0.17. The values of K_{OH}^{mix} have also been determined from surface potential under experimental conditions of kinetic study. The values of K_{OH}^{mix} obtained from kinetic data are almost similar to the corresponding K_{OH}^{mix} values obtained from surface potential measurements.[47]

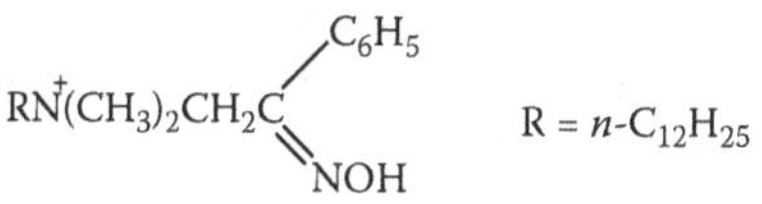

$$RN^{+}(CH_3)_2CH_2C\overset{\displaystyle C_6H_5}{\underset{\displaystyle NOH}{\Big\backslash}} \qquad R = n\text{-}C_{12}H_{25}$$

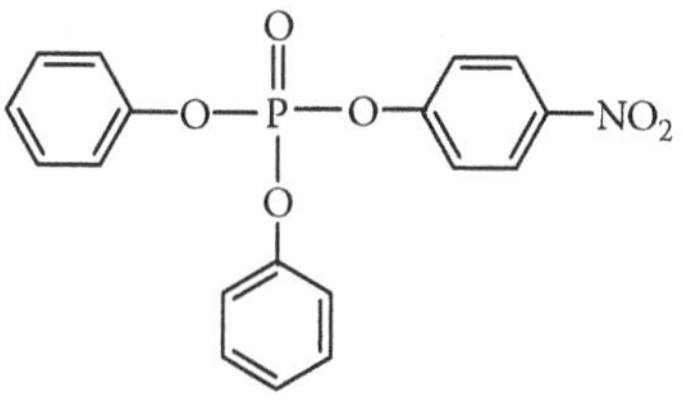

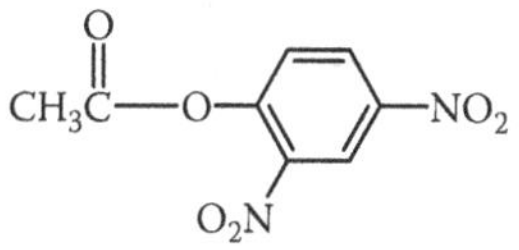

DHDBr

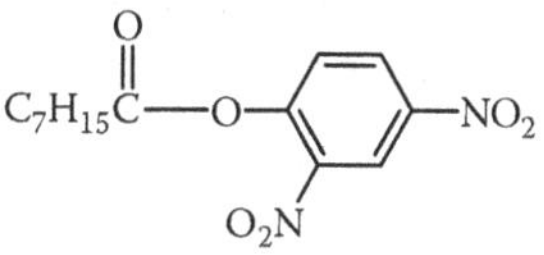

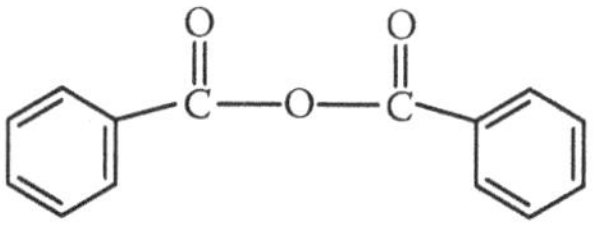

SCHEME 5.2

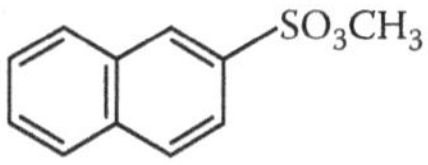

The values of K_{OH}^{mix} at different values of $X_{Brij\ 97}$ (= X_{sm2}) as summarized in Table 5.1 fit reasonably well to Equation 5.4 where S = OH and sm1 = CTABr. The nonlinear least-squares calculated values of $K_{sm2/sm1}$ and K_{OH}^{sm1} are 56 ± 7 M^{-1} and 2.9 ± 0.8, respectively. The calculated value of K_{OH}^{sm1} (= 56 M^{-1}) is not exactly the same as the one (= 69 M^{-1}) obtained experimentally at X_{ms1} = 1 (Table 5.1). The observed data have also been treated with the following empirical equation:

$$K_{OH}^{sm1} / K_{OH}^{mix} = \delta + K_{sm2/sm1} X_{sm2} \qquad (5.22)$$

where δ and $K_{sm2/sm1}$ are empirical constants. The linear least-squares calculated values of δ and $K_{sm2/sm1}$ with $K_{OH}^{sm1} = 69$ M^{-1} are 1.27 ± 0.05 and 3.4 ± 0.1, respectively.

Recently, Davies and Foggo[83] studied the effects of mixed anionic (SDS)/nonionic ($C_{12}E_{23}$) micelles on the rate of reaction of m-chloroperbenzoic acid and iodide. The observed data have been treated using a combined multiple micellar pseudophase (MMPP) model and transition state pseudoequilibrium constant approach. It is interesting to note that proportionately weighted nonlinear regression was used for the kinetic model fitting to a kinetic equation of nine disposable parameters, and it is almost certain that the reliability of the values of calculated disposable parameters from a kinetic equation derived based upon a kinetic model decreases with an increase in the number of such disposable parameters.

The effects of mixed CTABr–SDS micelles on pseudo-first-order rate constants (k_{obs}) for the reactions of HO$^-$ with phenyl benzoate (PB) showed the mixed micelles behavior as cationic in nature when MR ≤ 0.3 and anionic in nature when MR ≥ 4.0, where MR = $[SDS]_T/[CTABr]_T$.[84] The values of k_{obs} could not

TABLE 5.1
Effects of Mole Fraction ($X_{Brij\ 97}$) of Brij 97 on CTABr Micellar Binding Constant (K_{OH}) of HO$^-$ in Mixed CTABr/Brij 97 Micellar Solutions

$X_{Brij\ 97}$ [a]	$K_{OH}^{mix,\ b}$ M^{-1}	$K_{OH}^{mix}{}_{,calcd}^{c}$ M^{-1}	Y^d	Y_{calcd}^{e} M^{-1}	$K_{OH}^{mix,\ f}$
0.0	69 ± 2.8		1.00		69 ± 2.8
0.09	44 ± 1.8	44	1.57	1.58	38 ± 1.5
0.33	28 ± 1.1	29	2.40	2.40	33 ± 1.3
0.50	27 ± 1.1	23	2.98	2.98	28 ± 1.1
0.67	20 ± 0.8	19	3.57	3.56	22 ± 0.9
0.83	12 ± 0.5	17	4.11	4.11	11 ± 0.4

[a] $X_{Brij\ 97} = [Brij\ 97]_T/([Brij\ 97]_T + [CTABr]_T)$.
[b] These values are obtained from Reference 47 where K_{OH}^{mix}, values were determined kinetically.
[c] Calculated from Equation 5.4 with $K_{OH}^{sml} = 56 \pm 7$ M^{-1} and $K_{sm2/sm1} = 2.9 \pm 0.8$.
[d] $Y = K_{OH}^{sml}/K_{OH}^{mix}$ with $K_{OH}^{sml} = 69$ M^{-1}.
[e] Calculated from Equation 5.22 with $\delta = 1.27 \pm 0.05$ and $K_{sm2/sm1} = 3.4 \pm 0.1$.
[f] These values are obtained from Reference 47 where K_{OH}^{mix}, values were determined by using surface potential.

be obtained within the MR range of > 0.3 to < 4.0 because of the appearance of turbidity into the reaction mixtures under such conditions.

The values of k_{obs} for the reactions of HO^- with PB were found to be almost independent of $[C_{12}E_{23}]_T$ within its range 0.0 to 5×10^{-4} M in the presence of 0.006 to 0.030 M CTABr and 0.01 M NaOH.[53] The increase in $[C_{12}E_{23}]_T$ from $\geq$ 7×10^{-4} M caused the decrease in k_{obs} and this decrease became nearly independent of $[CTABr]_T$ at $\geq 25 \times 10^{-4}$ M $C_{12}E_{23}$. The values of k_{obs} remained significantly higher at $[CTABr]_T = 0.006, 0.01, 0.02,$ and 0.03 M than at $[CTABr]_T = 0.0$ under the presence of $[C_{12}E_{23}]_T$ in the range of > 5×10^{-4} to 0.02 M. These observations show that mixed CTABr/$C_{12}E_{23}$ micelles behave like pure CTABr micelles at $[C_{12}E_{23}]_T \leq 5 \times 10^{-4}$ M and the decrease in k_{obs} with the increase in $[CTABr]_T$ from 0.006 to 0.03 M at a constant $[C_{12}E_{23}]_T$ is due to dilution effect on $[HO_M^-]$. The decrease in k_{obs} with increase in $[C_{12}E_{23}]_T$ within its range of $\geq 7 \times 10^{-4}$ to $\leq 25 \times 10^{-4}$ M at a constant $[CTABr]_T$ is attributed to the decrease in $[HO_M^-]$ in the micellar environment of PB_M molecules due to increase in both the fractional micellar ionization (α) and the volume of the micellar pseudophase under such conditions. The values of k_{obs} became independent of $[CTABr]_T$ within its range 0.006 to 0.03 M at a constant $[C_{12}E_{23}]_T$ within its range 25×10^{-4} to 250×10^{-4} M, which could be due to insignificant dilution effect on $[HO_M^-]$ owing to increase in $[CTABr]_T$ at a constant $[C_{12}E_{23}]_T$ compared to the decreasing effects of $[C_{12}E_{23}]_T$ on $[HO_M^-]$ at a constant $[CTABr]_T$. The results on the effects of mixed CTABr/$C_{12}E_{23}$ micelles on the rates of alkaline hydrolysis of PB also revealed the different locations of phenolate ions, phenol, and PB molecules in the mixed micellar pseudophase.[53] The quantitative interpretation of these kinetic data could not be achieved in terms of any one of the existing micellar kinetic models. However, the effects of $[C_{16}E_{20}]_T$ on k_{obs} for alkaline hydrolysis of PB have been explained quantitatively in terms of PP model coupled with an empirical equation[85] as described in Chapter 3.

Pseudo-first-order rate constants (k_{obs}) for the reactions of HO^- with 4-nitrophthalimide (4-NPTH) showed an initial monotonic decrease with the increase in $[C_{12}E_{23}]_T$ from 0.0 to < 0.005 M followed by an almost monotonic increase with the increase in $[C_{12}E_{23}]_T$ from 0.01 to ≤ 0.07 M in the presence of 0.006 M CTABr and 0.03 M NaOH.[86] The hydrolysis of 4-NPTH ceased or became too slow to monitor kinetically at 0.006 M CTABr, 0.01 M NaOH, and ≥ 0.04 M $C_{12}E_{23}$. Similar observations were obtained at 0.01 and 0.02 M CTABr. The absence of detectable hydrolysis under such typical reaction conditions was attributed to kinetically insignificant concentration of HO^- in the micellar environment of micellized 4-NPTH.

The initial monotonic decrease in k_{obs} with the increase in $[C_{12}E_{23}]_T$ from 0.0 to < 0.005 M is due to the increase in both the volume of mixed micelles and the value of α, which, in turn, decrease the effective concentration of HO_M^-. These effects are expected to decrease the effective concentrations of all the counterions, HO^-, Br^-, and 4-NPT^-. However, the fact that the hydrophobicity of these counterions varies in the order $HO^- < Br^- <$ 4-NPT^- makes the least hydrophobic ion, HO^-, to move first from the cationic surface of a less hydrophilic region to a relatively more hydrophilic region, including aqueous pseudophase. Once such decreasing effects of $[C_{12}E_{23}]_T$ on $[HO_M^-]$ and $[Br_M^-]$ are leveled off, then the further increase in $[C_{12}E_{23}]_T$ causes the movement of 4-NPT$^-$ ions from the cationic surface of a less hydrophilic region to a relatively more hydrophilic micellar region. The values of k_{obs} for alkaline hydrolysis of 4-NPTH, at 0.01 M and 0.03 M NaOH in the absence of micelles, are nearly 4- to 7-fold larger than k_{obs} at 0.006 to 0.020 M CTABr in the absence of $C_{12}E_{23}$. These results form the basis for the conclusion that the increase in $[C_{12}E_{23}]_T$ from an optimum value causes the transfer of 4-NPT$^-$ ions from the less hydrated cationic surface to a more hydrated region of mixed micelle.

The increase in $[C_{12}E_{23}]_T$ at a constant $[CTABr]_T$ decreases the effective concentrations of counterions, HO$^-$, Br$^-$, and 4-NPT$^-$ at the cationic micellar surface, which, in turn, implies that it decreases the CTABr micellar binding constants of these counterions. Such an effect of $[C_{12}E_{23}]_T$ on CTABr binding constant of 4-NPT$^-$ is treated with an empirical equation (Equation 3.32 with replacement of [MX] by $[C_{12}E_{23}]_T$ and $K_{X/S}$ by $K_{C12E23/S}$), which, in conjunction with Equation 3.2 (i.e., PP model) as described in Chapter 3, leads to Equation 5.23.

$$k_{obs} = \frac{k_0' + \theta K[C_{12}E_{23}]_T}{1 + K[C_{12}E_{23}]_T} \qquad (5.23)$$

where

$$k_0' = \frac{k_W + k_M K_S^0[D_n]}{1 + K_S^0[D_n]} \qquad (5.24)$$

with $k_W = k_{obs}$ at $[D_n] = [C_{12}E_{23}]_T = 0$ and k_W and k_M represent pseudo-first-order rate constant,

$$\theta = k_W \qquad (5.25)$$

with $k_W = k_{obs}$ at $[D_n] = 0$ $[C_{12}E_{23}]_T \neq 0$, and

$$K = K_{C12E23/S}/(1 + K_S^0[D_n]) \qquad (5.26)$$

The observed data, obtained only at 0.006 M CTABr and within $[C_{12}E_{23}]_T$ range 0.01 to 0.07 M at 0.03 M NaOH and 0.01 to 0.03 M at 0.01 M NaOH, showed reasonably good fit to Equation 5.23 and the calculated values of k_0', θ and K are respectively $(1.96 \pm 0.69) \times 10^{-4}$ sec^{-1}, $(16.3 \pm 0.2) \times 10^{-4}$ sec^{-1}, and 21.4 ± 7.3 M^{-1} at 0.03 M NaOH and $(1.95 \pm 0.13) \times 10^{-4}$ sec^{-1}, $(7.93 \pm 1.26) \times 10^{-4}$ sec^{-1}, and 17.2 ± 6.4 M^{-1} at 0.01 M NaOH. The calculated values of K are almost independent of [NaOH], which is in agreement with Equation 5.26.[86] At 0.01 M NaOH, a low increase (~14%) and the absence of an increase in k_{obs} with the increase in $[C_{12}E_{23}]_T$ from 0.01 to 0.03 M at 0.01 and 0.02 M CTABr, respectively, may be the consequence of low values of K under such conditions as predicted by Equation 5.26. It is evident from Equation 5.25 that θ may be considered to be a constant in the data treatment with Equation 5.23 only if θ is independent of $[C_{12}E_{23}]_T$ at $[CTABr]_T = 0$. But the values of k_{obs} decrease from 2.50×10^{-4} to 1.79×10^{-4} sec^{-1} and to 1.31×10^{-4} sec^{-1} with respective increase in $[C_{12}E_{23}]_T$ from 0.01 to 0.03 M and to 0.07 M at 0.03 M NaOH.[86] Thus, although the observed data fit to Equation 5.23 is satisfactory in terms of residual errors ($= k_{obsi} - k_{calcdi}$ where k_{obsi} and k_{calcdi} represent experimentally determined and calculated rate constants at the i-th value of $[C_{12}E_{23}]_T$, respectively), the calculated values of some of the kinetic parameters k'_0, θ, and K may not be reliable.

Equation 5.23 predicts that $k'_0 = k_{obs}$ at $[C_{12}E_{23}]_T = 0$ and $[D_n] \neq 0$. However, the values of k_{obs} at $[C_{12}E_{23}]_T = 0$ and $[D_n] \approx 0.006$ M are nearly 2-fold larger than k'_0.[87] This is a mathematical paradox that one should be aware of while discussing the data treatment with a kinetic equation. In actuality k'_0 should be equal to k_{obs} at an optimum hypothetical value of $[C_{12}E_{23}]_T$ where the decrease in k_{obs} is maximum in the absence of a rate increasing effect of $[C_{12}E_{23}]_T$. However, this can never happen practically because the increase in $[C_{12}E_{23}]_T$ (1) decreases k_{obs} owing to transfer of HO$^-$ ions from the micellar environment of micellized 4-NPT$^-$ ions to the aqueous phase at lower values of $[C_{12}E_{23}]_T$ and (2) increases k_{obs} owing to transfer of 4-NPT$^-$ ions from the micellar environment of low $[H_2O]$ to one of high $[H_2O]$ at higher values of $[C_{12}E_{23}]_T$.

REFERENCES

1. (a) Holland, P.M., Rubingh, D.N. *Mixed Surfactant Systems*, Vol. 501, Holland, P.M., Rubingh, D.N., Eds. ACS Symposium Series, American Chemical Society, Washington, D.C., **1992**, pp. 1–30, chap. 1, and references cited therein. (b) Hoffmann, H., Possnecker, G. The mixing behavior of surfactants. *Langmuir* **1994**, *10*(2), 381–389.

2. Rosen, M.J. In *Phenomena in Mixed Surfactant Systems*, Vol. 311, Scamehorn, J.F., Ed. ACS Symposium Series, American Chemical Society, Washington, D.C., **1986**, p. 144.

3. Shiloach, A., Blankschtein, D. Predicting micellar solution properties on binary surfactant mixtures. *Langmuir* **1998**, *14*(7), 1618–1636.

4. Christian, S.D., Scamehorn, J.F. *Solubilization in Surfactant Aggregate*. Marcel Dekker: New York, 1995.

5. Shiloach, A., Blankschtein, D. Measurements and prediction of ionic/nonionic mixed micelle formation and growth. *Langmuir* **1998**, *14*(25), 7166–7182.

6. Islam, M.N., Okano, T., Kato, T. Surface phase behavior of a mixed system of anionic — nonionic surfactants studied by Brewster angle microscopy and polarization modulation infrared reflection — adsorption spectroscopy. *Langmuir* **2002**, *18*(26), 10068–10074.

7. Yoshida, K., Dubin, P.L. Complex formation between polyacrylic acid and cationic/nonionic mixed micelles: effect of pH on electrostatic interaction and hydrogen bonding. *Colloids Surf. A* **1998**, *147*(1–2), 161–167.

8. Palous, J.L., Turmine, M., Latellier, P. Mixtures of nonionic and ionic surfactants: determination of mixed micelle composition using cross-differentiation relations. *J. Phys. Chem.* **1998**, *102*(30), 5886–5890.

9. Ganesh, K.N., Mitra, P., Balasubramanian, D. Solubilization sites of aromatic optical probes in micelles. *J. Phys. Chem.* **1982**, *86*(22), 4291–4293.

10. Nilsson, P.G., Lindman, B. Mixed micelles of nonionic and ionic surfactants: a nuclear magnetic resonance self-diffusion and proton relaxation study. *J. Phys. Chem.* **1984**, *88*(22), 5391–5397.

11. Matsubara, H., Muroi, S., Kameda, M., Ikeda, N., Ohta, A., Aratono, M. Interaction between ionic and nonionic surfactants in the adsorbed film and micelle. 3. Sodium dodecyl sulfate and tetraethylene glycol monooctyl ether. *Langmuir* **2001**, *17*(25), 7752–7757.

12. Griffiths, P.C., Whatton, M.L., Abbott, R.J., Kwan, W., Pitt, A.R., Howe, A.M., King, S.M., Heenam, R.K. Small-angle neutron scattering and fluorescence studies of mixed surfactants with dodecyl tails. *J. Colloid Interface Sci.* **1999**, *215*(1), 114–123.

13. Ghosh, S., Moulik, S.P. Interfacial and micellization behaviors of binary and ternary mixtures of amphiphiles (Tween-20, Brij-35, and sodium dodecyl sulfate) in aqueous medium. *J. Colloid Interface Sci.* **1998**, *208*(2), 357–366.

14. Abe, M., Tsubaki, N., Ogino, K. Solution properties of mixed surfactant system. V. The effect of alkyl groups in anionic surfactant on surface tension of anionic–nonionic surfactant systems. *J. Colloid Interface Sci.* **1985**, *107*(2), 503–508.

15. Alargova, R.G., Kochijashky, I.I., Sierra, M.L., Kwetkat, K., Zana, R. Mixed micellization of dimeric (gemini) surfactants and conventional surfactants. *J. Colloid Interface Sci.* **2001**, *235*(1), 119–129.

16. Desai, T.R., Dixit, S.G. Interaction and viscous properties of aqueous solutions of mixed cationic and nonionic surfactants. *J. Colloid Interface Sci.* **1996**, *177*(2), 471–477.

17. Esumi, K., Miyazaki, M., Arai, T., Koide, Y. Mixed micellar properties of a cationic Gemini surfactant and a nonionic surfactant. *Colloids Surf. A* **1998**, *135*(1–3), 117–122.

18. Malliaris, A., Binana-Limbela, W., Zana, R. Fluorescence probing studies of surfactant aggregation in aqueous solutions of mixed ionic micelles. *J. Colloid Interface Sci.* **1986**, *110*(1), 114–120.

19. Zana, R., Muto, Y., Esumi, K., Meguro, K. Mixed micelles formation between alkyltrimethylammonium bromide and alkane-α,ω-bis(trimethylammonium) bromide in aqueous solution. *J. Colloid Interface Sci.* **1988**, *123*(2), 502–511.

20. Furuya, H., Moroi, Y., Sugihara, G. Micelle formation of binary mixtures of dodecylammonium perfluoro carboxylates. *Langmuir* **1995**, *11*(3), 774–778.

21. Attwood, D., Mosquera, V., Novas, L., Sarmiento, F. Micellization in binary mixtures of amphiphilic drugs. *J. Colloid Interface Sci.* **1996**, *179*(2), 478–481.

22. Moulik, S.P., Haque, M.E., Jana, P.K., Das, A.R. Micellar properties of cationic surfactants in pure and mixed states. *J. Phys. Chem.* **1996**, *100*(2), 701–708.

23. Lopez-Fontan, J.L., Suarez, M.J., Mosquera, V., Sarmiento, F. Micellar behavior of n-alkyl sulfates in binary mixed systems. *J. Colloid Interface Sci.* **2000**, *223*(2), 185–189.

24. Attwood, D., Patel, H.K. Mixed micelles of alkyltrimethylammonium bromides and chlorohexidine digluconate in aqueous solution. *J. Colloid Interface Sci.* **1989**, *129*(1), 222–230.

25. Velazquez, M.M., Garcia-Mateos, I., Lorente, F., Valero, M., Rodriquez, L.J. Fluorescence studies on the characterization of mixed micelles. *J. Mol. Liq.* **1990**, *45*(1–2), 95–100.

26. Bucci, S., Fagotti, C., de Giorgio, V., Piazza, R. Small-angle neutron-scattering study of ionic-nonionic mixed micelles. *Langmuir* **1991**, *7*(5), 824–826.

27. Guering, P., Nilsson, P.G., Lindman, B. Mixed micelles of ionic and nonionic surfactants: quasielastic light scattering and NMR self-diffusion studies of $C_{12}E_5$-SDS micelles. *J. Colloid Interface Sci.* **1985**, *105*(1), 41–44.

28. Garamus, V.M. Study of mixed micelles with varying temperature by small-angle neutron scattering. *Langmuir* **1997**, *13*(24), 6388–6392.

29. Clint, J.H. Micellization of mixed nonionic surface active agents. *J. Chem. Soc., Faraday Trans. 1.* **1975**, *71*(6), 1327–1334.

30. Holland, P.M., Rubingh, D.N. Nonideal multicomponent mixed micelle model. *J. Phys. Chem.* **1983**, *87*(11), 1984–1990.

31. Rubingh, D.N. In *Solution Chemistry of Surfactants*, Mittal, K., Ed. Plenum: New York, **1979**, pp. 337–362.

32. Motomura, K., Yamanaka, M., Aratono, M. Thermodynamic consideration of the mixed micelle of surfactants. *Colloid Polym. Sci.* **1984**, *262*(12), 948–955.

33. Georgiev, G.S. Markov chain model of mixed surfactant systems. Part 1. New expression for the non-ideal interaction parameter. *Colloid Polym. Sci.* **1996**, *274*(1), 49–58.

34. Nagarajan, R. Molecular theory for mixed micelles. *Langmuir* **1985**, *1*(3), 331–341.

35. Nagarajan, R. Micellization, mixed micellization and solubilization: the role of interfacial interactions. *Adv. Colloid Interface Sci.* **1986**, *26*(2–4), 205–264.

36. Puwada, S., Blankschtein, D. Thermodynamic description of micellization, phase behavior, and phase separation of aqueous solutions of surfactant mixtures. *J. Phys. Chem.* **1992**, *96*(13), 5567–5579.

37. Puwada, S., Blankschtein, D. Theoretical and experimental investigations of micellar properties of aqueous solutions containing binary mixtures of nonionic surfactants. *J. Phys. Chem.* **1992**, *96*(13), 5579–5592.

38. Eads, C.D., Robosky, L.C. NMR studies of binary surfactant mixture thermodynamics: molecular size model for asymmetric activity coefficients. *Langmuir* **1999**, *15*(8), 2661–2668.

39. Huang, L., Somasundaran, P. Theoretical model and phase behavior for binary surfactant mixtures. *Langmuir* **1997**, *13*(25), 6683–6688.

40. Junquera, E., Aicart, E. Mixed micellization of dodecylethyldimethylammonium bromide and dodecyltrimethylammonium bromide in aqueous solution. *Langmuir* **2002**, *18*(24), 9250–9258.

41. McDermott, D.C., Lu, J.R., Lee, E.M., Thomas, R.K., Rennie, A.R. Study of the adsorption from aqueous solution of hexaethylene glycol monododecyl ether on silica substrates using the technique of neutron reflection. *Langmuir* **1992**, *8*(4), 1204–1210.

42. Bergstrom, M., Skov Pedersen, J. Structure of pure SDS and DTAB micelles in brine determined by small-angle neutron scattering (SANS). *Phys. Chem. Chem. Phys.* **1999**, *1*(18), 4437–4446.

43. Baglioni, P., Dei, L., Rivara-Minten, E., Kevan, L. Mixed micelles of SDS/$C_{12}E_6$ and DTAC/$C_{12}E_6$ surfactants. *J. Am. Chem. Soc.* **1993**, *115*(10), 4286–4290.

44. Garamus, V.M. Formation of mixed micelles in salt-free aqueous solutions of sodium dodecyl sulfate and $C_{12}E_6$. *Langmuir* **2003**, *19*(18), 7214–7218.

45. Bastiat, G., Grassl, B., Khoukh, A., Francois, J. Study of sodium dodecyl sulfate-poly(propylene oxide) methylacrylate mixed micelles. *Langmuir* **2004**, *20*(14), 5759–5769.

46. Gao, H.-C., Zhao, S., Mao, S.-Z., Yuan, H.-Z., Yu, J.-Y., Shen, L.-F., Du, Y.-R. Mixed micelles of polyethylene glycol (23) lauryl ether with ionic surfactants studied by proton 1D and 2D NMR. *J. Colloid Interface Sci.* **2002**, *249*, 200–208.

47. Zakharova, L., Valeeva, F., Zakharov, A., Ibragimova, A., Kudryavtseva, L., Harlampidi, H. Micellization and catalytic activity of the cetyltrimethylammonium bromide — Brij 97 — water mixed micellar system. *J. Colloid Interface Sci.* **2003**, *263*, 597–605.

48. Fernandez, G., Rodriguez, A., Graciani, M. del M., Munoz, M., Moya, M.L. Study of the reaction methyl 4-nitrobenzenesulfonate + Cl$^-$ in mixed hexadecyltrimethylammonium chloride-triton X-100 micellar solutions. *Int. J. Chem. Kinet.* **2003**, *35*(2), 45–51.

49. Munoz, M., Rodriguez, A., Graciani, M. del M., Moya, M.L. Micellar medium effects on the hydrolysis of phenyl chloroformate in ionic, zwitterionic, nonionic, and mixed micellar solutions. *Int. J. Chem. Kinet.* **2002**, *34*, 445–451.

50. Khan, M.N. Effects of mixed anionic and cationic surfactants on rate of transesterification and hydrolysis of esters. *J. Colloid Interface Sci.* **1996**, *182*, 602–605.

51. Khan, M.N., Arifin, Z. Kinetics and mechanism of intramolecular general base-catalyzed methanolysis of phenyl salicylate in the presence of cationic micelles. *Langmuir* **1996**, *12*(2), 261–268.

52. Herrington, K.L., Kaler, E.W., Miller, D.D., Zasadzinski, J.A., Chiruvolu, S. Phase behavior of aqueous mixtures of dodecyltrimethylammonium bromide (DTAB) and sodium dodecyl sulfate (SDS). *J. Phys. Chem.* **1993**, *97*(51), 13792–13802.

53. Khan, M.N., Ismail, E., Yusoff, M.R. Effects of pure non-ionic and mixed non-ionic-cationic surfactants on the rates of hydrolysis of phenyl salicylate and phenyl benzoate in alkaline medium. *J. Phys, Org. Chem.* **2001**, *14*, 669–676.

54. Khan, M.N., Arifin, Z., Wahab, I.B., Ali, S.F.M., Ismail, E. Effects of cationic micelles on the rate of intramolecular general base-catalyzed reaction of ionized phenyl salicylate (PS$^-$) with tris(hydroxymethyl)aminomethane (Tris). *Colloids Surf. A* **2000**, *163*(2–3), 271–281.

55. Khan, M.N., Arifin, Z. Effects of cationic micelles on rates and activation parameters of intramolecular general base-catalyzed hydrolysis of ionized salicylate esters. *J. Colloid Interface Sci.* **1996**, *180*, 9–14.

56. Blasko, A., Bunton, C.A., Toledo, E.A., Holland, P.M., Nome, F. S_N2 reactions of a sulfonate ester in mixed cationic/phosphine oxide micelles. *J. Chem. Soc. Perkin Trans. 2.* **1995**, 2367–2373.

57. Joshi, H.M., Nagar, T.N. Kinetics of oxidation of rosaniline hydrochloride by potassium peroxydisulphate in mixed micelles of binary surfactant system. *Asian J. Chem.* **2002**, *14*(3–4), 1763–1765.

58. Ghosh, K.K., Pandey, A. Kinetics of hydrolysis of hydroxamic acid in mixed micelles of binary surfactant systems. *J. Indian Chem. Soc.* **1999**, *76*(4), 191–194.

59. Wright, S., Bunton, C.A., Holland, P.M. Binding of bromide ion to mixed cationic-nonionic micelles. In *Mixed Surfactant Systems*, Vol. 501, Holland, M., Rubingh, D.N., Eds. ACS Symposium Series, American Chemical Society, Washington, D.C., **1992**, pp. 227–233, chap. 13.

60. (a) Bunton, C.A., Savelli, G. Organic reactivity in aqueous micelles and similar assemblies. *Adv. Phys. Org. Chem.* **1986**, *22*, 213–309. (b) Larsen, J.W., Tepley, L.B. Effect of aqueous alcoholic solvents on counterion binding to CTAB [cetyltrimethylammonium bromide] micelles. *J. Colloid Interface Sci.* **1974**, *49*(1), 113–118. (c) Mackay, R.A. Reactions in microemulsions: the ion-exchange model. *J. Phys. Chem.* **1982**, *86*(24), 4756–4758. (d) Bunton, C.A., de Buzzaccarini, F. Quantitative treatment of bromide ion nucleophilicity in a microemulsion. *J. Phys. Chem.* **1982**, *86*(25), 5010–5014.

61. Vangeyte, P., Leyh, B., Auvray, L., Grandjean, J., Misselyn-Bauduin, A.-M., Jerome, R. Mixed self-assembly of poly(ethylene oxide)-b-poly(ϵ-caprolactone) copolymers and sodium dodecyl sulfate in aqueous solution. *Langmuir* **2004**, *20*(21), 9019–9028.

62. Bunton, C.A., Wright, S. S_N2 Reactions of a sulfonate ester in mixed cationic/nonionic micelles. *Langmuir* **1993**, *9*(1), 117–120.

63. Foroudian, H.J., Bunton, C.A., Holland, P.M., Nome, F. Nucleophilicity of bromide ion in mixed cationic/sulfoxide micelles. *J. Chem. Soc. Perkin Trans. 2.* **1996**, 557–561.

64. Blasko, A., Bunton, C.A., Toledo, E.A., Holland, P.M., Nome, F. S_N2 Reactions of a sulfonate ester in mixed cationic/phosphine oxide micelles. *J. Chem. Soc. Perkin Trans. 2.* **1995**, 2367–2373.

65. (a) Zana, R. Ionization of cationic micelles: effect of the detergent structure. *J. Colloid Interface Sci.* **1980**, *78*(2), 330–337. (b) Lianos, P., Zana, R. Micelles of tetradecyltrialkylammonium bromides with fluorescent probes. *J. Colloid Interface Sci.* **1982**, *88*(2), 594–598. (c) Van Nieuwkoop, J., Snoei, G. Conductivity measurements in single-phase microemulsions of the system sodium dodecyl sulfate/1-butanol/water/heptane. *J. Colloid Interface Sci.* **1985**, *103*(2), 417–435.

66. (a) Bacaloglu, R., Bunton, C.A., Ortega, F. Micellar enhancements of rates of S_N2 reactions of halide ions: the effect of head group size. *J. Phys. Chem.* **1989**, *93*(4), 1497–1502. (b) Chaudhuri, A., Romsted, L.S. Simultaneous determination of counterion, alcohol, and water concentrations at a three-component microemulsion interface using product distributions from a dediazoniation reaction. *J. Am. Chem. Soc.* **1991**, *113*(13), 5052–5053. (c) Chaudhuri, A., Laughlin, J.A., Romsted, L.S., Yao, J. Arenediazonium salts: new probes of the interfacial compositions of association colloids. 1. Basic approach, methods, and illustrative applications. *J. Am. Chem. Soc.* **1993**, *115*(18), 8351–8361. (d) Chaudhuri, A., Romsted, L.S., Yao, J. Arenediazonium salts: new probes of the interfacial compositions of association colloids. 2. Binding constants of butanol and hexanol in aqueous three-component cetyltrimethylammonium bromide microemulsions. *J. Am. Chem. Soc.* **1993**, *115*(18), 8362–8367.

67. (a) Bunton, C.A., Wright, S., Holland, P.M., Nome, F. S_N2 reactions of a sulfonate ester in mixed cationic/nonionic micelles. *Langmuir* **1993**, *9*(1), 127–120. (b) Bertoncini, C.R.A., Nome, F., Cerichelli, G., Bunton, C.A. Effect of 1-butanol upon S_N2 reactions in cationic micelles: a quantitative treatment. *J. Phys. Chem.* **1990**, *94*(15), 5875–5878. (c) Bertoncini, C.R.A., Neves, M. de F.S., Nome, F., Bunton, C.A. Effects of 1-butanol-modified micelles on S_N2 reactions in mixed-ion systems. *Langmuir* **1993**, *9*(5), 1274–1279.

68. (a) Bunton, C.A., Gan, L.-H., Moffatt, J.R., Romsted, L.S., Savelli, G. Reactions in micelles of cetyltrimethylammonium hydroxide: test of the pseudophase model for kinetics. *J. Phys. Chem.* **1991** *85*(26), 4118–4125. (b) Stadler, E., Zanette, D.,

Rezende, M.C., Nome, F. Kinetic behavior of cetyltrimethylammonium hydroxide: the dehydroclorination of 1,1,1-trichloro-2,2-bis(p-chlorophenyl)ethane and some of its derivatives. *J. Phys. Chem.* **1984**, *88*(9), 1892–1896. (c) Neves, M. de F.S., Zanette, D., Quina, F., Moretti, M.T., Nome, F. Origin of the apparent breakdown of the pseudophase ion-exchange-model for micellar catalysis with reactive counterion surfactants. *J. Phys. Chem.* **1989**, *93*(4), 1502–1505.

69. Bacaloglu, R., Bunton, C.A., Cerichelli, G., Ortega, F. Micellar effects upon rates of S_N2 reactions of chloride ion. II. Effects of cationic headgroups. *J. Phys. Chem.* **1990**, *94*(12), 5068–5073.

70. (a) Bunton, C.A., Moffatt, J.R. Ionic competition in micellar reactions: a quantitative treatment. *J. Phys. Chem.* **1986**, *90*(4), 538–541. (b) Bunton, C.A., Moffatt, J.R. Micellar effects upon substitutions by nucleophilic anions. *J. Phys. Chem.* **1988**, *92*(10), 2896–2902. (c) Bunton, C.A., Moffatt, R.R. A quantitative treatment of micellar effects in moderately concentrated hydroxide ion. *Langmuir* **1992**, *8*(9), 2130–2134. (d) Blasko, A., Bunton, C.A., Cerichelli, G., McKenzie, D.C. A nuclear magnetic resonance study of ion exchange in cationic micelles: success and failures of models. *J. Phys. Chem.* **1993**, *97*(43), 11324–11331. (e) Rodenas, E., Dolcet, C., Valiente, M. Simulations of micelle-catalyzed bimolecular reaction of hydroxide ion with a cationic substrate using the nonlinearized Poisson–Boltzmann equation. *J. Phys. Chem.* **1990**, *94*(4), 1472–1477. (f) Dolcet, C., Rodenas, E. Hydroxide ion specific adsorption on cetyltrimethylammonium bromide micelles explains kinetic data. *Colloids Surf.* **1993**, *75*, 39–50.

71. (a) Romsted, L.S. Micellar effects on reaction rates and equilibria. In *Surfactants in Solution*, Vol. 2, Mittal, K.L., Lindman, B., Eds. Plenum Press: New York, **1984**, pp. 1015–1068. (b) Germani, R., Ponti, P., Savelli, G., Spreti, N., Bunton, C.A., Moffatt, J.R. Modeling of micellar effects upon substitution reactions with moderately concentrated hydroxide ion. *J. Chem. Soc. Perkin Trans. 2.* **1989**, 401–405.

72. Bunton, C.A., Foroudian, H.J., Gillitt, N.D. Effects of headgroup structure on dephosphorylation of p-nitrophenyl diphenyl phosphate by functional oximate comicelles. *Langmuir* **1999**, *15*(4), 1067–1074.

73. Frescura, V.L.A., Marconi, D.M.O., Zanette, D., Nome, F., Blasko, A., Bunton, C.A. Effects of sulfobetaine-sodium dodecanoate micelles on deacylation and indicator equilibrium. *J. Phys. Chem.* **1995**, *99*(29), 11494–11500.

74. (a) Fernandez, M.S., Fromherz, P. Lipoid pH indicators as probes of electrical potential and polarity in micelles. *J. Phys. Chem.* **1977**, *81*(18), 1755–1761. (b) Zanette, D., Leite, M.R., Reed, W., Nome, F. Intrinsic basicity constant of 10-phenyl-10-(hydroxyamino)decanoate in aqueous solutions of hexadecyltrimethylammonium bromide: effect of salts and detergent concentration. *J. Phys. Chem.* **1987**, *91*(8), 2100–2102. (c) Romsted, L.S., Zanette, D. Quantitative treatment of indicator equilibria in micellar solutions of sodium decyl phosphate and sodium lauryl sulfate. *J. Phys. Chem.* **1988**, *92*(16), 4690–4698.

75. Marconi, D.M.O., Frescura, V.L.A., Zanette, D., Nome, F., Bunton, C.A. Nucleophilically assisted deacylation in sodium dodec anoate and dodecyl sufate micelles. *J. Phys. Chem.* **1994**, *98*(47), 12415–12419.

76. Lee, B.S., Nome, F. Effects of sulfobetaine–sodium decyl phosphate mixed micelles on deacylation and indicator equilibrium. *Langmuir* **2000**, *16*(26), 10131–10136.

77. Froehner, S.J., Nome, F., Zanette, D., Bunton, C.A. Micellar-mediated general acid catalyzed acetal hydrolysis. Reactions in comicelles. *J. Chem. Soc. Perkin Trans. 2.* **1996**, 673–676.

78. Freire, L., Iglesias, E., Bravo, C., Leis, J.R., Peña, M.E. Physicochemical properties of mixed anionic-non-ionic micelles: effects on chemical reactivity. *J. Chem. Soc. Perkin Trans. 2.* **1994**, 1887–1894.

79. (a) Romsted, L.S. A general kinetic theory of rate enhancements for reactions between organic substrates and hydrophilic ions in micellar systems. In *Micellization, Solubilization, Microemulsions*, Vol. 2, Mittal, K.L., Ed. Plenum Press: New York, **1977**, pp. 309–530. (b) Romsted, L.S., Bunton, C.A., Yao, J. Micellar catalysis, a useful misnomer. *Curr. Opin. Colloid Interface Sci.* **1997**, *2*(6), 622–628.

80. Iglesias, E., Montenegro, L. Kinetic investigations of the interaction between sodium dodecyl sulfate and the nonionic surfactants C_mE_n. Electrical conductivity and fluorescence probe measurements. *Phys. Chem. Chem. Phys.* **1999**, *1*(20), 4865–4874.

81. Iglesias, E. Ethyl cyclohexanone-2-carboxylate in aqueous micellar solutions. 1. Ester hydrolysis in cationic and nonionic micelles. *J. Phys. Chem. B* **2001**, *105*(42), 10287–10294.

82. Treiner, C. The thermodynamics of micellar solubilization of neutral solutes in aqueous binary surfactant systems. *Chem. Soc. Rev.* **1994**, *23*(5), 349–356.

83. Davies, D.M., Foggo, S.J. Kinetic treatment of the reaction of *m*-chloroperbenzoic acid and iodide in mixed anionic/non-ionic micelles. *J. Chem. Soc., Perkin Trans. 2.* **1998**, 247–251.

84. Khan, M.N. Effects of mixed anionic and cationic surfactants on rate of transesterification and hydrolysis of esters. *J. Colloid Interface Sci.* **1996**, *182*, 602–605.

85. (a) Khan, M.N., Ismail, E. Effects of non-ionic and mixed non-ionic-cationic micelles on the rate of aqueous cleavages of phenyl benzoate and phenyl salicylate in alkaline medium. *J. Phys. Org. Chem.* **2004**, *17*, 376–386. (b) Khan, M.N., Ismail, E. Effects of non-ionic and mixed cationic-non-ionic micelles on the rate of alkaline hydrolysis of phthalimide. *J. Phys. Org. Chem.* **2002**, *15*, 374–384.

86. Khan, M.N., Ismail, E. Effects of nonionic and mixed cationic–nonionic micelles on the rate of alkaline hydrolysis of 4-nitrophthalimide. *J. Colloid Interface Sci.* **2001**, *240*, 636–639.

87. Khan, M.N., Zunoliza, A. Kinetics and mechanism of alkaline hydrolysis of 4-nitrophthalimide in the absence and presence of cationic micelles. *Int. J. Chem. Kinet.* **2001**, *33*(7), 407–414.

6 Metallomicelles: Effects on Reaction Rates

6.1 INTRODUCTION

Metallomicellar catalysis is a micellar catalytic process that has a greater effect on the rate of a reaction in the presence than in the absence of metal ions. Metal ions, commonly known as Lewis acids, are general acids. If the rate of a reaction is sensitive to general acid catalysis, the metal ions, under the appropriate reaction conditions such as medium polarity, pH, etc., can catalyze the rate of such a reaction. Metal-ion-catalyzed hydrolysis of phosphate or carboxylate esters may be due to (1) coordination of metal ion to $P = O$ or $C = O$ group (Lewis acid or general acid or electrophilic catalysis), (2) charge neutralization, (3) activation of a nucleophilic water molecule by the decrease of its pK_a (nucleophilic catalysis), and (4) energetically assisted departure of a leaving group (general acid catalysis).

Micelle-forming surfactants with special head groups that could complex metal ions through electrovalent bonding interaction may cause very large rates of acceleration because these metallomicelles become so-called induced functional metallomicelles. A functional surfactant carries a reactive head group and, consequently, a functional micelle increases the rate of a bimolecular reaction owing to significant loss of degree of translational motion of one of the reactants. Similarly, a reactive counterion micelle consists of reactive counterions, and if the micellar surface is saturated with counterions, the concentration of the ionic reactant (i.e., counterion) at the ionic micellar surface should be constant, and the mole ratio of reactive ions to micellar head groups will be given by $\beta = 1 - \alpha$. Then, the rate of such a bimolecular reaction will depend only on the distribution of the substrate (i.e., other neutral reactant) between aqueous phase and micellar phase.

Almost all the kinetic studies on metallomicellar-mediated reactions have been carried out in the presence of either nonfunctional metallomicelles in which micellar head groups are incapable of acting as ligands for metal–ligand complex formation or induced functional metallomicelles in which micellar head groups act as effective ligands for metal–ligand complex formation.

6.2 NONFUNCTIONAL METALLOMICELLES

Zeng et al.[1–13] have carried out an extensive study on the effects of pure ionic and nonionic micelles on various metal-complex-catalyzed hydrolyses of activated esters. Recently, they developed the ternary complex kinetic (TCK) model to give quantitative or semiquantitative explanations of the observed kinetic data on metallomicellar-mediated reactions. However, this model, which is related to the pseudophase (PP) model, needs careful attention as described in the following text.

6.2.1 THE TCK MODEL FOR NONFUNCTIONAL METALLOMICELLAR-MEDIATED REACTIONS[1]

The TCK model is based on the following reaction scheme (Scheme 6.1):

$$n\,(L)_W + m\,(M)_W \xrightleftharpoons{K_M} (M_mL_n)_M$$

$$(M_mL_n)_M + (S)_W \xrightleftharpoons{K_T} (M_mL_nS)_M$$

$$(S)_W \xrightarrow{k_0'} P$$

$$(M_mL_nS)_M \xrightarrow{k_{N'M}} P$$

SCHEME 6.1

where subscripts W and M represent bulk water phase and metallomicellar phase, respectively; L, M, and S are ligand, metal ion, and neutral substrate such as ester, respectively; K_M is the association constant between m metal ions and n ligands; K_T is the association constant between a binary complex (M_mL_n) and a neutral substrate S; $k_{N'M}$ and k_0' are the apparent first-order rate constants for product formation in the metallomicellar phase and in the bulk water phase, respectively. The apparent first-order rate constant (k_0') for the product formation in bulk water phase is expressed by

$$k_0' = k_0 + k_M[M_W] + k_L[L_W] \tag{6.1}$$

where k_0 is the pseudo-first-order rate constant due to the buffer used to maintain the pH of the reaction medium, k_M and k_L are the second-order rate constant due to respective metal ion and ligand, $[M_W]$ and $[L_W]$ are the concentrations of metal ion and ligand in the bulk water phase, respectively. It is apparent from Scheme 6.1 that

$$K_M = ([M_mL_n])_M/([M_W]^m[L_W]^n) \tag{6.2}$$

where $([M_mL_n])_M$ is the concentration of binary complex in the metallomicellar phase and

$$K_T = ([M_mL_nS])_M/\{([M_mL_n])_M[S_W]\} \tag{6.3}$$

where $([M_mL_nS])_M$ and $[S_W]$ are the concentrations of substrate (S) in the metallomicellar phase and bulk water phase, respectively. The total concentrations of ligand ($[L]_T$), metal ion ($[M]_T$), and substrate ($[S]_T$) are expressed by Equation 6.4, Equation 6.5, and Equation 6.6, respectively.

$$[L]_T = [L_W] + n([M_mL_n])_M \tag{6.4}$$

$$[M]_T = [M_W] + m([M_mL_n])_M \tag{6.5}$$

$$[S]_T = [S_W] + n([M_mL_nS])_M \tag{6.6}$$

Experimentally determined rate law (rate = $k_{obs}[S]_T$), Scheme 6.1, and Equation 6.6 can lead to Equation 6.7

$$k_{obs} = \frac{k_0' + \{k_{NM}'([M_mL_nS])_M / [S_W]\}}{1 + \{([M_mL_nS])_M / [S_W]\}} \tag{6.7}$$

Equation 6.2 and Equation 6.7 with rearrangement can lead to Equation 6.8

$$\frac{1}{k_{obs} - k_0'} = \frac{1}{K_T(k_{NM}' - k_0')([M_mL_n])_M} + \frac{1}{k_{NM}' - k_0'} \tag{6.8}$$

Inserting Equation 6.4 and Equation 6.5 into Equation 6.2 and neglecting the high-order terms of $([M_mL_n])_M$, we have

$$([M_mL_n])_M = \frac{K_M[M]_T^m[L]_T^n}{1 + n^2K_M[M]_T^m[L]_T^{n-1} + m^2K_M[L]_T^n[M]_T^{m-1}} \tag{6.9}$$

Equation 6.8 and Equation 6.9 can lead to Equation 6.10

$$\frac{1}{k_{obs} - k_0'} = \frac{1}{K_T(k_{NM}' - k_0')K_M[M]_T^m[L]_T^n} + \frac{n^2}{K_T(k_{NM}' - k_0')[L]_T} +$$
$$\frac{m^2}{K_T(k_{NM}' - k_0')[M]_T} + \frac{1}{k_{NM}' - k_0'} \tag{6.10}$$

Equation 6.10 can have different forms depending on the nature of a particular reaction. For example, when n = 1 and m = 1, Equation 6.10 may be written as

$$\frac{1}{k_{obs} - k_0'} = \frac{1}{K_T(k_{NM}' - k_0')}\left(1 + \frac{1}{K_M[M]_T}\right)\frac{1}{[L]_T} +$$

$$\frac{1}{K_T(k_{NM}' - k_0')[M]_T} + \frac{1}{k_{NM}' - k_0'} \tag{6.11}$$

For a particular reaction system in which n = 1 and m = 2, Equation 6.10 may be written as

$$\frac{1}{k_{obs} - k_0'} = \frac{1}{K_T(k_{NM}' - k_0')}\left(1 + \frac{1}{K_M[M]_T^2}\right)\frac{1}{[L]_T} +$$

$$\frac{4}{K_T(k_{NM}' - k_0')[M]_T} + \frac{1}{k_{NM}' - k_0'} \tag{6.12}$$

For a particular reaction system in which n = 2 and m = 1, Equation 6.10 may be written as

$$\frac{1}{k_{obs} - k_0'} = \frac{1}{K_T(k_{NM}' - k_0')}\left(1 + \frac{1}{K_M[L]_T^2}\right)\frac{1}{[M]_T} +$$

$$\frac{4}{K_T(k_{NM}' - k_0')[L]_T} + \frac{1}{k_{NM}' - k_0'} \tag{6.13}$$

The values of k_{NM}', K_M, and K_T can be calculated from one of Equation 6.11 to Equation 6.13 by manipulating the reaction conditions for a particular reaction. However, although the TCK model is based upon the essential feature of the PP model and has been used extensively by Zeng et al.,[14–21] it contains the following assumptions that do not exist either in the PP model or in any one of its various extended forms as discussed in Chapter 3:

1. The free ligands (L), metal ions (M), and neutral substrate molecules (S) do not exist in the micellar phase, i.e., $[L_M] = [M_M] = [S_M] = 0$ where subscript M represents the micellar phase.
2. The concentrations of binary complex (M_mL_n) and ternary complex (M_mL_nS) in the bulk water phase are zero, i.e., $[(M_mL_n)_W] = [(M_mL_nS)_W] = 0$, where subscript W stands for the bulk water phase.

The rate of hydrolysis of 4-nitrophenyl picolinate or 4-nitrophenyl 2-pyridine carboxylate (1) has been studied at constant pH (maintained by buffers) in the presence of three ligands, N,N,N',N'-tetra(2-hydroxyethyl)-1,3-diaminopropane (2), N,N,N',N'-tetra(2-hydroxyethyl)-1,10-diaminodecane (3), and N,N,N',N'-tetra(2-hydroxyethyl)-1,4-diaminoxylene (4) as well as three metal ions, Zn(II),

TABLE 6.1
Apparent First-Order Rate Constants (k_{obs}) and Metal Ion Catalytic Factor (Y) for the Hydrolysis 1 at pH 7.00, 25°C[a]

System	$10^3 k_{obs}/sec^{-1}$ [b]	Y^c (for Zn^{2+})	Y^c (for Ni^{2+})	Y^c (for Co^{2+})
Buffer	0.0132	51	34	27
2 + Buffer	0.0226	14	17	11
3 + Buffer	0.0595	14	27	13
4 + Buffer	0.0457	15	28	14
Buffer + CTABr	0.0213	15	9	8
2 + Buffer + CTABr	0.0134	23	19	17
3 + Buffer + CTABr	0.0462	13	6	11
4 + Buffer + CTABr	0.0142	41	40	31
Buffer + Brij35	0.0239	19	19	17
2 + Buffer + Brij35	0.0337	21	25	19
3 + Buffer + Brij35	0.0513	26	29	25
4 + Buffer + Brij35	0.0402	22	29	23

[a] Data from Jiang, F., Jiang, B., Yu, X., Zeng, X. Metallomicellar catalysis: effects of Bridge-connecting ligands on the hydrolysis of PNPP catalyzed by Zn(II), Co(II), and Ni(II) complexes of ethoxy-diamine ligands in micellar solution. *Langmuir* **2002**, *18*(18), 6769–6774; reaction conditions: 0.01 M Tris buffer, μ = 0.1 M (KNO$_3$), [CTABr] = 0.01 M, [Brij35] = 0.001 M, [1] = 5 × 10^{-5} M, and [ligand] = [metal ion (M^{2+})] = 0.001 M.
[b] Values of k_{obs} at [M^{2+}] = 0.
[c] Y = k_{obs} (at a [M^{2+}])/k_{obs} (at [M^{2+}] = 0).

Ni(II), and Co(II). Effects of 0.01 M CTABr and 0.001 M Brij35 on the rate of hydrolysis of **1** have also been studied at constant pH in the absence and presence of ligands (**2**, **3**, and **4**) and metal ions {Zn(II), Ni(II), and Co(II)}.[1] The observed catalytic effects of metal ions, obtained under a variety of reaction conditions, are summarized in Table 6.1. The stoichiometry of metal complexes has been determined by kinetic versions of Job plots, which involve plots of apparent first-order rate constants (k_{obs}) vs. mole fraction of a ligand or metal ion at a constant total concentration of ligand and metal ion. Such plots show 2:1 and 1:2 complexes (metal or ligand) in the presence of CTABr and Brij35 micelles, respectively, and the proposed structures of these complexes are shown in Figure 6.1 and Figure 6.2.[1] The formation of 1:2 and 2:1 complexes of the ligand to metal ions in respective CTABr and Brij35 micelles has been reported in the related studies.[22,23]

1

2 : X = CH$_2$CH$_2$CH$_2$
3 : X = CH$_2$(CH$_2$)$_8$CH$_2$
4 : X = H$_2$C—⟨benzene⟩—CH$_2$

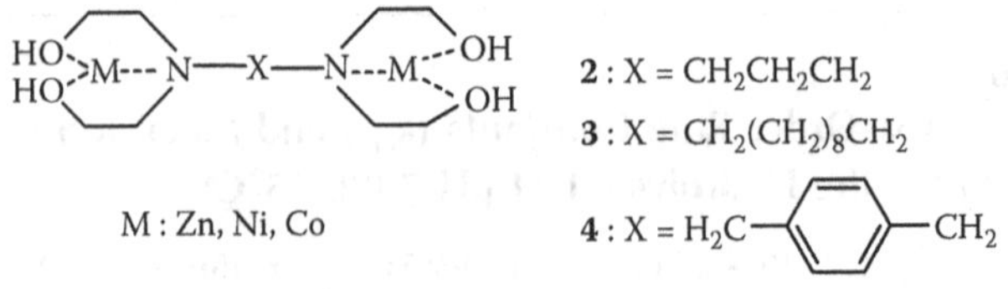

FIGURE 6.1 The structure of complex in CTABr micellar solution.

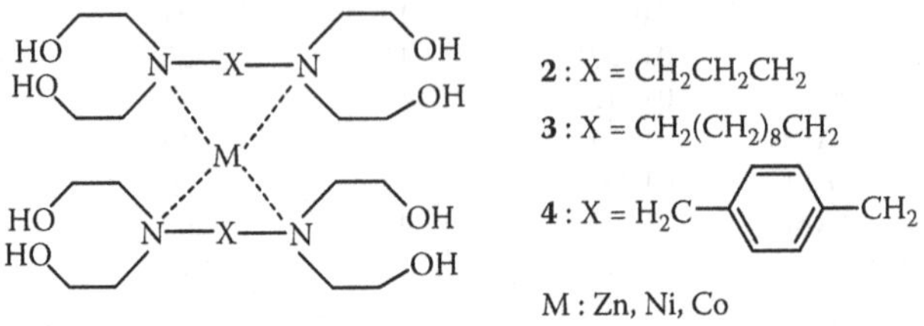

FIGURE 6.2 The structure of complex in Brij35 micellar solution.

It is apparent from Table 6.1 that the metal ion (Zn^{2+}, Ni^{2+}, and Co^{2+}) catalytic efficiency, which is measured by the Y values $\{Y = k_{obs}$ (at a $[M^{2+}])/k_{obs}$ (at $[M^{2+}] = 0)\}$, is significantly larger in the absence than in the presence of either ligands (**2**, **3**, and **4**) or both ligands + CTABr, as well as ligands + Brij35 micelles. However, in a CTABr micellar solution, both the ligands and metal–ligand complexes show little inhibition on the rate of hydrolysis of **1** when compared with the rate enhancement by the ligands or metal ions only. The nearly 20-fold rate enhancement due to the presence of Brij35 micelles is surprising, whereas a similar rate enhancement owing to the presence of CTABr micelles may be attributed to increased pH at the cationic micellar surface through ion-exchange HO^-/Br^-.[1]

The effects of metallo-CTABr and metallo-Brij35 micelles on k_{obs} for hydrolysis of **1** at pH 7.00 have been analyzed using Equation 6.12 and Equation 6.13, respectively. But the calculated values of k'_{NM}, K_M, and K_T do not show a plausible order of magnitude. For example, the values of K_M and K_T are expected to be lower in cationic CTABr micelles than in nonionic Brij35 micelles due to unfavorable electrostatic interaction between cationic micellar head groups and complexes M_mL_n and M_mL_nS. But most of the K_M and K_T values do not fit to this prediction. Even similar or larger values of k'_{NM} in Brij35 micelles compared to those in CTABr micelles are surprising.[1]

6.2.1.1 Effect of pH on k'_{NM}

It is a well-established fact that in the absence of micelles, the pK_a of a water molecule bound to a metal ion through an electrovalent bond is smaller by several pK units than that of the metal-unbound or free water molecule. Thus, it is obvious that one must obtain k_{NM} values dependent on pH. Jiang et al.[1] suggested a mechanism as shown by Scheme 6.2 to explain k'_{NM}–pH profile.

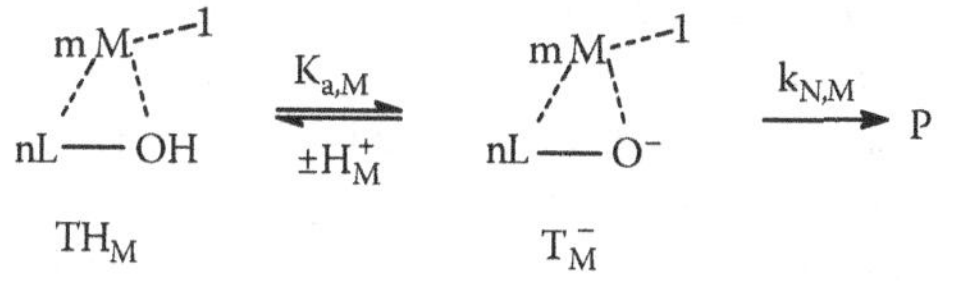

$$\underset{TH_M}{\overset{mM^{\cdots\text{-}1}}{\underset{nL-\!\!\!-OH}{\bigwedge}}} \quad \overset{K_{a,M}}{\underset{\pm H_M^+}{\rightleftharpoons}} \quad \underset{T_M^-}{\overset{mM^{\cdots\text{-}1}}{\underset{nL-\!\!\!-O^-}{\bigwedge}}} \quad \overset{k_{N,M}}{\longrightarrow} \; P$$

SCHEME 6.2

In Scheme 6.2, TH_M and T_M are the micellized nonionized and ionized ternary complex, respectively, TH_M is nonreactive and $k_{N,M}$ is the first-order rate constant for product formation from reactive T_M^-. It may be noted that Scheme 6.2 ignores the effects of pH on association constants K_M and K_T as described in Scheme 6.1. The rate law: rate = $k'_{NM} [(M_mL_nS)_M]$ and Scheme 6.2 can lead to Equation 6.14

$$k'_{NM} = \frac{k_{N,M}K_{a,M}}{[H_M^+] + K_{a,M}} \tag{6.14}$$

where $K_{a,M} = ([T_M^-][H_M^+])/([TH_M])$, $([M_mL_nS])_M = [TH_M] + [T_M^-]$, and $k_{N,M}$ is the first-order rate constant of the unimolecular conversion of T_M^- to products (P). The following linearized rearranged form of Equation 6.14 is used to calculate $k_{N,M}$ and $K_{a,M}$ from the calculated values of k'_{NM} at different pH.

$$1/k'_{NM} = 1/k_{N,M} + [H_M^+]/(k_{N,M}K_{a,M}) \tag{6.15}$$

Although the plots of $1/k'_{NM}$ vs. $[H^+]$ for all three ligands are linear[1], the data treatment using Scheme 6.2 involves an inherent assumption that $[H_W^+] \approx [H_M^+]$, which may not be correct, at least in cationic micelles (CTABr). It is interesting and surprising to note that most of the values of $k_{N,M}$ and $pK_{a,M}$ in CTABr micelles are not appreciably different from the corresponding values in Brij35 micelles.[1]

Effects of cationic (cetylpyridinium chloride, CPC) and anionic (SDS) micelles on the rate of reaction of chromium(VI) oxidation of formaldehyde have been studied in the presence and absence of picolinic acid.[24] Cationic micelles (CPC) inhibit whereas anionic micelles (SDS) catalyze the reaction rates that could be attributed to electrostatic interactions between reactants (cationic metal ions and catalyst H^+) and ionic head groups of ionic micelles. Experimentally determined kinetic data on these metallomicellar-mediated reactions have been explained by different kinetic models such as pseudophase ion-exchange (PIE) model, Menger's enzyme-kinetic-type model, and Piszkiewicz's cooperativity model (Chapter 3). The rate of oxidation of proline by vanadium(V) with water acting as nucleophile is catalyzed by aqueous micelles.[25] Effects of anionic micelles (SDS) on the rate of N-bromobenzamide-catalyzed oxidation of ethanol, propanol, and n-butanol in acidic medium reveal the presence of premicellar catalysis that has been rationalized in light of the positive cooperativity model.[26]

The anionic surfactant (SDS) accelerates whereas the cationic surfactant (CPC) retards the rate of picolinic-acid-promoted Cr(VI) oxidation of the hexitols to the respective aldohexoses in aqueous acidic media.[27]

In aqueous Brij35 micellar solution, the copper(II) complex of macrocyclic Schiff base can only catalyze the hydrolysis of 1 by the mechanism that involves the nucleophilic attack of external hydroxide ion on the carbonyl carbon of substrate (ester), whereas the zinc(II) complex of same ligand can accelerate the hydrolysis of 1 more strongly than that of 4-nitrophenyl acetate by the intramolecular nucleophilic attack of zinc-bound hydroxide ion on carbonyl carbon of esters.[28] The catalytic activity of Zn(II) complex is close to or higher than that of Cu(II) complex. The rate constants for the catalytic hydrolysis of bis(4-nitrophenyl) phosphate by complexes [(bpya)Cu]Cl$_2$ and [(bpya)Zn]Cl$_2$ (where bpya = 2,2$'$-dipyridylamine) in Brij35 micellar solution at 25°C and pH 7.02 are 1.2×10^6 times and 1.5×10^5 times higher than those for the spontaneous hydrolysis, respectively.[29]

The effects of cationic micelles (CTABr and cetylpyridinium bromide, CPBr) on the rate of chromium(VI) oxidation of glycolic acid have been explained by a model in which the reaction rate depends on the concentration of both reactants in the micellar pseudophase and some added inorganic salts (NaCl, NaBr, NaNO$_3$, and Na$_2$SO$_4$) reduce the micellar catalysis by excluding glycolic acid from the micellar reaction site.[30] The rate of picolinic-acid-catalyzed oxidation of dimethyl sulfoxide to dimethyl sulfone by chromium(VI) in aqueous acidic media is catalyzed by sodium dodecyl sulfate, whereas cetylpyridinium chloride retards the reaction continuously. The observed micellar effects have been explained by pseudophase ion-exchange model. The Piszkiewicz cooperativity model has been applied to determine the kinetic parameters, and it indicates the existence of catalytically productive submicellar aggregates.[31]

The effects of copper metallomicelles on apparent second-order rate constants (k$_{2ap}$) for Diels–Alder reactions have been studied by carrying out a limited number of kinetic runs (Table 6.2) that could allow for providing only a qualitative interpretation. Both Cu^{2+}-CTABr and Cu^{2+}-C$_{12}$E$_7$ metallomicelles show mild inhibitory effects on k$_{2ap}$ for bimolecular reactions of 5 with 6a, 6b, and 6c in the presence of a single constant [surfactant] (= 7.8 mM) above its CMC and constant [Cu(NO$_3$)$_2$] (= 0.10 mM). Inhibitions that are more significant with C$_{12}$E$_7$ than with CTABr are attributed to different locations of micellized reactants, diene (5) and dienophiles (6), in the micellar pseudophase.[32]

6a : X = H
6b : X = CH$_2$SO$_3$Na
6c : X = CH$_2$NMe$_3$Br

Pseudo-first-order rate constants (k$_{obs}$) for the reaction between metal ruthenium complex [Ru(NH$_3$)$_5$pz]$^{2+}$ (pz = pyrazine) and oxidant S$_2$O$_8^{2-}$ (peroxydisul-

TABLE 6.2
Effect of Micelles of CTABr and $C_{12}E_7$ on the Apparent Metal Ion and Metallomicellar Catalytic Factor (Y) for the Reaction between 5 and 6a, 6b, and 6c, Respectively, at 25°C[a]

Medium	6a	6b	6c
Water + 10 mM Cu(NO$_3$)$_2$	79×10^3	124×10^3	175×10^3
Water + 10 mM Cu(NO$_3$)$_2$ + CTABr[b]	29×10^3	11×10^3	131×10^3
Water + 10 mM Cu(NO$_3$)$_2$ + $C_{12}E_7$[b]	45×10^3	77×10^3	122×10^3

[a] From Otto, S., Engberts, J.B.F.N., Kwak, J.C.T. Million-fold acceleration of a Diels–Alder reaction due to combined Lewis acid and micellar catalysis in water. *J. Am. Chem. Soc.* **1998**, *120*(37), 9517–9525; Y = k_{2ap}/k_{2ap}^s where k_{2ap}^s (= 1.40×10^{-5} M^{-1} sec^{-1}) is the apparent second-order rate constant for the reaction in pure actonitrile solvent.[32]

[b] The concentration of surfactant is 7.8 mM above the CMC of the particular surfactant.

fate ion) with ratio $[S_2O_8^{2-}]/[Ru(NH_3)_5pz]^{2+} > 10$, obtained in the presence of CTACl micelles, first decrease on increasing $[CTACl]_T$, reaching a minimum, and then increase as the $[CTACl]_T$ increases.[33] It is concluded that the increase in k_{obs} with increase in $[CTACl]_T$ is due to a change in CTACl micellar binding constant (K) of counterion ($S_2O_8^{2-}$). The initial monotonic decrease in k_{obs} with increase in $[CTACl]_T$ until $[CTACl]_T = 0.0096$ M has been found to fit to Equation 6.16

$$k_{obs} = \frac{k_W + k_M K[D_n]}{1 + K[D_n]} \qquad (6.16)$$

which is similar to Equation 3.13 (Chapter 3) with $[D_n] = [CTACl]_T - CMC$, $[R]_T = [S_2O_8^{2-}]_T$, $k_W = k_W^2[R]_T$, $K = K_R$, and $k_M = k_M^{mr}K_S[R]_T$, where K_S is the CTACl micellar-binding constant of complex $[Ru(NH_3)_5pz]^{2+}$ and $1 \gg K_S[D_n]$. The calculated values of K_R, k_W, and k_M are 546 M^{-1}, 3.5×10^{-3} sec^{-1}, and ~ 0, respectively. The calculated value of k_W (= 3.5×10^{-3} sec^{-1}) is nearly 4-fold smaller than k_W ($\sim 14.0 \times 10^{-3}$ sec^{-1}) determined experimentally at $[CTACl]_T = 0$. These observations are rationalized in terms of specific effects of premicellar monomers or premicellar aggregates on the rate of reaction.[33]

The counterion binding with ionic micelles is generally described in terms of two alternative approaches: the first one is the widely used pseudophase ion-exchange model (Chapter 3, Subsection 3.3.7) and the second one, less commonly used, is to write the counterion binding constant in terms of an ionic micellar surface potential (Ω) (Chapter 3, Section 3.4). The value of K in Equation 6.16 is expected to remain independent of $[CTACl]_T$ as long as the degree of association of counterions (β) of ionic micelle-forming surfactant and, thus, the surface potential (Ω) of the micelles, remains a constant.[34,35] At higher surfactant concentrations, there is a condensation of counterions on the surface of the micelles, which causes a decrease of the surface potential[33] and, consequently, a decrease of K. This decrease in K corresponds obviously to a change of free energy of the reaction process. This free energy change (ΔG) can be written as the sum of two contributions: (1) a nonelectrostatic or intrinsic (ΔG_{nel}) and (2) an electrostatic contribution (ΔG_{el}). Thus

$$\Delta G = \Delta G_{nel} + \Delta G_{el} \qquad (6.17)$$

where ΔG_{el} can be expressed as

$$\Delta G = z\kappa F\Omega \qquad (6.18)$$

where z is the charge of the ion ($S_2O_8^{2-}$) whose binding to the ionic micellar surface is described by K ($z = -2$), κ is an empirical parameter that takes into account the location of the ion at the surface, which is not necessarily the same as the location of the probe used in the determination of Ω; that is, κ gives the fraction of surface potential (determined with a given probe) that decreases ΔG_{el} for another probe, and F is the Faraday constant.

If K_0 (nonelectrostatic binding constant) is defined as

$$K_0 = \exp(-\Delta G_{nel}/RT) \qquad (6.19)$$

it follows from Equation 6.17 and Equation 6.18 that

$$K = K_0 \exp(-(z\kappa F\Omega)/RT) \qquad (6.20)$$

In order to prove that the increase in k_{obs} with increase in $[CTACl]_T$ beyond 9.6×10^{-3} M CTACl is due to decrease in Ω under such experimental conditions, the values of K at different $[D_n]$ have been calculated from Equation 6.16 with $k_M = 0$ or $k_W >> k_M K [D_n]$ and $k_W = 3.5 \times 10^{-3}$ sec^{-1}. These calculated values of K fit reasonably well to Equation 6.20 with $- (z\kappa F\Omega)/RT \approx 0.08$, where the values of Ω at different desired $[CTACl]_T$ have been determined from the changes in pK_a of a suitable indicator (heptadecylumbelliferone).[33]

The values of pseudo-first-order rate constants (k_{obs}) for the hydrolysis of **1** at pH 7.03, 25°C and in the presence of water, water + 1×10^{-4} M Cu(NO$_3$)$_2$, water + 0.01 M CTABr, and water + 1×10^{-4} M Cu(NO$_3$)$_2$ + 0.01 M CTABr are 1.7×10^{-5}, 1.98×10^{-2}, 1.93×10^{-5}, and 3.78×10^{-2} sec^{-1}, respectively.[36] These observations show copper-ion catalytic factor of ~1100- and 2200-fold in the absence and presence of CTABr micelles, respectively. The nearly 2-fold larger value of k_{obs} in the presence of CTABr micelles clearly demonstrates that the catalytic species in the micellar-mediated hydrolysis is Cu^{2+}–OH$^-$, which is plausible because of the higher apparent pH at cationic micellar surface than the pH of the bulk water phase due to occurrence of ion-exchange HO⁻/Br⁻. However, the rate of a metal-ion- and cationic metal-complex-catalyzed reaction should be inhibited by the cationic micelles if the cationic metal complex is devoid of appreciable hydrophobicity and the substrate is significantly hydrophobic. For example, the oxidations of ferrocene (FcH) and *n*-butylferrocene (FcBu) by ferric salts (nitrate and bromide) are strongly inhibited by the aqueous CTABr and CTANO₃, whereas the rates of these redox reactions are strongly catalyzed by anionic micelles of SDS, and the kinetics can be fitted to a model in which the reaction rate depends on the concentration of both reactants in the micellar pseudophase. Some added salts reduce the micellar catalysis by excluding ferric ions from the micelle. The rates of oxidation of FcH and FcBu by ferricyanide ions, $[Fe(CN)_6]^{3-}$, are inhibited by anionic micelles of SDS.[37a] The rates of monoalkylation of Hg^{2+} in dilute acid by alkyl aquobis-(dimethyl glyoximato) cobalt(III) and the related propane derivatives are catalyzed by anionic micelles of SDS, whereas nonionic micelles of Igepal do not catalyze the rates of these reactions.[37b]

6.2.2 METALLOMICELLAR-MEDIATED INORGANIC REACTIONS

Micellar-mediated inorganic reactions involve (1) metal ion–ligand complex formation and (2) metal ion–ligand complex acting as either a catalyst or a reactant. In these reactions, if the ligand is hydrophilic and the reaction rates are not significantly sensitive to ionic strength, then such reactions are expected to occur in the micellar microreaction environment in which the water activity is almost similar to that of the bulk water solvent and, consequently, under such conditions, $k_M \approx k_W$. But if the metallomicellar-mediated reaction involves a reactant that is sufficiently hydrophobic, then such a hydrophobic reactant is expected to reside deep inside the micelle, causing an almost complete separation from other hydrophilic reactant or catalyst (metal ion–ligand complex) and under such conditions, $k_M \approx 0$.

Although the effect of micelles on the inorganic reaction rates has not been studied as extensively as that on organic reaction rates, an attempt at the study of the kinetics of the rates of inorganic reactions in the presence of pure and mixed micelles was made as early as the late 1970s. Reinsborough and Robinson[38] studied the effects of SDS micelles on the rate of reaction between aqueous Ni^{2+} and 2,2′-bipyridyl and 4,4′-dimethyl-2,2′-bipyridyl ligands. The mechanism of the complex formation at the micellar surface is the same as in aqueous solution but the rate is considerably enhanced because of the concentration effect of the micelles. The effect of micelles on the reaction rate is quantitatively explained in terms of Berezin's pseudophase model of micelle (Chapter 3, Section 3.4) that, under limiting conditions of $k_W^2 = k_M^2$ and $1 \ll K_S K_R[D_n]$, gives Equation 6.21

$$k_{obs} = k_W^2[R]_T/\{V_M[D_n][1 + (K_R[D_n])^{-1}][1 + (K_S[D_n])^{-1}]\} \qquad (6.21)$$

where R is the metal-ion reactant and S is the ligand reactant. In the observed kinetic data analysis in terms of Equation 6.21, generally, V_M, K_R, and K_S are considered to be unknown kinetic parameters. No significant partitioning of the ligand occurs between the interior and surface of the micelle. The rate-determining step is the release of a solvating H_2O molecule from the metal ion, and the rate constant for this process is similar to that in aqueous solutions.[38]

Since 1979, Equation 6.21 has been used to explain the observed data quantitatively in various studies on the effects of both pure and mixed micelles on the rates of metal ion–ligand complex formation.[39-43] In the mixed micellar systems, $[D_n] = [Surf]_T - CMC^{mix}$, where $[Surf]_T$ is the total concentration of surfactant that includes both surfactants and CMC^{mix} is the critical micelle concentration of mixed micelles. Metal ion–ligand complex formation reactions are generally reversible reactions. However, it is assumed that the backward reaction step in the metal ion complexation is of little kinetic consequence, especially in the presence of micelles. Pronounced rate enhancement or inhibition of Ni^{2+}-ligand complexations is often observed at surfactant concentrations much below the CMC, and the results are interpreted in terms of Ni^{2+}-surfactant micelles as the agents responsible for the rate changes in dilute surfactant solution.[44]

The kinetic data on the rates of hydrolysis of the chloropentaamminecobalt(III) cation, studied at 25°C and 0.05 M ionic strength in aqueous mixed micelles of anionic SDS and nonionic n-dodecylpenta(oxyethylene glycol) monoether ($C_{12}E_5$) surfactants over a wide range of total surfactant concentration ($[Surf]_T = [SDS]_T + [C_{12}E_5]_T$) and SDS mole fraction (X_{SDS}), have shown that interactions of the cationic cobalt complex with mixed micelles cause a decrease of the reaction rate with increasing $[Surf]_T$ (at constant X_{SDS}) and X_{SDS} (at constant $[Surf]_T$).[45] The experimentally determined second-order rate constants ($k_{2obs}/M^{-1}sec^{-1}$; first order with respect to each reactant, complex, and HO^-) have been satisfactorily treated with a kinetic equation similar to Equation 3.2 (Chapter 3). However, the reaction is apparently a bimolecular one in which HO^- is one of the two reactants and, consequently, the proposed reaction scheme[45] is similar to Scheme 3.3 (Chapter 3) with $K_R = 0$, where R represents HO^-.

6.3 INDUCED FUNCTIONAL METALLOMICELLES

Surfactant molecules with head groups acting as effective ligands can act as induced functional surfactant molecules in the presence of complex-forming metal ions. Such metallomicelles could cause an unusually large rate enhancement if one or both reactants in a bimolecular reaction became bound to the metal ion through electrovalent bonding interaction. Such rate enhancements are generally attributed to induced intramolecularity as well as to metal ion effect on either nucleophilicity or electrophilicity, or both nucleophilicity and electrophilicity of the reactants bound to metal ions. However, the magnitude of such metallomicellar catalytic effect depends on the choice of the reference reaction in which the induced intramolecularity due to metallomicelle formation does not exist.

Pseudo-first-order rate constants (k_{obs}) for the aqueous cleavage **1** at pH 7.03 and 25°C are 1.7×10^{-5} sec^{-1}, 6.8×10^{-4} sec^{-1}, 1.93×10^{-5} sec^{-1}, 1.92×10^{-3} sec^{-1}, 3.78×10^{-2} sec^{-1}, and 4.91 sec^{-1} in the aqueous buffer of N-ethylmorpholine-HNO$_3$, aqueous buffer + N-methyl-2-(hydroxymethyl)imidazole (**7**), aqueous buffer + 0.01 M CTABr, aqueous buffer + 0.01 M CTABr + 1×10^4 M **8** {where **8** = N-dodecyl-2-(hydroxymethyl)imidazole}, aqueous buffer + 0.01 M CTABr + 1×10^{-4} M Cu^{2+}, and aqueous buffer + 0.01 M CTABr + 1×10^{-4} M **8** + 1×10^{-4} M Cu^{2+}, respectively.[36] Although these observations are not sufficient to provide a definite quantitative explanation, they are interesting enough to discuss, at least qualitatively. The nearly 40-fold increase in k_{obs} due to the presence of 1 $\times 10^{-4}$ M **7** shows the occurrence of intramolecular general-base-assisted nucleophilic attack by primary alcoholic oxygen of **7** at carbonyl carbon of ester in the rate-determining step as shown in transition state **TS$_1$**. Thus, it may not be unreasonable to assume a similar rate enhancement ($\sim$ 40-fold) due to the presence of 1×10^{-4} M **8**. The value of k_{obs} is increased $\sim$ 100-fold due to the presence of 0.01 M CTABr and 1×10^{-4} M **8** that shows an $\sim$ 2.5-fold increase in k_{obs} due to the presence of 0.01 M CTABr. An impressive rate acceleration of $\sim 2.5 \times 10^5$-fold is obtained in the presence of 1×10^{-4} M **8** + 1×10^{-4} M Cu^{2+} that is $\sim$ 35-fold larger than k_{obs} obtained in the presence of 1×10^{-4} M **7** + 1×10^{-4} M Cu^{2+} at the same pH. Thus, the metallomicelar effect is only $\sim$ 35-fold, which may be attributed to both micellar reaction medium and volume effect. The nearly 7×10^3-fold rate acceleration in the presence of 1×10^{-4} M **8** + 1×10^{-4} M Cu^{2+} is explained through the involvement of the proposed transition state **TS$_2$** in the rate-determining step of the aqueous cleavage of **1**.[36]

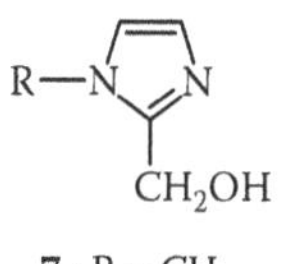

7 : R = CH$_3$
8 : R = C$_{12}$H$_{25}$

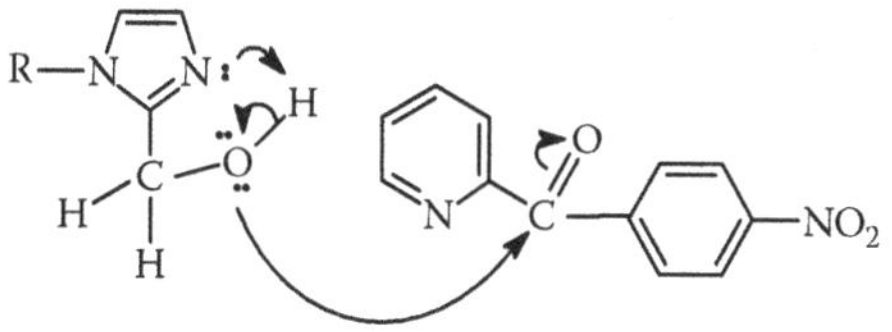

TS$_1$

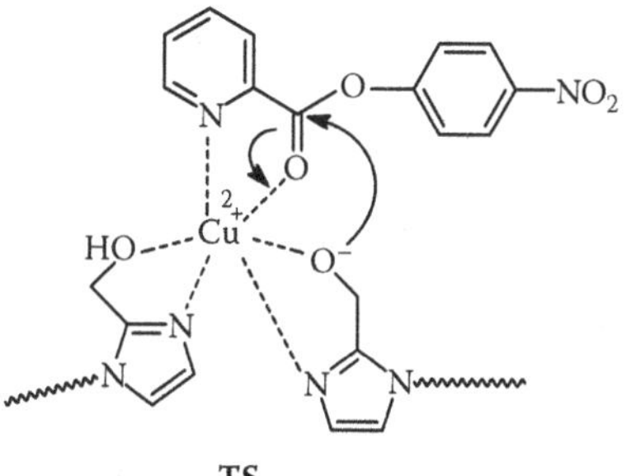

TS$_2$

The values of k_{obs} for the cleavage of **1** in the buffer of pH 7.03 at 25°C and $[8] = [Cu^{2+}] = 1 \times 10^{-4}$ M become independent of $[CTABr]_T$ at ≥ 0.005 M. Thus, at 0.01 M CTABr, the substrate **1** is almost completely micelle bound. Pseudo-first-order rate constants (k_{obs}), at a constant $[8]$ and pH, increase monotonically with increasing $[Cu^{2+}]$. These observations have been analyzed by assuming a reaction mechanism as shown in Scheme 6.3[36]

$$nL + M \xrightleftharpoons{K_M} M.L_n$$

$$S + M.L_n \xrightarrow{k_N'} P$$

$$S \xrightarrow{k_0'} P$$

SCHEME 6.3

where a metal ion (M) forms a complex (M.L$_n$) with n ligands (L) with an association constant K_M. The complex reacts with substrate S (i.e., **1**) with an apparent second-order rate constant k_N' to give products P. The products P are also formed through k_0'-step without involving an M.L$_n$ complex. Thus,

$$k_0' = k_0 + k_L[L]_T + k_M[M]_T \tag{6.22}$$

where subscript T represents total concentration. The observed rate law: rate = $k_{obs}[S]_T$ and Scheme 6.3 with n = 1 and n = 2 can lead to Equation 6.23 and Equation 6.24, respectively.

$$k_{obs} = k_0' + \frac{k_N' K_M [L]_T [M]_T}{1 + K_M([L]_T + [M]_T)} \tag{6.23}$$

$$k_{obs} = k_0' + \frac{k_N' K_M [L]_T^2 [M]_T}{1 + K_M[L]_T^2 + 4K_M[L]_T[M]_T} \tag{6.24}$$

It should be noted that the derivation of Equation 6.23 and Equation 6.24 from Scheme 6.3 involves approximations whose validity should be justified.[36] In the presence of micelles, these equations also indirectly assume that either $[S_W] = [L_W] = [M_W] = 0$ or $[S_M] = [L_M] = [M_M] = 0$, where subscripts W and M stand for aqueous pseudophase and micellar pseudophase, respectively. These assumptions need to be justified also. Equation 6.25, which is the rearranged form of Equation 6.23, has been used to calculate k_N' and K_M.[36]

$$\frac{1}{k_{obs} - k_0'} = \frac{1}{[M]_T} \left(\frac{1 + K_M[L]_T}{k_N' K_M[L]_T^2} \right) + \frac{1}{k_N'[L]_T} \qquad (6.25)$$

Pseudo-first-order rate constants (k_{obs}) for the aqueous cleavage of **1** in the presence of Zn^{2+} or Cu^{2+} at constant pH, mixed micelles of CTABr, and a hydrophobic bis-imidazole ligand, 1-dodecyl-4-hydroxymethyl-α-(1-dodecyl-2-imidazolyl)-2-imidazolemethanol (**9**), follow Equation 6.25.[46] The kinetic and product analysis indicated that the reaction proceeds through the transacylation from **1** to the hydroxyl group of ligand–metal ion complexes and that the active complexes undergoing transacylation are a 2:1 and a 1:1 (ligand:metal ion) complex for Zn^{2+} and a 1:1 (ligand:metal ion) complex for Cu^{2+}, respectively. For the ligand (**9**), the transacylation is highly site selective, i.e., it occurs predominantly on the secondary hydroxyl group with Zn^{2+}, but on the primary hydroxyl group with Cu^{2+}. Such site selectivity has been discussed in terms of the coordination structure of the active metal–ligand complexes.

The rate of cleavage of **1** has been studied in the absence and presence of metallomicelles of surfactant with tridentate ligand head groups (**10** and **11**) at a constant pH, but the observed kinetic data are not sufficient to provide a quantitative explanation of these data in terms of a micellar kinetic model.[47] However, these kinetic data may provide a qualitative assessment of catalytic factor (Y) due to aqueous metal ion catalysis (Y^M), metal ion–tridentate ligand complex catalysis (Y^{ML}), and metal ion–tridentate ligand surfactant complex catalysis (Y^{MLS}). Thus, $Y^M = (k_{obs}^M - k_0)/k_0$, $Y^{ML} = (k_{obs}^{ML} - k_0)/k_0$, and $Y^{MLS} = (k_{obs}^{MLS} - k_0)/k_0$, where k_0 is the pseudo-first-order rate constant in the presence of buffer of constant pH, and k_{obs}^M, k_{obs}^{ML}, and k_{obs}^{MLS} represent the pseudo-first-order rate constants in the presence of metal ion, metal ion–ligand complex, and metal ion–ligand surfactant complex, respectively, at a constant pH. Experimentally determined value of k_0 and calculated values of Y^X with X = M, L, LS, ML, and MLS for M = Zn^{2+} and Cu^{2+}, L = **12**, **13**, and LS = **10**, **11** are summarized in Table 6.3.

It is evident from Table 6.3[47] that the most impressive rate enhancements are due to metal ion–water complex catalysis: ~ 100- and 14000-fold rate acceleration due to Zn^{2+}–H_2O complex and Cu^{2+}–H_2O complex, respectively. Catalytic effects due to Zn^{2+}-**12** or **13** and Cu^{2+}-**12** or **13** complexes are ~ 1.6- to 5-fold and ~ 10-fold, respectively. Similarly, the catalytic effects due to Zn^{2+}-**10** or **11** and Cu^{2+}-**10** or **11** complexes are ~ 2-fold and 5-fold, respectively, which could be easily explained in terms of merely usual concentration and medium effects of the micromicellar reaction environment.

Unlike metallomicelles, functional surfactant micelles compared to nonfunctional surfactant micelles exhibit rate enhancement of many orders. The reason for such apparent characteristic difference between the catalytic efficiency of metallomicelles and functional surfactant micelles is the choice of the reference reactions used to evaluate the catalytic efficiency of metallomicelles and func-

TABLE 6.3
Apparent Ligand, Metal Ion, and
Metallomicellar Catalytic Factor (Y) for the
Hydrolysis 1 at pH 7.25 and 25°C

Ligand	Metal Ion	Y^a
H_2O	None	0
H_2O	Zn^{2+}	100
H_2O	Cu^{2+}	14,400
12	None	0.3
10	None	1.0
12	Zn^{2+}	500
12	Cu^{2+}	144,000
10	Zn^{2+}	1100
10	Cu^{2+}	763,000
13	None	0.4
11	None	0.8
13	Zn^{2+}	164
13	Cu^{2+}	130,000
11	Zn^{2+}	390
11	Cu^{2+}	1,600,000

[a] $Y = (k_{obs} - k_0)/k_0$ where k_0 and k_{obs} are pseudo-first-order rate constants in the presence of aqueous buffer and metal ion or ligand, or metal ion + ligand, respectively, and the value of k_0 in aqueous buffer of pH 7.25 is 2.8×10^{-5} sec^{-1}.

tional surfactant micelles. Induced intramolecularity of reaction generally does not exist in nonfunctional surfactant micellar-mediated reactions that are considered to be the reference reactions for evaluation of the functional surfactant micellar catalytic efficiency. Whereas, in the evaluation of metallomicellar catalytic efficiency, the reference reactions involve metal ion–ligand complexes (devoid of hydrophobic moieties in ligand molecules), which contain a high degree of induced intramolecularity and, consequently, such reference reactions progress with a huge rate enhancement.

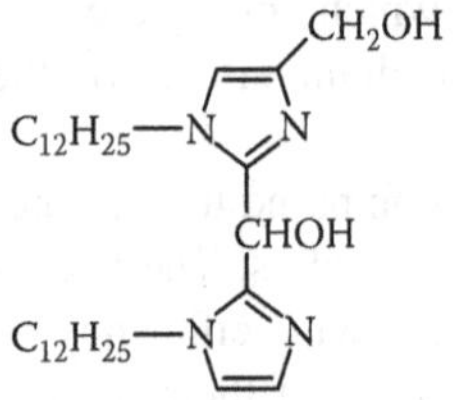

9

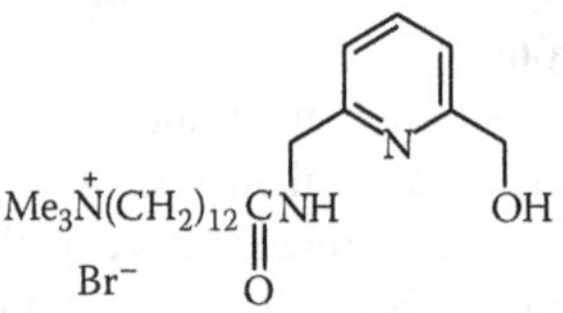

10

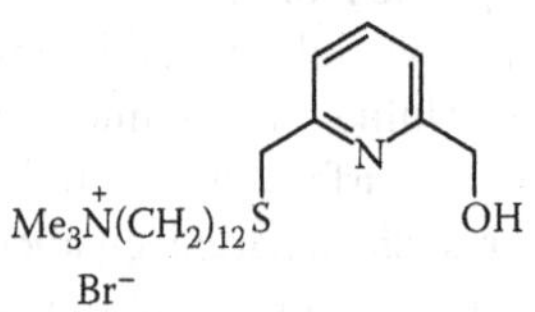

11

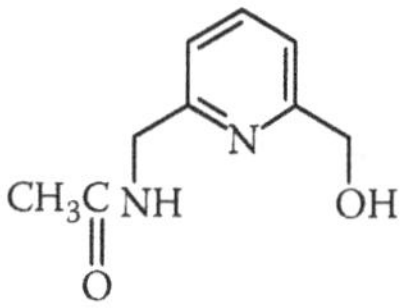

12 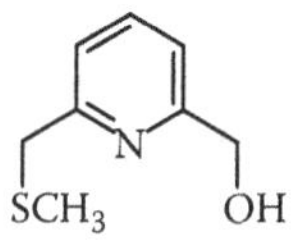   13

Hydrophobic ligands **14** to **16** form metallomicelles in the presence of Cu(II) ions that are catalytically active in the cleavage of the 4-nitrophenyl esters of acetic, hexanoic, and dodecanoic acids and 4-nitrophenyl diphenyl phosphate.[23] The ligand with free hydroxyl (**14a**) is more effective in the cleavage of carboxylate esters than the ligands with methylated alcoholic group (**14b**) or devoid of it (**16**), whereas the opposite behavior is observed in the cleavage of phosphate triester. The quantification of metallomicellar catalytic factor is not possible owing to lack of data on the catalytic effects of Cu(II) complexes of hydrophilic head group ligands on the cleavage of these esters under similar experimental conditions.[23]

Menger et al.[48] studied the hydrolysis of 4-nitrophenyl diphenyl phosphate (**17**) in the presence of micelles of a long-chain chelate of cupric ion (**18**) and found that the rate of hydrolysis occurs $>10^5$ times faster at 1.5×10^{-3} M **18** and pH 6.0 than in the absence of catalyst (**18**). But **18**-promoted hydrolyses are only >200 times faster than those catalyzed by an equivalent concentration of cupric ion complexed with tetramethylethylenediamine. Significantly large rate enhancement (> 10^5-fold) is attributed to multiple effects including induced intramolecular reaction between micellized ester (**17**) and HO⁻ of metallomicellar head group [Cu(L)(HO)]⁺. The hydroxide ion loosely associated with the cationic metallomicellar surface cannot be considered to be the catalytically active nucleophile, because the values of pseudo-first-order rate constants (k_{obs}) increase only from 4.1 $\times 10^{-2}$ to 5.8×10^{-2} sec⁻¹ with the increase in pH from 6.0 to 8.3 at 1.5×10^{-3} M **18**.[48]

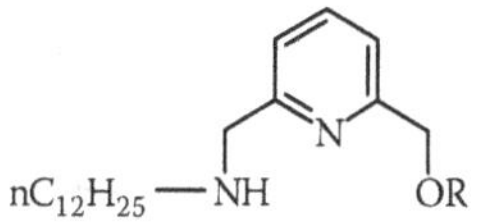

14a : R = H
14b : R = CH₃
                15 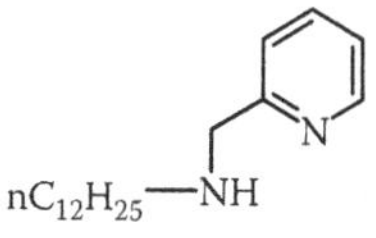   16

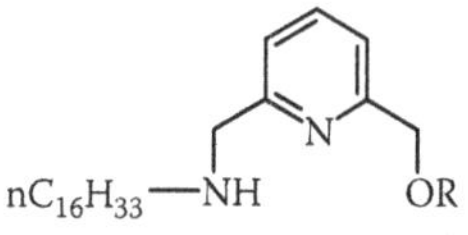

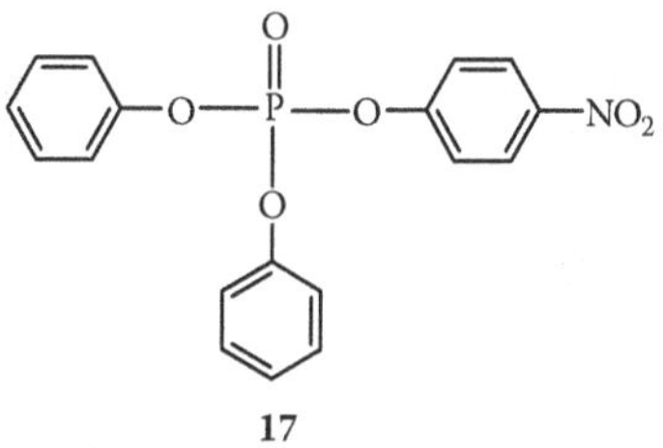

17 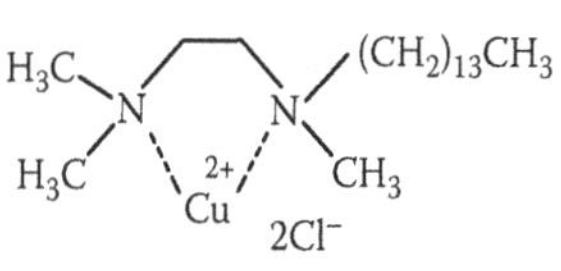   18

TABLE 6.4

Apparent Rate Enhancements (k_{rel}) for the Hydrolysis of Amides and Esters at pH 7.0 and 31°C[a]

[Cu(L)]Cl$_2$	19	20	21	22	4-NPA	2-NPA	25
None	1[b]	1[c]	1[d]	1[e]	1[f]	1[g]	1[h]
0.001 M	43	25	5050	26	2.6		44
0.001 M + 0.001 M CTABr	18	22	3210				37
0.001 M + 0.001 M Triton	38	22					40
0.002 M	46	48	7000	58	5.4	4.1	45
0.002 M + 0.001 M CTABr	29	40					
0.002 M + 0.001 M Triton	45	43					41

[a] $k_{rel} = k_{obs}/k_0$ where k_{obs} is the pseudo-first-order rate constant in the presence of aqueous buffer and [Cu(L)]Cl$_2$ or [Cu(L)]Cl$_2$ + inert surfactant and $k_0 = k_{obs}$ at [[Cu(L)]Cl$_2$] = 0.
[b] $k_0 = 2.17 \times 10^{-5}$ sec^{-1}.
[c] $k_0 = 1.62 \times 10^{-5}$ sec^{-1}.
[d] $k_0 = 2.0 \times 10^{-6}$ sec^{-1}.
[e] $k_0 = 1.7 \times 10^{-5}$ sec^{-1}.
[f] $k_0 = 1.0 \times 10^{-5}$ sec^{-1}.
[g] $k_0 = 1.34 \times 10^{-5}$ sec^{-1}.
[h] $k_0 = 6.6 \times 10^{-5}$ sec^{-1}.

Most of the studies on the effects of metallomicelles on the rate of hydrolysis of esters involve so-called activated esters in which nucleophilic attack is the rate-determining step. The effects of copper-containing metallomicelles (formed from both copper(II)-hydrophobic ligand complex as well as from hydrophobic ligand [L = N,N,N'-trimethyl-N'-tetradecylethylenediamine] containing free Cu^{2+} ions) on the rate of hydrolysis of amides **19** and **20** as well as activated esters **21**, **22**, and **23** have been studied at pH 7.0 and 31°C.[49] The apparent rate enhancements (k_{rel}) of Cu^{2+}(L) metallomicelles on the rate of hydrolysis of **19** to **23**, 2-nitrophenyl acetate (2-NPA), and 4-nitrophenyl acetate (4-NPA) under various reaction conditions are summarized in Table 6.4.[49] The actual rate enhancements due to metallomicelles, Cu^{2+}(L), under various reaction conditions are not possible to estimate because of the lack of pseudo-first-order rate constants (k_{obs}) for hydrolysis of **19** to **23**, 2-NPA, and 4-NPA in the presence of Cu^{2+}–N,N,N',N'–tetramethylethylenediamine complex. However, the values of k_{rel} for **19** and **20** are almost same at 0.002-M Cu^{2+}(L) and 0.002 M Cu^{2+}(L) + 0.001-M Triton. The presence of 0.001-M CTABr comicelles has no effect on k_{rel} for **20** but decreases k_{rel} from 46 to 29 for **19** (Table 6.4), which may be attributed to larger cationic mixed micellar affinity of anionic **19** than that of neutral **20**.

The similarity in the magnitudes of k_{rel} for hydrolysis of **19** and **20** at pH 7.0 may be understood in terms of a general mechanism of these reactions as shown in Scheme 6.4. The alkaline hydrolysis of trifluoroacetanilides involves solvent assisted C–N bond breaking as the rate-determining step[50] and, consequently, k_{obs}

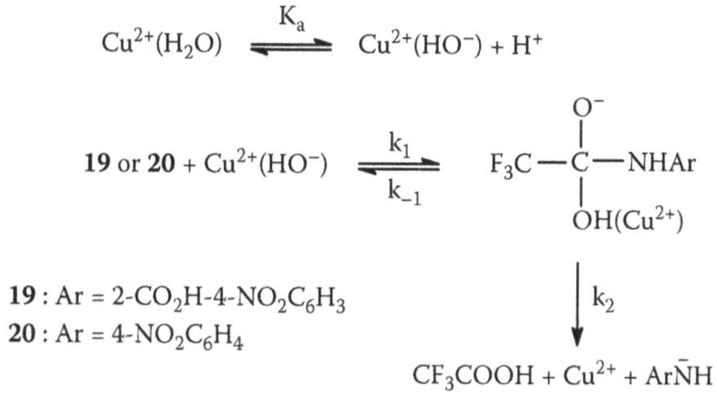
Cu2+(H2O)
Ka
Cu2+(HO−) + H+
19 or 20 + Cu2+(HO−)
k1
k−1
F3C—C—NHAr
O−
OH(Cu2+)
19 : Ar = 2-CO2H-4-NO2C6H3
20 : Ar = 4-NO2C6H4
k2
CF3COOH + Cu2+ + ArN̄H

SCHEME 6.4

$= k_1k_2/k_{-1}$. The value of k_1 for **19** is expected to be significantly larger than that for **20** because of the presence of induced intramolecularity in k_1-step for **19** as shown in transition state $\mathbf{TS_3}$. Although it is difficult to predict the relative magnitudes of k_{-1} for **19** and **20**, it seems that k_{-1} for **19** may not be very different from k_{-1} for **20**. The value of k_2 for **20** is conceivably significantly larger than that of **19** due to the probable internal hydrogen bonding as shown in $\mathbf{TS_4}$. Thus, the conclusions — k_1 (**19**) > k_1 (**20**), k_{-1} (**19**) ≈ k_{-1} (**20**), and k_2 (**19**) < k_2 (**20**) — predict that k_1k_2/k_{-1} for **19** may be similar to k_1k_2/k_{-1} for **20** provided k_1k_2 (**19**) ≈ k_1k_2 (**20**). The hydrolysis of esters generally follow a stepwise addition–elimination mechanism as shown in Scheme 6.4, in which k_1-step is the rate-determining one for activated esters[51] and hence $k_{obs} = k_1$ for **21**, **22**, 2-NPA, and 4-NPA. Induced intramolecular nucleophilic attack in the rate-determining step for hydrolysis of **21** as shown in $\mathbf{TS_5}$ causes k_{rel} for **21** ~ 200-fold larger than k_{rel} for **22** under similar experimental conditions.[49]

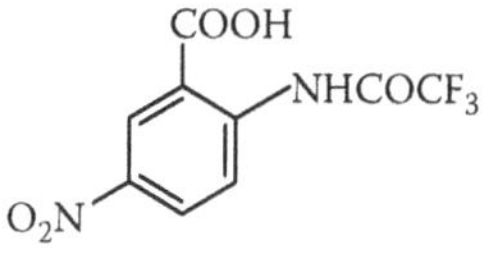
COOH
NHCOCF3
O2N

19

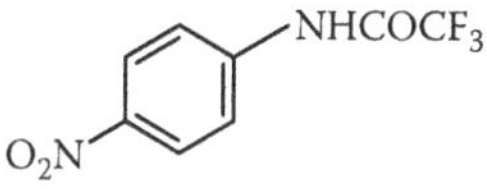
NHCOCF3
O2N

20

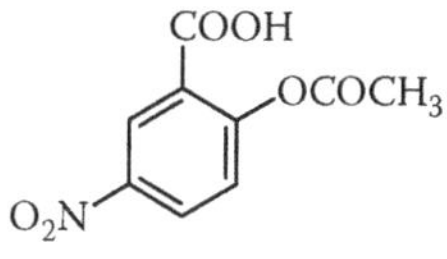
COOH
OCOCH3
O2N

21

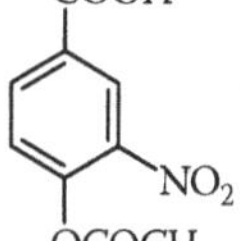
COOH
NO2
OCOCH3

22

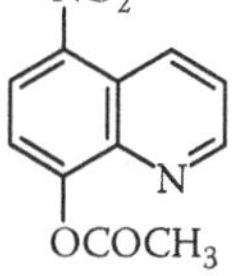
NO2
N
OCOCH3

23

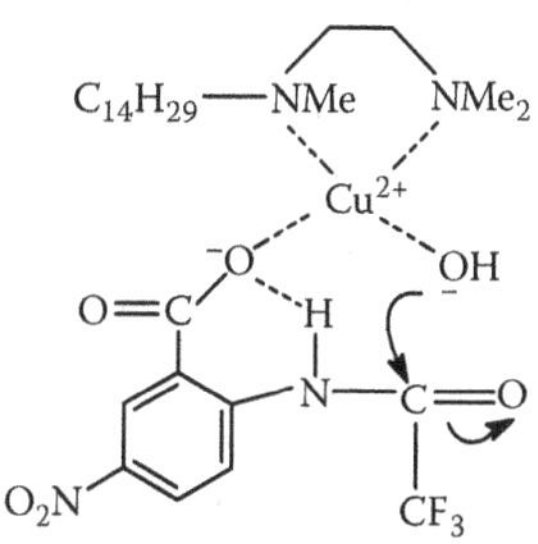
C14H29—NMe
NMe2
Cu2+
OH
O=C
H
N
C=O
O2N
CF3

$\mathbf{TS_3}$

$$TS_4 \qquad\qquad TS_5$$

Hydrolysis of several esters of picolinic acid with leaving groups of pK_a ranging from 3.56 to 16.0 has been studied at 25°C, pH 6.3, and 7.5 in the presence of Cu^{2+} complexes of water, hydroxyl-functionalized surfactant 14, and 6-((methylamino)methyl)-2-(hydroxymethyl)pyridine, 24.[52] The rate enhancements due to H_2O–Cu^{2+}, 24–Cu^{2+}, and 14–Cu^{2+} complexes over the pure buffer for hydrolysis of 25a are 1.5×10^3, 4.2×10^4, and 1.6×10^6, respectively; for 25b, 6.2×10^3, 1.7×10^4, and 7.8×10^5, respectively; and for 25c, 5.9×10^4, 1.1×10^4, and 8.4×10^3, respectively, under the standard quite low concentrations ([Cu^{2+}] = [ligand] = 2×10^{-4} M) that were used. These data show that the metalomicellar rate enhancements are only 38, 46, and 0.8 for 25a, 25b, and 25c, respectively. An interesting observation of this study is that as the leaving group of the esters becomes increasingly poorer, the hydroxyl of the tridentate ligand becomes less and less effective up to the case of the trifluoroethyl picolinate (25c) that reacts virtually at the same rate in the presence of metallomicelles 14a-Cu^{2+} or 14b-Cu^{2+}.[52]

Apparently, the hydroxyl group of the tridentate ligand is the effective nucleophile as long as the rate-determining step is the formation of tetrahedral intermediate (T_1 or T_2, Scheme 6.5), i.e., when $k_2^a \gg k_{-1}^a$ (route a, Scheme 6.5). For nonactivated esters where $k_2^a \ll k_{-1}^a$, the effectiveness of the hydroxyl group of ligand as nucleophile vanishes, and the hydroxide ion becomes the effective nucleophile (route b, Scheme 6.5) and, consequently, k_2^b/k_{-1}^b becomes larger than k_2^a/k_{-1}^a. The change in the reaction path from route a to route b with the change in the rate determining from k_1^a-step to k_1^b-step may be explained in terms of the effect of the nucleophilic group on the stability of the transition states of these rate-determining steps. The k_1^a-step involves a nucleophile that is devoid of hydrogen attached to the nucleophilic site whereas the nucleophile in k_1^b-step carries a hydrogen attached to the nucleophilic site. The departure of the leaving group (OL) in k_2^b-step is assisted by the internal 1,3-proton transfer whereas such assistance in k_2^a-step is not available. Similar conclusions have been ascertained in the observed efficient hydrolytic reactivity and nonreactivity of nonionized and ionized phthalamic and N-substituted phthalamic acids, respectively.[53]

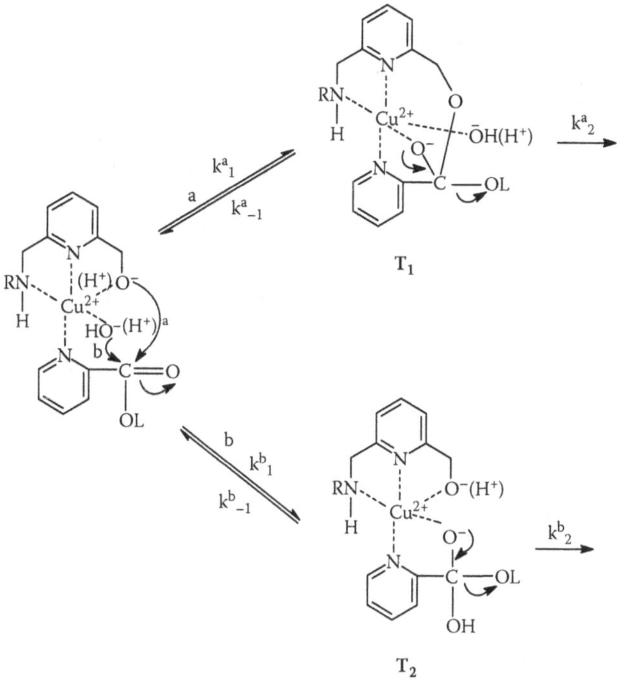

SCHEME 6.5

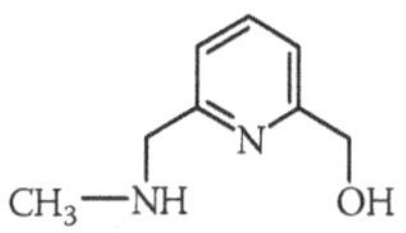

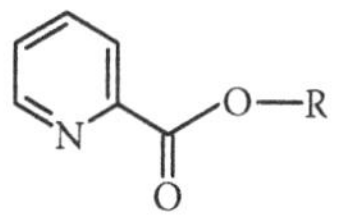

25a : R = 2,4-dinitrophenyl
25b : R = 4-nitrophenyl
25c : R = trifluoroethyl

In an attempt to discover the effects of chiral hydrophobic ligands on enantiose-lective cleavage of α-amino acid esters, Scrimin et al. [54] studied the kinetics of the aqueous cleavage of 4-nitrophenyl esters of α-amino acids — phenyl alanine (PhePNP), phenylglycine (PhgPNP), and leucine (LeuPNP) — in the presence of several chiral hydrophobic ligands (**26a,b, 27a–d**) and metal ions {Cu(II), Zn(II), and Co(II)} at constant pH (= 5.50 or 6.25 or 7.00) and temperature (= 25°C). Apparent rate enhancements up to ~ 130-fold due to mixed CTABr-Cu(II) metallomicellar catalysis and enantioselectivity ratios ranging from 3.2 to 11.6 have been observed under typical reaction conditions. The chiral ligand reacts faster with the enantiomeric ester of opposite absolute configuration, i.e., the catalyzed hydrolysis is favored between (R)-ligand and (S)-substrate/ester (or *vice versa*). The catalytic hydrolysis of esters due to metallomicelar catalysis and mixed CTABr-metallomicellar catalysis involves (1) the formation of a ternary (ligand–metal ion–substrate) complex, (2) within such a complex, a nucleophilic attack of the ligand hydroxyl on the substrate to give a transacylation intermediate, and (3) the metal ion promoted hydrolysis of the transacylation intermediate with a relatively fast turnover of the catalyst.[54]

Most of the metallomicellar-mediated reactions involve either hydrophobic bidentate or tridentate ligands. Two hydrophobic tetradentate ligands (**28, 29**) and two hydrophobic tridentate ligands (**30, 31**) have been recently used to discover the effects of Cu(II) complexes of these ligands on the rate of hydrolysis of 4-nitrophenyl hexanoate (PNPH) and 4-nitrophenyl diphenyl phosphate (PNPDPP) at pH 7.6, 25°C, and 0.005-M CTABr.[55] Pseudo-first-order rate constants (k_{obs}) are larger by 53-, 41-, 2.5-, and 4.5-fold for PNPH as well as 33-, 11-, 5.5-, and 5.3-fold for PNPDPP in the presence of 5×10^{-4} M metallomicellar complexes **28**–Cu^{2+}, **29**–Cu^{2+}, **30**–Cu^{2+}, and **31**–Cu^{2+}, respectively, compared to k_{obs} at [ligand] = 0, [CTABr] = 0.01 M, and [CuCl$_2$] = 5×10^{-4} M. Significantly larger reactivity of **28** and **29** over **30** or **31** may be attributed in part to the higher nucleophilicity of the Cu^{2+}-bound –CH$_2$O$^-$ group relative to the metal-bound hydroxyl group, which is plausible because the respective second-order rate constants for the reactions of 4-nitrophenyl acetate[56a] and aspirin[56b] are 58-fold and 68-fold larger with CH$_3$O$^-$ than those with HO$^-$. However, the estimation of metallomicellar catalytic effect is not possible due to lack of data for the hydrolysis of these esters in the presence of **32**–Cu^{2+}, **33**–Cu^{2+}, **34**–Cu^{2+}, and **35**–Cu^{2+} complexes.

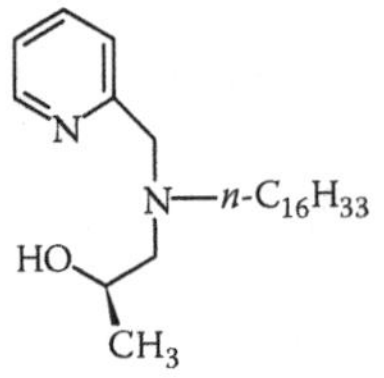

26a

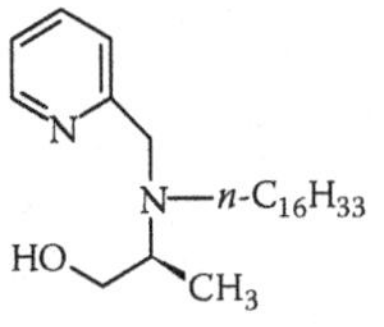

26b

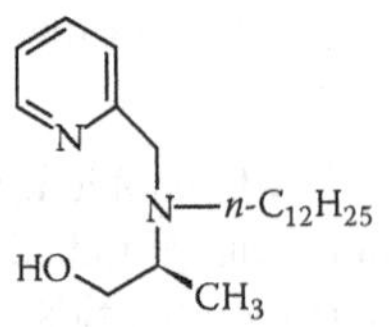

27a

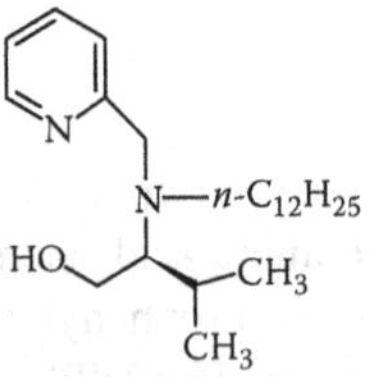

27b

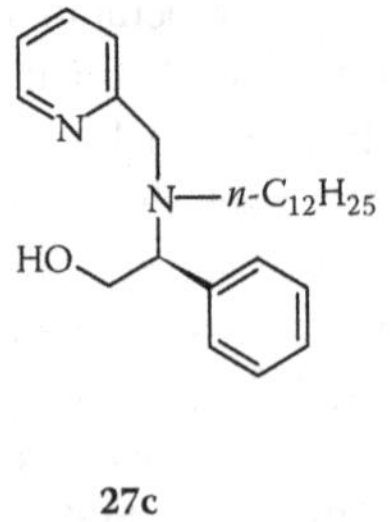

27c

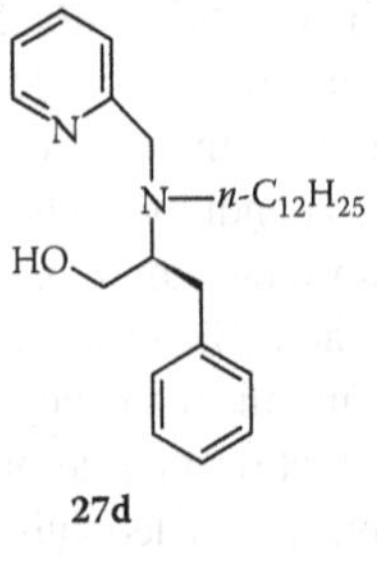

27d

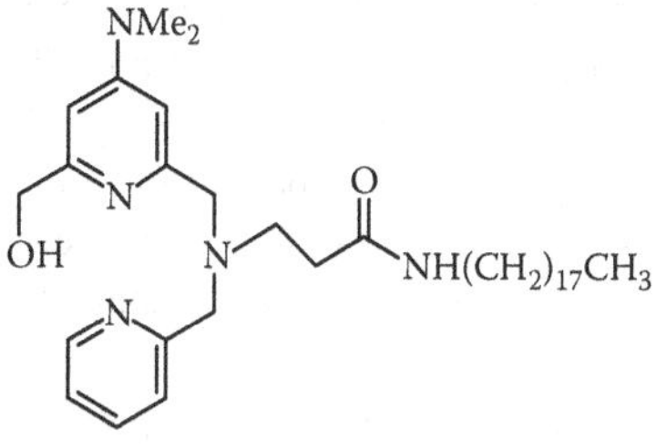

28

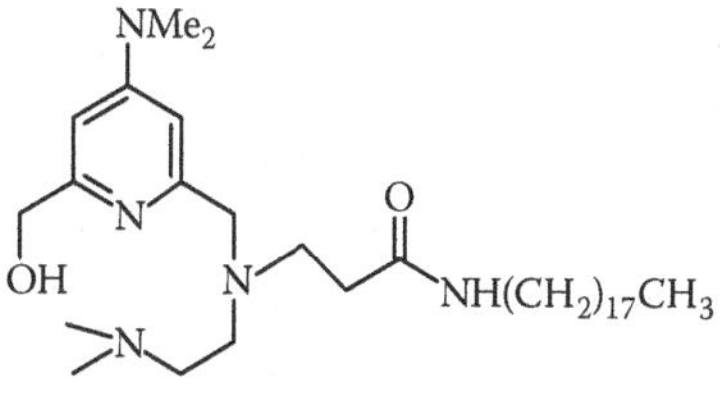

29

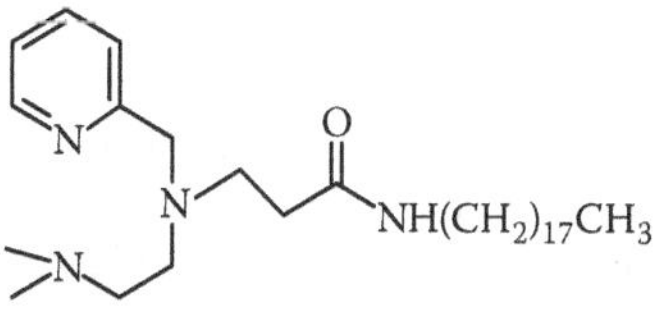

30

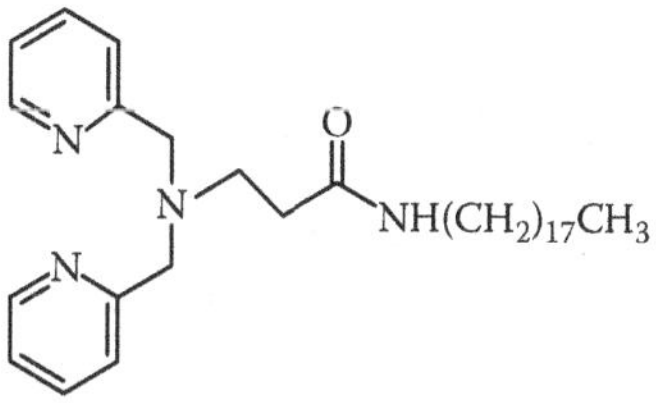

31

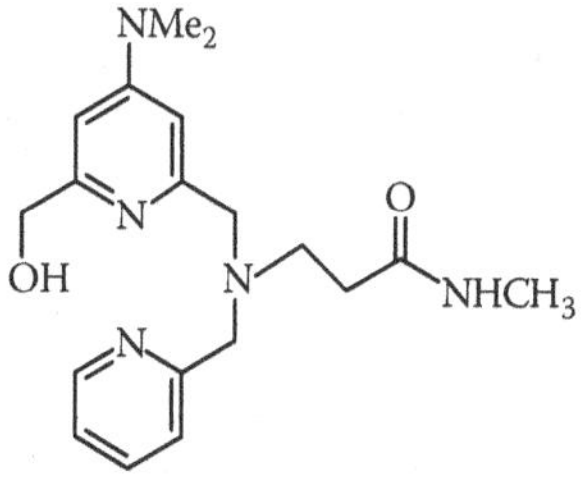

32

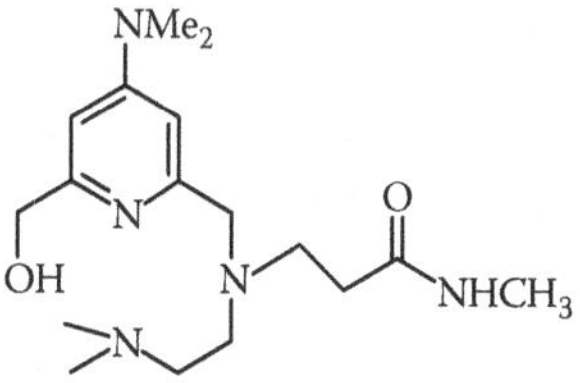

33

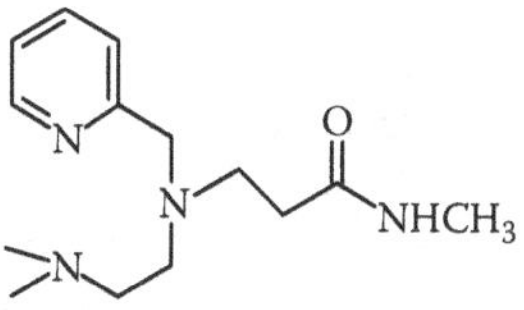

34

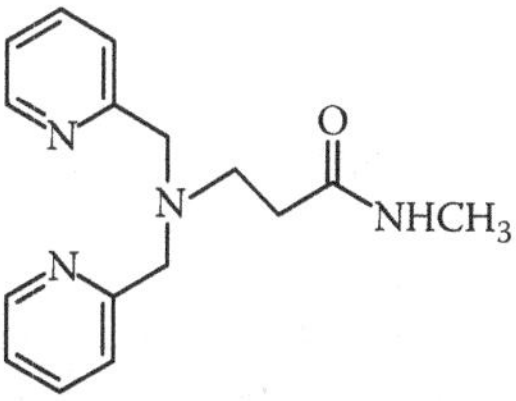

35

The studies on the effects of micelles and mixed micelles on the rate of metal ion–ligand complex formation (i.e., the rate of inorganic reactions), as well as the effects of metallomicelles and induced functional metallomicelles on the rate of organic reactions have been carried out extensively during the last nearly five decades. Induced functional metallomicelles are expected to provide impressive rate enhancements, because rate enhancements of several orders have been observed in several functional surfactant micellar-mediated reactions. Thus, induced functional metallomicellar effects on the reaction rates are still considered to be of great interest.[57–65] However, in all these studies, special and unusual expected metallomicellar effects on reaction rates are lacking. Scrimin et al.[66] have conclusively showed that the apparent impressive rate enhancements due to induced functional metallomicellar catalysis, reported by a few researchers[67–69], are chimeras that disappear when transfer equilibria between the aqueous and micellar pseudophase are taken into account.[66,70]

REFERENCES

1. Jiang, F., Jiang, B., Yu, X., Zeng, X. Metallomicellar catalysis: effects of Bridge-connecting ligands on the hydrolysis of PNPP catalyzed by Zn(II), Co(II), and Ni(II) complexes of ethoxy-diamine ligands in micellar solution. *Langmuir* **2002**, *18*(18), 6769–6774.

2. Jiang, B., Xie, F., Xie, J., Jiang, W., Hu, C., Zeng, X. Investigation of carboxylic acid ester hydrolyses catalyzed by micellar copper(II) complexes. *J. Dispersion Sci. Technol.* **2004**, *25*(2), 139–147.

3. Yu, X., Jiang, B., Cheng, S., Huang, Z., Zeng, X. Comparative reactivities of metal cation-catalyzed hydrolysis of *p*-nitrophenyl picolinate in micellar solutions. *J. Dispersion Sci. Technol.* **2003**, *24*(6), 761–765.

4. Huang, Z., Cheng, S., Zeng, X. Catalytic hydrolysis of *p*-nitrophenyl acetate by copper ion complexes of diamine-based ligands in CTAB micellar solution. *J. Dispersion Sci. Technol.* **2003**, *24*(2), 213–218.

5. Yan, X., Zeng, X., Cheng, S., Li, Y., Xie, J. Catalytic hydrolysis of *p*-nitrophenyl picolinate by copper(II) and zinc(II) complexes of N-(2-deoxy-β-D-glucopyrano-syl-2-salicylaldimino). *Int. J. Chem. Kinet.* **2002**, *34*(6), 345–350.

6. Xiang, Y., Zeng, X., Jiang, B., Xie, J. Metallomicellar catalysis: catalytic hydrolysis of *p*-nitrophenyl picolinate by bis-{N-(2-deoxy-β-D-glucopyranosyl-2-[3-car-boxylsalicylaldimino])}M₂(II) (M = Cu, Zn, Co) in CTAB micellar solution. *J. Dispersion Sci. Technol.* **2001**, *22*(5), 453–459.

7. Xiang, Y., Zeng, X., Cheng, S., Li, Y., Xie, J. Accelerated cleavage of *p*-nitrophenyl picolinate catalyzed by copper(II) and zinc(II)complexes of D-glucosamine Schiff base in micellar solution. *J. Dispersion Sci. Technol.* **2000**, *21*(7), 857–867.

8. Xie, J., Cheng, S., Du, J., Zeng, X. Catalytic hydrolysis of bis(*p*-nitrophenyl) phosphate by lipophilic diaquo(tetraamine) cobalt(III) in micellar solution. *J. Dispersion Sci. Technol.* **2001**, *22*(4), 337–341.

9. Xiang, Q., Xiang, Y., Nan, Z., Zeng, X., Xie, J. Hydrolysis of *p*-nitrophenyl picolinate catalyzed by copper(II) and zinc(II) binuclear complexes coordinating tripeptide in micellar solution. *J. Dispersion Sci. Technol.* **2001**, *22*(1), 103–109.

10. Yan, X., Zeng, X., Cheng, S., Li, Y., Xie, J. Metallomicellar catalysis cleavage of *p*-nitrophenyl picolinate catalyzed by binuclear metal complexes coordinating tripeptide in CTAB micellar solution. *J. Colloid Interface Sci.* **2001**, *235*(1), 114–118.

11. Cheng, S., Zeng, X. Catalytic hydrolysis of *p*-nitrophenyl picolinate catalyzed by dioxocyclam zinc(II) complexes in micellar solution. *J. Dispersion Sci. Technol.* **2000**, *21*(5), 655–662.

12. Zeng, X., Zhang, Y., Yu, X., Tian, A. Metallomicellar catalysis: cleavage of *p*-nitrophenyl picolinate in copper(II) coordinating N-myristoyl-N-(β-hydroxyethyl)ethylenediamine in CTAB micelles. *Langmuir* **1999**, *15*(5), 1621–1624.

13. Cheng, S., Meng, X., Chen, Y., Wang, Q., Zeng, X., Xie, J. Catalase mimic with Fe(II) metallomicelle. *J. Dispersion Sci. Technol.* **2000**, *21*(2), 199–207.

14. Jiang, F., Du, J., Yu, X., Bao, J., Zeng, X. Metallomicellar catalysis: accelerated hydrolysis of BNPP by copper(II), zinc(II), and nickel(II) complexes of long alkanol-imidazole in CTAB micellar solution. *J. Colloid Interface Sci.* **2004**, *273*(2), 497–504.

15. Jiang, B., Xiang, Y., Du, J., Xie, J., Hu, C., Zeng, X. Hydrolysis of *p*-nitrophenyl picolinate catalyzed by divalent metal ion complexes containing imidazole groups in micellar solution. *Colloids Surf. A* **2004**, *235*(1–3), 145–151.

16. Jiang, F., Jiang, B., Chen, Y., Yu, X., Zeng, X. Metallomicellar catalysis: effects of bridge-connecting ligands on the hydrolysis of ONPP catalyzed by Cu(II) complexes of ethoxy-diamine ligands in micellar solution. *J. Mol. Catalysis A* **2004**, *210*(1–2), 9–16.

17. Jiang, F., Jiang, B., Cao, Y., Meng, X., Yu, X., Zeng, X. Metallomicellar catalysis. *Colloids Surf. A* **2005**, *254*(1–3), 91–97.

18. Cheng, S., Zeng, X., Meng, X., Yu, X. Metallomicellar catalysis hydrolysis of *p*-nitrophenyl picolinate catalyzed by copper(II), nickel(II), and zinc(II) complexes of long alkylpyridine ligands in micellar solution. *J. Colloid Interface Sci.* **2000**, *224*(2), 333–337.

19. Yan, X., Jiang, B., Zeng, X., Xie, J. Metallomicellar catalysis: catalytic cleavage of *p*-nitrophenyl picolinate by Cu²⁺ complex of 4-chloro-2,6-bis[[(2-hydroxyethyl)amino]methyl]phenol in micellar solution. *J. Colloid Interface Sci.* **2002** *247*(2), 366–371.

20. Cheng, S., Yu, X., Zeng, X. Hydrolysis of *p*-nitrophenyl picolinate catalyzed by zinc(II) or nickel(II) complexes of long alkylpyridines with hydroxyl groups in micellar solution. *J. Dispersion Sci. Technol.* **1999**, *20*(7), 1821–1830.

21. Zeng, X., Cheng, S., Yu, X., Huang, Z. Hydrolysis of *p*-nitrophenyl picolinate catalyzed by copper(II) complexes of long alkylpyridines with hydroxyl groups in micellar solution. *J. Dispersion Sci. Technol.* **1999**, *20*(6), 1581–1593.

22. Tagaki, W., Ogino, K., Fujita, T., Yosbida, T., Nishi, K. Inaba, Y. Hydrolytic metalloenzyme models: catalysis in the hydrolysis of *p*-nitrophenyl 2-pyridine-carboxylate by copper and zinc ion complexes of anionic surfactants having functional imidazole and hydroxyl moieties. *Bull. Chem. Soc. Jpn.* **1993**, *66*(1), 140–147.

23. Scrimin, P., Tecilla, P., Tonellato, U. Metallomicelles as catalysts of the hydrolysis of carboxylic and phosphoric acid esters. *J. Org. Chem.* **1991**, *56*(1), 161–166.

24. Mondal, S.K., Das, M., Kar, D., Das, A.K. Micellar effect on the reaction of chromium(VI) oxidation of formaldehyde in the presence and absence of picolinic acid in aqueous acidic media: a kinetic study. *Indian J. Chem. Sect. A* **2001**, *40A*(4), 352–360.

25. Tiwari, M., Pandey, A. Kinetics of oxidation of praline by vanadium(V) in micellar system. *Oxidation Commun.* **2004**, *27*(1), 133–139.

26. Pare, B., Bhagwat, V.W., Radhakrishnan, A., Shastry, V.R. Kinetics and mechanism of micellar-catalyzed autoxidation of ethanol, propanol and n-butanol by N-bromobenzamide. *J. Indian Chem. Soc.* **2003**, *80*(8), 787–789.

27. Saha, B., Das, M., Das, A.K. Micellar effects on the reaction of Cr(V) oxidation of hexitols in the presence and absence of picolinic acid in aqueous acidic media. *J. Chem. Res. (S)* **2003**, 658–661.

28. Kou, X., Cheng, S., Du, J., Yu, X., Zeng, X. Catalytic hydrolysis of carboxylic acid esters by Cu(II) and Zn(II) complexes containing a tetracoordinate macrocyclic Schiff base ligand in Brij35 micellar solution. *J. Mol. Catal. A* **2004**, *210*(1–2), 23–29.

29. Du, J., Chen, M., Wu, Y., Zeng, X., Liu, Y., Shunzo, Y., Yoshimi, S. Metallomicellar catalysis: hydrolysis of phosphodiester with Cu(II) and Zn(II) complexes in micellar solution. *J. Dispersion Sci. Technol.* **2003**, *24*(5), 683–689.

30. Kabir-ud-Din, Hartani, K., Khan, Z., Micellar catalysis on the redox reaction of glycolic acid with chromium(VI). *Int. J. Chem. Kinet.* **2001**, *33*(6), 377–386.

31. Das, A.K., Mondal, S.K., Kar, D., Das, M. Micellar effects on the reaction of picolinic acid catalyzed chromium(VI) oxidation of dimethyl sulfoxide in aqueous acidic media: a kinetic study. *Int. J. Chem. Kinet.* **2001**, *33*(3), 173–181.

32. Otto, S., Engberts, J.B.F.N., Kwak, J.C.T. Million-fold acceleration of a Diels–Alder reaction due to combined Lewis acid and micellar catalysis in water. *J. Am. Chem. Soc.* **1998**, *120*(37), 9517–9525.

33. Lopez-Cornejo, P., Perez, P., Garacia, F., De la Vega, R., Sanchez, F. Use of the pseudophase model in the interpretation of reactivity under restricted geometry conditions: an application to the study of the $[Ru(NH_3)_5pz]^{2+}$ + $S_2O_8^{2-}$ electron-transfer reaction in different microheterogeneous systems. *J. Am. Chem. Soc.* **2002**, *124*(18), 5154–5164.

34. Bernas, A., Grand, D., Hautecloque, S., Giannotti, C. Interfacial electrical potential in micelles: its influence on N,N,N,N-tetramethylbenzidine cation decay. *J. Phys. Chem.* **1986**, *90*(23), 6189–6194 and references cited therein.

35. Grand, D., Hautecloque, S. Electron transfer from nucleophilic species to N,N,N,N-tetramethylbenzidine cation in micellar media: effect of interfacial electrical potential on cation decay. *J. Phys. Chem.* **1990**, *94*(2), 837–841.

36. Tagaki, W., Ogino, K. Tanaka, O., Machiya, K., Kashihara, N., Yoshida, T. Hydrolytic metalloenzyme models: micellar effects on the activation of the hydroxyl groups of N-alkyl-2-(hydroxymethyl)imidazole ligands by Cu^{2+} in the transacylation of *p*-nitrophenyl picolinate. *Bull. Chem. Soc. Jpn.* **1991**, *64*(1), 74–80.

37. (a) Bunton, C.A., Cerichelli, G. Micellar effects upon electron transfer from ferrocenes. *Int. J. Chem. Kinet.* **1980**, *12*, 519–533. (b) Allen, R.J., Bunton, C.A. Micellar catalysis of the alkylation of mercuric ions by alkyl cobalt(III) complexes. *Bioinorg. Chem.* **1976**, *5*, 311–324.

38. Reinsborough, V.C., Robinson, B.H. Micellar catalysis of metal complex formation. Part 2. Kinetics of the reaction between aquated nickel(2+) and various neutral bidentate ligands in the presence of sodium dodecyl sulfate micelles in aqueous solution. *J. Chem. Soc., Faraday Trans. 1.* **1979**, *75*(11), 2395–2405.

39. Hicks, J.R., Reinsborough, V.C. Effects of added cations on rate enhancements for the nickel(II)-PADA reaction in micellar solutions. *Can. J. Chem.* **1984**, *62*(5), 990–994.

40. Favaro, Y.L., Reinsborough, V.C. Micellar catalysis in mixed anionic/cationic surfactant systems. *Can. J. Chem.* **1994**, *72*(12), 2443–2446.

41. Fletcher, P.D., Hicks, J.R., Reinsborough, V.C. Rate enhancements for the nickel(II)-PADA complexation in mixed anionic micelles. *Can. J. Chem.* **1983**, *61*(7), 1594–1597.

42. Hicks, J.R., Reinsborough, V.C. Rate enhancement of the nickel(II)-PADA complex formation in sodium alkanesulfonate micellar solutions. *Aust. J. Chem.* **1982**, *35*(1), 15–19.

43. Reinsborough, V.C., Stultz, T.D.M., Xiang, X. Rate enhancement of nickel(II)-PADA complex formation in mixed sodium perfluorooctanoate/octanesulfonate micellar solutions. *Aust. J. Chem.* **1990**, *43*(1), 11–19.

44. Drennan, C.E., Hughes, R.J., Reinsborough, V.C., Soriyan, O.O. Rate enhancement of nickel(II) complexation in dilute anionic surfactant solutions. *Can. J. Chem.* **1998**, *76*(2), 152–157.

45. Calvaruso, G., Cavasino, F.P., Sbriziolo, C., Turco Liveri, M.L. Interaction of the chloropentaamminecobalt(III) cation with mixed micelles of anionic and non-ionic surfactants: a kinetic study. *J. Chem. Soc., Faraday Trans.* **1995**, *91*(7), 1075–1079.

46. Ogino, K., Kashihara, N., Ueda, T., Isaka, T., Yoshida, T., Tagaki, W. Hydrolytic metalloenzyme models: metal ion dependent site-selective acylation of hydroxyl groups of bis-imidazole ligands catalyzed by Zn^{2+} and Cu^{2+} in the reaction with *p*-nitrophenyl 2-pyridinecarboxylate in a cationic surfactant micelle. *Bull. Chem. Soc. Jpn.* **1992**, *65*(2), 373–384.

47. Fornasier, R., Scrimin, P., Tecilla, P., Tonellato, U. Bolaform and classical cationic metallomicelles as catalysts of the cleavage of *p*-nitrophenyl picolinate. *J. Am. Chem. Soc.* **1989**, *111*(1), 224–229.

48. Menger, F.M., Gan, L.H., Johnson, E., Durst, D.H. Phosphate ester hydrolysis catalyzed by metallomicelles. *J. Am. Chem. Soc.* **1987**, *109*(9), 2800–2803.

49. Broxton, T.J., Nasser, A. Micellar catalysis of organic reactions. Part 37. A comparison of the catalysis of ester and amide hydrolysis by copper-containing micelles. *Can. J. Chem.* **1997**, *75*, 202–206.

50. Schowen, R.L., Hopper. C.R., Bazikian, C.M. Amide hydrolysis. V. Substituent effects and solvent isotope effects in the basic methanolysis of amides. *J. Am. Chem. Soc.* **1972**, *94*(9), 3095–3097.

51. Hine, J., Khan, M.N. Internal amine-assisted attack of alcoholic hydroxyl group on esters: serine esterase models. *Indian J. Chem.* **1992**, *31B*, 427–435 and references cited therein.

52. Scrimin, P., Tecilla, P., Tonellato, U. Leaving group effect in the cleavage of picolinate esters catalyzed by hydroxyl-functionalized metallomicelles. *J. Org. Chem.* **1994**, *59*(1), 18–24.

53. Khan, M.N., Arifin, A. Kinetics and mechanism of intramolecular carboxylic acid participation in the hydrolysis of N-methoxyphthalamic acid. *Org. Biomol. Chem.* **2003**, *1*, 1404–1408 and references cited therein.

54. Scrimin, P., Tecilla, P., Tonellato, U. Chiral lipophilic ligands. 1. Enantioselective cleavage of α-amino acid esters in metallomicellar aggregates. *J. Org. Chem.* **1994**, *59*(15), 4194–4201.

55. Bhattacharya, S., Snehalatha, K., Kumar, V.P. Synthesis of new Cu(II)-chelating ligand amphiphiles and their esterolytic properties in cationic micelles. *J. Org. Chem.* **2003**, *68*(7), 2741–2747.

56. (a) Jencks, W.P., Gilchrist, M. Nonlinear structure-reactivity correlations: the reactivity of nucleophilic reagents toward esters. *J. Am. Chem. Soc.* **1968**, *90*(10), 2622–2637. (b) Khan, M.N., Gleen, P.C., Arifin, Z. Effects of inorganic salts and mixed aqueous-organic solvents on the rates of alkaline hydrolysis of aspirin. *Indian J. Chem.* **1996**, *35A*, 758–765.

57. Scrimin, P., Ghirlanda, G., Tecilla, P., Moss, R.A. Comparative reactivities of phosphate ester cleavages by metallomicelles. *Langmuir* **1996**, *12*(26), 6235–6241.

58. Scrimin, P., Tecilla, P., Tonellato, U., Valle, G., Veronese, A. A zinc(II)-organized molecular receptor as a catalyst for the cleavage of amino acid esters. *J. Chem. Soc., Chem. Commun.* **1995**, 1163–1164.

59. Scrimin, P., Tecilla, P., Tonellato, U., Valle, G., Veronese, A. Metal ions cooperativity in the catalysis of the hydrolysis of a β-amino ester by a macrocyclic dinuclear Cu(II) complex. *Tetrahedron* **1995**, *51*(2), 527–538.

60. Scrimin, P., Tecilla, P., Tonellato, U., Vendrame, T. Aggregate structure and ligand location strongly influence copper(II) binding ability of cationic metallosurfactants. *J. Org. Chem.* **1989**, *54*(25), 5988–5991.

61. Weijnen, J.G.J., Koudijis, A., Engbersen, J.F.J. Synthesis of chiral 1,10-phenanthroline ligands and the activity of metal-ion complexes in the enantoselective hydrolysis of N-protected amino acid esters. *J. Org. Chem.* **1992**, *57*(26), 7258–7265.

62. Bracken, K., Moss, R.A., Ragunathan, K.G. Remarkably rapid cleavage of a model phosphodiester by complexed ceric ions in aqueous micellar solutions. *J. Am. Chem. Soc.* **1997**, *119*(39), 9323–9324.

63. Claudia, S., Rossi, P., Felluga, F., Formaggio, F., Palumbo, M., Tecilla, P., Toniolo, C., Scrimin, P. Dinuclear Zn^{2+} complexes of synthetic heptapeptides as artificial nucleases. *J. Am. Chem. Soc.* **2001**, *123*(13), 3169–3170.

64. Yamada, K., Takahashi, Y., Yamamura, H., Araki, S., Saito, K., Kawai, M. Phosphodiester bond cleavage mediated by a cyclic β-sheet peptide-based dinuclear zinc(II) complex. *Chem. Commun.* **2000**, 1315–1316.

65. Bhattacharya, S., Snehalatha, K., George, S.K. Synthesis of some copper(II)-chelating (dialkylamino)pyridine amphiphiles and evaluation of their esterolytic capacities in cationic micellar media. *J. Org. Chem.* **1998**, *63*(1), 27–35.

66. Scrimin, P., Tecilla, P., Tonellato, U., Bunton, C.A. Nucleophilic catalysis of hydrolyses of phosphate and carboxylate esters by metallomicelles: facts and misconceptions. *Colloids Surf. A* **1998**, *144*, 71–79.

67. Gellman, S.H., Petter, R., Breslow, R. Catalytic hydrolysis of a phosphate trimester by tetracoordinated zinc complexes. *J. Am. Chem. Soc.* **1986**, *108*(9), 2388–2394.

68. Kimura, E., Hashimoto, H., Koioke, T. Hydrolysis of lipophilic esters catalyzed by a zinc(II) complex of a long alkyl-pendant macrocyclic tetraamine in micellar solution. *J. Am. Chem. Soc.* **1996**, *118*(45), 10963–10970.

69. Weijnen, J.G.J., Engbersen, J.F.J. Catalytic hydrolysis of phosphate esters by metallocomplexes of 1,10-phenanthroline derivatives in micellar solution. *Recl. Trav. Chim. Pays-Bas* **1993**, *112*(6), 351–357.

70. Bunton, C.A., Scrimin, P., Tecilla, P. Source of catalysis of dephosphorylation of *p*-nitrophenyl diphenyl phosphate by metallomicelles. *J. Chem. Soc., Perkin Trans. 2.* **1996**, 419–425.

7 Chemical Kinetics and Kinetic Parameters

7.1 INTRODUCTION

Let us begin with the very fundamentals of reaction kinetics. As reaction kinetics deals with chemical reactions, the very basic questions that a beginner in this area of research can pose may include, (1) what is a chemical reaction? (2) why does a chemical reaction occur? (3) how does a chemical reaction occur? (4) why do we need to know the answers to these questions? and (5) what is the role of chemical kinetics in explaining these questions?

7.1.1 WHAT IS A CHEMICAL REACTION?

The transformation of atoms or molecules or atoms + molecules from one stable state (ionic or nonionic or ionic + nonionic) to another stable state (ionic or nonionic or ionic + nonionic) in a reaction medium is termed a *chemical reaction*. For example, consider the following aqueous reactions:

$$H^+ + Cl^- \rightleftharpoons HCl \tag{7.1}$$

$$H^+ + HO^- \rightleftharpoons H_2O \tag{7.2}$$

$$H^+ + NH_3 \rightleftharpoons NH_4^+ \tag{7.3}$$

$$HCOOH + CH_3NH_2 \rightleftharpoons HCCO^{-+}NH_3CH_3 \tag{7.4}$$

$$CH_3COCl + H_2O \rightleftharpoons CH_3COOH + HCl \tag{7.5}$$

$$nCO_2 + nH_2O \rightleftharpoons (CH_2O)_n + nO_2 \tag{7.6}$$

7.1.1.1   One-Step or Elementary Chemical Reactions

Such chemical reactions, also called *concerted chemical reactions*, occur in a single reaction step, i.e., bond formation and bond cleavage in such reactions involve a single transition state (the highest energy point on the lowest energy path in a chemical transformation from reactants to products is called *transition state* or *activated state*). A hypothetical example of an irreversible one-step chemical reaction is

$$R \longrightarrow P$$

7.1.1.2   Multistep Chemical Reactions

Such chemical reactions, also called *stepwise reactions*, involve two or more one-step chemical reactions. For example, a hypothetical irreversible multistep chemical reaction is represented as

$$R \longrightarrow I_1 \longrightarrow I_2 \longrightarrow I_3 \longrightarrow P$$

where I_1, I_2, and I_3 represent chemical intermediates on the reaction path for the overall reaction R = P.

7.1.1.3   Chemical Intermediate

A chemical configuration or entity, which can exist on the reaction path for a period greater than 10^{-13} sec is considered a chemical intermediate.[1]

7.1.2   WHY DOES A CHEMICAL REACTION OCCUR?

The answer to this very fundamental question certainly depends on the individual's basic perception of a chemical reaction, but I attempt to answer this question as follows.

It seems that the most fundamental single law of nature that the entire universe, including the most fundamental particles (the building blocks or the sub-building blocks of the known universe), ought to follow is to attain the most stable state through the least energy path.[2]

Since the dawn of civilization, it has been customary for humans to inquire into and seek explanations about unexplained phenomena encountered directly or indirectly. Even today, there are several unanswered and unexplained questions such as these: Why are we here? Why and where do we go from here at the time of the so-called death? Why are all these powerful events beyond our control? Unlike atoms, molecules, and macromolecules, what makes us think, reason out, and pose such questions? Is death considered the end of the essence of life? Why are reactants nonedible and products edible in some reactions and *vice versa*, although these simple reactions involve only sharing, exchange, or donation of a pair of electrons among the two atoms, which constitute the reaction sites? Some unanswered questions, such as why energy and charge cannot be created and destroyed, are beyond our ability to understand and find answers to. The essence of such a complex question is generally coined as the *natural law*.

7.1.3 How Does a Chemical Reaction Occur?

The answer to this question is actually what we call the reaction mechanism. A *reaction mechanism* is defined as the detailed description of the electronic or both electronic and nuclear reorganization during the course of a chemical reaction. For example,

Reaction: $HO^- + CH_3CH_2Br = CH_3CH_2OH + Br^-$
The mechanism for the above reaction is shown in Scheme 7.1.

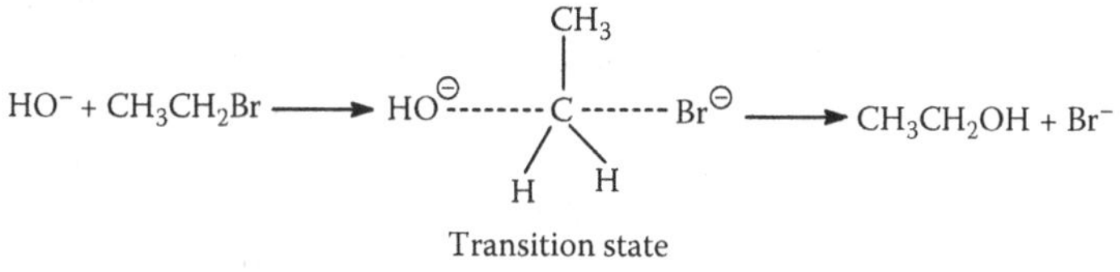

Transition state

SCHEME 7.1

Reaction: $NH_3 + CH_3COCl = CH_3CONH_2 + HCl$
The mechanism for the above reaction is shown in Scheme 7.2.

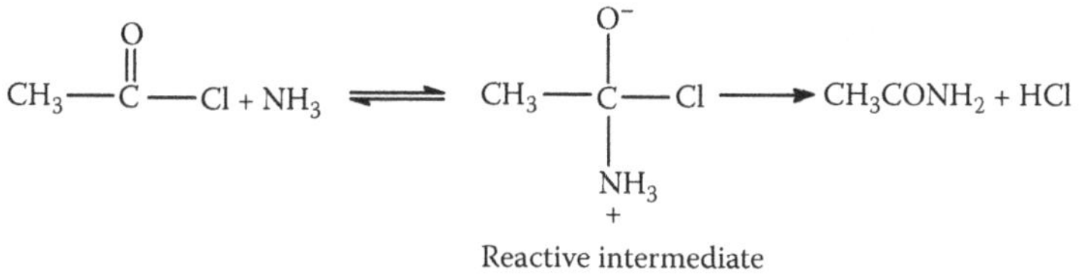

Reactive intermediate

SCHEME 7.2

7.1.4 Why Do We Need to Know the Answers to These Questions?

The world and the universe we live in, the things we see around us, and we ourselves are essentially made up through chemical reactions. If we want to improve the essence of our survival, we need to know the answers to these very basic questions. We may be the most intelligent inhabitants of Earth, capable of reasoning about and evaluating the events that happen around us, yet we actually know very little about chemical reactions at the molecular level. For instance, aspirin (chemical name: acetyl salicylic acid) has been known to humans as a painkiller for more than 150 years, but even today no one knows why and how it relieves pain. This is true of almost all the drugs that we use. We cannot make our survival safer without knowing, with absolute certainty, why and how these reactions occur. The chemical approach, which is the ideal approach to answering such questions, is extremely difficult. Probably that is why only a very few enthusiastic scientists dare to enter this area of research, which could provide answers to how a reaction occurs at the most refined molecular level.

7.1.5 WHAT IS THE ROLE OF CHEMICAL KINETICS IN ANSWERING THESE QUESTIONS?

Chemical kinetics deals with the dynamics of the chemical conversion or transformation of reactants to products in chemical reactions. It gives very fine details of the reaction mechanism that probably cannot be given by any other chemical or physical technique. It gives the exact reaction period for 1, 2, 3, ..., n half-lives of a reaction. It also gives a quantitative measure of the effects of various factors, such as catalysts, reaction medium, temperature, pH, and inert salts, on the reaction rates. Kinetic study on reaction rates has been an extremely important and essential tool in diagnosing the fine details of the mechanism of molecular transformation from one stable state (i.e., reactant state) to the other (i.e., product state). Its importance in the study of reaction rates dates back to at least 1850.[3]

Kinetic study of micellization of surfactant molecules, micellar solubilization of solubilizate molecules, and micellar-mediated chemical reaction rates has been an area of considerable interest since the early 1950s. Such studies have helped us understand, at least at the qualitative level, the dynamic behavior of micelle, its properties such as water penetration into micelle, micellar surface roughness, ionic strength of ionic micellar surface, polarity/dielectric constant of the micellar environment, and adsorption sites for solubilizates of different hydrophobicity. The occurrence of ion exchange between counterions at the ionic micellar surface was probably first detected by kinetic study.[4] Kinetic studies on micellar-mediated reactions (both organic and inorganic) have indirectly shown that the micellar environment is not perfectly homogeneous in terms of water concentration, polarity/dielectric constant, ionic strength (for ionic micelles), and distribution and micellar location of solubilizates of different hydrophobicity and molecular structure.[5]

7.2 FUNDAMENTALS OF CHEMICAL KINETICS

7.2.1 RATE OF CHEMICAL REACTION

The rate of a chemical reaction is intuitively and empirically defined as

$$\text{rate} \propto \kappa \times \text{frequency of effective collisions between}$$
$$\text{reactant atoms or molecules (R)} = k\,[R] \qquad (7.7)$$

where κ is the fraction of effective collisions between reactant atoms or molecules (R) that occur in unit time and lead to the product formation (thus, $\kappa < 1$ and > 0), [R] is the concentration of R, and k is the proportionality constant, which is known as *rate constant* for the reaction. The magnitude of rate constant, k, is the measure of the velocity or rate of reaction. The value of k can never be negative. *Effective collision* is defined as a collision that provides sufficient energy to colliding reactant atoms or molecules to surmount the transition state. Thus, logically the fraction of the frequency of effective collisions between reactant atoms or molecules (R) should be proportional to the concentration of reactant ([R]) and not to the concentration of product ([P]). However, it seems amazing that although both reactant (R) and product (P) molecules exist together in the same reaction-vessel, the rate of reaction is proportional to the frequency of effective collisions between reactant molecules only or R molecules and the wall of the reaction vessel and not between R and P molecules. It does not mean that collisions between R and P molecules do not occur. It simply means that these collisions can never be effective.

The early study[3] on the effect of temperature on rate constant k led to the empirical relationship as expressed by Equation 7.8

$$k = B \exp(-A'/T) \qquad (7.8)$$

where B and A' are empirical constants. In 1884, some theoretical significance to Equation 7.8 was given by van't Hoff,[3] who argued on the basis of the effect of temperature on equilibrium constants. This idea was extended by Arrhenius[3] and was successfully applied by him to the data of a number of reactions; on account of his work, Equation 7.9 is usually referred to as Arrhenius equation.

$$k = A \exp(-E/RT) \qquad (7.9)$$

where A is an empirical constant as is B in Equation 7.8, but E is now referred to as energy of activation.

In 1935, Eyring[6] presented a clear and more advanced theoretical formulation of Equation 7.8 that could lead to Equation 7.10, which does not contain any empirical constant.

$$k = (k_B T/h) \exp(\Delta S^*/R) \exp((-\Delta H^*/RT) \qquad (7.10)$$

In Equation 7.10, all the symbols have their usual meanings, but for unimolecular reactions, ΔH^* is related to E of Equation 7.9 by the relationship $\Delta H^* = E - RT$.

7.2.2 Concept of Energy Barrier or the Energy of Activation

The reactant state and product state correspond to two valleys (scientifically known as *potential wells*) of, most likely, different depths. These valleys are

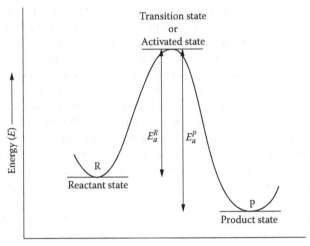

FIGURE 7.1 The energy barrier for an exothermic one-step reaction.

separated by a high hill or mountain. The height of the hill or mountain with respect to the reactant valley (state) is defined as the *energy of activation* (E_a). This situation may best be explained pictorially as in Figure 7.1, where R represents a reactant state, P the product state for the chemical transformation from R to P (the forward reaction), E_a^R stands for the energy of activation for the conversion of R to P, and E_a^P the energy of activation for the conversion of P to R (the backward reaction). The solid line in Figure 7.1 represents the least energy path for transformation of R to P and P to R.

The energy barrier is almost zero for reactions between the fundamental particles, because the fundamental particles are the most unstable entities and, consequently, their state energies are almost similar to their corresponding transition state energies. Thus, $E_a^R \approx 0$ for the reactions between fundamental particles and such reactions are the so-called pure irreversible reactions. The rates of some atomic reactions are diffusion controlled and proceed with virtually no energy barrier.

7.2.3 WHY REACTIONS WITH MODERATELY STABLE REACTANTS INVOLVE ENERGY BARRIERS

The conversion of reactant state into product state in a chemical reaction requires electronic reorganization or perturbation in the reactant (atoms or molecules), which could take place by absorption of energy from the external energy source or by collision among the reactant molecules. The absorption or intake of external energy by reactant atoms or molecules makes them relatively unstable and, consequently, they acquire a relatively high energy state. Once this instability in reactant state reaches a critical unstable state (we call it *transition state* or *activated state*), the product formation occurs.

7.2.4 Rate Law and Order of a Reaction

The kinetic equation, which shows the relationships between the rate (v) of a reaction and the concentrations of reactants, and in some cases products also, is called the *rate law*. For example, the rate law for the reaction

$$A + B = P \qquad (7.11)$$

is expressed as

$$v = \alpha \, [A][B]^2/(\beta + \delta[P]) \qquad (7.12)$$

where α, β, and δ are empirical constants and $v = -d[A]/dt = -d[B]/dt = d[P]/dt$. The empirical constants, α, β, and δ are generally the functions of rate constants and equilibrium constants. The rate law is considered to be the backbone of the use of chemical kinetics in understanding the reaction mechanisms. The rate law (represented by Equation 7.12) shows that the reaction depicted by Equation 7.11 involves more than one elementary reaction step and, consequently, follows a complex reaction mechanism. A rate law, sometimes referred to as a rate expression or a rate equation, is always determined experimentally and, hence, a rate equation is an empirical equation.

The *order of a reaction* is defined as the sum of the exponents appearing in the rate law. For elementary reactions, the order of reaction is equal to the sum of the coefficients of the left-hand side of a written balanced chemical equation. For a reaction that obeys a complex rate law, such as Equation 7.12, the reaction order cannot be defined, and such reactions are said to have no simple reaction order.

7.2.5 Integrated Form of Rate Law

The rate laws with simple reaction order, such as first-order Equation 7.13 and second-order Equation 7.14, can be easily integrated to get the respective integrated forms, Equation 7.15 and Equation 7.16, of rate laws.

$$-d[A]/dt = k[A] \qquad (7.13)$$

$$-d[A]/dt = k[A][B] \qquad (7.14)$$

$$[A] = [A_0]e^{-kt} \qquad (7.15)$$

$$\{1/([B_0] - [A_0])\} \ln\{[A_0][B]/[B_0][A]\} = kt \qquad (7.16)$$

In Equation 7.15 and Equation 7.16, $[A_0]$ and $[B_0]$ represent respective concentrations of reactants A and B at reaction time t = 0. The values of [A] and [B] in Equation 7.16 are related by the relationship $[A] = [B] + [A_0] - [B_0]$ and also, $[A] = [A_0] - [X]$ as well as $[B] = [B_0] - [X]$, where [X] refers to the concentrations of A and B, which have reacted at reaction time t.

In the special case in which the concentrations of A and B at t = 0 are identical (i.e., $[A_0] = [B_0] = [R_0]$), the rate law is given by Equation 7.17, in which [R] is the remaining concentration of A or B at any time t. The integrated rate law is given by Equation 7.18, which is also valid for second-order reaction involving a single reactant such as A and, in this case, the elementary reaction is represented as 2A → P with second-order rate constant k.

$$-d[R]/dt = k[R]^2 \qquad (7.17)$$

$$1/[R] - 1/[R_0] = kt \qquad (7.18)$$

The integrated form of rate law shows the relationship between the concentration of either reactant or product as a function of rate constant and reaction time (t). It is generally easier and convenient to verify experimentally the integrated form of rate law (i.e., Equation 7.15), than the rate law (i.e., Equation 7.13).

7.2.6 EXPERIMENTAL RATE LAW

The dependence of the rate of a reaction upon the concentration of the reaction components (i.e., reactants), determined experimentally by using physical or chemical techniques, constitutes the so-called experimental rate law. Thus, an experimental rate law, which lacks theoretical justification, is an empirical relationship, and the reliability of such a relationship remains unquestionable within the domain of the reliability of experimental techniques used to determine it. For example, the experimental rate laws for reactions Equation 7.19 and Equation 7.20 are given by Equation 7.21 and Equation 7.22, respectively.

$$CH_3COCH_3 + Br_2 \overset{k}{=} CH_3COCH_2Br + HBr \qquad (7.19)$$

$$HCOOCH_3 + HO^- \overset{k}{=} HCOO^- + CH_3OH \qquad (7.20)$$

$$v = k\,[CH_3COCH_3][H^+] \qquad (7.21)$$

$$v = k\,[HCOOCH_3][HO^-] \qquad (7.22)$$

7.2.7 DETERMINATION OF THE EXPERIMENTAL RATE LAW

Experimental determination of the rate law is the most important and perhaps the most fundamental aspect of the use of reaction kinetics in diagnosing the reaction mechanisms. There are several methods used for the determination of the experimental rate law, and these methods are critically and well described in several books on reaction kinetics. The most convenient and commonly used method involves the UV-visible spectrophotometric technique, and an attempt is made to describe this particular method in some detail in this chapter.

7.2.7.1 One-Step First-Order Reactions

Consider a simple one-step first-order reaction,

$$R \xrightarrow{k} P \qquad (7.23)$$

which obeys integrated rate law as expressed by Equation 7.15 where k is the first-order rate constant and P represents the products. If reactant R absorbs more strongly than P at a working wavelength (λ) within UV-visible region, then Equation 7.24 can be easily derived from Equation 7.15 by replacing $[A_0]$ by $[R_0]$

$$A_{obs} = \delta_{app}[R_0]e^{-kt} + A_\infty \qquad (7.24)$$

where A_{obs} is the observed absorbance of reaction mixture at any reaction time t, δ_{app} is an apparent molar extinction coefficient of the reaction mixture, and $A_\infty = A_{obs}$ at t = ∞. Under such experimental conditions, $\delta_{app} = \delta_R - \delta_P$ and $A_\infty = \delta_P$ $[R_0]$ where δ_R and δ_P represent the molar extinction coefficients of R and P, respectively, at the working wavelength (λ). These relationships also show that $\delta_{app}[R_0] = A_0 - A_\infty$ where $A_0 = A_{obs}$ at t = 0.

If P absorbs more strongly than R at λ, then Equation 7.15 can lead to Equation 7.25 with $[R_0] \equiv [A_0]$.

$$A_{obs} = \delta_{app}[R_0](1 - e^{-kt}) + A_0 \qquad (7.25)$$

In Equation 7.25, $\delta_{app} = \delta_P - \delta_R$, $A_0 = \delta_R [R_0]$, and $\delta_{app} [R_0] = A_\infty - A_0$. The rearrangement of Equation 7.25 gives Equation 7.26

$$A_\infty - A_{obs} = (A_\infty - A_0)e^{-kt} \qquad (7.26)$$

where $A_\infty > A_0$ and $A_\infty > A_{obs}$. The rearrangement of Equation 7.24 also gives Equation 7.26, where $A_\infty < A_0$ and $A_\infty < A_{obs}$.

The linearized forms of Equation 7.24 and Equation 7.25 have been very often used to calculate k either by graphical procedure (i.e., the plot of ln ($A_{obs} - A_\infty$) or ln ($A_\infty - A_{obs}$) vs. t) or by using the linear least-squares technique. The serious problem in using the linearized form of Equation 7.24 or Equation 7.25 is concerned with the fact that the statistical reliability of $A_{obs} - A_\infty$ or $A_\infty - A_{obs}$ decreases as $A_{obs} \rightarrow A_\infty$ and, consequently, $A_{obs} - A_\infty$ or $A_\infty - A_{obs} \rightarrow 0$ and $\ln(A_{obs} - A_\infty)$ or $\ln(A_\infty - A_{obs}) \rightarrow$ an unrealistic large negative number. Thus, the use of the linearized form of Equation 7.24 or Equation 7.25 suffers from the disadvantages of placing a very high emphasis on the values of A_{obs} as $A_{obs} \rightarrow A_\infty$ and being very sensitive to even small errors in A_{obs} values under such conditions. Thus, it is strongly recommended that a kinetic equation in its most original form be used — not a simplified or modified form such as the linearized form of a nonlinear kinetic equation used with the help of logarithms.

The use of the linearized form of the nonlinear equation Equation 7.24 or Equation 7.25 in the determination of the rate constant k requires the exact value of A_∞ under a typical reaction condition of kinetic run, which is sometimes difficult to obtain, especially when the rate of reaction is very slow or the reaction under investigation is not a simple one-step irreversible reaction. There is no perfect, decisive, and completely error-free method to determine an exact value of A_∞. Some experimental approaches have been described by Jencks[7] for the determination of a reliable value of A_∞. The necessary and basic requirement in these approaches is that the reaction must obey the first-order rate law within the reaction period of at least 10 half-lives (i.e., time required for 99.9% completion of the reaction). This requirement is difficult to achieve with complete certainty even with moderately slow reactions.

7.2.7.1.1 The Iterative Method and Linear Least-Squares Regression Analysis of Kinetic Data

An iterative method of obtaining a reliable value of A_∞ may be described as follows. But this method requires a computer and a linear least-squares software program in order to facilitate the computational process. A linear least-squares software program in BASIC is given in Appendix A. The values of k, $\ln (A_0 - A_\infty)$ {when $A_\infty < A_0$} or $\ln (A_\infty - A_0)$ {when $A_\infty > A_0$}, and least squares, $\Sigma\, d_i^2$, where $d_i = \ln(A_{obs,i} - A_\infty) - \ln(A_{calcd,i} - A_\infty)$ {when $A_\infty < A_0$} or $\ln(A_\infty - A_{obs,i}) - \ln(A_\infty - A_{calcd,i})$ {when $A_\infty > A_0$} with $A_{obs,i}$ and $A_{calcd,i}$ representing respective observed and least-squares calculated values of absorbance at the i_{th} reaction time (t_i), are calculated from Equation 7.27 (a linearized form of Equation 7.24 where $A_\infty < A_0$) or Equation 7.28 (a linearized form of Equation 7.25 where $A_\infty > A_0$), for a presumed value of A_∞ by using linear least-squares technique.

$$\ln(A_{obs} - A_\infty) = \ln(A_0 - A_\infty) - kt \qquad (7.27)$$

$$\ln(A_\infty - A_{obs}) = \ln(A_\infty - A_0) - kt \qquad (7.28)$$

The values of Σd_i^2 are obtained at different presumed values of A_∞ and the best among these values of A_∞ is the one for which the value of Σd_i^2 is minimum. This method has been used to calculate k for two typical sets of observed data as summarized in Table 7.1 and Table 7.2. The observed data in Table 7.1 represent

TABLE 7.1
The Values of Observed Absorbance (A_{obs}) at 280 nm as a Function of Reaction Time (t) for Methanolysis of Ionized Phenyl Salicylate (PS⁻)

t/sec	A_{obs}	$A_{calcd}{}^a$	$A_{calcd}{}^a$	$A_{calcd}{}^a$	$A_{calcd}{}^a$	$A_{calcd}{}^a$	$A_{calcd}{}^a$	$A_{calcd}{}^a$	$A_{calcd}{}^a$
25	0.399	0.694[b]	0.676[c]	0.312[d]	0.357[e]	0.399[f]	0.422[g]	0.449[h]	0.465[i]
35	0.470	0.699	0.684	0.413	0.444	0.472	0.488	0.506	0.518
40	0.505	0.702	0.688	0.457	0.482	0.505	0.517	0.532	0.541
50	0.561	0.707	0.696	0.533	0.548	0.562	0.570	0.579	0.584
60	0.610	0.712	0.703	0.595	0.603	0.611	0.615	0.619	0.622
70	0.655	0.717	0.711	0.646	0.649	0.652	0.653	0.654	0.655
80	0.690	0.722	0.718	0.688	0.688	0.687	0.686	0.685	0.684
90	0.718	0.727	0.724	0.723	0.720	0.717	0.714	0.711	0.709
100	0.741	0.731	0.731	0.751	0.746	0.742	0.739	0.734	0.731
110	0.760	0.736	0.737	0.774	0.769	0.763	0.759	0.754	0.751
120	0.778	0.740	0.743	0.793	0.787	0.781	0.777	0.772	0.768
180	0.844	0.764	0.774	0.854	0.849	0.844	0.841	0.836	0.833
240	0.869	0.784	0.799	0.872	0.871	0.868	0.866	0.864	0.863
360	0.880	0.815	0.835	0.880	0.880	0.880	0.880	0.881	0.882
2160	0.897	0.897	0.897						
$10^3 k/sec^{-1}$ =		2.69 ± 0.39	3.79 ± 0.34	19.7 ± 0.6	18.1 ± 0.3	16.5 ± 0.1	15.5 ± 0.2	14.2 ± 0.4	13.3 ± 0.5
$-10\ln(A_\infty - A_0)$ =		15.22 ± 2.22	14.14 ± 1.98	0.709 ± 0.885	1.94 ± 0.41	3.16 ± 0.11	3.87 ± 0.31	4.74 ± 0.57	5.31 ± 0.72
$10^3 \Sigma d_i^2$ =		7842	6222	513.5	109.4	8.563	63.26	212.7	337.4

Note: Data analysis with Equation 7.28. Reaction condition: $[R_0] = 3.2 \times 10^{-4}$ M where R = phenyl salicylate, [KOH] = 0.01 M, [KCl] 0.14 M, CH_3OH = 80% v/v, H_2O = 19.2% v/v, CH_3CN = 0.8% v/v, and 30°C.

[a] Calculated from Equation 7.28 with kinetic parameters listed in Table 7.1. [b] A_∞ = 0.8980. [c] A_∞ = 0.8971. [d] A_∞ = 0.8805. [e] A_∞ = 0.881. [f] A_∞ = 0.882. [g] A_∞ = 0.883. [h] A_∞ = 0.885. [i] A_∞ = 0.887.

TABLE 7.2
The Values of Observed Absorbance (A_{obs}) at 300 nm as a Function of Reaction Time (t) of a Reaction Mixture Containing Phthalimide (PTH) and 0.50-M Trimethylamine Buffer of pH 10.95

t/sec	A_{obs}	A_{calcd}[a]	A_{calcd}[a]	A_{calcd}[a]	A_{calcd}[a]	A_{calcd}[a]	A_{calcd}[a]	A_{calcd}[a]	A_{calcd}[a]
30	0.616	0.614[b]	0.615[c]	0.615[d]	0.620[e]	0.625[f]	0.629[g]	0.640[h]	0.481[i]
50	0.599	0.598	0.599	0.599	0.604	0.607	0.611	0.621	0.475
60	0.591	0.591	0.591	0.592	0.595	0.599	0.602	0.611	0.472
90	0.568	0.568	0.568	0.568	0.561	0.574	0.577	0.583	0.464
120	0.547	0.546	0.546	0.546	0.549	0.550	0.552	0.557	0.455
180	0.504	0.504	0.505	0.505	0.505	0.506	0.506	0.508	0.439
210	0.485	0.485	0.485	0.485	0.485	0.485	0.485	0.485	0.431
240	0.467	0.467	0.467	0.466	0.466	0.465	0.464	0.463	0.423
300	0.431	0.431	0.431	0.431	0.429	0.427	0.426	0.422	0.408
420	0.368	0.369	0.368	0.368	0.364	0.362	0.359	0.353	0.379
600	0.290	0.291	0.291	0.290	0.286	0.282	0.279	0.270	0.340
1140	0.143	0.143	0.143	0.143	0.140	0.137	0.135	0.129	0.244
1800	0.060	0.060	0.060	0.060	0.061	0.062	0.063	0.064	0.163
10500	0.001								0.001
10^3k/sec^{-1} =		1.31 ± 0.00	1.31 ± 0.00	1.32 ± 0.00	1.39 ± 0.01	1.45 ± 0.01	1.50 ± 0.02	1.64 ± 0.04	0.611 ± 0.037
$-10\ln(A_0 - A_\infty)$ =		4.46 ± 0.01	4.47 ± 0.00	0.01	0.05	4.51 ± 0.08	4.50 ± 0.12	4.44 ± 0.23	7.13 ± 1.08
$10^4\,\Sigma\,d_i^2$ =		0.399	0.150	0.241	17.52	60.55	130.2	449.7	16468

Note: Data analysis with Equation 7.27. Reaction condition: $[R_0]$ = 3.2 × 10^{-4} M where R = phthalimide, ionic strength = 1 M, H_2O = 98.4% v/v, CH_3CN = 1.6% v/v, and 30°C.

[a] Calculated from Equation 7.27 with kinetic parameters listed in Table 7.2. [b] A_∞ = −0.001. [c] A_∞ = 0.0. [d] A_∞ = 0.001. [e] A_∞ = −0.001. [f] A_∞ = 0.009. [g] A_∞ = 0.015. = 0.020. [h] A_∞ = 0.030. [i] A_∞ = 0.0.

the rate of formation of phenolate ions in the reaction of ionized phenyl salicylate with methanol at 0.01-M KOH, 0.15 M ionic strength, and 30°C.[8] Similarly, the values of A_{obs} vs. t as shown in Table 7.2 are obtained for the aqueous cleavage of phthalimide in the presence of constant total buffer concentration of trimethylamine.[9] The rates of reactions have been monitored for the reaction period of 50 half-lives and 20 half-lives for reactions described by observed data (A_{obs} vs. t) in Table 7.1 and Table 7.2, respectively. The use of either Equation 7.27 or Equation 7.28 yields a reliable fit of observed data with reasonable and plausible values of A_{∞} only for the reaction period of < 10 half-lives. The observed data in Table 7.1 and Table 7.2 could not fit to Equation 7.28 and Equation 7.27, respectively, when the observed data at > 20 half-lives are included in the data analysis.

The hydrolysis and alkylaminolysis products of phthalimide are phthalamic acid and N-substituted phthalamide, respectively, and these products show an insignificant absorption at 300 nm. The values of molar extinction coefficients of phthalamic acid and phthalic acids are same as 40 M^{-1} cm^{-1} at 300 nm.[10] Thus, the calculated value of A_{∞} for the reaction mixtures containing 0.50-M trimethylamine at pH 10.95 ($A_{\infty} \approx 0.0$, Table 7.2) is plausible in view of insignificant absorption of phthalamic acid at 300 nm.

The observed kinetic data summarized in Table 7.1 and Table 7.2 follow the simple first-order rate law for less than nine half-lives of the reactions. However, the observed kinetic data for the reaction of methylamine with phthalimide at 0.08-M CH$_3$NH$_2$ buffer of pH 10.93 do not follow strictly a first-order rate law if the data analysis includes all the observed data points within observed t range 20 to 2220 sec, as is evident from the A_{calcd} values in Table 7.3. But, the rate of methylaminolysis of phthalimide follows the first-order rate law for the reaction period of < ~ 10 half-lives, as is evident from the data analysis shown in Table 7.3. Similarly, the rate of 2-hydroxyethylaminolysis of phthalimide at 0.20-M 2-hydroxyethylamine buffer of pH 9.84 follows the first-order rate law for the reaction period of < ~ 3.5 half-lives. The chemical reasons for the deviation of observed data points from the first-order rate law in these reactions are described in detail elsewhere.[11]

In case the value of A_{∞} cannot be obtained with certainty, the classical Guggenheim method[12] can be used to overcome this problem. But the Guggenheim method also suffers from a serious limitation, i.e., it must be certain that the reaction obeys the first-order rate law until t = ∞ (i.e., the reaction time at which $A_{obs} = A_{\infty}$). Thus, this method may yield an erroneous value of k if it is used for the reactions of primary amines, such as methylamine, with phthalimide. However, the observed kinetic data for such reactions may be easily and relatively accurately analyzed by the iterative method (Table 7.3) rather than other conventional approaches and the Guggenheim method.

7.2.7.1.2 The Nonlinear Least-Squares Regression Analysis of Kinetic Data

The exact solution of a nonlinear kinetic equation, such as Equation 7.24 and Equation 7.26, cannot be obtained by any conventional or nonconventional

TABLE 7.3
The Values of Observed Absorbance (A_{obs}) at 300 nm as a Function of Reaction Time (t) of a Reaction Mixture Containing Phthalimide (PTH) and 0.08-M Methylamine Buffer of pH 10.93

t/sec	A_{obs}	A_{calcd}[a]	A_{calcd}[a]	A_{calcd}[a]	A_{calcd}[a]	A_{calcd}[a]	A_{calcd}[a]	A_{calcd}[a]	A_{calcd}[a]
20	0.570	0.299[b]	0.299[c]	0.307[d]	0.445[e]	0.503[f]	0.534[g]	0.568[h]	0.657[i]
30	0.480	0.297	0.296	0.302	0.402	0.436	0.454	0.474	0.524
35	0.430	0.296	0.294	0.300	0.383	0.408	0.420	0.434	0.471
40	0.400	0.295	0.293	0.299	0.366	0.382	0.390	0.400	0.425
45	0.370	0.293	-0.292	0.297	0.349	0.358	0.364	0.370	0.386
50	0.345	0.292	-0.290	0.295	0.334	0.337	0.340	0.343	0.352
60	0.300	0.290	-0.287	0.291	0.307	0.300	0.299	0.298	0.297
70	0.270	0.288	-0.285	0.217	0.283	0.271	0.267	0.264	0.257
80	0.240	0.286	-0.282	0.283	0.262	0.246	0.241	0.236	0.227
90	0.220	0.283	-0.279	0.280	0.244	0.226	0.220	0.215	0.204
100	0.200	0.281	-0.277	0.276	0.229	0.210	0.204	0.199	0.188
180	0.145	0.264	-0.257	0.249	0.161	0.152	0.150	0.148	0.144
300	0.140	0.241	-0.230	0.216	0.136	0.139	0.140	0.140	0.140
600	0.130	0.191	-0.177	0.157					
1080	0.120	0.133	-0.123	0.108					
2220	0.076	0.059	-0.069	0.075					
10^3k/sec^{-1} =		0.806 ± 0.178	1.16 ± 0.21	1.73 ± 0.21	14.4 ± 1.3	20.2 ± 1.0	22.5 ± 0.8	24.8 ± 0.5	29.7 ± 0.6
$-10\ln(A_0 - A_\infty)$ =		12.2 ± 1.15	13.7 ± 1.32	14.1 ± 1.	8.67 ± 1.47	6.02 ± 1.15	4.79 ± 0.87	3.55 ± 0.56	0.65 ± 0.67
$10^3 \Sigma d_i^2$ =		2249	2995	3248	1346	820.7	470.1	196.0	280.1

Note: Data analysis with Equation 7.27. Reaction condition: $[R_0] = 3.2 \times 10^{-4}$ M where R = phthalimide, ionic strength = 1 M, H_2O = 98.4% v/v, CH_3CN = 1.6% v/v, and 30°C.

[a] Calculated from Equation 7.27 with kinetic parameters listed in Table 7.3. [b] $A_\infty = 0.010$. [c] $A_\infty = 0.050$. [d] $A_\infty = 0.070$. [e] $A_\infty = 0.130$. [f] $A_\infty = 0.139$. [g] $A_\infty = 0.139$. [h] $A_\infty = 0.1395$. [i] $A_\infty = 0.1399$.

approach. Such equations can be solved only by approximation techniques, and among such techniques, the nonlinear least-squares technique is the most commonly used to calculate unknown kinetic parameters such as rate constants. But the nonlinear least-squares technique cannot be easily used without the aid of a computer and an appropriate program. In the absence of these facilities, one has no choice except to use a linearized form of such a nonlinear equation. However, the process of linearization of a nonlinear equation adds unavoidable uncertainty to the calculated kinetic parameters. This complication generally arises because of the shift in the statistical importance of observed data from more reliable to less reliable with change in data analysis from a nonlinear to a linearized form of the nonlinear equation. For example, in the nonlinear least-squares data analysis using Equation 7.24, the statistical importance of A_{obs} decreases as the value of A_{obs} decreases with the increase in reaction time t, whereas the reverse is true in the linear least-squares data analysis using Equation 7.27 because absolute value of $\ln (A_{obs} - A_\infty)$ increases as $(A_{obs} - A_\infty)$ decreases with increasing t.

In order to compare the relative reliability of the observed data fit to the nonlinear kinetic equations, Equation 7.24 and Equation 7.25, and their respective linearized forms — Equation 7.27 and Equation 7.28 — the observed data summarized in Table 7.1, Table 7.2, and Table 7.3 have been used to calculate kinetic parameters, k, δ_{app}, and A_0 or A_∞ using the nonlinear least-squares technique. A computer program in BASIC for the nonlinear least-squares data analysis is given in Appendix B. The calculated kinetic parameters (k, δ_{app}, and A_0 or A_∞) using these observed data (A_{obs} vs. t) under reaction time range ($t_1 - t_q$ with subscripts 1 and q representing 1st and q-th reaction time in a kinetic run) of different percentages of the completion of the reaction are summarized in Table 7.4 to Table 7.6. A critical look at the calculated kinetic parameters and the values of calculated absorbance (A_{calcd}) shows that the reliability of the data fit to Equation 7.24 and Equation 7.27 or to Equation 7.25 and Equation 7.28 is almost the same if the observed data at ≤ 10 half-lives are considered for data analysis. But the observed data at ≤ 20 to 50 half-lives could not yield a reliable and acceptable fit to Equation 7.27 and Equation 7.28 (Table 7.1 and Table 7.2) whereas they fit to Equation 7.24 and Equation 7.25 (Table 7.4 and Table 7.5), respectively, with almost the same precision as the observed data at ≤ 10 half-lives. Furthermore, it is evident from Table 7.3 and Table 7.6 that the best value of A_∞ is 0.1395 to 0.1394 for the data treatment with Equation 7.27 at $t \leq 10.7$ half-lives. But the values of k are appreciably sensitive to the selected value of A_∞, which is evident from the change in k from 22.5×10^{-3} to 29.7×10^{-3} sec^{-1} with change in A_∞ from 0.1390 to 0.1399. Such sensitivity of k to A_∞ is less evident for the observed data in Table 7.2, where data analysis includes A_{obs} values at $t \leq 3.4$ half-lives.

The linear least-squares and nonlinear least-squares data analysis of the same set of observed data within the half-life range of ~ 0.6 to 8.6 as shown in Table 7.1 and Table 7.4 show that the best values of A_∞ are 0.882 and 0.881, respectively. The calculated values of k at different selected values of A_∞ in Table 7.1 show that an insignificant change in A_∞ (from 0.8805 to 0.8830) causes a significant

TABLE 7.4

The Values of Observed Absorbance (A_{obs}) at 280 nm as a Function of Reaction Time (t) for Methanolysis of Ionized Phenyl Salicylate (PS⁻)

t/sec	A_{obs}	$A_{calcd}{}^a$	$A_{calcd}{}^a$	$A_{calcd}{}^a$	$A_{calcd}{}^a$	$A_{calcd}{}^a$	$A_{calcd}{}^a$
25	0.399	0.400	0.398	0.397	0.398	0.399	0.399
35	0.470	0.472	0.472	0.472	0.472	0.471	0.471
40	0.505	0.504	0.505	0.505	0.504	0.504	0.504
50	0.561	0.561	0.562	0.563	0.562	0.561	0.561
60	0.610	0.610	0.611	0.612	0.611	0.610	
70	0.655	0.651	0.652	0.653	0.653		
80	0.690	0.686	0.687	0.687	0.689		
90	0.718	0.716	0.717	0.717	0.719		
100	0.741	0.741	0.742	0.741			
110	0.760	0.763	0.763	0.762			
120	0.778	0.782	0.781	0.779			
180	0.844	0.847	0.844				
240	0.869	0.872	0.868				
360	0.880	0.885	0.880				
2160	0.897	0.887					
$10^3 \ k/sec^{-1}$ =		16.1 ± 0.3	16.5 ± 0.2	17.3 ± 0.4	15.9 ± 0.6	15.8 ± 1.6	15.3 ± 4.0
$\delta_{app}/M^{-1} \ cm^{-1}$ =		2277 ± 22	2282 ± 11	2280 ± 12	2319 ± 20	2310 ± 83	2344 ± 275
A_0 =		0.158 ± 0.008	0.151 ± 0.004	0.141 ± 0.007	0.155 ± 0.007	0.158 ± 0.012	0.153 ± 0.025

Note: Data analysis with Equation 7.25.

Reaction condition is described in Table 7.1.

[a] Calculated from Equation 7.25 with kinetic parameters listed in Table 7.4.

TABLE 7.5
The Values of Observed Absorbance (A_{obs}) at 300 nm as a Function of Reaction Time (t) of a Reaction Mixture Containing Phthalimide (PTH) and 0.50-M Trimethylamine Buffer of pH 10.95

t/sec	A_{obs}	A_{calcd}[a]	A_{calcd}[a]	A_{calcd}[a]	A_{calcd}[a]	A_{calcd}[a]	A_{calcd}[a]	A_{calcd}[a]
30	0.616	0.615	0.615	0.615	0.615	0.615	0.615	0.616
50	0.599	0.599	0.599	0.599	0.599	0.599	0.599	0.599
60	0.591	0.591	0.591	0.591	0.591	0.591	0.591	0.591
90	0.568	0.568	0.568	0.568	0.568	0.568	0.568	0.568
120	0.547	0.546	0.546	0.546	0.546	0.546	0.546	0.546
180	0.504	0.505	0.505	0.505	0.505	0.505	0.505	0.504
210	0.485	0.485	0.485	0.485	0.485	0.485	0.485	0.485
240	0.467	0.466	0.466	0.466	0.466	0.466	0.466	0.467
300	0.431	0.431	0.431	0.431	0.431	0.431	0.431	
420	0.368	0.368	0.368	0.368	0.368	0.368		
600	0.290	0.290	0.290	0.290	0.290			
1140	0.143	0.143	0.143	0.143				
1800	0.060	0.060	0.060					
10500	0.001	0.001	0.000					
$10^3\,k/sec^{-1}$ =		1.32 ± 0.00	1.32 ± 0.00	1.33 ± 0.01	1.32 ± 0.02	1.35 ± 0.04	1.40 ± 0.09	1.51 ± 0.16
$\delta_{app}/M^{-1}\,cm^{-1}$ =		1997 ± 1	1997 ± 2	1998 ± 5	1970 ± 20	1906 ± 45	1791 ± 99	1994 ± 152
A_{∞} =		0.001 ± 0.001	0.001 ± 0.001	0.001 ± 0.002	0.031 ± 0.007	0.068 ± 0.015	0.002 ± 0.032	0.010 ± 0.049

Note: Data analysis with Equation 7.24. Reaction condition is described in Table 7.2.

[a] Calculated from Equation 7.24 with kinetic parameters listed in Table 7.5.

TABLE 7.6
The Values of Observed Absorbance (A_{obs}) at 300 nm as a Function of Reaction Time (t) of a Reaction Mixture Containing 3.2×10^{-4} M Phthalimide (PTH) and 0.08-M Methylamine Buffer of pH 10.93

t/sec	A_{obs}	A_{calcd}^{a}	A_{calcd}^{a}	A_{calcd}^{a}	A_{calcd}^{a}	A_{calcd}^{a}	A_{calcd}^{a}	A_{calcd}^{a}
20	0.570	0.563	0.566	0.567	0.568	0.570	0.571	0.571
30	0.480	0.475	0.475	0.475	0.475	0.475	0.474	0.474
35	0.430	0.438	0.437	0.437	0.437	0.436	0.435	0.435
40	0.400	0.405	0.403	0.403	0.402	0.401	0.400	0.4000
45	0.370	0.375	0.373	0.373	0.372	0.371	0.370	0.370
50	0.345	0.348	0.347	0.346	0.345	0.344	0.344	0.344
60	0.300	0.303	0.302	0.301	0.301	0.300	0.301	
70	0.270	0.267	0.266	0.266	0.266	0.266	0.269	
80	0.240	0.238	0.238	0.238	0.238	0.239		
90	0.220	0.215	0.216	0.216	0.217	0.219		
100	0.200	0.196	0.198	0.199	0.200	0.203		
180	0.145	0.133	0.142	0.145	0.148			
300	0.140	0.121	0.132	0.137	0.140			
600	0.130	0.120	0.132	0.136				
1080	0.120	0.120	0.132					
2220	0.076	0.120						
10^3 k/sec^{-1} =		22.0 ± 1.3	23.5 ± 0.5	24.0 ± 0.4	24.4 ± 0.4	25.5 ± 0.9	27.5 ± 1.8	26.8 ± 5.2
δ_{app}/M^{-1} cm^{-1} =		2151 ± 92	2169 ± 37	2177 ± 28	2183 ± 24	2195 ± 27	2192 ± 32	2197 ± 75
A_{∞} =		0.120 ± 0.070	0.132 ± 0.003	0.136 ± 0.002	0.1394 ± 0.0025	0.148 ± 0.006	0.166 ± 0.015	0.159 ± 0.056

Note: Data analysis with Equation 7.24. Reaction condition is described in Table 7.3.

a Calculated from Equation 7.24 with kinetic parameters listed in Table 7.6.

change in k (from 19.7×10^{-3} to 15.5×10^{-3} sec^{-1}) and Σd_i^2. Similarly, it is also evident from Table 7.3 and Table 7.6 that the best values of A_∞ are 0.1395 and 0.1394. But the values of k are appreciably sensitive to the values of A_∞ in the data treatment with Equation 7.27, which is evident from the change in k from 22.5×10^{-3} to 29.7×10^{-3} sec^{-1} with a change in A_∞ from 0.1390 to 0.1399.

The effects of the use of A_{obs} values within different reaction time ranges $t_{0.6}$ to t_n, where subscripts 0.6 and n stand for t at 0.6 and n half-lives of the reaction, respectively, on the nonlinear least-squares calculated kinetic parameters using Equation 7.25, show that the values of k, δ_{app}, and A_0 are affected by ±6, ±2, and ±5%, respectively, with n changing from 50 to 1.1 (Table 7.4). However, the calculated values of k, δ_{app}, and A_0 become increasingly unreliable when the observed data cover reaction times for < 1 half-life. The effects of use of A_{obs} values within a different t range ($t_{0.06} - t_n$) on k, δ_{app}, and A_∞, calculated from Equation 7.24 reveal that the variations in these respective kinetic parameters are $1.32 \times 10^{-3} - 1.51 \times 10^{-3}$ sec^{-1}, 1997 to 1791 M^{-1} cm^{-1}, and 0.001 to 0.068 with change in n from 20 to 0.5 (Table 7.5). It is also evident from Table 7.4 and Table 7.5 that the values of kinetic parameters are almost unaffected by change in n from 50 to 1.4 (Table 7.4) and 20 to 1.1 (Table 7.5). The observed data shown in Table 7.6 are certain to obey the first-order rate law at reaction time t for ≤ 10 half-lives. However, the nonlinear least-squares treatment of the observed data with Equation 7.24 reveal that the change in n from 70 to 1.9 with the first data point at $t_{0.7}$ causes variation in k, δ_{app}, and A_∞ from 22.0×10^{-3} to 26.8×10^{-3} sec^{-1}, 2151 to 2197 M^{-1} cm^1, and 0.120 to 0.159, respectively (Table 7.6).

If the first-order or pseudo-first-order rate constant for a reaction under a specific reaction condition is 0.035 sec^{-1}, then such a reaction is completed almost 50% within 20 sec — an average minimum time to obtain first data point on the A_{obs} vs. t plot in a conventional UV-visible spectrophotometric technique. In order to test reliable and satisfactory fit of observed kinetic data for such reactions to Equation 7.24, the experimentally determined kinetic data (A_{obs} vs. t) for a typical kinetic run on the nucleophilic substitution reaction of pyrrolidine with phenyl salicylate are shown in Table 7.7 in which the rate of reaction remains strictly first order for the reaction period of 53 half-lives.[13] The observed kinetic data in Table 7.7 have been treated with Equation 7.24 using the nonlinear least-squares technique. The kinetic parameters (k, δ_{app}, and A_∞) calculated using different reaction time ranges (t_n to t_{53}, where subscripts n and 53 represent half-lives of the reaction and n < 53) with n = 0.8, 1.8, 2.4, and 3.6, are summarized in Table 7.7. The extent of reliability of the data fit to Equation 7.24 is evident from the values of A_{calcd} and standard deviations associated with the calculated kinetic parameters (Table 7.7). A similar calculation has been carried out on kinetic data of Table 7.1 using Equation 7.25, and the results obtained are summarized in Table 7.8.

The observed data summarized in Table 7.7 cover the reaction period of 0.8 to 52.5 half-lives. The kinetic parameters (k, δ_{app}, and A_∞) calculated from Equation 7.24 are almost unchanged with change in the selected t range of the observed data. The selected t ranges for the data treatment with Equation 7.24 are $t_{0.8}$–t_{53},

TABLE 7.7

The Values of Observed Absorbance (A_{obs}) at 350 nm as a Function of Reaction Time (t) of a Reaction Mixture Containing Phenyl Salicylate and 0.048-M Pyrrolidine at pH 11.92

t/sec	A_{obs}	A_{calcd}[a]	A_{calcd}[a]	A_{calcd}[a]	A_{calcd}[a]
20	0.571	0.572			
30	0.439	0.439			
35	0.385	0.386			
40	0.340	0.339			
45	0.300	0.298	0.301		
50	0.264	0.262	0.264		
55	0.233	0.231	0.232		
60	0.204	0.204	0.204	0.204	
65	0.180	0.181	0.180	0.180	
70	0.160	0.160	0.159	0.159	
80	0.122	0.127	0.125	0.125	
90	0.100	0.101	0.100	0.099	0.100
120	0.056	0.056	0.055	0.055	0.056
300	0.023	0.022	0.023	0.023	0.023
1320	0.023	0.023	0.023	0.023	0.023
10^3 k/sec^{-1} =		27.6 ± 0.2	28.6 ± 0.3	28.9 ± 0.6	28.2 ± 0.2
δ_{app}/M^{-1} cm^{-1} =		5977 ± 30	6300 ± 86	6403 ± 225	6093 ± 97
A_∞ =		0.022 ± 0.001	0.023 ± 0.001	0.023 ± 0.001	0.023 ± 0.000

Note: Data analysis with Equation 7.24.

Reaction condition: $[R_0] = 1.6 \times 10^{-4}$ M where R = phenyl salicylate, ionic strength = 1 M, H_2O = 99.2% v/v, CH_3CN = 0.8% v/v, and 30°C.

[a] Calculated from Equation 7.24 with kinetic parameters listed in Table 7.7.

$t_{1.9}$–t_{53}, $t_{2.5}$–t_{53}, and $t_{3.7}$–t_{53} where subscripts stand for half-lives of the reaction (Table 7.7). It is noteworthy that if the observed data are obtained only after nearly 90% completion of the reaction, where data follow Equation 7.24, the nonlinear least-squares treatment can yield reliable values of the kinetic parameters. The effects of the use of the observed data covering different half-lives (n), with final observed data point at 50 half-lives, on the kinetic parameters calculated from Equation 7.25 are described by the observed data shown in Table 7.1. The calculated values of k, δ_{app}, and A_0 vary from 16.1×10^{-3} to 13.5×10^{-3} sec^{-1}, 2277 to 1774 M^{-1} cm^{-1} and 0.158 to 0.324, respectively, with variation in the selected data coverage of 0.6 to 50 and 2.3 to 50 half-lives (Table 7.8). Although the variations in the kinetic parameters are small, they do show a regular variation which could be due to mutual error cancellation effects of the three kinetic parameters leaving residual errors (= $A_{obs\,i}$ – $A_{calcd\,i}$) very small (<1%). When the data analysis of observed data covering 2.3 to 50 half-lives also included the first

TABLE 7.8
The Values of Observed Absorbance (A_{obs}) at 280 nm as a Function of Reaction Time (t) for Methanolysis of Ionized Phenyl Salicylate (PS$^-$)

t/sec	A_{obs}	A_{calcd}^a	A_{calcd}^a	A_{calcd}^a	A_{calcd}^a	A_{calcd}^a	A_{calcd}^a
25	0.399	0.400					0.399
35	0.470	0.472					
40	0.505	0.504	0.507				
50	0.561	0.561	0.563				
60	0.610	0.610	0.610				
70	0.655	0.651	0.651	0.657			
80	0.690	0.686	0.686	0.689			
90	0.718	0.716	0.716	0.716			
100	0.741	0.741	0.741	0.740	0.741		
110	0.760	0.763	0.762	0.760	0.761		
120	0.778	0.782	0.781	0.778	0.778	0.779	0.776
180	0.844	0.847	0.846	0.843	0.842	0.841	0.844
240	0.869	0.872	0.871	0.871	0.870	0.869	0.871
360	0.880	0.885	0.885	0.887	0.887	0.887	0.886
2160	0.897	0.887	0.888	0.890	0.891	0.892	0.889
10^3 k/sec^{-1}	=	16.1 ± 0.3	15.9 ± 0.4	14.6 ± 0.6	13.9 ± 1.3	13.5 ± 2.3	15.4 ± 0.6
δ_{app}/M^{-1} cm^1	=	2277 ± 22	2248 ± 41	2026 ± 93	1872 ± 243	1774 ± 494	2253 ± 43
A_0	=	0.158 ± 0.008	0.168 ± 0.014	0.242 ± 0.031	0.292 ± 0.080	0.324 ± 0.161	0.168 ± 0.015

Note: Data analysis with Equation 7.25. Reaction condition is described in Table 7.1.

^{a} Calculated from Equation 7.25 with kinetic parameters listed in Table 7.8.

data point (A_{obs} at t = 25 sec, i.e., at 0.6 half-life), the values of k, δ_{app}, and A_0 change from 13.5×10^{-3} to 15.4×10^{-3} sec^{-1}, 1774 to 2253 M^{-1} cm^{-1}, and 0.324 to 0.168, respectively (Table 7.8). Thus, the data analysis with Equation 7.25 requires the inclusion of at least one data point obtained at < 1 half-life.

The following inferences may be extracted from the values of A_{obs}, A_{calcd}, and kinetic parameters (k, δ_{app}, and A_0 or A_∞) listed in Table 7.1 to Table 7.8:

1. In the use of linear Equation 7.27 or Equation 7.28, the observed data should be obtained at ≤ 10 half-lives. If the observed data cover ~ 10 half-lives, the fit of these data to Equation 7.27 or Equation 7.28 becomes highly sensitive to the value of A_∞ (a change of ~ 0.3% in the best value of A_∞ causes an appreciable change in k). However, if the observed data cover ≤ ~ 3.4 half-lives, a change of nearly 3% in the best value of A_∞ causes only ~ 25% change in k.

2. For the observed data covering ≤ ~ 10 half-lives, the iterative method provides the best value of A_∞ for the data treatment with Equation 7.27 or Equation 7.28, which is almost similar to the one obtained by the data treatment with Equation 7.24 or Equation 7.25, where A_∞ or A_0 is considered to be an unknown parameter.

3. The nonlinear least-squares technique can be safely used in the data treatment with Equation 7.24 or Equation 7.25 using observed data obtained until 10 to 70 half-lives. The precision of the data fit to these nonlinear equations remains essentially unchanged with a change of observed data coverage from 10 to 70 half-lives.

4. The effect of the observed data coverage from 0.1 to 1.1 and 0.1 to 50 half-lives has essentially no effect on the precision of the data fit to Equation 7.24 or Equation 7.25. In general, the reliable nonlinear least-squares treatment of observed data with Equation 7.25 requires that the observed data should contain at least one data point at < 0.7 half-life and one at > 1.5 half-lives.

5. The effect of the observed data coverage from 0.8 to 53 and 3.7 to 53 half-lives gives almost the same precision of the data fit to Equation 7.24.

7.2.7.2 Irreversible First-Order Consecutive Reactions

A multistep irreversible first-order reaction, as expressed by Equation 7.29, is also called an

$$R \xrightarrow{k_1} I_1 \xrightarrow{k_2} I_2 \xrightarrow{k_3} P \qquad (7.29)$$

irreversible first-order consecutive reaction if the values of k_1/k_2 and k_2/k_3 are not very different from those in the range 0.5 to 2.0, which is a purely arbitrary range. These values of k_1/k_2 and k_2/k_3 show that the differences in ground state stability of R and I_1, I_1 and I_2, as well as I_2 and P are not very large and, consequently, I_1 and I_2 cannot be considered to be highly reactive intermediates on the reaction path. Thus, one cannot apply the steady state approximation to the rates of change of concentration of I_1 and I_2 during the course of the reaction. In view of Equation 7.29, the rate laws for the rate of change of concentration of R, I_1, I_2, and P may be expressed as

$$d[R]/dt = -k_1[R] \tag{7.30}$$

$$d[I_1]/dt = k_1[R] - k_2[I_1] \tag{7.31}$$

$$d[I_2]/dt = k_2[I_1] - k_3[I_2] \tag{7.32}$$

$$d[P]/dt = k_3[I_2] \tag{7.33}$$

The integration of the simple first-order rate law, Equation 7.30, gives Equation 7.34.

$$[R] = [R_0]\exp(-k_1 t) \tag{7.34}$$

where $[R] = [R_0]$ at $t = 0$. There are different approaches for solving differential equations such as Equation 7.31 and Equation 7.32,[14] but, perhaps the simplest approach, under the conditions where $k_1 \neq k_2 \neq k_3$, may be described as follows. Equation 7.31 may be rearranged to give Equation 7.35

$$\exp(k_2 t)\{(d[I_1]/dt) + k_2[I_1]\} = \exp(k_2 t)k_1[R] \tag{7.35}$$

where $\exp(k_2 t)$ is obtained from the integration factor $\exp(\int k_2 dt)$. Equation 7.35 may be rewritten as

$$d\{[I_1]\exp(k_2 t)\}/dt = k_1[R_0]\exp((k_2 - k_1)t)$$

or

$$[I_1]\exp(k_2 t) = k_1[R_0]\int \exp((k_2 - k_1)t)$$
$$= \{k_1[R_0]/(k_2 - k_1)\}\exp((k_2 - k_1)t) + IC \tag{7.36}$$

where IC represents the integration constant. The application of boundary condition $[I_1] = 0$ at $t = 0$, gives $IC = -k_1[R_0]/(k_2 - k_1)$, which when introduced into Equation 7.36, gives Equation 7.37.

$$[I_1] = \{k_1[R_0]/(k_2 - k_1)\}\{\exp(-k_1 t) - \exp(-k_2 t)\} \tag{7.37}$$

Similarly, Equation 7.32 with integration factor $\exp(\int k_3 dt) = \exp(k_3 t)$ and boundary condition $[I_2] = 0$ at $t = 0$ gives Equation 7.38.

$$[I_2] = \frac{k_2 k_1 [R_0]}{(k_2 - k_1)(k_3 - k_1)(k_3 - k_2)} \tag{7.38}$$

$$[(k_3 - k_2)\exp(-k_1 t) - (k_3 - k_1)\exp(-k_2 t) + (k_2 - k_1)\exp(-k_3 t)]$$

Equation 7.30 to Equation 7.33 and boundary conditions $[R] = [R_0]$ at $t = 0$ and $[I_1] = [I_2] = [P] = 0$ at $t = 0$ can lead to Equation 7.39.

$$[P] = [R_0] - [R] - [I_1] - [I_2] \tag{7.39}$$

Under the typical reaction condition where $k_1 = k_2 = k_3 = k$, the integration of Equation 7.31 and Equation 7.32 gives Equation 7.40 and Equation 7.41, respectively (Hint: $\int x e^{ax} dx = (ax - 1)e^{ax}/a^2$).

$$[I_1] = kt[R_0] \exp(- kt) \tag{7.40}$$

$$[I_2] = [R_0] \exp(- kt)[1 - (1 + kt) \exp(- kt)] \tag{7.41}$$

The kinetic equations such as Equation 7.37, Equation 7.38, Equation 7.40, and Equation 7.41 are known as *transcendental equations*, whose direct solution cannot be obtained. Such a kinetic equation is generally solved by the use of approximation techniques such as Newton–Raphson iterative method[15] and non-linear least-squares method. But, these methods have limitations of a different nature. For instance, the nonlinear least-squares method, which is most commonly used in such kinetic studies, tends to provide less reliable values of calculated kinetic parameters with increase in the number of such parameters.

To illustrate, a chemical example in which the rate of a reaction follows an irreversible two-step first-order consecutive reaction path is the spectrophotometric kinetic study on the aqueous cleavage of N-methylphthalamic acid (R) in mixed H_2O–CH_3CN solvent.[16] The rate of hydrolysis of R at 0.03-M HCl has been carried out spectrophotometrically by monitoring the change in absorbance (A_{obs}) at 310 nm with a change in the reaction time (t). The hydrolysis in mixed acidic water–acetonitrile solvent follows the reaction scheme as expressed by Equation 7.42

$$B \xleftarrow{k_1} R \underset{-E}{\xrightarrow{k_2}} C \xrightarrow{k_3} D \tag{7.42}$$

where B, C, D, and E represent N-methylphthalimide, phthalic anhydride, phthalic acid, and $MeNH_2$, respectively. It is evident from Equation 7.42 that under the boundary conditions $t = 0$, $[R] = [R_0]$, and $[B] = [C] = [D] = [E] = 0$,

$$[D] = [R_0] - [R] - [B] - [C] \tag{7.43a}$$

$$[E] = [R_0] - [R] - [B] \tag{7.43b}$$

where [D], [R], [B], [C], and [E] represent the concentrations of the respective D, R, B, C, and E at any reaction time t. The value of A_{obs} of the reaction mixture at 310 nm may be given as

$$
\begin{aligned}
A_{obs} &= \delta_B[B] + \delta_R[R] + \delta_C[C] + \delta_D[D] + \delta_E[E] \\
&= (\delta_B - \delta_D - \delta_E)[B] + (\delta_R - \delta_D - \delta_E)[R] + (\delta_C - \delta_D)[C] + (\delta_D + \delta_E)[R_0] \\
&= \frac{(\delta_B - \delta_D - \delta_E)[R_0]k_1}{k_1 + k_2}[1 - \exp(-(k_1 + k_2)t)] + (\delta_R - \delta_D - \delta_E) \times \\
&\quad [R_0]\exp(-(k_1 + k_2)t)] \\
&\quad + \frac{(\delta_C - \delta_D)[R_0]k_2}{k_3 - (k_1 + k_2)}[\exp(-(k_1 + k_2)t) - \exp(-(k_3 t)] + (\delta_D + \delta_E)[R_0] \\
&= \left\{ \frac{(\delta_B - \delta_D - \delta_E)[R_0]k_1}{k_1 + k_2} - (\delta_R - \delta_D - \delta_E)[R_0] \right\}[1 - \exp(-(k_1 + k_2)t)] + \\
&\quad (\delta_R - \delta_D - \delta_E)[R_0] \\
&\quad + \frac{(\delta_C - \delta_D)[R_0]k_2}{k_3 - (k_1 + k_2)}[\exp(-(k_1 + k_2)t) - \exp(-(k_3 t)] + (\delta_D + \delta_E)[R_0]
\end{aligned}
$$

or

$$
\begin{aligned}
A_{obs} &= [R_0]\left\{ \frac{E_1 k_1}{k_1 + k_2} + \delta_D + \delta_E \right\}[1 - \exp(-(k_1 + k_2)t)] \\
&\quad + \frac{[R_0]E_2 k_2}{k_3 - (k_1 + k_2)}[\exp(-(k_1 + k_2)t) - \exp(-(k_3 t)] + \tag{7.44} \\
&\quad A_0 \exp(-(k_1 + k_2)t)
\end{aligned}
$$

where $E_1 = (\delta_B - \delta_D - \delta_E)$, $E_2 = (\delta_C - \delta_D)$, and $A_0 = \delta_R[R_0]$.

Under the typical reaction conditions where [C] is insignificant compared with [R] + [B] + [D] + [E] during the course of the reaction (i.e., when $k_3 \gg k_2$), Equation 7.44 reduces to Equation 7.45

$$A_{obs} = [R_0]E_1'[1 - \exp(-k_{obs}t)] + A_0' \tag{7.45}$$

where $k_{obs} = k_1 + k_2$, $E_1' = E_1 k_1/(k_1 + k_2)$, and $A_0' = [R_0]\{ \delta_D + \delta_E + (\delta_R - \delta_D - \delta_E) \exp(-(k_1 + k_2)t\} \approx [R_0]\delta_R \approx A_0$, if $\delta_R \approx \delta_D + \delta_E$.

The observed data, obtained at an acetonitrile content of $\leq 50\%$ v/v where $k_3/k_2 > 30$, fit to Equation 7.45 where k_{obs}, E_1', and A_0' have been considered to be unknown parameters in the data analysis and the derived values of k_1 and k_2 change from 2.18×10^{-6} to 1.10×10^{-6} sec^{-1} and 0.82×10^{-5} to 2.86×10^{-5} sec^{-1}, respectively, with change in the acetonitrile content from 2 to 50% v/v.[16] But at least in acetonitrile content of $\geq 60\%$, the observed data could not fit to Equation 7.45 because under such conditions, [C] cannot be ignored compared with [R] + [B] + [D] and, consequently, the observed data have been found to fit to Equation 7.44 where k_1, k_2, and A_0 have been treated as unknown parameters in the data analysis. The values of δ_B, δ_C, δ_D, and k_3 have been obtained by using authentic samples of B, C, and D. The calculated values of k_1 and k_2 vary from 1.08×10^{-6} to 1.93×10^{-6} sec^{-1} and 3.48×10^{-5} to 4.06×10^{-5} sec^{-1}, respectively, whereas A_0 values remain insignificant (<0.010) with change in the acetonitrile content from 60 to 80% v/v.[16]

Perhaps it is noteworthy that the plot at 60% v/v CH_3CN in Figure 7.2 shows an apparent monotonic change in A_{obs} with reaction time t. Such plots are generally the characteristic features of reactions involving only one rate-determining step and, consequently, the observed data should fit to a kinetic equation such as Equation 7.45. Therefore, the observed data at 60% v/v CH_3CN have also been

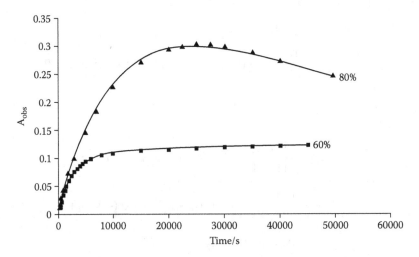

FIGURE 7.2 Plots of absorbance (A_{obs}) at 310 nm against the reaction time (s) for a mixed aqueous solution of N –methylphthalamic acid (4×10^{-4} M) in 60 and 80% v/v CH_3CN, 0.03-M HCl at 30°C. The solid lines are drawn through the calculated data points using Equation 7.44 and kinetic parameters $10^6 k_1 = 1.08 \pm 0.05$ sec^{-1}, $10^5 k_2 = 3.48 \pm 0.03$ sec^{-1}, $A_0 = 0.007 \pm 0.001$, $E_1 = 1042$ M^{-1} cm^{-1}, $E_2 = 254$ M^{-1} cm^{-1}, and $10^5 k_3 = 38.8$ sec^{-1} for 60% CH_3CN and $10^6 k_1 = 2.21 \pm 0.08$ sec^{-1}, $10^5 k_2 = 4.91 \pm 0.24$ sec^{-1}, $A_0 = 0.002 \pm 0.010$, $E_1 = 864$ M^{-1} cm^{-1}, $E_2 = 171$ M^{-1} cm^{-1}, and $10^5 k_3 = 6.1$ sec^{-1} for 80% CH_3CN.

tried to fit to Equation 7.45, and the calculated values of k_{obs}, E_1', and A_0' are $(31.9 \pm 1.0) \times 10^{-5}$ sec^{-1}, 27.4 ± 0.3 M^{-1} cm^{-1}, and 0.009 ± 0.001, respectively. The fit of the observed data to Equation 7.45 may be considered satisfactory in terms of A_{calcd} values as shown in Table 7.9. But the values of k_{obs} and E_1 give k_1 and k_2 as 8.1×10^{-6} sec^{-1} and 31.1×10^{-5} sec^{-1}, respectively, which are abnormally high compared with the corresponding values at 50% v/v CH_3CN (k_1 = 1.10×10^{-6} sec^{-1} and $k_2 = 2.86 \times 10^{-5}$ sec^{-1}) and 70% v/v CH_3CN ($k_1 = 1.08 \times 10^{-6}$ sec^{-1} and $k_2 = 3.29 \times 10^{-5}$ sec^{-1}). However, the treatment of the observed data at 60% v/v CH_3CN with Equation 7.44 gives k_1, k_2, and A_0 as $(1.08 \pm 0.05) \times 10^{-6}$ sec^{-1}, $(3.48 \pm 0.03) \times 10^5$ sec^{-1}, and 0.007 ± 0.000, respectively, which are plausible in view of the corresponding values at 50 and 70% v/v CH_3CN. The values of A_{calcd} (Table 7.9) show also the reliable fit of these observed data to Equation 7.44.

It is also noteworthy that the value of $k_3/k_2 \approx 10$ at 60% v/v CH_3CN and, consequently, one might tend to believe that the rate of reaction should follow a simple first-order rate law at 60% v/v CH_3CN because k_3 is ~ 10-fold larger than k_2. But the practical reality is that although the observed data do follow an apparent first-order rate law, the calculated kinetic parameters are very unreliable, and the observed data fit equally well to a kinetic equation derived for a two-step irreversible first-order consecutive reaction. Thus, it appears that the reliable fit of observed data to a kinetic equation is a necessary but not sufficient requirement for the correctness of the kinetic equation. The reliability of the calculated kinetic parameters from the kinetic equation must also be tested if possible.

The values of A_{obs} vs. t at 80% v/v CH_3CN clearly reveal a maximum (Figure 7.2) which shows the presence of an intermediate (C) on the reaction path. These observed data fit reasonably well to Equation 7.44 as is evident from A_{calcd} values shown in Table 7.9.

7.3 EMPIRICAL KINETIC EQUATIONS

The algebraic relationship between experimentally determined rate constants (k) as a function of factors that affect the reaction rate, such as the concentration of reaction ingredients, including catalysts and temperature, is defined as *empirical kinetic equation*. The validity of an empirical kinetic equation is solely supported by experimental observations and, thus, its authenticity is beyond any doubt as far as a reliable data fit to the empirical equation is concerned. However, the nature and the values of calculated empirical parameters or constants remain obscure until the empirical kinetic equation is justified theoretically or mechanistically. The experimental determination of the empirical kinetic equations is considered to be the most important aspect of the use of kinetic study in the mechanistic diagnosis of the reactions. The classical and perhaps the most important empirical kinetic equation, determined by Hood[3] in 1878, is Equation 7.8.

The rate of a reaction is generally affected by all or some of the following factors: temperature, reaction medium, the concentration of reactants and cata-

TABLE 7.9
The Values of Observed Absorbance (A_{obs}) at 310 nm as a Function of Reaction Time (t) for Hydrolysis of N-Methylphthalamic Acid

CH₃CN/% v/v =	60			80		
t/sec	A_{obs}	A_{calcd}^a	A_{calcd}^b	t/sec	A_{obs}	A_{calcd}^c
110	0.012	0.013	0.011	110	0.018	0.017
195	0.015	0.016	0.014	194	0.021	0.020
389	0.021	0.022	0.021	500	0.031	0.030
500	0.024	0.025	0.025	1000	0.046	0.046
695	0.031	0.031	0.031	2000	0.075	0.075
890	0.037	0.036	0.037	3000	0.102	0.102
1195	0.045	0.044	0.045	5000	0.147	0.148
1500	0.052	0.051	0.052	7000	0.185	0.186
2000	0.062	0.061	0.062	10010	0.227	0.229
2500	0.071	0.069	0.071	15010	0.271	0.273
3000	0.078	0.077	0.078	19988	0.294	0.294
3500	0.083	0.083	0.083	22490	0.300	0.298
4000	0.088	0.088	0.088	24990	0.302	0.299
4500	0.092	0.093	0.092	27495	0.301	0.298
5000	0.096	0.096	0.096	30000	0.298	0.295
6000	0.101	0.102	0.101	35000	0.287	0.286
8000	0.107	0.110	0.107	40000	0.272	0.274
10010	0.110	0.114	0.111	49510	0.246	0.249
15010	0.114	0.118	0.114			
19990	0.116	0.118	0.116			
24990	0.118	0.119	0.118			
30000	0.120	0.119	0.119			
35000	0.121	0.119	0.120			
40000	0.122	0.119	0.125			
45010	0.124	0.119	0.121			

Note: Data analysis with Equation 7.44 and Equation 7.45.

Reaction condition: $[R_0] = 4.0 \times 10^{-3}$ M where R = N-methylphthalamic acid generated from N-methylphthalimide, HCl] = 0.03 M and 30°C.

[a] Calculated from Equation 7.45 with $10^4 k_{obs} = 3.19 \pm 0.10$ sec⁻¹, $E_1' = 27.4 \pm 0.3$ M^{-1} cm⁻¹, and $A_0' = 0.009 \pm 0.001$.

[b] Calculated from Equation 7.44 with $10^6 k_1 = 1.08 \pm 0.05$ sec⁻¹, $10^5 k_2 = 3.48 \pm 0.03$ sec⁻¹, $A_0 = 0.007 \pm 0.001$, $E_1 = 1042$ M^{-1} cm⁻¹, $E_2 = 254$ M^{-1} cm⁻¹, and $10^5 k_3 = 38.8$ sec⁻¹.

[c] Calculated from Equation 7.44 with $10^6 k_1 = 2.21 \pm 0.08$ sec⁻¹, $10^5 k_2 = 4.91 \pm 0.24$ sec⁻¹, $A_0 = 0.002 \pm 0.010$, $E_1 = 864$ M^{-1} cm⁻¹, $E_2 = 171$ M^{-1} cm⁻¹, and $10^5 k_3 = 6.1$ sec⁻¹.

lysts, ionic strength, and specific salt. The effects of reaction medium, ionic strength, and concentrations of catalysts and reactants on reaction rates provide vital information about the fine details of reaction mechanism. Empirical kinetic equations, based on the effects of the concentration of salts or additives, organic cosolvent in mixed aqueous-organic solvents, reactants, and catalysts on the reaction rates, may be described as follows.

7.3.1 EFFECTS OF ORGANIC COSOLVENT IN MIXED AQUEOUS SOLUTION ON THE REACTION RATES

The theoretical and, consequently, quantitative explanation of the effects of mixed aqueous–organic solvents on the rates of reactions is extremely difficult, if not impossible. Highly dynamic structural features of such solvent systems and various possible solute–solvent interactions both at ground state as well as transition state pose a considerable amount of difficulty and uncertainty in achieving even just a reasonable theoretical model, which could describe the effects of mixed aqueous–organic solvents on the reaction rates. Under such circumstances, generally empirical kinetic equations are used, which could provide only a qualitative explanation for the effects of mixed aqueous–organic solvents on the reaction rates. There are several multiparameter empirical kinetic equations, which are used depending upon the nature of both the reaction and the reaction medium.[17] The only danger in using multiparameter empirical kinetic equations is the fact that the reliability of the values of the calculated empirical parameters or constants decreases as the number of empirical parameters or constants increases. It is, therefore, advisable to use an empirical kinetic equation, which contains the least number of empirical parameters or constants.

Pseudo-first-order rate constants (k_{obs}) for alkaline hydrolysis of N-(4′-methoxyphenyl)phthalimide in mixed OS – H_2O solvents fit to Equation 7.46) within OS content range of

$$k_{obs} = k_0 \exp(-\Psi X) \tag{7.46}$$

2 to 50% v/v for OS = CH_3CN and 10 to 80% v/v for OS = 1,4-dioxan. In Equation 7.46, k_0 and Ψ are empirical constants and X represents % v/v content of organic cosolvent (OS). The value of k_0 can be compared with k_{obs} obtained at X = 0. The values of k_{obs} for intramolecular carboxylic group-catalyzed cleavage of N-(4′-methoxyphenyl)phthalamic acid at 0.05 M HCl also fit to Equation 7.46 within OS content range of 10 to 80 for OS = 1,4-dioxan.[18]

Pseudo-first-order rate constants (k_{obs}) for the intramolecular general base-catalyzed nucleophilic substitution reactions of ionized phenyl salicylate with ROH, R = CH_3 [19], CH_3CH_2 [20], $HOCH_2CH_2$ [21], and ROH = glucose [22], in mixed aqueous solvents within the ROH content or concentration range of varying values, have been found to fit to the following empirical equation:

$$k_{obs} = \frac{\alpha[ROH]_T}{1 + \beta[ROH]_T} \qquad (7.47)$$

where $[ROH]_T$ represents total concentration of ROH as well as α and β are empirical constants. Equation 7.47 is applicable even in the presence of inert inorganic and organic salts, but it fails in the presence of cationic micelles.[19,21]

7.3.2 EFFECTS OF [SALTS] OR [ADDITIVES] ON THE REACTION RATES

Effects of the inert inorganic salts on the rate constants (k) for the reactions involving ionic reactants are generally explained in terms of the Debye–Huckel or extended Debye–Huckel theory.[22] In actuality, the extended Debye–Huckel theory involves an empirical term, which makes the theory a semiempirical theory. However, there are many reports in which the effects of salts on k of such ionic reactions cannot be explained by the Debye–Huckel theory.[23,24] For instance, pseudo-first-order rate constants (k_{obs}) for the reaction of HO$^-$ with acetyl salicylate ion (aspirin anion) show a fast increase at low salt concentration followed by a slow increase at high concentration of several salts. But the lowest salt concentration for each salt remains much higher than the limiting concentration (0.01 M for salts such as M$^+$X$^-$) above which the Debye–Huckel theory is no longer valid. These k_{obs} values fit reasonably well to Equation 7.48

$$k_{obs} = \frac{k_0 + \alpha[M_kX_n]}{1 + \beta[M_kX_n]} \qquad (7.48)$$

where $k_0 = k_{obs}$ at $[M_kX_n] = 0$, α and β are empirical constants, and $[M_kX_n]$ represents the total concentration of salt M_kX_n.[25]

Rate-decreasing effects of the concentration of caffeine and theophylline-7-acetate (the additives) have been observed for the alkaline hydrolysis phenyl and substituted phenyl benzoates.[26]

Pseudo-first-order rate constants for the alkaline hydrolysis of these esters in the presence of a series of varying concentrations of caffeine and theophylline-7-acetate obey Equation 7.48 with $[M_kX_n]$ replaced by [caffeine] or [theophylline-7-acetate].

7.3.3 EFFECTS OF [CATALYSTS] AND [REACTANTS] ON THE REACTION RATES

A catalyst changes the rate of a reaction compared to the rate of the same reaction under the same reaction conditions, but in the absence of catalyst. The effect of the catalyst on the reaction rate might be caused by either the catalyst providing a different microreaction environment compared with the bulk reaction medium or the catalyst acting as a reactant in or before the rate-determining step followed by the recovery of the catalyst in a fast reaction step after the rate-determining step. Sometimes, a catalyst might change the rate of a reaction owing to the formation of a reactive or nonreactive molecular complex between the catalyst and reactants through weak molecular interaction forces such as dispersion forces, hydrogen bonding, dipole–dipole, and ion–dipole interaction forces. A catalyst may not necessarily be different from a reactant molecule. Intermolecular general acid and general base catalysis, in the reactions of primary and secondary amines with esters, imides, amides, and related compounds, generally involve two amine molecules, one acting as a nucleophile and the other acting as a catalyst.

7.3.3.1 Effects of [Micelles] on the Reaction Rates

In the very early phase of the kinetic studies on the effects of [micelles] on the reaction rates, it has been observed that an empirical kinetic equation similar to Equation 7.48 with replacement of $[M_kX_n]$ by $[Surf]_T$ – CMC (where $[Surf]_T$ and CMC represent total surfactant concentration and critical micelle concentration, respectively) is applicable in many micellar-mediated reactions.[27–29] But the plots of k_{obs} vs. ($[Surf]_T$ – CMC) for alkaline hydrolysis of some esters reveal maxima when surfactants are cationic in nature.[28,30,31] Similar kinetic plots have also been observed in the hydrolysis of methyl orthobenzoate in the presence of anionic micelles.[32] Bunton and Robinson[33] suggested a semiempirical equation similar to Equation 7.49, which could explain the presence of maxima in the plots of k_{obs} vs. ($[Surf]_T$ – CMC). In Equation 7.49), μ is an empirical constant,

$$k_{obs} = \frac{k_W + k_M K_R [D_n]}{1 + K_R [D_n] + \mu [D_n]^2} \tag{7.49}$$

subscripts W and M refer to the water pseudophase and micellar pseudophase, respectively, $[D_n] = [Surf]_T - CMC$, K_R is the micellar-binding constant of reactant (R), i.e., $K_R = [R_M]/([R_W][D_n])$, and k_W and k_M represent pseudo-first-order rate constants for the cleavage of R in the water pseudophase and micellar pseudophase, respectively. Equation 7.49 is similar to Equation 7.48 with $k_W = k_0$, $k_M K_R = \alpha$, $K_R = \beta$, and $\mu = 0$.

7.3.3.2 Effects of [Inert Salts] on k_{obs} for Ionic Micellar-Mediated Semiionic Reactions

The effects of inert inorganic and organic salts on k_{obs} for alkaline hydrolysis of phenyl benzoate[34] and phthalimide, as evident from observed data for two typical sets of observations shown in Table 7.10 and Table 7.11, have been explained in terms of the semiempirical kinetic equation Equation 7.50, which

TABLE 7.10
Pseudo-First-Order Rate Constants (k_{obs}) for Hydrolysis of Phenyl Benzoate at Different [NaCl] in the Presence of 0.01-M NaOH and 0.007-M CTABr

[NaCl] M	$10^3 k_{obs}$ sec^{-1}	$10^3 k_{calcd}^a$ sec^{-1}	$10^3 k_{calcd}^a$ sec^{-1}	$10^3 k_{calcd}^a$ sec^{-1}	$10^3 k_{calcd}^a$ sec^{-1}	$10^3 k_{calcd}^a$ sec^{-1}	$10^3 k_{calcd}^a$ sec^{-1}
0.0	15.4 ± 0.1[b]						
0.0	15.8 ± 0.1						
0.001	14.1 ± 0.1	13.4	13.9	14.3	13.8	13.8	13.7
0.002	12.4 ± 0.2	12.3	12.5	12.7	12.5	12.4	12.3
0.003	11.3 ± 0.2	11.4	11.5	11.5	11.4	11.3	11.2
0.004	9.19 ± 0.19	10.6	10.6	10.5	10.5	10.4	10.4
0.007	8.84 ± 0.14	8.84	8.68	8.51	8.64	8.59	8.56
0.010	7.94 ± 0.09	7.64	7.44	7.24	7.42	7.42	7.44
0.020	5.84 ± 0.09	5.46	5.27	5.12	5.33	5.43	5.59
0.030	4.75 ± 0.05	4.40	4.27	4.18	4.37	4.53	4.77
0.050	3.69 ± 0.05	3.36	3.32	3.31	3.46	3.69	4.01
0.100	2.45 ± 0.02	2.43	2.50	2.57	2.68	2.98	
0.200	1.67 ± 0.01	1.91	2.04	2.16	2.25		
0.300	1.33 ± 0.01	1.73	1.88	2.02			
$10^3 k_0$/sec^{-1}	=	14.8	15.6	16.4	15.6	15.6	15.6
10^3 F k_{obs}^{salt}/sec^{-1}	=	1.35 ± 0.34[b]	1.55 ± 0.34[b]	1.74 ± 0.37[b]	1.79 ± 0.39[b]	2.17 ± 0.45[b]	2.64 ± 0.55[b]
K/M^{-1}	=	114 ± 11	139 ± 14	167 ± 17	145 ± 15	156 ± 18	170 ± 22
$10^6 \Sigma d_i^2$	=	3.130	3.453	4.343			

Note: Reaction condition: [Phenyl benzoate]$_0$ = 2 × 10^{-4} M, 35°C, the aqueous reaction mixture in each kinetic run contains 2% v/v CH$_3$CN and CTABr represents cetyltrimethylammonium bromide.

[a] Calculated from Equation 7.50 with empirical kinetic parameters, F k_{obs}^{salt} and K, listed in Table 7.10. [b] Error limits are standard deviations.

TABLE 7.11
Pseudo-First-Order Rate Constants (k_{obs}) for Hydrolysis of Phthalimide at Different [M_kX_n] in the Presence of 0.02-M KOH and 0.006-M CTABr

M_kX_n = [M_kX_n] / M	(CH$_3$)$_4$NCl			[M_kX_n] / M	(CH$_3$)$_4$NBr		
	$10^4\,k_{obs}$ sec^{-1}	$10^4\,k_{calcd}^{a}$ sec^{-1}	$10^4\,k_{obs}^{salt,b}$ sec^{-1}		$10^4\,k_{obs}$ sec^{-1}	$10^4\,k_{calcd}^{b}$ sec^{-1}	$10^4\,k_{obs}^{salt,b}$ sec^{-1}
0.02	3.05 ± 0.05[c]	3.15	21.5	0.02	4.74 ± 0.11[c]	4.28	21.7
0.04	3.72 ± 0.06	4.02	21.8	0.04	5.97 ± 0.08	5.97	21.8
0.05	4.55 ± 0.08	4.52	22.7	0.05	6.63 ± 0.17	6.90	22.7
0.06	5.01 ± 0.12	4.84	22.1	0.06	7.53 ± 0.21	7.44	22.1
0.08	5.52 ± 0.12	5.40	21.1	0.08	8.53 ± 0.27	8.24	20.9
0.10	5.83 ± 0.20	6.09	21.4	0.10	9.16 ± 0.27	8.85	20.0
0.15	7.13 ± 0.11	7.39	21.1	0.15	10.8 ± 0.4	10.3	19.3
0.20	7.98 ± 0.20	8.05	19.7	0.20	10.4 ± 0.4	11.7	19.7
0.30	9.36 ± 0.13	8.95	18.0	0.40	13.1 ± 0.2	12.9	17.3
0.40	10.2 ± 0.10	9.64	17.2	0.60	13.6 ± 0.1	14.0	17.2
0.60	9.90 ± 0.24	10.4	16.1	0.80	13.5 ± 0.5	12.6	14.7
				1.00	12.1 ± 0.1	12.4	14.0
$10^4\,k_0$/sec^{-1} =	2.21				2.21		
F =	0.97 ± 0.06[c]				1.01 ± 0.04[c]		
K/M^{-1} =	2.64 ± 0.33				5.88 ± 0.60		

Note: Reaction condition: [Phthalimide]$_0$ = $2 \times 10^{-4}\,M$, 35°C, the aqueous reaction mixture in each kinetic run contains 2% v/v CH$_3$CN and CTABr represents cetyltrimethylammonium bromide.

[a] Calculated from Equation 7.50 with empirical kinetic parameters, F and K, listed in Table 7.11.
[b] The values of k_{obs}^{salt} were obtained experimentally at different [M_kX_n] in the absence of CTABr.
[c] Error limits are standard deviations.

is discussed in detail in Chapter 3. In Equation 7.50, k_{obs} represents a pseudo-first-order rate constant

$$k_{obs} = \frac{k_0 + Fk_{obs}^{salt}K[M_kX_n]}{1 + K[M_kX_n]} \qquad (7.50)$$

for alkaline hydrolysis of phenyl benzoate where reactants are HO^- and phenyl benzoate (S) under a typical set of reaction conditions, $k_0 = \{k_W + k_M^{mr} K_{OH}^0 K_S [D_n]\}/\{(1 + K_{OH}^0 [D_n])(1 + K_S [D_n])\}$ with various symbols as described in Scheme 3.8 and Equation 3.11 (Chapter 3) where $R = HO^-$ and $k_W = k_{obs}$ at $[D_n] = [M_kX_n] = 0$; $F = \mu/k_{obs}^{salt}$ where μ is an empirical constant; $k_{obs}^{salt} = k_{obs}$ at the typical value of $[M_kX_n]$ with $[D_n] = 0$; and $K = K_{X/OH}/(1 + K_{OH}^0 [D_n])$ where $K_{X/OH}$ is an empirical constant derived from empirical equation: $K_{OH} = K_{OH}^0/(1 + K_{X/OH} [M_kX_n])$. If the values of k_{obs} are independent of $[M_kX_n]$, then $F k_{obs}^{salt}$ should be independent of $[M_kX_n]$ and, consequently, it should be treated as a constant at a constant value of $[D_n]$. The value of $\mu(1 + K_S[D_n])/k_{obs}^{salt}$ gives the measure of the fraction of micellized hydroxide ion (HO_M^-) transferred from the micellar pseudophase to the aqueous pseudophase by the limiting concentration of X^{k-} (the *limiting concentration of a salt*, M_kX_n, is defined as its specific concentration at which the rate of reaction becomes independent of $[M_kX_n]$).

Pseudo-first-order rate constants (k_{obs}) for hydrolysis of phenyl benzoate at 0.01 M NaOH, 0.007 M CTABr, 35°C, 98% v/v H_2O, 2% v/v CH_3CN, and varying concentration of NaCl (Table 7.10) fit to Equation 7.50 where $F k_{obs}^{salt}$ and K are treated as unknown parameters, whereas k_0 is a known parameter. The values of k_{obs}^{salt} are almost independent of [NaCl] at ≤ 0.3 M NaCl and $[CTABr]_T = 0$. The calculated value of $F k_{obs}^{salt}$ ($= k_{obs}^{salt}/(1 + K_S [D_n])$ only when $\mu(1 + K_S[D_n])/k_{obs}^{salt} = 1$) can be easily validated by determining the values of k_{obs}^{salt} in the absence of micelles (D_n) and K_S separately.[34] In order to test the sensitivity of the calculated values of $F k_{obs}^{salt}$ and K as well as k_{calcd} to k_0 slightly different values of k_0 have been used to calculate $F k_{obs}^{salt}$, K, and k_{calcd} by the use of nonlinear least-squares method, and these results are summarized in Table 7.10. The experimentally determined average value of k_0 is 0.0156 sec^{-1} at 0.007-M CTABr, 0.01-M NaOH, 35°C, 98% v/v H_2O, and 2% v/v CH_3CN. The change in k_0 from 0.0156 sec^{-1} to 0.0148 sec^{-1} (~5% decrease) causes decrease in $F k_{obs}^{salt}$, K, and Σd_i^2 (residual error, $d_i = k_{obs\ i} - k_{calcd\ i}$) by 13, 18, and 9%, respectively. Similarly, ~ 5% increase in k_0 causes increase in $F k_{obs}^{salt}$, K, and Σd_i^2 by 12, 20, and 26%, respectively. In the data analysis with Equation 7.50, the change in the observed data coverage from [NaCl] range of 0.001–0.300 M to 0.001–0.200, 0.001 to 0.100, and 0.001 to 0.050 M increases $F k_{obs}^{salt}$ and K by 15 and 4%, 40 and 12%, and 70 and 22%, respectively (Table 7.10).

In Equation 7.50, for pH-independent hydrolysis of phthalimide where phthalimide exists in almost 100% ionized form (S^-), $k_0 = (k_W + k_M K_S^0 [D_n])/(1 + K_S^0 [D_n])$ with various symbols as described in Scheme 3.1 and Equation 3.2 (Chapter 3) and $k_W = k_{obs}$ at $[D_n] = [M_kX_n] = 0$; $k_{obs}^{salt} = k_{obs}$ at the typical value of $[M_kX_n]$

with $[D_n] = 0$; $F = \theta/k_{obs}^{salt}$ where θ is an empirical constant and the value of F gives the measure of the fraction of micellized anionic phthalimide (S_M^-) transferred from the micellar pseudophase to the aqueous pseudophase by the limiting concentration of X^{k-} (the *limiting concentration* of a salt, M_kX_n, is defined as its specific concentration at which the rate of reaction becomes independent of $[M_kX_n]$) provided CTABr micellar binding constant of water is < 1, and K = $K_{X/S}/(1 + K_S^0 [D_n])$ where $K_{X/S}$ is an empirical constant derived from the empirical equation $K_S = K_S^0/(1 + K_{X/S} [M_kX_n])$.

The values of k_{obs} for pH-independent hydrolysis of phthalimide at 0.02-M KOH, 0.006 M CTABr, 35°C, 98% v/v H_2O, 2% v/v CH_3CN, and different concentrations of $(CH_3)_4NCl$ and $(CH_3)_4NBr$ as shown in Table 7.11 obey Equation 7.50. The values of k_{obs}^{salt} have been found to be weakly sensitive to $[(CH_3)_4NCl]$ and $[(CH_3)_4NBr]$ at $[CTABr]_T = 0$. The nonlinear least-squares calculated values of F and K are summarized in Table 7.11.

7.3.3.3 Effects of the Concentration of Nucleophiles/Catalysts on the Rate of Nucleophilic Cleavage of Imides, Amides, and Esters

Pseudo-first-order rate constants (k_{obs}) for the nucleophilic reaction of piperidine with phthalimide at different total piperidine concentration ($[Pip]_T$) in 100% v/v CH_3CN solvent obey the empirical kinetic equation, Equation 7.51,

$$k_{obs} = \frac{\Psi[Pip]_T^2}{1 + \Phi[Pip]_T} \qquad (7.51)$$

where Ψ and Φ are empirical constants.[35] The nonlinear least-squares calculated values of Ψ and Φ are $(4.39 \pm 0.72) \times 10^{-3}$ M^{-2} sec^{-1} and 1.32 ± 0.47 M^{-1}, respectively. However, the values of k_{obs} vary linearly with $[Pip]_T$ at a constant content of acetonitrile in mixed water–acetonitrile solvents within the acetonitrile content range 2 to 80% v/v.[35]

Kinetic studies on the nucleophilic cleavage of phthalimide in the buffers of 2-hydroxyethylamine and 2-methoxyethylamine reveal that the values of k_{obs} follow a nonlinear kinetic equation similar to Equation 7.51 with replacement of $[Pip]_T$ by $[Buf]_T$ where $[Buf]_T = [Am] + [AmH^+]$ and Am represents 2-hydroxyethylamine or 2-methoxyethylamine.[9]

Pseudo-first-order rate constants (k_{obs}) for aminolysis of amide,[36] imide,[37] and ester[38] groups in buffers of amines follow Equation 7.52

$$k_{obs} = \alpha + \beta[Buf]_T + \Phi[Buf]_T^2 \qquad (7.52)$$

where α, β, and Φ are empirical constants and $[Buf]_T$ represents total amine buffer concentration. The values of k_{obs} for hydrolysis of amides[39a,40] and β-sultams[41] containing no acidic group of $pK_a \leq \sim 25$ follow Equation 7.53 under highly alkaline medium. In Equation 7.53, A_1 and A_2 are empirical constants.

$$k_{obs} = A_1 [HO^-] + A_2[HO^-]^2 \qquad (7.53)$$

But the values of k_{obs} for hydrolysis amides,[39] under highly alkaline medium, follow Equation 7.54, whereas Equation 7.55 has been found to explain k_{obs} vs. [HO] profiles for alkaline hydrolysis of imide[42] and ester[43] groups of substrates containing an acidic group of $pK_a \leq \sim 15$.

$$k_{obs} = \frac{A_1[HO^-]}{1 + A_3[HO^-]} \qquad (7.54)$$

$$k_{obs} = \frac{A_1[HO^-] + A_2[HO^-]^2}{1 + A_3[HO^-]} \qquad (7.55)$$

In Equation 7.54 and Equation 7.55, A_1, A_2, and A_3 represent empirical constants.

7.4 REACTION MECHANISM, THEORETICAL RATE LAW, AND THEORETICAL KINETIC EQUATIONS

Empirical kinetic equations for dynamic processes such as reaction rates very often form the basis of theoretical developments that show the fine details of the mechanisms of reactions. Perhaps the most classical example of an empirical kinetic equation is Equation 7.8, which was discovered experimentally in 1878. But a satisfactory theoretical justification for Equation 7.8 was provided by Eyring in 1935, which provides the physicochemical meanings of the empirical constants, A' and B, of Equation 7.8. Empirical kinetic equations, such as Equation 7.47 to Equation 7.55, obtained as the functions of concentrations of reactants, catalysts, inert salts, and solvents, provide vital information regarding the fine details of reaction mechanisms. The basic approach in using kinetics as a tool for elucidation of the reaction mechanism consists of (1) experimental determination of empirical kinetic equation, (2) proposal of a plausible reaction mechanism, (3) derivation of the rate law in view of the proposed reaction mechanism (such a derived rate law is referred to as theoretical rate law), and (4) comparison of the derived rate law with experimentally observed rate law, which leads to the so-called theoretical kinetic equation. The theoretical kinetic equation must be similar to the empirical kinetic equation with definite relationships between empirical constants and various rate constants and equilibrium constants used in the proposed reaction mechanism.

It should be noted that the kinetic approach used to elucidate the fine details of reaction mechanisms may be thought to be the best, and even a necessary, approach, but it is not always sufficient. Additional experimental evidence should be used to strengthen the correctness of the proposed reaction mechanism. Sometimes, two or more alternative mechanisms can lead to the same theoretical rate law or theoretical kinetic equation with, of course, different constant parameters, which means that the empirical kinetic equation cannot differentiate between these alternative mechanisms. Under such circumstances, other appropriate physicochemical approaches are needed to differentiate between alternative reaction mechanisms. An attempt is made in this section of the chapter to give some representative mechanistic examples in which detailed reaction mechanisms are established based on empirical kinetic equations.

7.4.1 Mechanistic Implication of Equation 7.47

An extensive kinetic study on methanolysis of ionized phenyl salicylate (PS$^-$), as described elsewhere[8,44], reveals that the overall reaction involves PS$^-$ and CH_3OH as the reactants. Thus, the apparent rate law for methanolysis of PS$^-$ may be given as

$$\text{rate} = k_2[CH_3OH][PS^-]_T \qquad (7.56)$$

where k_2 represents a second-order rate constant and $[PS^-]_T$ is the total concentration of ionized phenyl salicylate. The observed rate law rate $= k_{obs}[PS^-]_T$ and Equation 7.56 predict that k_{obs} should vary linearly with total concentration of methanol, $[CH_3OH]_T$, with essentially zero intercept, provided $[CH_3OH]$ is directly proportional to $[CH_3OH]_T$. But the rate constants k_{obs} for methanolysis of PS$^-$ follow Equation 7.47, where ROH = CH_3OH. The most plausible mechanistic explanation for the fit of k_{obs} values to Equation 7.47 is the self-association of methanol molecules in the mixed CH_3OH-H_2O solvents.[45] Thus, the total concentration of methanol, $[CH_3OH]_T$, is considered to be the sum of the concentrations of monomeric, dimeric, trimeric, ..., and $n$meric methanol where the formation of dimeric, trimeric, ..., and $n$meric methanol occurs by a stepwise association of CH_3OH molecules. In view of this proposal, the $[CH_3OH]_T$ should be given by Equation 7.57.

$$[CH_3OH]_T = [CH_3OH] + 2[(CH_3OH)_2] + 3[(CH_3OH)_3]$$
$$+ \ldots + n[(CH_3OH)_n] \qquad (7.57)$$

For purely mathematical convenience, the assumption of equal association constant for one more addition of CH_3OH molecule to an aggregate changes Equation 7.57 to Equation 7.58

$$[CH_3OH]_T = [CH_3OH]\{1 + 2(K_A[CH_3OH]) + 3(K_A[CH_3OH])^2$$

$$+ \ldots + n(K_A[CH_3OH])^{n1}\} \tag{7.58}$$

where K_A represents the association constant for dimerization of CH_3OH (i.e., $K_A = [(CH_3OH)_2]/[CH_3OH]^2$). Equation 7.58 can lead to Equation 7.59 if $K_A[CH_3OH] \ll 1$.

$$[CH_3OH] = \frac{[CH_3OH]_T}{1 + 2K_A[CH_3OH]_T} \tag{7.59}$$

The observed rate law rate = $k_{obs}[PS^-]_T$ and Equation 7.56 and Equation 7.59 can yield Equation 7.60.

$$k_{obs} = \frac{k_2[CH_3OH]_T}{1 + 2K_A[CH_3OH]_T} \tag{7.60}$$

The theoretical kinetic equation Equation 7.60 is similar to the empirical kinetic equation Equation 7.47 with ROH = CH_3OH, $\alpha = k_2$, and $\beta = 2K_A$.

7.4.2 MECHANISTIC IMPLICATION OF EQUATION 7.48

Rate of alkaline hydrolysis of acetyl salicylate ion (AS$^-$) is proportional to both [HO$^-$] and [AS$^-$] at a constant ionic strength. Pseudo-first-order rate constants (k_{obs}) at a constant [NaOH] and different concentrations of inert salt M_kX_n (= NaCl, Na_2CO_3, KCl, and $BaCl_2$) fit to Equation 7.48. These observations have been explained in terms of ion-pair formation mechanism as shown in Scheme 7.3

$$AS^- + M^{n+} \underset{}{\overset{K_A}{\rightleftharpoons}} AS^- \ldots M^{n+} \quad n = 1 \text{ or } 2$$

$$AS^- + HO^- \xrightarrow{k_0^2} \text{Salicylate ion} + \text{Acetate ion}$$

$$AS^- \ldots M^{n+} + HO^- \xrightarrow{k_s^2} \text{Salicylate ion} + \text{Acetate ion}$$

SCHEME 7.3

The observed rate law: rate = k_{obs} [AS$^-$]$_T$, (where [AS$^-$]$_T$ = [AS$^-$] + [AS$^-$ ... M^{n+}]) and Scheme 7.3 can lead to Equation 7.61

$$k_{obs} = \frac{k_0 + k_sK_A[M_kX_n]}{1 + K_A[M_kX_n]} \tag{7.61}$$

where $k_0 = k_0^2$ [HO$^-$] and $k_s = k_s^2$ [HO$^-$]. Equation 7.61 is similar to Equation 7.48 with $\alpha = k_s K_A$ and $\beta = K_A$.

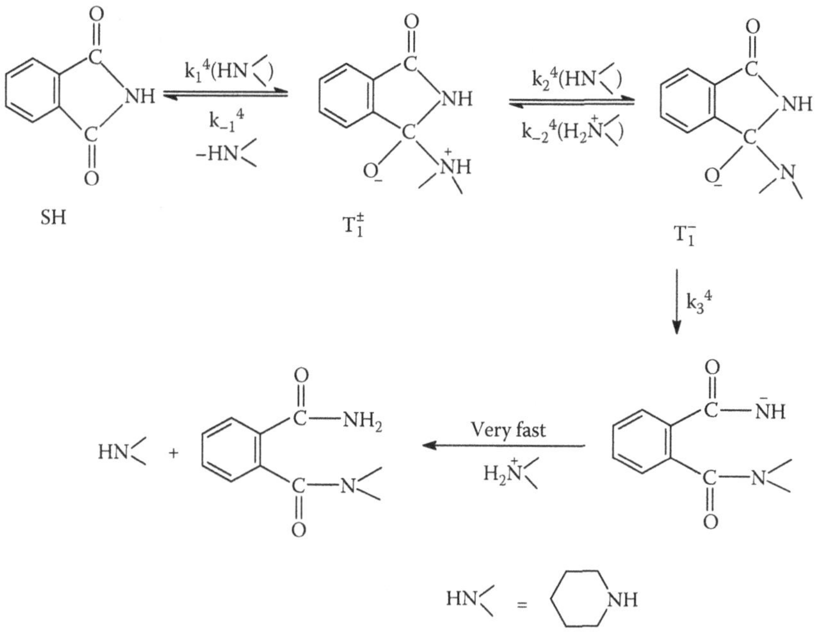

SCHEME 7.4

7.4.3 Mechanistic Implication of Equation 7.51

The nucleophilic cleavage of phthalimide (SH) at varying concentrations of piperidine in pure acetonitrile solvent has been explained empirically by Equation 7.51. There are at least three different probable mechanisms that could explain Equation 7.51. The first and most plausible mechanism, which is quite common in aqueous solvents, is shown in Scheme 7.4. The observed rate law rate = k_{obs} [SH] (where [SH] represents total concentration of phthalimide) and Scheme 7.4 lead to Equation 7.62

$$k_{obs} = \frac{K_1^4 k_2^4 f_a^2 [Pip]_T^2}{1 + f_{aH}(k_{-2}^4 / k_3^4)[Pip]_T} \tag{7.62}$$

where $f_a = K_a^{Pip}/(a_H + K_a^{Pip})$ with K_a^{Pip} representing the ionization constant of piperidinium ion (PipH$^+$), $f_{aH} = 1 - f_a$, $[Pip]_T = [Pip] + [PipH^+]$, and $K_1^4 = k_1^4/k_{-1}^4$. Equation 7.62 is similar to Equation 7.51 with $\Psi = K_1^4 k_2^4 f_a^2$ and $\Phi = f_{aH} (k_{-2}^4/k_3^4)$.

The second plausible mechanism, which could also provide a mechanistic explanation for Equation 7.51, involves the equilibrium formation of reactive complex between SH and Pip in pure acetonitrile (an aprotic solvent) followed by an irreversible reaction between reactive complex (SH•Pip) and Pip as shown in Scheme 7.5. The observed rate law and Scheme 7.5 can lead to Equation 7.63 provided $[Pip]_T = [Pip] + [SH•Pip] \approx [Pip]$ for $[Pip] >> [SH•Pip]$.

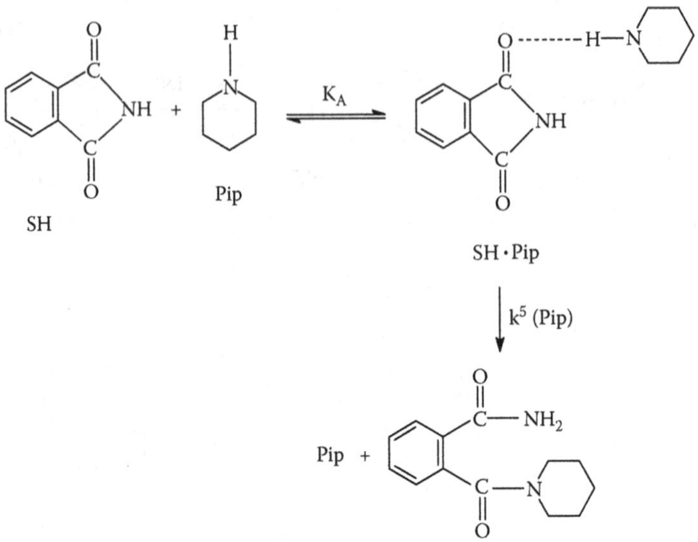

SCHEME 7.5

$$k_{obs} = \frac{k^5 K_A [Pip]_T^2}{1 + K_A [Pip]_T} \tag{7.63}$$

Equation 7.63 is similar to Equation 7.51 with $\Psi = k^5 K_A$ and $\Phi = K_A$.

The third possible mechanism, which could also explain Equation 7.51, is shown in Scheme 7.6, which involves self-association of Pip molecules in acetonitrile solvent.

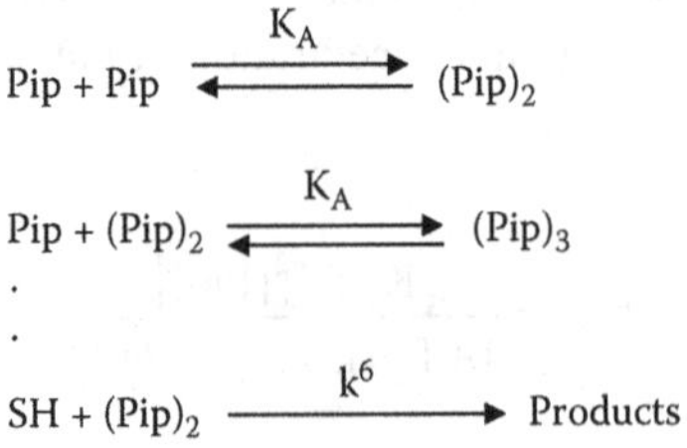

$$Pip + Pip \xrightarrow{\quad K_A \quad} (Pip)_2$$

$$Pip + (Pip)_2 \xrightarrow{\quad K_A \quad} (Pip)_3$$

$$\cdot$$
$$\cdot$$

$$SH + (Pip)_2 \xrightarrow{\quad k^6 \quad} \text{Products}$$

SCHEME 7.6

Scheme 7.6 reveals that the total concentration of piperidine ($[Pip]_T$) may be expressed as

$$[Pip]_T = [Pip] + 2[(Pip)_2] + 3[(Pip)_3] \ldots$$

$$[Pip]_T = [Pip]\{1 + 2(K_A[Pip]) + 3(K_A[Pip])^2 \ldots \tag{7.64}$$

Equation 7.64 can lead to Equation 7.65 if $K_A[Pip] \ll 1$.

$$[Pip] = \frac{[Pip]_T}{1 + 2K_A[Pip]_T} \tag{7.65}$$

The observed rate law: rate = $k_{obs}[SH]$, Scheme 7.6 and Equation 7.65 can yield Equation 7.66 provided $(2K_A[Pip]_T)^2 \ll 1 + 4K_A[Pip]_T$.

$$k_{obs} = \frac{k^6 K_A[Pip]_T^2}{1 + 4K_A[Pip]_T} \tag{7.66}$$

The theoretical kinetic equation, Equation 7.66, is similar to the empirical kinetic equation, Equation 7.51, with $\Psi = k^6 K_A$ and $\Phi = 4K_A$.

Although all three reaction mechanisms as shown by Scheme 7.4, Scheme 7.5, and Scheme 7.6 give theoretical kinetic equations, which are similar to empirical kinetic equations (Equation 7.51), reaction mechanisms represented by Scheme 7.5 and Scheme 7.6 have been ruled out based upon reasons described in details elsewhere.[35]

Kinetic studies on nucleophilic cleavage of phthalimide (SH) in the buffers of RNH_2 (= 2-hydroxyethylamine and 2-methoxyethylamine) reveal nonlinear plots of k_{obs} vs. $[Buf]_T$ (total buffer concentration) at constant pH, which fit to Equation 7.51 with replacement of $[Pip]_T$ by $[Buf]_T$. These observations have been explained in terms of a brief reaction mechanism as shown in Scheme 7.7 in

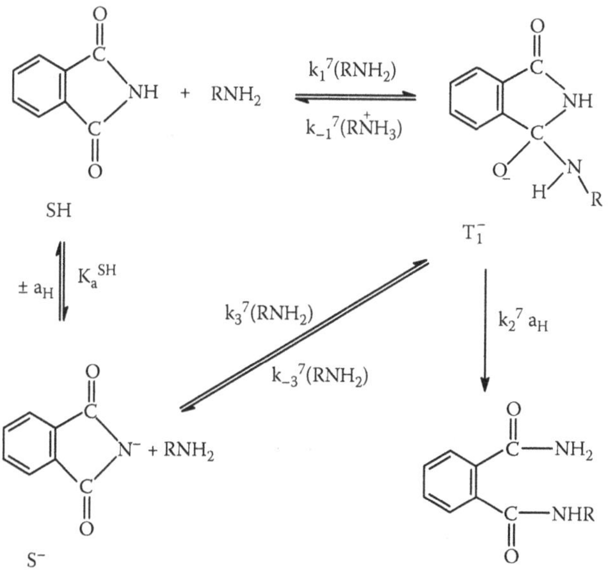

SCHEME 7.7

which the reactions of primary amine with nonionized (SH) and ionized phthalimide (S$^-$) involve the usual intermolecular general-base-catalyzed stepwise mechanism and preassociation stepwise mechanism, respectively.[9] The observed rate law rate = $k_{obs}[PT]_T$ where $[PT]_T = [SH] + [S^-]$ and Scheme 7.7 give Equation 7.67

$$k_{obs} = \frac{k_2^7 a_H (k_1^7 f_a^2 f_{SH} + k_3^7 f_a^2 f_{S-})[Buf]_T^2}{k_2^7 a_H + (k_{-1}^7 f_{aH} + k_{-3}^7 f_a)[Buf]_T} \qquad (7.67)$$

where $[Buf]_T = [RNH_2] + [RNH_3^+]$, $f_a = K_a^{Am}/(a_H + K_a^{Am})$ with $K_a^{Am} = a_H[RNH_2]/[RNH_3^+]$, $f_{SH} = a_H/(a_H + K_a^{SH})$ with $K_a^{SH} = [S^-]a_H/[SH]$, $f_{S-} = 1 - f_{SH}$, and $f_{aH} = 1 - f_a$. Equation 7.67 is similar to Equation 7.51 with $[Pip]_T = [Buf]_T$, $\Psi = k_1^7 f_a^2 f_{SH} + k_3^7 f_a^2 f_{S-}$, and $\Phi = (k_{-1}^7 f_{aH} + k_{-3}^7 f_a)/k_2^7 a_H$.

7.4.4 MECHANISTIC IMPLICATION OF EQUATION 7.52

Pseudo-first-order rate constants (k_{obs}) for the nucleophilic cleavage of phthalimide in the presence of buffer solutions of pyrrolidine follow Equation 7.52.[46] These observed data may be explained by a brief reaction mechanism as shown by Scheme 7.8, in which SH and S$^-$ represent respective nonionized and ionized phthalimide and Am represents a secondary amine in base form.

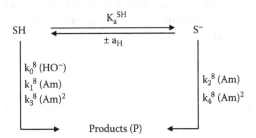

P = N-substituted phthalamide + phthalamate ion

SCHEME 7.8

Scheme 7.8 can lead to a rate law that is expressed by Equation 7.68

$$\text{rate} = k_0^8[HO^-][SH] + k_1^8[Am][SH] + k_2^8[Am][S^-] + k_3^8[Am]^2[SH] +$$
$$k_4^8[Am]^2[S^-]$$
$$= (k_0^8 f_{SH} HO^- + k_1^8 f_a f_{SH}[Buf]_T + k_2^8 f_a f_{S-}[Buf]_T + \qquad (7.68)$$
$$k_3^8 f_a^2 f_{SH}[Buf]_T^2 + k_4^8 f_a^2 f_{S-}[Buf]_T^2)[PT]_T$$

where $[PT]_T = [SH] + [S^-]$, $[Buf]_T = [Am] + [AmH^+]$, $f_a = K_a^{Am}/(a_H + K_a^{Am})$ with $K_a^{Am} = [Am]a_H/[AmH^+]$, $f_{S-} = K_a^{SH}/(a_H + K_a^{SH})$ with $K_a^{SH} = [S^-]a_H/[SH]$ and $f_{SH} = 1 - f_{S-}$. The observed rate law rate = $k_{obs}[PT]_T$ and Equation 7.68 yield Equation 7.69,

$$k_{obs} = k_0{}^8f_{SH}[HO^-] + (k_1{}^8f_{SH} + k_2{}^8f_{S^-})f_a[Buf]_T +$$
$$(k_3{}^8f_{SH} + k_4{}^8f_{S^-})f_a{}^2[Buf]_T{}^2 \tag{7.69}$$

which is similar to Equation 7.52 with $\alpha = k_0{}^8f_{SH}[HO^-]$, $\beta = (k_1{}^8f_{SH} + k_2{}^8f_{S^-})f_a$, and $\Phi = (k_3{}^8f_{SH} + k_4{}^8f_{S^-})f_a{}^2$.

7.4.5 Mechanistic Implication of Equation 7.53

Biechler and Taft[39a] reported that in the aqueous alkaline hydrolysis of a series of N-methylanilides, $RCON(CH_3)C_6H_5$, pseudo-first-order rate constants (k_{obs}) strictly follow Equation 7.53. In order to explain the appearance of $A_2[HO^-]^2$ term in Equation 7.53, these authors, perhaps for the first time, suggested the existence of the highly reactive oxydianionic tetrahedral intermediate $T_2{}^{2-}$ on the reaction path in the alkaline hydrolytic cleavage of amide bond. This report became a catalyst for a number of researchers to go into establishing such an intermediate in the related reactions, and during the last nearly 5 decades, a series of papers have appeared in which such an unusual intermediate ($T_2{}^{2-}$) has been reported to exist on the reaction path.[47] In order to give a mechanistic explanation to Equation 7.53, Biechler and Taft[39a] suggested a reaction mechanism as shown in Scheme 7.9 where the $k_1{}^9$ step and $k_2{}^9$ step are rate-determining steps. The observed rate law rate = $k_{obs}[Sub]$ and Scheme 7.9 give Equation 7.70,

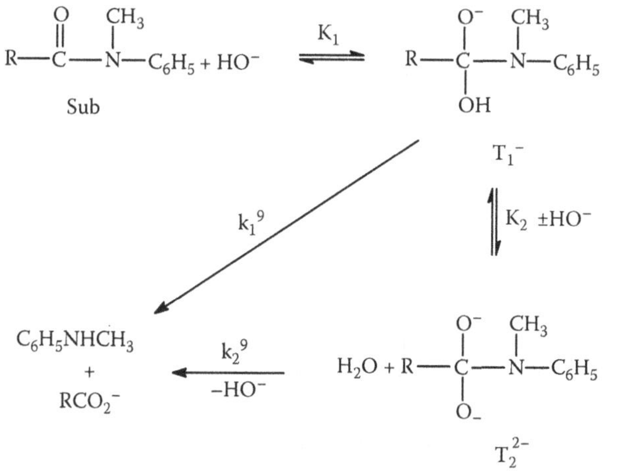

SCHEME 7.9

$$k_{obs} = k_1{}^9K_1[HO^-] + k_2{}^9K_1K_2[HO^-]^2 \tag{7.70}$$

which is similar to Equation 7.53 with $A_1 = k_1{}^9K_1$ and $A_2 = k_2{}^9K_1K_2$.

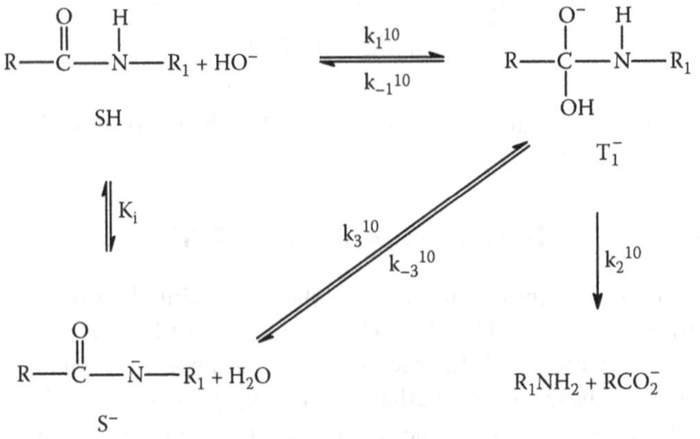

Trifluoroacetanilide, R = CF$_3$ and R$_1$ = C$_6$H$_5$

N - (*o*-carboxybenzoyl)carbamate, R = *o*-CO$_2^-$ C$_6$H$_4$ and R$_1$=CO$_2$C$_2$H$_5$

SCHEME 7.10

7.4.6 MECHANISTIC IMPLICATION OF EQUATION 7.54

Kinetic studies on the hydrolytic cleavage of the amide bond in trifluoroacetanilide[39a] and *N*-(*o*-carboxybenzoyl)carbamate[39b] under highly alkaline medium reveal that k_{obs}-[HO$^-$] profiles follow Equation 7.54. These observations may be explained by a general reaction mechanism as described in Scheme 7.10. The observed rate law rate = k_{obs}[Sub] and Scheme 7.10 can lead to Equation 7.71,

$$k_{obs} = \frac{k_2^{10}(k_1^{10} + k_3^{10}K_i')[HO^-]}{(1 + K_i'[HO^-])(k_{-1}^{10} + k_{-3}^{10})} \tag{7.71}$$

provided the k_2^{10} step is the rate-determining step. In Equation 7.71, $K_i' = K_i/[H_2O]$ = K_a^{SH}/K_w with $K_a^{SH} = [S^-]a_H/[SH]$ and $K_w = a_H\, a_{OH}$. Equation 7.71 is similar to Equation 7.54 with $A_1 = k_2^{10}(k_1^{10} + k_3^{10}K_i')/(k_{-1}^{10} + k_{-3}^{10})$ and $A_3 = K_i'$.

7.4.7 MECHANISTIC IMPLICATION OF EQUATION 7.55

The kinetic study on the hydrolytic cleavage of phenyl salicylate (PS) reveals that the pseudo-first-order rate constants (k_{obs}) obtained within hydroxide ion concentration range of 0.05 to 2.0 *M* at 35°C follow Equation 7.55.[48] The occurrence of a plausible reaction mechanism that could explain Equation 7.55 is shown in Scheme 7.11. The derived or theoretical rate law, based on Scheme 7.11, and observed rate law rate = k_{obs}[PS]$_T$ (where [PS]$_T$ = [SH] + [S$^-$], i.e., the total concentration of PS) give Equation 7.72

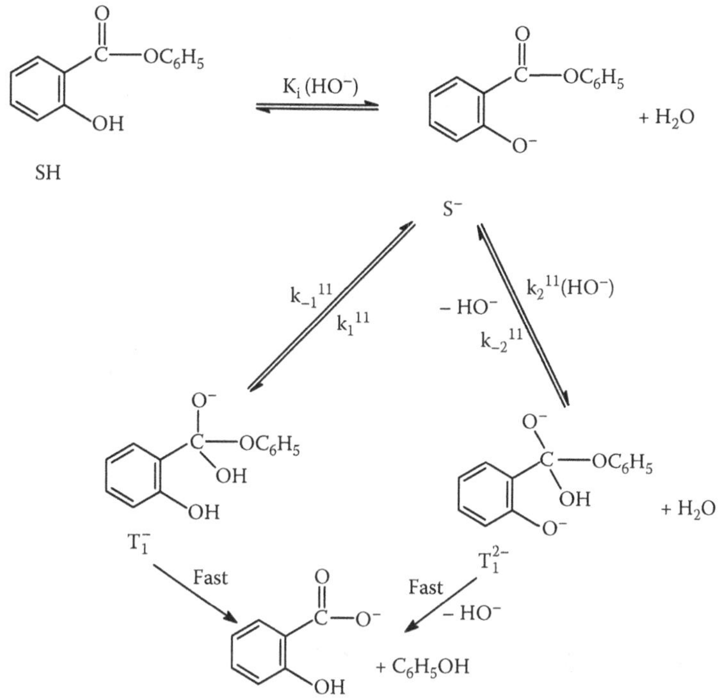

SCHEME 7.11

$$k_{obs} = \frac{k_1^{11}K_i'[H_2O][HO^-] + k_2^{11}K_i'[HO^-]^2}{1 + K_i'[HO^-]} \tag{7.72}$$

where $K_i' = K_a^{SH}/K_w$. Equation 7.72 is similar to Equation 7.55 with $A_1 = k_1^{11}K_i'[H_2O]$, $A_2 = k_2^{11} K_i'$, and $A_3 = K_i'$.

An alternative mechanism involving nonionized (SH) rather than ionized phenyl salicylate (S^-) as a reactant is described in Scheme 7.12 where T_1^- and T_1^{2-} are highly reactive transient intermediates.[43] Scheme 7.12 and observed rate law give Equation 7.73

$$k_{obs} = \frac{k_1^{12}[HO^-](k_2^{12} + k_3^{12}K[HO^-])}{(1 + K_i'[HO^-])(k_{-1}^{12} + k_2^{12} + k_3^{12}K[OH^-])} \tag{7.73}$$

where $K_i' = K_a^{SH}/K_w$. Equation 7.73 could be reduced to Equation 7.55 only if it is assumed that $k_{-1}^{12} \gg k_2^{12}$ and k_3^{12} K [HO$^-$]). These assumptions necessitate that both the k_2^{12} step and k_3^{12} step should be rate-determining steps, but this is not correct because the pK_a of leaving group in the k_2^{12} step and k_3^{12} step is

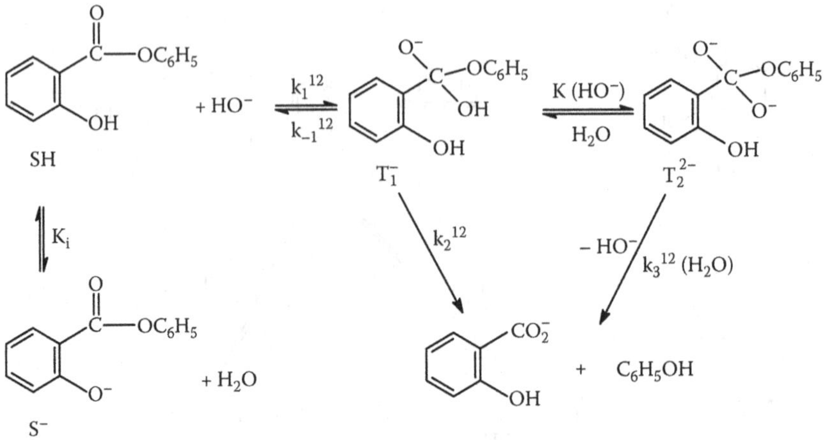

SCHEME 7.12

smaller by ~ 6 pK units than that in the k_{-1}^{12} step. Consequently, the mechanism in Scheme 7.12 is not correct.

7.5 RELIABILITY OF DERIVED OR CALCULATED KINETIC PARAMETERS

The observed or primary kinetic data are described as those kinetic data determined directly by experiment, and the treatment of such data with an empirical or theoretical kinetic equation involves either a linear (Equation 7.74) or a nonlinear equation (Equation 7.75), which shows the variation of dependent variable Y with independent variable X. The dependent variable may be the concentration or equivalent of any physical property such as absorbance, A_{obs}, of either reactant or product that has been monitored as a function of an independent variable such as reaction time, t. In kinetic studies, the rate constants, under specific reaction conditions, are obtained from primary kinetic data and, therefore, such experimentally determined rate constants may be considered secondary kinetic data. Dependent variables may be experimentally determined rate constants (also called *observed rate constants* or *secondary kinetic data*), which have been determined as functions of independent variables such as the concentrations of reaction components (reactants and catalysts), temperature, and reaction medium. The reliability of the calculated kinetic parameters from a kinetic equation using primary and secondary kinetic data may be described as follows.

7.5.1 KINETIC EQUATIONS INVOLVING PRIMARY KINETIC DATA (Y vs. X)

Let us consider the following kinetic equations:

$$Y = \theta + \Phi X \tag{7.74}$$

$$Y = \alpha \exp(-\beta X) + \psi \tag{7.75}$$

where θ, Φ, α, β, and ψ represent kinetic parameters or constants. If Y represents the concentration of the reaction components, and X the reaction time, t, then θ and Φ or α, β, and ψ should be determined under the reaction conditions where the change in Y, (= ΔY), between X = 0 and X = F (a final attained value of X in a typical experiment) (i.e., $|\Delta Y| = |Y_0| - |Y_F|$) must be large so that $|\Delta Y|$ /($|Y_0| - |Y_\infty|$) (where Y = Y_∞ at X = ∞) could be ≥ 0.75. Furthermore, if Y represents a physical property such as absorbance of the reaction components, then $|Y_0| - |Y_\infty|$ should preferably be $\geq \sim 0.1$ and $\leq \sim 1.5$ absorbance units (Abs) provided a conventional UV-visible spectrophotometer is used where photometric accuracy is generally ± 0.002 Abs at 0.5 Abs. The kinetic parameters, Φ, α, β, and ψ must be positive numbers, because these parameters represent either concentration or rate constant or a mixed function of both concentration and rate constant. If, for example, the calculated value of any one of Φ, α, β, and ψ is negative, then the data analysis with Equation 7.74 and Equation 7.75 becomes unreliable. Sometimes it does happen that the value of ψ is negative but associated with standard deviation of more than 100%. Under such circumstances, the value of ψ may be considered to be zero for a reliable data analysis with Equation 7.75.

7.5.2 KINETIC EQUATIONS INVOLVING SECONDARY KINETIC DATA (Y vs. X)

Let us consider Equation 7.76,

$$k_{obs} = \frac{k_W + k_M K_S [D_n]}{1 + K_S [D_n]} \tag{7.76}$$

which relates the variation of experimentally determined rate constants, k_{obs} (= Y), with micelles concentration, $[D_n]$ (= X), for a micellar-mediated reaction. Standard deviations associated with the calculated kinetic parameters (such as k_M and K_S from Equation 7.76) and the residual errors ($d_i = k_{obs,i} - k_{calcd,i}$, where $k_{calcd,i}$ represents the ith calculated rate constant using Equation 7.76 with calculated kinetic parameters k_M and K_S) may not be considered the sole criteria for a good or bad fit of observed data, k_{obs} vs. $[D_n]$, to a specific kinetic equation, Equation 7.76. The calculated value of k_M from a set of observed data (k_{obs} vs. $[D_n]$) may be considered reliable if $k_M K_S [D_n]/(k_W + k_M K_S [D_n]) \geq 0.5$ under the specific reaction conditions in which the observed data have been used to calculate k_M from Equation 7.76. Similarly, the calculated value of K_S may be reliable if k_M $K_S [D_n]/(k_W + k_M K_S [D_n]) \geq 0.5$ and $K_S[D_n] \geq 0.5$ even when $k_W >> k_M K_S [D_n]$ under the specific reaction conditions. It is also important to note that in order to calculate the unknown kinetic parameters k_M and K_S from Equation 7.76, it is probably better to use the q number of k_{obs} at the q number of $[D_n]$ than the q number of k_{obs} at the q/2 or q/3 number of $[D_n]$. These criteria are not developed by statistical methods, but are the outcome of intuition and experimentation.

To give an experimental justification for some of the points mentioned, let us analyze the experimentally determined pseudo-first-order rate constants (k_{obs}) at different total hydrazine buffer concentration $(Buf]_T)$ for the cleavage of phthalimide at pH 8.21.[50] These kinetic data as summarized in Table 7.12 have been treated with Equation 7.77

$$k_{obs} = k_0 + k_n[Buf]_T + k_b[Buf]_T^2 \qquad (7.77)$$

where k_0, k_n, and k_b represent respective apparent rate constants for hydrolysis and uncatalyzed and buffer-catalyzed hydrazinolysis of phthalimide. Both k_n and k_b are functions of rate constants and pH. The least-squares calculated values of k_0, k_n, and k_b are shown in Table 7.13. Although the fit of k_{obs} values to Equation 7.77 is considered to be good in view of residual errors, d_i, because the maximum d_i value is < 2% (Table 7.12), the calculated negative value of k_0 (= -1.61×10^{-3} sec^{-1}, Table 7.12) is physically/chemically meaningless. The most reliable value of k_0 at pH 8.21 is 6×10^{-5} sec^{-1}, which shows that the maximum contribution of k_0 compared to $k_n [Buf]_T + k_b [Buf]_T^2$ in Equation 7.77 is < 3%.[49] Thus, the calculated negative value of k_0 is merely due to the insignificant contribution of k_0 in Equation 7.77 under the experimental conditions of the study. Under such circumstances, k_0 should not be treated as an unknown parameter and, hence, a relatively more reliable data treatment is to use Equation 7.78 with $k_0 = 6 \times 10^{-5}$ sec^{-1}, which has been determined under experimental conditions in which contribution due to terms such as $k_n [Buf]_T + k_b [Buf]_T^2$ was insignificant compared with k_0.

$$k_{obs} - k_0 = k_n[Buf]_T + k_b[Buf]_T^2 \qquad (7.78)$$

The values of k_{obs} at different $[Buf]_T$ (Table 7.12) have been treated with Equation 7.78 and the least-squares calculated values of k_n and k_b are summarized in Table 7.13. Although the values of the calculated parameters k_n and k_b are positive and associated with low standard deviations, the values of k_{calcd} show the systematic positive deviations of 29 and 9% compared with the corresponding values of k_{obs} at $[Buf]_T = 0.025$ and 0.0375 M, respectively. This shows that the data fit to Equation 7.78 is unsatisfactory. The apparent second-order rate constants k_{2obs} (= $(k_{obs} - k_0)/[Buf]_T$ where $10^5 k_0 = 6.0$ sec^{-1}) have been treated with Equation 7.79

$$k_{2obs} = \frac{\alpha[Buf]_T}{1 + \beta[Buf]_T} \qquad (7.79)$$

where α and β are empirical constants, which could be the function of rate constants and pH. The values of k_{2obs} at different $[Buf]_T$ have been used to

TABLE 7.12

Pseudo-First-Order Rate Constants (k_{obs}) for Hydrolytic Cleavage of Phthalimide in Buffers of Hydrazine at pH = 8.21 ± 0.02 and 30°C

$[Buf]_T$ [a] M	$10^3\,k_{obs}$ sec^{-1}	$10^3\,k_{calcd}$ [b] sec^{-1}	$10^3\,k_{calcd}$ [c] sec^{-1}	$10^3\,k_{calcd}$ [c] sec^{-1}	$10^3\,k_{calcd}$ [c] sec^{-1}	$10^3\,k_{2obs}$ $M^{-1}\,sec^{-1}$	$10^3\,k_{2calcd}$ [d] $M^{-1}\,sec^{-1}$	$10^3\,k_{2calcd}$ [e] $M^{-1}\,sec^{-1}$
0.0250	2.06 ± 0.04 [f]	2.02	2.66			80.0	82.3	79.9
0.0375	3.83 ± 0.04	3.90	4.17	4.26		101	101	101
0.0500	5.85 ± 0.06	5.82	5.80	5.91	6.03	116	113	114
0.0625	7.73 ± 0.04	7.78	7.58	7.68	7.80	123	122	124
0.0750	9.84 ± 0.09	9.78	9.48	9.57	9.68	130	129	130
0.1000	13.9 ± 0.1	13.9	13.7	13.8	13.8	138	139	139
0.1250	18.2 ± 0.1	18.2	18.5	18.4	18.3	145	146	145

Note: Reaction condition: $[Phthalimide]_0 = 2 \times 10^{-4}\,M$, ionic strength 1.0 M and the aqueous reaction mixture in each kinetic run contains 1% v/v CH_3CN.

[a] $[Buf]_T = [NH_2NH_2] + [NH_2NH_3^+]$.

[b] Calculated from Equation 7.77 with kinetic parameters listed in Table 7.13.

[c] Calculated from Equation 7.78 with $10^5\,k_0 = 6.0\,sec^{-1}$ and kinetic parameters listed in Table 7.13.

[d] Calculated from Equation 7.79 with kinetic parameters listed in Table 7.13.

[e] Calculated from Equation 7.80 with kinetic parameters listed in Table 7.13.

[f] Error limits are standard deviations.

TABLE 7.13

Least-Squares Calculated Kinetic Parameters from Equation 7.77 to Equation 7.80 Using Data in Table 7.12

$10^3 k_0$ (sec^{-1})	$10^3 k_n$ (M^{-1} sec^{-1})	$10^3 k_b$ (M^2 sec^{-1})	α (M^2 sec^{-1})	β (M^{-1})	$10^3 k_n$ (M^{-1} sec^{-1})
-1.61 ± 0.11^b	$142 \pm 3^{a,b}$	$132 \pm 22^{a,b}$	$14.4 \pm 4.3^{b,c}$	$61.9 \pm 0.12.2^{b,c}$	$-61.8 \pm 25.5^{b,c}$
	93.4 ± 6.6^d	430 ± 66^d	6.02 ± 0.23^e	32.2 ± 1.8^e	
	97.3 ± 5.2^f	394 ± 52^f			
	102 ± 4^g	353 ± 35^g			

a Kinetic parameters are calculated from Equation 7.77 with k_{obs} obtained within $[Buf]_T$ range 0.0250 – 0.1250 M.

b Error limits are standard deviations.

c Kinetic parameters are calculated from Equation 7.80.

d Kinetic parameters are calculated from Equation 7.78 with k_{obs} obtained within $[Buf]_T$ range 0.0250 to 0.1250 M.

e Kinetic parameters are calculated from Equation 7.79.

f Kinetic parameters are calculated from Equation 7.78 with k_{obs} obtained within $[Buf]_T$ range 0.0375 to 0.1250 M.

g Kinetic parameters are calculated from Equation 7.78 with k_{obs} obtained within $[Buf]_T$ range 0.0500 to 0.1250 M.

calculate α and β from Equation 7.79 by the use of the nonlinear least-squares technique, and these values of the calculated α and β are summarized in Table 7.13. The satisfactory fit of the k_{2obs} values to Equation 7.79 is evident from the standard deviations associated with the values of α and β and from the k_{2calcd} values listed in Table 7.12.

An attempt to fit the kinetic data (k_{2obs} vs. $[Buf]_T$) to Equation 7.80 resulted in an apparent satisfactory fit of almost the same degree as with Equation 7.79 in view of residual errors d_i (Table 7.12).

$$k_{2obs} = k_n + \frac{\alpha[Buf]_T}{1 + \beta[Buf]_T} \qquad (7.80)$$

But a significantly large negative value of k_n ($= -(61.8 \pm 25.5) \times 10^{-3}$ M^{-1} sec^{-1}, Table 7.13) associated with a standard deviation of $> 40\%$ shows that apparent satisfactory fit of the kinetic data to Equation 7.80 in view of k_{2calcd} values is fortuitous and, consequently, calculated values of k_n, α, and β are unreliable.

It is perhaps worthy to note that the least-squares treatment of k_{obs} with Equation 7.78 at $[Buf]_T \geq 0.050$ M gives an apparently satisfactory fit in view of k_{calcd} values summarized in Table 7.12. The maximum value of the residual error is 3%, which is similar to the one obtained with Equation 7.79 by using k_{2obs} values at $[Buf]_T \geq 0.025$ M. Thus, a researcher might end up believing that the k_{obs} values fit to Equation 7.78 reasonably well if, for some reasons, k_{obs} values were not obtained at $[Buf]_T < 0.050\ M$. Such a kinetic ambiguity may be resolved by using an external buffer whose base component cannot act as an effective nucleophile such as carbonate buffer. The values of k_{obs} obtained at different $[Buf]_T$ (where $[Buf]_T$ = total concentration of external buffer) in the presence of a constant concentration of hydrazine should reveal a linear plot of k_{obs} vs. $[Buf]_T$ if the kinetic data follow Equation 7.78. But a linear plot of k_{obs} vs. $[Buf]_T$ is unlikely to exist if the kinetic data follow Equation 7.79. Such an experimental approach showed that the rate of hydrazinolysis of phthalimide followed Equation 7.79.[49]

Another experimental example, which could demonstrate the importance of the so-called empirical criteria described in the beginning of this section, is the kinetic study of alkaline hydrolysis of 4-nitrophthalimide (NPT) in the absence and presence of cationic micelles (cetyltrimethylammonium bromide, CTABr micelles) at 35°C.[42] The pH-independent and pH-dependent rate of alkaline hydrolysis of NPT in the absence and, presumably, in the presence of micelles involve the reaction of HO⁻ with nonionized (SH) and ionized (S⁻) NPT, respectively.[42] Pseudo-first-order rate constants (k_{obs}) for hydrolytic cleavage of NPT at 0.01-M NaOH, 35°C and varying total concentrations of CTABr ($[CTABr]_T$), as listed in Table 7.14, have been treated with Equation 7.76 where $[D_n] = [CTABr]_T$ – CMC, K_S is CTABr micellar binding constant of NPT, k_W and k_M represent pseudo-first-order rate constants for hydrolysis of NPT in the aqueous pseudophase and micellar pseudophase, respectively. The nonlinear least-squares calculated values of k_M, K_S, and least squares ($\sum d_i^2$) are shown in Table 7.14. The values of k_{calcd} are comparable with the corresponding k_{obs} values within the domain of the acceptable residual errors d_i (Table 7.14).

The occurrence of ion exchange in ionic micellar-mediated ionic or semiionic reactions has been unequivocally established (References 12, 13, 16, 23, 24, 36, 39 cited in Chapter 3). The expected kinetically effective ion exchange in the present reacting system is Br⁻/HO⁻.[42] Pseudophase ion exchange (PIE) model can lead to Equation 7.81

$$k_{obs} = \frac{k_{0,W} + k_{OH,W}[HO_W^-] + (k_{0,M} + k_{OH,M}^{mr} m_{OH})K_S[D_n]}{1 + K_S[D_n]} \tag{7.81}$$

where $k_{0,W}$ and $k_{OH,W}$ represent respective first- and second-order rate constants for the reactions of HO_W^- with SH_W and S_W^-, $k_{OH,M}^{mr} = k_{OH,M}/V_M$, $k_{OH,M}$ is second-order rate constant for the reaction of S_M^- with HO_M^-, V_M is the molar volume of the micellar reaction region, $k_{0,M}$ represents first-order rate constant for pH-independent hydrolysis of NPT in the micellar pseudophase, and $m_{OH} = [HO_M^-]/[D_n]$. The kinetic parameters $k_{0,M}$, $k_{OH,M}^{mr}$, and K_S have been calculated with known values of $k_{0,W}$ and $k_{OH,W}$, determined experimentally under the same reaction conditions with $[CTABr]_T = 0$, and these results are summarized in Table 7.14. Although the data treatment with Equation 7.81 gives $\sum d_i^2$, which is significantly lower than that obtained with Equation 7.76, the negative value of $k_{OH,M}^{mr}$ with standard deviation of greater than 50% shows that the data treatment

TABLE 7.14
Pseudo-First-Order Rate Constants (k_{obs}) for Hydrolytic Cleavage of 4-Nitrophthalimide in the Presence of CTABr at 0.01-M NaOH and 35°C

10^4 [CTABr]$_T$ M	10^4 k_{obs} sec^{-1}	10^4 k_{calcd}[a] sec^{-1}	10^4 k_{calcd}[b] sec^{-1}	10^4 k_{calcd}[c] sec^{-1}	m_{OH}[d]
0.2	22.6 ± 0.2[e]				
0.4	22.6 ± 0.4				
0.6	22.6 ± 0.3				
1.0	22.6 ± 0.3				
2.0	22.0 ± 0.5				
3.0	18.8 ± 0.6	17.6	18.0	17.5	0.509
4.0	13.0 ± 0.3	12.6	12.8	12.6	0.469
5.0	9.60 ± 0.20	10.1	10.1	10.1	0.438
7.0	6.83 ± 0.05	7.62	7.26	7.63	0.390
10.0	5.32 ± 0.06	5.98	5.43	5.99	0.341
20.0	4.19 ± 0.05	4.26	3.76	4.26	0.252
60.0	3.77 ± 0.07	3.20	3.26	3.20	0.139
100.0	3.51 ± 0.05	3.00	3.32	3.00	0.0996
200.0	2.80 ± 0.03	2.85	3.46	2.85	0.0604
10^8 Σd_i^2	=	3.586	2.109	3.697	
10^4 k_M/sec^{-1}	=	2.71 ± 0.38[e]			
10^4 $k_{0,M}$/sec^{-1}	=		3.80 ± 0.64[e]	2.70 ± 0.39[e]	
10^4 $k_{OH,M}$/sec^{-1}	=		8.9 ± 4.8	0	
K_S/M^{-1}	=	6770 ± 680	4900 ± 860	6710 ± 683	

Notes: Reaction condition: [NPT]$_0$ = 1 × 10^{-4} M and the aqueous reaction mixture in each kinetic run contains 1% v/v CH$_3$CN.

[a] Calculated from Equation 7.76 with 10^4 CMC = 2.5 M, 10^4 k_W 22.6 sec^{-1} and the calculated values of k_M and K_S listed in Table 7.14.

[b] Calculated from Equation 7.81 with 10^4 CMC = 2.5 M, K_{Br}^{OH} = 20, β = 0.8, 10^4 $k_{0,W}$ = 17.9 sec^{-1}, 10^4 $k_{OH,W}$ = 463 M^{-1} sec^{-1}, and the calculated values of $k_{0,M}$, $k_{OH,M}$, and K_S listed in Table 7.14.

[c] Calculated from Equation 7.81 with 10^4 CMC = 2.5 M, K_{Br}^{OH} = 20, β = 0.8, 10^4 $k_{0,W}$ = 17.9 sec^{-1}, 10^4 $k_{OH,W}$ = 463 M^{-1} sec^{-1}, $k_{OH,M}$ = 0 and the calculated values of $k_{0,M}$ and K_S listed in Table 7.14.

[d] m_{OH} = [HO$_M^-$]/[D$_n$].

[e] Error limits are standard deviations.

with Equation 7.81 is not reliable and the value of $k_{OH,M}{}^{mr} m_{OH}$ is insignificant compared to other terms in the numerator of Equation 7.81. The contribution of $k_{OH,W}$ [$HO_W{}^-$] is only nearly 20% compared with $k_{0,W}$ at 0.01-M NaOH, and it is highly unlikely that the value of $k_{0,W}/k_{OH,W}$ may be appreciably different from $k_{0,M}/k_{OH,M}$. Thus, relatively more reliable data treatment with Equation 7.81 is expected with $k_{OH,M}{}^{mr} = 0$. The calculated values of $k_{0,M}$, K_S, and $\sum d_i^2$ using Equation 7.81 with $k_{OH,M}{}^{mr} = 0$ (Table 7.14) show that the statistical reliability of data fit is almost as good as with Equation 7.76. It is apparent that in the present reaction system, the occurrence of ion exchange Br$^-$/HO$^-$ is kinetically ineffective and, consequently, it is preferable to use the pseudophase micellar (PM) model (i.e., Equation 7.76) rather than the PIE model (i.e., Equation 7.81), because the PM model involves fewer assumptions compared to the PIE model.

REFERENCES

1. Jencks, W.P. Ingold lecture: how does a reaction choose its mechanism? *Chem Soc, Rev.* **1981**, *10*(3), 345–375.
2. Khan, M.N. What We Know and What We Think We Know about Chemical Catalysis. Inaugural Lecture at University of Malaya on February 18, **2005**.
3. Laidler, K.J. *Chemical Kinetics*, 2nd ed. Tata McGraw-Hill, Bombay-New Delhi, **1965** and the references cited therein.
4. (a) Bunton, C.A., Robinson, L. Micellar effects upon the reaction of *p*-nitrophenyl diphenyl phosphate with hydroxide and fluoride ions. *J. Org. Chem.* **1969**, *34*(4), 773–780. (b) Dunlap, R.B., Cordes, E.H. Secondary valence force catalysis. VI. Catalysis of hydrolysis of methyl orthobenzoate by sodium dodecyl sulfate. *J. Am. Chem. Soc.* **1968**, *90*(16), 4395–4404.
5. Cordes E.H. Kinetics of organic reactions in micelles. *Pure Appl. Chem.* **1978**, *50*(7), 617–625 and references cited therein.
6. (a) Eyring, H. Activated complex in chemical reactions. *J. Chem. Phys.* **1935**, *3*, 107–115. (b) Wynne-Jones, W.F.K., Eyring, H. Absolute rate of reactions in condensed phases. *J. Chem. Phys.* **1935**, *3*, 492–502.
7. Jencks, W.P. *Catalysis in Chemistry and Enzymology*, McGraw-Hill, New York, **1969**.
8. Khan, M.N. Kinetics and mechanism of transesterification of phenyl salicylate. *Int. J. Chem. Kinet.* **1987**, *19*, 757–776.
9. Khan, M.N., Ohayagha, J.E. Kinetics and mechanism of the cleavage of phthalimide in buffers of tertiary and secondary amines: evidence of intramolecular general acid-base catalysis in the reactions of phthalimide with secondary amines. *J. Phys. Org. Chem.* **1991**, *4*, 547–561.
10. Blackburn, R.A.M., Capon, B., McRitchie, A.C. The mechanism of hydrolysis of phthalamic acid and N-phenylphthalamic acid: the spectrophotometric detection of phthalic anhydride as an intermediate. *Bioorg. Chem.* **1977**, *6*, 71–78.
11. Khan, M.N. Kinetic studies on general base-catalyzed cleavage of phthalimide (PTH) in 2-hydroxyethylamine and 2-methoxyethylamine buffers. *Int. J. Chem. Kinet.* **1996**, *28*, 421–431.
12. Guggenheim, E.A. The determination of the velocity constant of a unimolecular reaction. *Phil. Mag.* (1798–1977), **1926**, *2*, 538–613.

13. Khan, M.N. Intramolecular general base catalysis and the rate-determining step in the nucleophilic cleavage of ionized phenyl salicylate with primary and secondary amines. *J. Chem. Soc. Perkin Trans. 2.* **1989**, 199–208.

14. Ritchie, C.D. *Physical Organic Chemistry — The Fundamental Concepts*, 2nd ed. Marcel Dekker, New York, 1990.

15. Khan, M.N., Khan, A.A. Kinetics and mechanism of hydrolysis of succinimide under highly alkaline medium. *J. Org. Chem.* **1975**, *40*(12), 1793–1794.

16. Arrifin, A., Khan, M.N. Unexpected rate retardation in the formation of phthalic anhydride from N-methylphthalamic acid in acidic H_2O-CH_3CN medium. *Bull. Korean Chem. Soc.* **2005**, *26*(7), 1037–1043.

17. Reichardt, C. *Solvents and Solvent Effects in Organic Chemistry*, 2nd ed. VCH, Weinheim, **1988**.

18. Leng, S.Y., Ariffin, A., Khan, M.N. Effects of mixed aqueous-organic solvents on the rate of intramolecular carboxylic group-catalyzed cleavage of N-(4-methoxyphenyl)phthalamic acid. *Int. J. Chem. Kinet.* **2004**, *36*, 316–325.

19. Khan, M.N., Arifin, Z. Kinetics and mechanism of intramolecular general base-catalyzed methanolysis of ionized phenyl salicylate in the presence of cationic micelles. *Langmuir* **1996**, *12*(2), 261–268.

20. Khan, M.N., Audu, A.A. Kinetic probe to study the structural behavior of the mixed aqueous-organic solvents: salt effects on the kinetics and mechanism of intramolecular general base-catalyzed ethanolysis of ionized phenyl salicylate. *Int. J. Chem. Kinet.* **1990**, *22*, 37–57.

21. Khan, M.N., Arifin, Z. Effects of cationic micelles on rate of intramolecular general base-catalyzed ethanediolysis of ionized phenyl salicylate (PS). *Langmuir* **1997**, *13*(25), 6626–6632.

22. Khan, M.N. Kinetic probe to study the structure of the mixed aqueous-organic solvents: kinetics and mechanism of nucleophilic reactions of *n*-propanol and D(+)-glucose with ionized phenyl salicylate. *J. Phys. Chem.* **1988**, *92*(22), 6273–6278.

23. Frost, A.A., Pearson, R.G. *Kinetics and Mechanism*, 2nd ed. John Wiley & Sons, New York, **1961**.

24. Holba, V., Benko, J., Okalova, K. Effect of ionic strength and dielectric constant on base hydrolysis of monomethyl phthalate. *Collect. Czech. Chem. Commun.* **1978**, *43*(6), 1581–1587.

25a. Bag, B.C., Das, M.N. Ion association and specific solvent effects on rates of alkaline hydrolysis of monomethyl succinate. *Indian J. Chem.* **1982**, *21A*, 1035–1039.

25b. Khan, M.N., Gleen, P.C., Arifin, Z. Effects of inorganic salts and mixed aqueous-organic solvents on the rates of alkaline hydrolysis of aspirin. *Indian J. Chem.* **1996**, *35A*, 758–765.

26. Pirinccioglu, N., Williams, A. Studies of reactions within molecular complexes: alkaline hydrolysis of substituted phenyl benzoates in the presence of xanthines. *J. Chem. Soc., Perkin Trans. 2.* **1998**, 37–40.

27. Bruice, T.C., Katzhendler, J., Fedor, L.R. Nucleophilic micelles. II. Effect on the rate of solvolysis of neutral, positively, and negatively charged esters of varied chain length when incorporated into nonfunctional and functional micelles of neutral, positive, and negative charge. *J. Am. Chem. Soc.* **1968**, *90*(5), 1333–1348.

28. Bunton, C.A., Robinson, L. Electrolyte and micellar effects upon the reaction of 2,4-dinitrofluorobenzene with hydroxide ion. *J. Org. Chem.* **1969**, *34*(4), 780–785.

29. Menger, F.M., Portnoy, C.E. Chemistry of reactions proceeding inside molecular aggregates. *J. Am. Chem. Soc.* **1967**, *89*(18), 4698–4703.

30. Bunton, C.A., Robinson, L., Stam, M. The hydrolysis of *p*-nitrophenyl diphenyl phosphate catalyzed by a nucleophilic reagent. *J. Am. Chem. Soc.* **1970**, *92*(25), 7393–7400.

31. Romsted, L.R., Cordes, E.H. Secondary valence force catalysis. VII. Catalysis of hydrolysis of *p*-nitrophenyl hexanoate by micelle-forming cationic detergents. *J. Am. Chem. Soc.* **1968**, *90*(16), 4404–4409.

32. Dunlap, R.B., Cordes, E.H. Secondary valence force catalysis. VIII. Catalysis of hydrolysis of methyl orthobenzoate by anionic surfactants. *J. Phys. Chem.* **1969**, *73*(2), 361–370.

33. Bunton, C.A., Robinson, L. Micellar effects upon nucleophilic aromatic and aliphatic substitution. *J. Am. Chem. Soc.* **1968**, *90*(22), 5972–5979.

34. Khan, M.N. Mechanisms of catalysis in micellar systems. *Encyclopedia of Surface and Colloid Science*, Hubbard, A., Ed. Marcel Dekker, New York, **2002**, pp. 3178–3191.

35. Khan, M.N. Effect of CH_3CN-H_2O and CH_3CN solvents on rate of reaction of phthalimide with piperidine. *J. Phys. Org. Chem.* **1999**, *12*, 187–193.

36. (a) Page, M.I., Proctor, P. Mechanism of β-lactam ring opening in cephalosporins. *J. Am. Chem. Soc.* **1984**, *106*(13), 3820–3825. (b) Fox, J.P., Jencks, W.P. General acid and general base catalysis of the methoxyaminolysis of 1-acetyl-1,2,4-triazole. *J. Am. Chem. Soc.* **1974**, *96*(5), 1436–1449.

37. Khan, M.N. Kinetics and mechanism of aminolysis of phthalimide and *N*-substituted phthalimides: evidence for the occurrence of intramolecular general acid-base catalysis in the reactions of ionized phthalimides with primary amines. *J. Chem. Soc. Perkin Trans. 2.* **1990**, 435–444.

38. (a) Satterthwait, A.C., Jencks, W.P. The mechanism of the aminolysis of acetate esters. *J. Am. Chem. Soc.* **1974**, *96*(22), 7018–7031. (b) Bruice, P.Y., Bruice, T.C. Aminolysis of substituted phenyl quinoline-8- and -6-carboxylates with primary and secondary amines: involvement of proton-slide catalysis. *J. Am. Chem. Soc.* **1974**, *96*(17), 5533–5542.

39. (a) Biechler, S.S., Taft, R.W., Jr. The effect of structure on kinetics and mechanism of the alkaline hydrolysis of anilides. *J. Am. Chem. Soc.* **1957**, *79*, 4927–4935. (b) Khan, M.N. Co-operative catalysis of the cleavage of an amide by neighboring carboxy group in alkaline hydrolysis. *J. Chem. Soc., Perkin Trans. 2.* **1989**, 233–237.

40. (a) Menger, F.M., Donohue, J.A. Base-catalyzed hydrolysis of *N*-acylpyrroles: a measurable acidity of a steady-state tetrahedral intermediate. *J. Am. Chem. Soc.* **1973**, *95*(2), 432–437. (b) Young, J.K., Pazhanisamy, S., Schowen, R.L. Energetics of carbonyl addition and elimination: kinetic manifestations of acyl substituent effects in alkaline hydrolysis. *J. Org. Chem.* **1984**, *49*(22), 4148–4152 and references cited therein.

41. Baxter, N.J., Rigoreau, L.J.M., Laws, A.P., Page, M.I. Reactivity and mechanism in the hydrolysis of β-sultams. *J. Am. Chem. Soc.* **2000**, *122*(14), 3375–3385.

42. Khan, M.N., Abdullah, Z. Kinetics and mechanism of alkaline hydrolysis of 4-nitrophthalimide in the absence and presence of cationic micelles. *Int. J. Chem. Kinet.* **2001**, *33*, 407–414.

43. Khan, M.N., Gambo, S.K. Intramolecular catalysis and the rate-determining step in the alkaline hydrolysis of ethyl salicylate. *Int. J. Chem. Kinet.* **1985**, *17*, 419–428.

44. Khan, M.N., Audu, A.A. Kinetic probe to study the structural behavior of the mixed aqueous-organic solvents: effects of the inorganic and organic salts on the kinetics and mechanism of intramolecular general base-catalyzed methanolysis of ionized phenyl salicylate. *J. Phys. Org. Chem.* **1992**, *5*, 129–141.

45. Khan, M.N. Kinetic probe to study the structural behavior of the mixed aqueous-organic solvents and micelles: kinetics and mechanism of intramolecular general base-catalyzed alkanolysis of phenyl salicylate in the presence of anionic micelles. *Int. J. Chem. Kinet.* **1991**, *23*, 837–852.

46. Khan, M.N., Ohayagha, J.E. Kinetic studies of the reactions of phthalimide with ammonia and pyrrolidine. *J. Phys. Org. Chem.* **1994**, *4*, 518–524.

47. (a) DeWolfe, R.H., Newcomb, R.C. Hydrolysis of formanilides in alkaline solutions. *J. Org. Chem.* **1971**, *36*(25), 3870–3878. (b) Brown, R.S., Bennet, A.J., Siebocka-Tilk, H. Recent perspectives concerning the mechanism of H_3O^+- and hydroxide-promoted amide hydrolysis. *Acc. Chem. Res.* **1992**, *25*(11), 481–488.

48. Khan, M.N., Olagbemiro, T.O., Umar, U.Z. The kinetics and mechanism of hydrolytic cleavage of phenyl salicylate under highly alkaline medium. *Tetrahedron* **1983**, *39*(5), 811–814.

49. Khan, M.N. Kinetic evidence for the occurrence of a stepwise mechanism in hydrazinolysis of phthalimide. *J. Org. Chem.* **1995**, *60*(14), 4536–4541.

Appendix A

```
10      PRINT "LINEAR LEAST SQUARES COMPUTER PROGRAM IN
        BASIC"
30      PRINT "NO. OF PARAMETERS = ";
40      INPUT N
50      PRINT "NO. OF POINTS";
60      INPUT K
70      DIM J(K,N),L(N,K),E(K,1),C(K,1),O(K,3),B(K,1),V(1,K),W(K,1)
71      DIM D(K),G(K),H(K)
80      DIM M(N,N),X(N,1),T(N,1),U(1,1),F(N,N),S(N,1)
90      K1=10^(-7.53)
100     K2=10^(-10.18)
190     FOR I=1 TO K
210     READ Z1
215     O(I,1) =Z1
220     NEXT I
230     FOR I= 1 TO K
250     READ Z2
255     O(I,2) =Z2
256     NEXT I
257     FOR I=1 TO K
258     READ Z3
259     O(I,3)=Z3
260     NEXT I
261     DATA.04,.08,.1,.15,.3
262     DATA.000843,.00163,.002,.0025,.0035
263     DATA.2,.2,.2,.2,.2
270     IF N < 2 THEN 320
300     IF N = 2 THEN 370
310     IF N = 3 THEN 430
312     IF N = 4 THEN 460
314     IF N = 5 THEN 472
320     FOR I=1 TO K
330     J(I,1) = O(I,1)
340     B(I,1) = O(I,2)
350     NEXT I
351     P=0.0
```

```
352    FOR I=1 TO K
353    P1=O(I,2)/O(I,1)
354    W(I,1)=P1
355    P=P+P1
356    NEXT I
357    T(1,1)=P/K
358    P2=0!
359    FOR I=1 TO K
360    P2=P2+((W(I,1)-T(1,1))*(W(I,1)-T(1,1)))
361    NEXT I
362    S(1,1)=SQR(P2/(K-1))
365    GOTO 1110
370    FOR I=1 TO K
380    J(I,1) =1
390    J(I,2) =O(I,1)
400    B(I,1) = O(I,2)
410    NEXT I
420    GOTO 490
430    FOR I=1 TO K
440    J(I,1)= 1
450    J(I,2)=O(I,1)
454    J(I,3)=O(I,1)*O(I,1)
456    B(I,1)=O(I,2)
457    NEXT I
458    GOTO 490
460    FOR I=1 TO K
461    J(I,1)=1
462    J(I,2)=1.449E-14/O(I,1)
464    J(I,3)=K1*O(I,1)*O(I,3)/(O(I,1)^2+O(I,1)*K1+K1*K2)
466    J(I,4)=K1*K2*O(I,3)/(O(I,1)^2+O(I,1)*K1+K1*K2)
468    B(I,1)= O(I,2)
470    NEXT I
471    GOTO 490
472    FOR I=1 TO K
473    D(I)=10^(-7.93)/(O(I,1)+10^(-7.93))
475    G(I)=10^(-5.97)/(O(I,1)+10^(-5.97))
477    H(I)=10^(-9.45)/(O(I,1)+10^(-9.45))
480    J(I,1)=D(I)*G(I)
482    J(I,2)=(1-D(I))*G(I)
483    J(I,3)=D(I)*(1-G(I))*H(I)*A(I)
484    J(I,4)=D(I)*(1-G(I))*(1-H(I))*A(I)
485    J(I,5)=(1-D(I))*(1-G(I))*(1-H(I))*A(I)
486    PRINT J(I,1),J(I,2),J(I,3),J(I,4),J(I,5)
488    B(I,1)=O(I,2)
489    NEXT I
```

```
490     FOR I=1 TO K
500     FOR M=1 TO N
510     L(M,I)=J(I,M)
520     NEXT M
530     NEXT I
540     FOR I=1 TO N
550     FOR J=1 TO N
560     F(I,J)=0.0
570     X(I,1)=0.0
580     T(I,1)=0.0
590     NEXT J
600     NEXT I
610     FOR M=1 TO N
620     FOR JJ=1 TO N
630     FOR I=1 TO K
640     F(M,JJ)=L(M,I)*J(I,JJ) + F(M,JJ)
650     NEXT I
660     NEXT JJ
670     NEXT M
680     FOR KK=1 TO N
690     FOR J= 1 TO N-1
700     M(KK,J)=F(KK,J+1)/F(KK,1)
710     NEXT J
845     M(KK,N)=1/F(KK,1)
850     IF KK=1 THEN 890
855     I=1
860     R1=F(I,1)
865     FOR J=1 TO N-1
870     M(I,J)=F(I,J+1)-R1*M(KK,J)
875     NEXT J
880     M(I,N)=-R1*M(KK,N)
885     IF KK=2 THEN 915
890     I=2
895     R2=F(I,1)
896     FOR J=1 TO N-1
897     M(I,J)=F(I,J+1)-R2*M(KK,J)
900     NEXT J
905     M(I,N)=-R2*M(KK,N)
910     IF KK=3 THEN 955
915     IF N=2 THEN 1004
920     I=3
925     R3=F(I,1)
930     FOR J=1 TO N-1
935     M(I,J)=F(I,J+1)-R3*M(KK,J)
940     NEXT J
```

```
945    M(I,N)=-R3*M(KK,N)
950    IF KK=4 THEN 990
955    IF N=3 THEN 1004
960    I=4
965    R4=F(I,1)
970    FOR J=1 TO N-1
975    M(I,J)=F(I,J+1)-R4*M(KK,J)
980    NEXT J
983    M(I,N)=-R4*M(KK,N)
986    IF KK=5 THEN 1004
990    IF N=4 THEN 1004
992    I=5
994    R5=F(I,1)
996    FOR J=1 TO N-1
998    M(I,J) =F(I,J+1)-R5*M(KK,J)
1000   NEXT J
1002   M(I,N)=-R5*M(KK,N)
1004   FOR I=1 TO N
1006   FOR J=1 TO N
1008   F(I,J)=M(I,J)
1010   NEXT J
1012   NEXT I
1014   NEXT KK
1025   FOR M=1 TO N
1026   FOR I=1 TO K
1027   X(M,1)=X(M,1)+L(M,I)*B(I,1)
1028   NEXT I
1029   NEXT M
1030   FOR MM=1 TO N
1040   FOR I=1 TO N
1050   T(MM,1)=T(MM,1)+F(MM,I)*X(I,1)
1060   NEXT I
1070   NEXT MM
1080   IF N=2 THEN 1160
1100   IF N = 3 THEN 1210
1104   IF N = 4 THEN 1227
1106   IF N = 5 THEN 1235
1110   FOR I=1 TO K
1120   C(I,1)=T(1,1)*O(I,1)
1130   E(I,1)=O(I,2)-C(I,1)
1140   NEXT I
1150   GOTO 1350
1160   FOR I=1 TO K
1170   C(I,1)=T(1,1)+T(2,1)*O(I,1)
1180   E(I,1)=O(I,2)-C(I,1)
```

```
1190    NEXT I
1200    GOTO 1250
1210    FOR I=1 TO K
1220    C(I,1)=T(1,1)+T(2,1)*O(I,1)+T(3,1)*O(I,1)*O(I,1)
1224    E(I,1)=O(I,2)-C(I,1)
1225    NEXT I
1226    GOTO 1250
1227    FOR I=1 TO K
1228    Z7=O(I,1)^2+O(I,1)*K1+K1*K2
1229    C(I,1)=T(1,1)+T(2,1)*(1.449E 14/O(I,1))+T(3,1)*K1*O(I,1)*O(I,3)/
        Z7+T(4,1)*K1*K2*O(I,3)/Z7
1231    E(I,1)=O(I,2)-C(I,1)
1233    NEXT I
1234    GOTO 1250
1235    FOR I=1 TO K
1236    C1=T(1,1)*D(I)*G(I)+T(2,1)*(1-D(I))*G(I)+T(3,1)*D(I)*
        (1-G(I))*H(I)*A(I)
1237    C2=(T(4,1)*D(I)*(1-G(I))*(1-H(I))+T(5,1)*(1-D(I))*(1-G(I))*
        (1-H(I)))*A(I)
1238    C(I,1)=C1+C2
1240    E(I,1)=O(I,2)-C(I,1)
1241    PRINT C(I,1)
1242    NEXT I
1250    FOR I=1 TO K
1260    V(1,I)=E(I,1)
1270    NEXT I
1280    U(1,1)=0.0
1290    FOR I=1 TO K
1300    U(1,1)=U(1,1)+V(1,I)*E(I,1)
1310    NEXT I
1320    FOR I=1 TO N
1330    S(I,1)=SQR(U(1,1)*F(I,I)/(K-N))
1340    NEXT I
1345    IF N>1 THEN 1370
1350    PRINT "A1 =";T(1,1);"STD. =";S(1,1)
1360    GOTO 1415
1370    PRINT "C =";T(1,1), "STD. =";S(1,1)
1380    PRINT "A1 =";T(2,1), "STD =";S(2,1)
1390    IF N=2 THEN 1410
1400    PRINT "A2 =";T(3,1); "STD. =";S(3,1)
1402    IF N=3 THEN 1410
1403    PRINT "A3 =";T(4,1); "STD =";S(4,1)
1405    IF N=4 THEN 1410
1406    PRINT "A4 =";T(5,1); "STD =";S(5,1)
1410    PRINT "LEAST SQUARE VALUE =";U(1,1)
```

```
1415    PRINT"DO YOU WANT Ycalcd";
1416    INPUT Y
1420    PRINT".............................................................
1430    PRINT" Xobs Yobs Ycalcd %Res Error"
1440    FOR I=1 TO K
1450    PRINT O(I,1),O(I,2),C(I,1),100*E(I,1)/O(I,2)
1460    NEXT I
1470    END
```

Appendix B

```
10      PRINT "NONLINEAR LEAST SQUARES COMPUTER PROGRAM
        IN BASIC"
30      PRINT "NO. OF PARAMETERS = ";
40      INPUT N
50      PRINT "NO. OF DATA POINTS";
60      INPUT K
70      DIM J(K,N),L(N,K),E(K,1),C(K,1),O(K,2),B(K,1),V(1,K),W(N,1)
80      DIM M(N,N),X(N,1),T(N,1),U(1,1),F(N,N),S(N,1)
90      PRINT "NN =";
100     INPUT NN
190     FOR I=1 TO K
210     READ Z1
215     O(I,1) = Z1
220     NEXT I
230     FOR I= 1 TO K
250     READ Z2
255     O(I,2) = Z2
260     NEXT I
270     PRINT "INITIAL GUESS VALUE OF A1 =";
280     INPUT A1
290     PRINT "INITIAL GUESS VALUE OF A2 =";
294     INPUT A2
295     IF N=2 THEN 310
296     PRINT "INITIAL GUESS VALUE OF A3 =";
297     INPUT A3
298     IF N=3 THEN 310
299     PRINT "INITIAL GUESS VALUE OF A4 =";
300     INPUT A4
302     IF N=4 THEN 310
304     PRINT "INITIAL GUESS VALUE OF A5 =";
306     INPUT A5
310     PRINT "INITIAL CONC. OF SUBSTRATE =";
320     INPUT X0
330     T(1,1)=A1
340     T(2,1)=A2
342     IF N=2 THEN 360
```

```
346    T(3,1)=A3
350    IF N=3 THEN 360
352    T(4,1)=A4
354    IF N=4 THEN 360
356    T(5,1)=A5
360    PRINT "TOTAL NUMBER OF ITERATION =:
370    INPUT K2
380    DATA.004,.01,.04,.08,.1,.2,.4,.6,1,1.4,2
381    DATA 1.226,1.222,1.188,1.103,1.072,.865,.733,.647,.563,.571,.548
400    PRINT"ERROR CHECK =";
410    INPUT E0
420    PRINT"TOTAL NUMBER OF ITERATION ="K2
421    PRINT"ERROR CHECK ="E0
430    GOSUB 722
440    A1=T(1,1)
450    A2=T(2,1)
452    IF N=2 THEN 470
454    A3=T(3,1)
456    IF N=3 THEN 470
458    A4=T(4,1)
460    IF N=4 THEN 470
464    A5=T(5,1)
470    FOR I=1 TO K
471    V(1,I)=E(I,1)
472    NEXT I
480    U(1,1)=0.0
481    FOR I=1 TO K
482    U(1,1)=U(1,1)+V(1,I)*E(I,1)
483    NEXT I
490    FOR I=1 TO N
500    S(I,1)=SQR(U(1,1)*F(I,I)/(K-N))
510    NEXT I
520    PRINT"ITERATION NUMBER ="K1
530    PRINT"A1 ="T(1,1);"STANDARD DEVIATION ="S(1,1)
540    PRINT"A2 ="T(2,1);" STANDARD DEVIATION ="S(2,1)
550    IF N=2 THEN 590
560    PRINT"A3 ="T(3,1);" STANDARD DEVIATION ="S(3,1)
570    IF N=3 THEN 590
580    PRINT"A4 ="T(4,1);" STANDARD DEVIATION ="S(4,1)
585    IF N=4 THEN 590
587    PRINT"A5 =";T(5,1);" STANDARD DEVIATION ="S(5,1)
590    PRINT"LEAST SQUARES VALUE ="U(1,1)
600    PRINT"---------------------------------------------------------------------------"
610    FOR I=1 TO N
620    IF ABS(W(I,1)/T(I,1))<E0 THEN 640
```

```
630    IF ABS(W(I,1)/T(I,1))>E0 THEN 660
640    NEXT I
641    PRINT"DO YOU WANT LEAST SQUARES VALUE";
642    INPUT Y9
650    GOTO 680
660    K1=K1+1
670    IF K1<K2 THEN 430
680    PRINT" TIME A_OBS A_CALCD 100*(A_OBS - A_CALCD)/A_OBS)"
690    FOR I=1 TO K
700    PRINT O(I,1),O(I,2),C(I,1),100*(O(I,2)-C(I,1))/O(I,2)
710    NEXT I
720    STOP
722    IF NN>2 THEN 792
730    FOR I=1 TO K
740    J(I,1)=O(I,1)*A2*X0*EXP(-A1*O(I,1))
750    J(I,2)=X0*(1-EXP(-A1*O(I,1)))
760    IF N=2 THEN 785
765    J(I,3)=1
770    IF N=3 THEN 785
774    J(I,4)=--------
776    IF N=4 THEN 785
778    J(I,5)=-------------
785    C(I,1)=A2*X0*(1-EXP(-A1*O(I,1)))+A3
789    E(I,1)=O(I,2)-C(I,1)
790    NEXT I
791    GOTO 800
792    FOR I=1 TO K
793    J(I,1)=-O(I,1)*A2*X0*EXP(-A1*O(I,1))
794    J(I,2)=X0*EXP(-A1*O(I,1))
795    J(I,3)=1!
796    C(I,1)=A2*X0*EXP(-A1*O(I,1))+A3
797    E(I,1)=O(I,2)-C(I,1)
798    NEXT I
800    FOR I=1 TO K
801    FOR M=1 TO N
802    L(M,I)=J(I,M)
803    NEXT M
804    NEXT I
806    FOR I=1 TO N
807    FOR J=1 TO N
808    F(I,J)=0.0
809    X(I,1)=0.0
810    W(I,1)=0.0
811    NEXT J
812    NEXT I
```

```
813    FOR M=1 TO N
814    FOR JJ=1 TO N
815    FOR I=1 TO K
816    F(M,JJ)=L(M,I)*J(I,JJ) + F(M,JJ)
817    NEXT I
818    NEXT JJ
819    NEXT M
820    FOR KK=1 TO N
825    FOR J= 1 TO N-1
830    M(KK,J)=F(KK,J+1)/F(KK,1)
840    NEXT J
845    M(KK,N)=1/F(KK,1)
850    IF KK=1 THEN 890
855    I=1
860    R1=F(I,1)
865    FOR J=1 TO N-1
870    M(I,J)=F(I,J+1)-R1*M(KK,J)
875    NEXT J
880    M(I,N)=-R1*M(KK,N)
885    IF KK=2 THEN 915
890    I=2
895    R2=F(I,1)
896    FOR J=1 TO N-1
897    M(I,J)=F(I,J+1)-R2*M(KK,J)
900    NEXT J
905    M(I,N)=-R2*M(KK,N)
910    IF KK=3 THEN 955
915    IF N=2 THEN 1004
920    I=3
925    R3=F(I,1)
930    FOR J=1 TO N-1
935    M(I,J)=F(I,J+1)-R3*M(KK,J)
940    NEXT J
945    M(I,N)=-R3*M(KK,N)
950    IF KK=4 THEN 990
955    IF N=3 THEN 1004
960    I=4
965    R4=F(I,1)
970    FOR J=1 TO N-1
975    M(I,J)=F(I,J+1)-R4*M(KK,J)
980    NEXT J
983    M(I,N)=-R4*M(KK,N)
986    IF KK=5 THEN 1004
990    IF N=4 THEN 1004
992    I=5
```

```
994    R5=F(I,1)
996    FOR J=1 TO N-1
998    M(I,J) =F(I,J+1)-R5*M(KK,J)
1000   NEXT J
1002   M(I,N)=-R5*M(KK,N)
1004   FOR I=1 TO N
1006   FOR J=1 TO N
1008   F(I,J)=M(I,J)
1010   NEXT J
1012   NEXT I
1014   NEXT KK
1020   FOR M=1 TO N
1030   FOR I=1 TO K
1040   X(M,1)=X(M,1)+L(M,I)*E(I,1)
1050   NEXT I
1060   NEXT M
1070   FOR MM=1 TO N
1080   FOR I=1 TO N
1090   W(MM,1)=W(MM,1)+F(MM,I)*X(I,1)
1100   NEXT I
1110   NEXT MM
1120   FOR I=1 TO N
1130   T(I,1)=T(I,1)+W(I,1)
1140   NEXT I
1150   RETURN
1160   END
```

Author Index

Subject Index

S

Printed in the United States
by Baker & Taylor Publisher Services